CORROSION ENGINEERING

McGraw-Hill Series in Materials Science and Engineering

Brick, Pense, and Gordon: *Structure and Properties of Engineering Materials*
Dieter: *Engineering Design: A Materials and Processing Approach*
Dieter: *Mechanical Metallurgy*
Drauglis, Gretz, and Jaffee: *Molecular Processes on Solid Surfaces*
Flemings: *Solidification Processing*
Fontana: *Corrosion Engineering*
Gaskell: *Introduction to Metallurgical Thermodynamics*
Guy: *Introduction to Materials Science*
Kehl: *The Principles of Metallographic Laboratory Practice*
Leslie: *The Physical Metallurgy of Steels*
Rhines: *Phase Diagrams in Metallurgy: Their Development and Application*
Rozenfeld: *Corrosion Inhibitors*
Shewmon: *Transformations in Metals*
Smith: *Principles of Materials Science and Engineering*
Smith: *Structure and Properties of Engineering Alloys*
Vander Voort: *Metallography: Principles and Practice*
Wert and Thomson: *Physics of Solids*

CORROSION ENGINEERING

Third Edition

Mars G. Fontana
Regents' Professor and Chairman Emeritus
Department of Metallurgical Engineering
Fontana Corrosion Center
The Ohio State University
Executive Director Emeritus
Materials Technology Institute of the Chemical Process Industries, Inc.

Tata McGraw-Hill Publishing Company Limited
NEW DELHI

McGraw-Hill Offices
New Delhi New York St Louis San Francisco Auckland Bogotá Caracas Kuala Lumpur Lisbon London Madrid Mexico City Milan Montreal San Juan Santiago Singapore Sydney Tokyo Toronto

Tata McGraw-Hill

CORROSION ENGINEERING

Tata McGraw-Hill Edition 2005

Third reprint 2006
RXLCRDDKRCRDL

Reprinted in India by arrangement with The McGraw-Hill Companies, Inc., New York

Sales territories: India, Pakistan, Nepal, Bangladesh, Sri Lanka and Bhutan

Library of Congress Cataloging-in-Publication Data

Fontana, Mars Guy, 1910
Corrosion engineering
Bibliography: p.
Includes index.
1. Corrosion and anti-corrosives. I. Title.
TA418.74.F6 1986 620.1'1223 85-19713
ISBN 0-07-021463-8

ISBN 0-07-060744-3

Published by Tata McGraw-Hill Publishing Company Limited,
7 West Patel Nagar, New Delhi 110 008, and printed at
Pushp Print Services, Delhi 110 053

The McGraw·Hill Companies

ABOUT THE AUTHOR

Mars G. Fontana is Regents' Professor and Chairman Emeritus, Department of Metallurgical Engineering, The Ohio State University. He was the first Executive Director of the Materials Technology Institute of the Chemical Process Industries.

He is a graduate of the University of Michigan, from which he received a bachelor's degree in Chemical Engineering, M.S. and Ph.D. (1935) in metallurgical engineering, and an honorary Doctor of Engineering in 1975.

As a metallurgical engineer and supervisor (1934–1945) for the DuPont Company he pioneered industrial uses of nylon, Teflon, and other plastics. He joined the faculty of Ohio State in 1945 as a full professor and served as chairman of the Department of Metallurgical Engineering for 27 years. He established one of the first courses in the United States on corrosion in 1946. This same year he started the Corrosion Center, now the Fontana Corrosion Center. The Metallurgical Engineering Building was named the Mars G. Fontana Laboratories in his honor in May 1981.

He was elected to the National Academy of Engineering in 1967. He is an Honorary Member of the American Society for Metals, Cambell Lecturer in 1970, ASM Gold Medal, and Fellow, ASM, A.I.Ch.E., and A.I.M.E. He was President of the National Association of Corrosion Engineers in 1952 and received their Speller Award in Corrosion Engineering in 1956. He originated the use of high silicon iron anodes for cathodic protection and is the father of Alloy 20. He was the first editor of the journal *Corrosion*. He has received several teaching awards. He has been consultant to industry and government since 1945. He has published over 200 papers in recognized journals. He is a member of the Board of Directors of Worthington Industries (since 1973).

To Betty, Martha, Marybeth, Dave, Tom, Jeff, Steven, Brian, Sarah, Scott, Beth, Mike, Carley, Lauren, and Katie for all the time I spent away from them while preparing this third edition.

CONTENTS

PREFACE

This third edition maintains the unique approach of the previous editions. It is unique because corrosion data are presented in terms of *corrosives* or environments rather than in terms of materials. This approach saves thumbing through many chapters on materials to determine likely candidate materials for a given corrosion problem (e.g., sulfuric acid). *Isocorrosion* charts (invented by the author) present a quick look at candidates for a particular corrosive.

There are some exceptions to the above in Chapter 5, particularly when a material has outstanding characteristics for certain environments.

Corrosion testing is the *backbone* of corrosion engineering. Chapter 4 includes simple and advanced complicated tests. Description of corrosion tests for plastics and elastomers has been expanded. The effects of the "revolution" in electronic instrumentation are described. Many types of electronic instrumentation are mentioned and references are provided for in-depth study.

In response to requests to make the text more challenging to college students, some "cutting edge" items are included—for example, Section 9-12, "Mechanism of the Growth and Breakdown of Passive Films." Advanced testing techniques such as AC Impedance and Small Amplitude Cycle Voltammetry (SACV) will be used more and more in the future.

Many environments have been added, such as the pulp and paper industry and nuclear waste isolation, and also subjects such as fracture mechanics and laser alloying.

The need for more corrosion engineers to reduce the costs of corrosion is described. The enormous costs of product liability claims is emphasized, since producers must watch their p's and q's, particularly QC and QA—quality control and quality assurance.

Although this book was first written as a textbook, it has proved useful as a reference book. The reference aspect has been enhanced through provision of literature references for in-depth study. An improved index is presented.

Little attempt has been made to cover paints, cathodic protection, and water treatment comprehensively. These are more of an "art" (experience) than a science, and whole books have been written about them. References are provided. The novice should contact expert organizations in these fields, of which there are many.

This text covers practically all the important aspects of corrosion engineering and corrosion science, including noble metals, "exotic" metals, nonmetallics, coatings, mechanical properties, and corrosion testing, and includes modern concepts as well. This coverage eliminates some of the deficiencies of previous books on corrosion. The book is designed to serve many purposes: It can be used for undergraduate courses, graduate courses, intensive short courses, in-plant training, self-study, and as a useful reference text for plant engineers and maintenance personnel.

Professors in metallurgical engineering, materials engineering, materials science, chemical engineering, mechanical engineering, chemistry, or other physical science or engineering disciplines could teach a beginning course using this text without extensive background or much work in preparation. Section 2-15, "Study Sequence," suggests different procedures depending on the "needs" of the students, plant personnel, and others. This means that considerable flexibility exists for material to be covered or presented.

Many examples are presented to illustrate the causes and cures of corrosion problems. Case histories are helpful in engineering teaching. Descriptions, including mechanical properties, of materials are presented so that the reader will get the proper "feel" for materials.

A Solutions Manual is available as a separate booklet.

In order to keep the price of the book down, the second edition (1978) consisted of the addition of an update, Chapter 12. In retrospect, this was less than a brilliant idea. Accordingly, Chapter 12 disappears; its information is integrated into the other chapters.

The Materials Technology Institute of the Chemical Process Industries was established in June 1977 and I was the first executive director (now emeritus). Members of MTI pay dues and sponsor work by outside contractors on work of mutual interest. The main purpose is to provide the corrosion engineer or materials engineer with tools and information to do his job more effectively. I am grateful to the board of directors of MTI for permission to use as much of the information developed as I wished. A substantial amount of this information and references to MTI publications appear in the book.

This edition is based on my 50 years of experience in industry, teaching, and consulting. Much of my time was devoted to solving corrosion problems.

I am happy to say that a large number of former students are successful corrosion engineers, and a score of them are teaching corrosion courses.

I wish to gratefully acknowledge the assistance of my friends and colleagues with this revision for the third edition. These include David Bowers (Pulp and Paper Industry), Ron Latanision (Metallic Glasses), Digby Macdonald (Passivity Models, SACV, Electronic Instrumentation and review of Chapters 9 and 10), Mike McKubre (AC Impedance), Tom Murata (Sour Resistance, SR values), Tom Oettinger (Waste Treatment), Bob Rapp (High-Temperature Corrosion), Mike Streicher (Crevice Corrosion, CCI), John Stringer (Coal Conversion), and Dick Treseder (CO_2 Corrosion). All are experts in their particular fields of corrosion. I also acknowledge other friends, former students, and colleagues in industry who supplied data and photographs.

If this book results in the better education of many more people in the field of corrosion, particularly the young people in colleges and universities, and in a greater awareness of the cost and evils of corrosion as well as of the means for alleviating it, this book will have served its major purpose.

I would like to express my thanks for the many useful comments and suggestions provided by colleagues who reviewed this text during the course of its development, especially to Judith Todd, University of Southern California, and Ellis Verink, University of Florida.

Mars G. Fontana

CHAPTER

ONE

INTRODUCTION

1-1 Cost of Corrosion

Estimates of the annual cost of corrosion in the United States vary between \$8 billion and \$126 billion. I believe \$30 billion is the most realistic figure. In any case, corrosion represents a tremendous economic loss and much can be done to reduce it. These large dollar figures are not surprising when we consider that corrosion occurs, with varying degrees of severity, wherever metals and other materials are used. Several examples follow.

According to the *Wall Street Journal* (Sept. 11, 1981) cost to oil and gas producers is nearly \$2 billion. Costs are increasing because of deeper wells and more hostile environments—higher temperatures and corrosive sulfur gases (e.g., 500°F and hydrogen sulfide).

Corrosion of bridges is a major problem as they age and require replacement, which costs billions. The collapse (because of stress corrosion) of the Silver Bridge into the Ohio River cost 40 lives and millions of dollars. Corrosion of bridge decks costs about \$500 million. Proper design and use of cathodic protection reduces costs substantially. One large chemical company spent more than \$400,000 per year for corrosion maintenance in its sulfuric acid plants, even though the corrosion conditions were not considered to be particularly severe. Another spends \$2 million per year on painting steel to prevent rusting by a marine atmosphere. A refinery employing a new process developed a serious problem after just 16 weeks of operation; some parts showed a corrosion loss of as much as 1/8 inch. The petroleum industry spends a million dollars per day to protect underground pipelines. The paper industry estimates corrosion increases the cost of paper \$6 to \$7 per ton. Coal conversion to gas and oil involves high temperatures, erosive particles, and corrosive gases, thus presenting severe problems that must be solved.

Corrosion costs of automobiles—fuel systems, radiators, exhaust systems, and bodies—are in the billions. I personally incurred costs of $500 in refurbishing an automobile fuel system in which water had been mixed with gasoline! (A photograph of the gasoline tank is on the cover of *Materials Performance*, March 1982.) Approximately 3 million home water heaters are replaced every year. Corrosion touches all—inside and outside the home, on the road, on the sea, in the plant, and in aerospace vehicles.

Total annual costs of floods, hurricanes, tornadoes, fires, lightning, and earthquakes are less than the costs of corrosion. Costs of corrosion will escalate substantially during the next decade because of worldwide shortages of construction materials, higher energy costs, aggressive corrosion environments in coal conversion processes, large increases in numbers and scope of plants, and other factors.

"Political" considerations are also a factor. We depend largely on foreign sources for some metals: 90 percent for chromium (the main alloying element for stainless steel) and 100 percent for columbium (niobium) used in high-temperature alloys. Our sources could be shut off or the prices boosted. For example, during a recent crisis the price per pound of columbium jumped from $5 to $50.

Production of metals used for corrosion resistance and to replace corroded parts require large amounts of energy, thus compounding the nation's energy problems.

The most comprehensive study of the annual cost of *metallic* corrosion in the United States was conducted by the National Bureau of Standards (NBS) and Battelle Memorial Institute in response to a congressional directive. Results are published in a seven-part series. The first is, "NBS-Battelle Cost of Corrosion Study ($70 Billion) Part 1—Introduction," by J. H. Payer, W. K. Boyd, D. B. Dippold, and W. H. Fisher of Battelle (*Materials Performance*, May 1980). The other six parts appeared in subsequent issues of *Materials Performance* (June–November 1980). The figure of $70 billion* covers corrosion (in 1975) of metals (nonmetallics not included) and are costs incurred if corrosion *did not exist*; this amount has no practical significance, but it does emphasize the *magnitude* of the problem. Unfortunately, $70 billion has been simply stated as the "cost of corrosion" in later literature and is misleading (implying that $70 billion could be saved) because nothing can be done *economically* to reduce most of these costs. It is somewhat like asking how much you could save on your food budget if you *stopped eating*. However, the report states that about $10 billion could be saved if best, and presently known, practices to combat

*The U.S. Department of Commerce stated, "Corrosion will cost the United States an estimated 126 billion dollars in 1982." (*Materials Performance*, 57, Feb. 1983).

corrosion were applied. Chemical industry efforts involve high costs, but this industry is in the forefront with regard to utilizing corrosion control practices.

In fact our economy would be drastically changed if there were *no* corrosion. For example, automobiles, ships, underground pipelines, and household appliances would not require coatings. The stainless steel industry would essentially disappear and copper would be used only for electrical purposes. Most metallic plants, as well as consumer products, would be made of steel or cast iron.

Although corrosion is inevitable, its cost can be considerably reduced. For example, an inexpensive magnesium anode could double the life of a domestic hot water tank. Washing a car to remove road deicing salts is helpful. Proper selection of materials and good design reduce costs of corrosion. A good maintenance painting program pays for itself many times over. Here is where the *corrosion engineer* enters the picture and is effective—his or her primary function is to combat corrosion.

Aside from its direct costs in dollars, corrosion is a serious problem because it definitely contributes to the depletion of our natural resources. For example, steel is made from iron ore, and our domestic supply of high grade directly smeltable iron ore has dwindled. Another important factor concerns the world's supply of metal resources. The rapid industrialization of many countries indicates that the competition for and the price of metal resources will increase. The United States is no longer the chief consumer of mineral resources.

1-2 Corrosion Engineering

Corrosion engineering is the application of science and art to prevent or control corrosion damage economically and safely.

In order to perform their function properly, corrosion engineers must be well versed in the practices and principles of corrosion; the chemical, metallurgical, physical, and mechanical properties of materials; corrosion testing; the nature of corrosive environments; the availability and fabrication of materials; computers*; and design. They also must have the usual attributes of engineers—a sense of human relations, integrity, the ability to think and analyze, an awareness of the importance of safety, common sense, a sense of organization, and, of prime importance, a solid feeling for economics. In solving corrosion problems, the corrosion engineer must select the method that will maximize profits. One definition of economics is simply—"there is no free lunch."

The following articles offer insight into applications of computer

*See Chapter 4 for the revolution brought about by the introduction of electronic instrumentation in corrosion science and engineering.

technology in corrosion engineering: Thinking Machines (Artificial Intelligence) and the CPI, *Chem. Eng.* 45–51 (Sept. 20, 1982), which describes several examples including prediction of stress corrosion cracking; S. N. Smith and F. E. Rizzo, Computer Assisted Corrosion Engineering, *Materials Performance*, **19**:21–23 (Oct. 1980); and C. Edeleanu, The Effect of the Microprocessors on Corrosion Technology, *Materials Performance*, **22**: 82–83 (Oct. 1983).

In the past, relatively few engineers received educational training in corrosion. Most of the people then engaged in this field had chemical, electrical, or metallurgical backgrounds. Fortunately this picture has changed. From only three in 1946, now 65 U.S. universities and colleges (including the author's) offer formal courses in corrosion.* *Corrosion Engineering* is a popular textbook for these courses. What this all means is that now there are hundreds of engineers in the field who have had a formal course in corrosion. In the past, and even today, corrosion is often regarded as a "necessary evil" to be tolerated. *Ignorance* is the cause of many premature, unexpected, and expensive failures—ignorance even by people who should know better. For example, two vendors of sacrificial anodes describe their systems as anodic protection! Actually it is cathodic protection, which is completely different.

1-3 Definition of Corrosion

Corrosion is defined as the destruction or deterioration of a material because of reaction with its environment. Some insist that the definition should be restricted to metals, but often the corrosion engineers must consider both metals and nonmetals for solution of a given problem. For purposes of this book we include ceramics, plastics, rubber, and other nonmetallic materials. For example, deterioration of paint and rubber by sunlight or chemicals, fluxing of the lining of a steelmaking furnace, and attack of a solid metal by another molten metal (liquid metal corrosion) are all considered to be corrosion.

Corrosion can be fast or slow. Sensitized 18-8 stainless steel is badly attacked in hours by polythionic acid. Railroad tracks usually show slight rusting—not sufficient to affect their performance over many years. The famous iron Delhi Pillar in India was made almost 2000 years ago and is almost as good as new. It is about 32 feet high and 2 feet in diameter. It should be noted, however, that it has been exposed mostly to arid conditions.

Corrosion of metals could be considered as extractive metallurgy in reverse as illustrated by Fig. 1-1. Extractive metallurgy is concerned primarily with the winning of the metal from the ore and refining or alloying the metal

*I took a corrosion course in 1932 at the University of Michigan. If readers know of an earlier college course, I would like to hear about it.

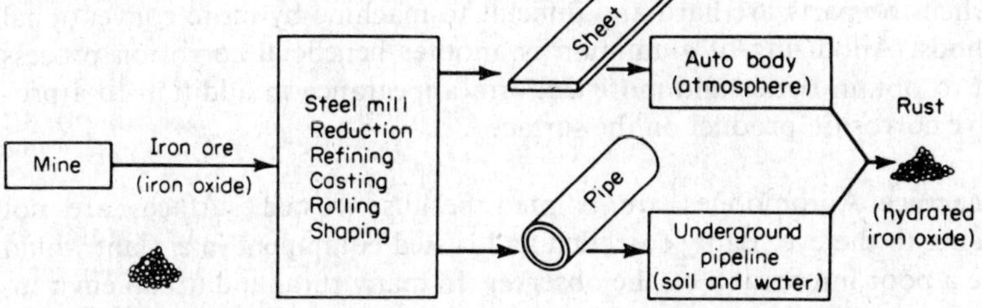

Figure 1-1 Metallurgy in reverse.

for use. Most iron ores contain oxides of iron, and rusting of steel by water and oxygen results in a hydrated iron oxide. *Rusting* is a term reserved for steel and iron corrosion, although many other metals form their oxides when corrosion occurs.

1-4 Environments

Practically all environments are corrosive to some degree. Some examples are air and moisture; fresh, distilled, salt, and mine waters; rural, urban, and industrial atmospheres; steam and other gases such as chlorine, ammonia, hydrogen sulfide, sulfur dioxide, and fuel gases; mineral acids such as hydrochloric, sulfuric, and nitric; organic acids such as naphthenic, acetic, and formic; alkalies; soils; solvents; vegetable and petroleum oils; and a variety of food products. In general, the "inorganic" materials are more corrosive than the "organics." For example, corrosion in the petroleum industry is due more to sodium chloride, sulfur, hydrochloric and sulfuric acids, and water, than to the oil, naphtha, or gasoline.

The trend in the chemical process industries toward higher temperatures and pressures has made possible new processes or improvements in old processes—for example, better yields, greater speed, and lower production costs. This also applies to power production, including nuclear power, missiles, and many other methods and processes. Higher temperatures and pressures usually involve more severe corrosion conditions. Many of the present-day operations would not have been possible or economical without the use of corrosion-resistant materials.

1-5 Corrosion Damage

Some of the deleterious effects of corrosion are described in the next few paragraphs. However, corrosion is beneficial or desirable in some cases. For example, *chemical machining* or chemical milling is widely used in aircraft and other applications. Unmasked areas are exposed to acid and excess metal is dissolved. This process is adopted when it is more economical or when the parts are hard and difficult to machine by more conventional methods. Anodizing of aluminum is another beneficial corrosion process used to obtain better and more uniform appearance in addition to a protective corrosion product on the surface.

Appearance Automobiles are painted because rusted surfaces are not pleasing to the eye. Badly corroded and rusted equipment in a plant would leave a poor impression on the observer. In many rural and urban environments it would be cheaper to make the metal thicker in the first place (corrosion allowance) than to apply and maintain a paint coating. Outside surfaces or trim on buildings are often made of stainless steel, aluminum, or copper for the sake of appearance. The same is true for restaurants and other commercial establishments. These are examples where service life versus dollars is not the controlling factor.

Maintenance and operating costs Substantial savings can be obtained in many types of plants through the use of corrosion-resistant materials of construction. One example is classic in this respect. A chemical plant effected an annual saving of more than $10,000 merely by changing the bolt material on some equipment from one alloy to another more resistant to the conditions involved. The cost of this change was negligible. In another case a waste acid recovery plant operated in the red for several months until a serious corrosion problem was solved. This plant was built to take care of an important waste disposal problem. Application of cathodic protection can cut leak rates in existing underground pipelines to practically nil with attendant large savings in repair costs. Maintenance costs are scrutinized because the labor picture accents the necessity for low-cost operation.

Close cooperation between the corrosion engineer and process and design personnel *before* a plant is built can eliminate or substantially reduce maintenance costs in many cases. Slight changes in the process sometimes reduce the corrosiveness of plant liquors without affecting the process itself, thus permitting the use of less expensive materials. These changes can often be made after the plant is in operation, but original preventive measures are more desirable. Corrosion difficulties can often be "designed out" of equipment, and the time to do this is in the original design of the plant.

When I started working for Du Pont in 1934, we had to write a report every month and indicate the dollars saved through our efforts. After a year or so we convinced management that *preventive* action was necessary and economical. The design people and the corrosion engineers collaborated at the inception of the project and much corrosion was "designed out" of the equipment. It is easier and cheaper to erase lines on a drawing than to repair or replace failed equipment in a plant.

Plant Shutdowns Frequently plants are shut down or portions of a process stopped because of unexpected corrosion failures. Sometimes these shutdowns are caused by corrosion involving no change in process conditions, but occasionally they are caused by changes in operating procedures erroneously regarded as incapable of increasing the severity of the corrosive conditions. It is surprising how often some minor change in process or the addition of a new ingredient changes corrosion characteristics completely. The production of a chemical compound vital to national defense is an example. To increase its production, the temperature of the cooling medium in a heat-exchanger system was lowered and the time required per batch decreased. Lowering the temperature of the cooling medium resulted, however, in more severe thermal gradients across the metal wall. They, in turn, induced higher stresses in the metal. Stress corrosion cracking of the vessels occurred quickly, and the plant was shut down with production delayed for some time.

Corrosion monitoring of a plant process is helpful in preventing unexpected corrosion failure and plant shutdown. This can be done by periodically examining corrosion specimens that are continually exposed to the process or by using a corrosion probe that continuously records the corrosion rate. Periodic inspection of equipment during scheduled downtimes can help prevent unexpected shutdown.

Contamination of product In many cases the market value of the product is directly related to its purity and quality. Freedom from contamination is a vital factor in the manufacture and handling of transparent plastics, pigments, foods, drugs, and semiconductors. In some cases a very small amount of corrosion, which introduces certain metal ions into the solution, may cause catalytic decomposition of a product, for example, in the manufacture and transporting of concentrated hydrogen peroxide or hydrazine.

Life of the equipment is not generally an important factor in cases where contamination or degradation of product is concerned. Ordinary steel may last many years, but more expensive material is used because the presence of rust is undesirable from the product standpoint.

Loss of valuable products No particular concern is attached to slight leakage of sulfuric acid to the drain, because it is a cheap commodity. However, loss of a material worth several dollars per gallon requires prompt corrective action. Slight losses of uranium compounds or solutions are hazardous and can be very costly. In such cases, utilization of more expensive design and better materials of construction are well warranted.

Effects on safety and reliability The handling of hazardous materials such as toxic gases, hydrofluoric acid, concentrated sulfuric and nitric acids, explosive and flammable materials, radioactive substances, and chemicals at high temperatures and pressures demands the use of construction materials that minimize corrosion failures. Stress corrosion of a metal wall separating the fuel and oxidizer in a missile could cause premature mixing, which could result in a loss of millions of dollars and in personal injury. Failure of a small component or control may result in failure or destruction of the entire structure. Corroding equipment can cause some fairly harmless compounds to become explosive. Economizing on materials of construction is not desirable if safety is risked.

Other health considerations are also important such as contamination of potable water. Corrosion products could make sanitizing of equipment more difficult. An interesting example here involves milk and other dairy product plants. The straight chromium stainless steels are satisfactory in old plants where much of the equipment is disassembled and sanitized by "dishpan" techniques. Newer plants use in-place cleaning and sanitizing which require more corrosive chemicals, particularly with regard to chloride ions and pitting. These solutions are circulated through the system without taking it apart thus saving many labor hours. These advances require use of more pit-resistant stainless steels, such as type 316 containing nickel and molybdenum.

Corrosion also plays an important part in medical metals used for hip joints, screws, plates, and heart valves. Reliability is, of course, of paramount importance here.

An *unusual* experience (*Chem. Eng.*, 28, March 19, 1984) emphasizes the importance of safety considerations. A large carbon steel vessel was cleaned, washed, and entered for maintenance. A workman was asphyxiated and died because the air became oxygen-deficient (about 1% O_2)—a situation "created by rapid rusting" of the empty steel vessel. If a second manhole had been opened, a natural draft would have changed the air.

Product liability There is an important and disturbing trend in this country toward putting the blame and legal responsibility on the producers or manufacturer of any item or piece of equipment that fails because of corrosion or for any other reason. The U.S. Department of Commerce has issued a report on the increase of product liability claims that points out that such claims have far outstripped inflation and are approaching medical malpractice insurance claims. One estimate indicates an average loss in 1965 from a product liability claim was $11,644. By 1973 this figure was $79,940, an increase of 686 percent. Lack of "contract," or "negligence," is no longer a defense.

A ridiculous example (to make the point) would be blaming the auto manufacturer if your car corroded because you drove it through a lake of hydrochloric acid! The car could be made of tantalum, but the cost would be astronomical, nobody would buy it, and then a disclaimer would have to be filed stating that hydrofluoric acid must not be present!

What this all means is that the manufacturer or producer of a product must make sure that it is made of proper materials, under good quality control, to a design that is as safe as possible, and the inspection must be critical. The corrosion engineer must be doubly sure that failure will not occur in the actual environment and should also be aware of the legal liability aspects. Passage of time is not a precluding factor; lawsuits resulted from failure of a bridge that had been in use for about 40 years.

The numbers listed in the third paragraph above have escalated tremendously (*Chem. Eng. Progr.*, p. 146, Mar. 1984). In the product liability area *alone*, jury awards "are now approaching $100 *billion* dollars per year." Corporate legal costs to defend suits are about $50 billion. One reason for this escalation is that there are roughly 600,000 attorneys in this country or about one lawyer for every 400 citizens. In Japan the corresponding number is one lawyer for every 16,000 people. In 1984 American law schools graduated about 35,000 lawyers—a number higher than the *total* number of American graduate students "in engineering, chemistry, physics and biology, combined!"

We are indeed a litigious society today.

1-6 Classification of Corrosion

Corrosion has been classified in many different ways. One method divides corrosion into low-temperature and high-temperature corrosion. Another separates corrosion into direct combination (or oxidation) and electrochemical corrosion. The preferred classification here is (1) *wet* corrosion and (2) *dry* corrosion.

Wet corrosion occurs when a liquid is present. This usually involves aqueous solutions or electrolytes and accounts for the greatest amount of corrosion by far. A common example is corrosion of steel by water. Dry corrosion occurs in the absence of a liquid phase or above the dew point of the environment. Vapors and gases are usually the corrodents. Dry corrosion is most often associated with high temperatures. An example is attack on steel by furnace gases.

The presence of even small amounts of moisture could change the corrosion picture completely. For example, dry chlorine is practically noncorrosive to ordinary steel, but moist chlorine, or chlorine dissolved in water, is extremely corrosive and attacks most of the common metals and alloys. The reverse is true for titanium—dry chlorine gas is more corrosive than wet chlorine.

1-7 Future Outlook

The future will place greater and greater demands on corrosion engineers. They must meet the challenge with their expertise and must exercise ingenuity to solve new problems. Energy considerations, materials shortages, and political aspects are relatively new complicating factors. The abnormal conditions of today will be normal tomorrow. In the past the emphasis has been on the development of "bigger and better alloys" and other materials; in the future, acceptable substitutes may be emphasized. For example, a Fe-6Cr-6Al alloy might be used instead of 18Cr-8Ni where the full corrosion resistance of the latter is not essential. New research tools are now available and better ones will be available later to aid in the study and understanding of corrosion and its prevention. Closer collaboration between corrosion engineers and corrosion scientists is a must. Greater collaboration between countries will occur.

Closer collaboration between corrosion engineers (and materials engineers) and design engineers is a must. The corrosion engineer must be a part of the design team from the beginning of the project. He should "sign off" on drawings and specifications. The corrosion and design engineers must understand fracture mechanics aspects and also inspection techniques including nondestructive examination.

There is a greater national awareness today than a decade ago. Witness the corrosion cost study resulting from a congressional directive (Sec. 1-1). This awareness will increase.*

M. H. Van de Voorde in "Materials for Advanced Energy Technologies—A European Viewpoint" (*J. Metals*, 19–23, July 1983) emphasizes the importance of many points made in this section and this chapter. For example, "Investment in materials research may be crucial for the survival of European energy supply and industrial innovation." Also, "Materials science must be reassessed and its great potential as a future profession must be acknowledged."

The Materials Technology Institute of the Chemical Process Industries was established in 1977. Consumers and producers alike are contributing funds for study of procedures to mitigate corrosion losses in areas of mutual interest. Some of the results are described later in this book. Other industry groups should form similar organizations. These combined efforts are more cost effective and productive than individual efforts.

A large number of plants using corrosive processes will be built in the future. These include coal conversion, power, refineries, synthetic fuel plants, oil and gas wells, thousands of miles of pipelines, and many other process plants. The number of environmental control systems will mushroom at great cost. In many cases corrosion problems will increase in severity.

There is a great clamor for universities and colleges to provide training in the field of corrosion. Closer alliance between universities and industries

*The federal budget for 1985 calls for over $1 billion for R and D in materials science and engineering (*Materials Performance*, 63, May 1984).

should occur. The best way to reduce corrosion costs is to have more practicing corrosion engineers.

The prospects of an interesting and rewarding career look bright for the corrosion engineer.

CHAPTER

TWO

CORROSION PRINCIPLES

2-1 Introduction

To view corrosion engineering in its proper perspective, it is necessary to remember that the choice of a material depends on many factors, including its corrosion behavior. Figure 2-1 shows some of the properties that determine the choice of a structural material. Although we are primarily concerned with the corrosion resistance of various materials, the final choice frequently depends on factors other than corrosion resistance. As mentioned in Chap. 1, the cost and the corrosion resistance of the material usually are the most important properties in most engineering applications requiring high chemical resistance. However, for architectural applications, appearance is often the most important consideration. Fabricability, which includes the ease of forming, welding, and other mechanical operations, must also be considered. In engineering applications, the mechanical behavior or strength is also important and has to be considered even though the material is being selected for its corrosion resistance. Finally, for many highly resistant materials such as gold, platinum, and some of the super-alloys, the availability of these materials frequently plays a deciding factor in whether or not they will be used. In many instances the delivery time for some of the exotic metals and alloys is prohibitive.

The engineering aspects of corrosion resistance cannot be overemphasized. Complete corrosion resistance in almost all media can be achieved by the use of either platinum or glass, but these materials are not practical in most cases.

Corrosion resistance or chemical resistance depends on many factors. Its complete and comprehensive study requires a knowledge of several fields of scientific knowledge as indicated in Fig. 2-2. Thermodynamics and

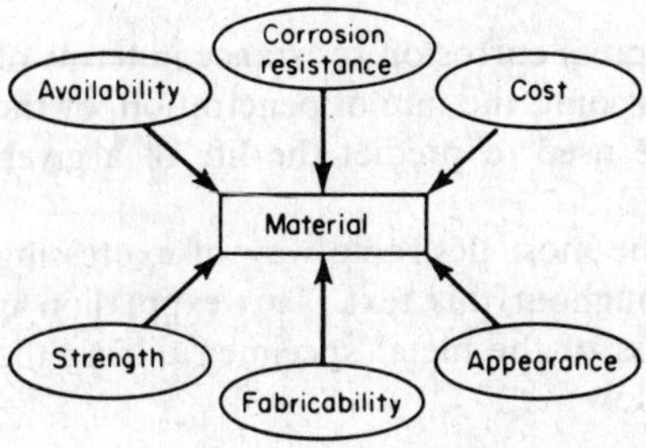

Figure 2-1 Factors affecting choice of an engineering material.

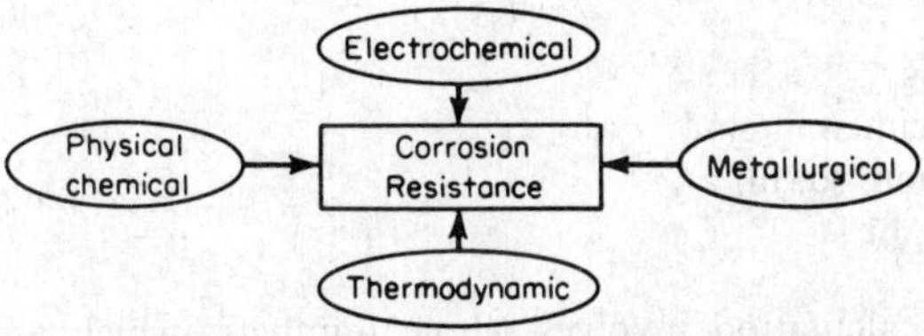

Figure 2-2 Factors affecting corrosion resistance of a metal.

electrochemistry are of great importance in understanding and controlling corrosion.

Thermodynamic studies and calculations indicate the spontaneous direction of a reaction. In the case of corrosion, thermodynamic calculations can determine whether or not corrosion is theoretically possible. Electrochemistry and its associated field, electrode kinetics, are introduced in this chapter and discussed in considerable detail in Chaps. 9 and 10.

Metallurgical factors frequently have a pronounced influence on corrosion resistance. In many cases the metallurgical structure of alloys can be controlled to reduce corrosive attack. Physical chemistry and its various disciplines are most useful for studying the mechanisms of corrosion reactions, the surface conditions of metals, and other basic properties.

In this chapter and the ones that follow, all of these disciplines that are important for the understanding and controlling of corrosion will be utilized. Since the rate of corrosion is of primary interest for engineering application, electrochemical theory and concepts will be considered in greater detail.

2-2 Corrosion Rate Expressions

Throughout this book, metals and nonmetals will be compared on the basis of their corrosion resistance. To make such comparisons meaningful, the rate of attack for each material must be expressed quantitatively. Corrosion rates have been expressed in a variety of ways in the literature; such as percent weight loss, milligrams per square centimeter per day, and grams per square inch per hour. These do not express corrosion resistance in terms of penetration. From an engineering viewpoint, the rate of penetration, or the thinning of a structural piece, can be used to predict the life of a given component.

The expression *mils per year* is the most desirable way of expressing corrosion rates and will be used throughout this text. This expression is readily calculated from the weight loss of the metal specimen during the corrosion test by the formula given below:

$$\text{mpy} = \frac{534W}{DAT}$$

where W = weight loss, mg
D = density of specimen, g/cm^3
A = area of specimen, sq. in.
T = exposure time, hr

This corrosion rate calculation involves whole numbers, which are easily handled.

Section 4-13 in Chap. 4 describes corrosion rate expressions in greater detail, including the metric system.

ELECTROCHEMICAL ASPECTS

2-3 Electrochemical Reactions

The electrochemical nature of corrosion can be illustrated by the attack on zinc by hydrochloric acid. When zinc is placed in dilute hydrochloric acid, a vigorous reaction occurs; hydrogen gas is evolved and the zinc dissolves, forming a solution of zinc chloride. The reaction is:

$$Zn + 2HCl \rightarrow ZnCl_2 + H_2 \tag{2.1}$$

Noting that the chloride ion is not involved in the reaction, this equation can be written in the simplified form:

$$Zn + 2H^+ \rightarrow Zn^{2+} + H_2 \tag{2.2}$$

Hence, zinc reacts with the hydrogen ions of the acid solution to form zinc ions and hydrogen gas. Examining the above equation, it can be seen that during the reaction, zinc is oxidized to zinc ions and hydrogen ions are reduced to hydrogen. Thus Eq. (2.2) can be conveniently divided into two reactions, the oxidation of zinc and the reduction of hydrogen ions:

Oxidation (anodic reaction)*	$Zn \rightarrow Zn^{2+} + 2e$	(2.3)
Reduction (cathodic reaction)	$2H^+ + 2e \rightarrow H_2$	(2.4)

An oxidation or anodic reaction is indicated by an increase in valence or a production of electrons. A decrease in valence charge or the consumption of electrons signifies a reduction or cathodic reaction. Equations (2.3) and (2.4) are partial reactions—both must occur simultaneously and at the same rate on the metal surface. If this were not true, the metal would spontaneously become electrically charged, which is clearly impossible. This leads to one of the most important basic principles of corrosion: *during metallic corrosion, the rate of oxidation equals the rate of reduction* (in terms of electron production and consumption).

The above concept is illustrated in Fig. 2-3. Here a zinc atom has been transformed into a zinc ion and two electrons. These electrons, which remain in the metal, are immediately consumed during the reduction of hydrogen ions. Figure 2-3 shows these two processes spatially separated for clarity. Whether or not they are actually separated or occur at the same point on the surface does not affect the above principle of charge conservation. In some corrosion reactions the oxidation reaction occurs uniformly on the surface, while in other cases it is localized and occurs at specific areas. These effects are described in detail in following chapters.

The corrosion of zinc in hydrocholoric acid is an electrochemical process. That is, any reaction that can be divided into two (or more) partial reactions of oxidation and reduction is termed electrochemical. Dividing corrosion or other electrochemical reactions into partial reactions makes them simpler to understand. Iron and aluminum, like zinc, are also rapidly

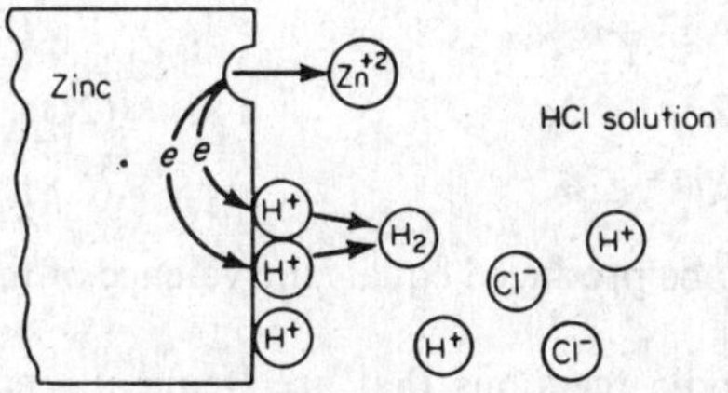

Figure 2-3 Electrochemical reactions occurring during corrosion of zinc in air-free hydrochloric acid.

*Until recently, corrosion theory usually has been based on the concept of local anode and cathode areas on metal surfaces. However, descriptions of corrosion phenomena based on modern electrode kinetic principles (mixed-potential theory) are more general since they apply to any corroding system and do not depend on assumptions regarding the distribution of anodic and cathodic reactions. It should be emphasized that these two methods of treating corrosion are not conflicting—they merely represent two different approaches. In this text, we have used electrode kinetic descriptions because of their greater simplicity and more general application.

corroded by hydrochloric acid. The reactions are:

$$Fe + 2HCl \rightarrow FeCl_2 + H_2 \tag{2.5}$$

$$2Al + 6HCl \rightarrow 2AlCl_3 + 3H_2 \tag{2.6}$$

Although at first sight these appear quite different, comparing the partial processes of oxidation and reduction indicates that reactions (2.1), (2.5), and (2.6) are quite similar. All involve the hydrogen ion reduction and they differ only in their oxidation or anodic reactions:

$$Zn \rightarrow Zn^{2+} + 2e \tag{2.3}$$

$$Fe \rightarrow Fe^{2+} + 2e \tag{2.7}$$

$$Al \rightarrow Al^{3+} + 3e \tag{2.8}$$

Hence, the problem of hydrochloric acid corrosion is simplified since in every case the cathodic reaction is the evolution of hydrogen gas according to reaction (2.4). This also applies to corrosion in other acids such as sulfuric, phosphoric, hydrofluoric, and water-soluble organic acids such as formic and acetic. In each case, only the hydrogen ion is active, the other ions such as sulfate, phosphate, and acetate do not participate in the electrochemical reaction.

When viewed from the standpoint of partial processes of oxidation and reduction, all corrosion can be classified into a few generalized reactions. The anodic reaction in every corrosion reaction is the oxidation of a metal to its ion. This can be written in the general form:

$$M \rightarrow M^{+n} + ne \tag{2.9}$$

A few examples are:

$$Ag \rightarrow Ag^{+} + e \tag{2.10}$$

$$Zn \rightarrow Zn^{2+} + 2e \tag{2.3}$$

$$Al \rightarrow Al^{3+} + 3e \tag{2.8}$$

In each case the number of electrons produced equals the valence of the ion.

There are several different cathodic reactions that are frequently encountered in metallic corrosion. The most common are:

Hydrogen evolution	$2H^{+} + 2e \rightarrow H_2$	(2.4)
Oxygen reduction (acid solutions)	$O_2 + 4H^{+} + 4e \rightarrow 2H_2O$	(2.11)
Oxygen reduction (neutral or basic solutions)	$O_2 + 2H_2O + 4e \rightarrow 4OH^{-}$	(2.12)
Metal ion reduction	$M^{3+} + e \rightarrow M^{2+}$	(2.13)

Metal deposition $$M^+ + e \rightarrow M \qquad (2.14)$$

Hydrogen evolution is a common cathodic reaction since acid or acidic media are frequently encountered. Oxygen reduction is very common, since any aqueous solution in contact with air is capable of producing this reaction Metal ion reduction and metal deposition are less common reactions and are most frequently found in chemical process streams. All of the above reactions are quite similar—they consume electrons.

The above partial reactions can be used to interpret virtually all corrosion problems. Consider what happens when iron is immersed in water or seawater which is exposed to the atmosphere (an automobile fender or a steel pier piling are examples). Corrosion occurs. The anodic reaction is:

$$Fe \rightarrow Fe^{2+} + 2e \qquad (2.7)$$

Since the medium is exposed to the atmosphere, it contains dissolved oxygen. Water and seawater are nearly neutral, and thus the cathodic reaction is:

$$O_2 + 2H_2O + 4e \rightarrow 4OH^- \qquad (2.12)$$

Remembering that sodium and chloride ions do not participate in the reaction, the overall reaction can be obtained by adding (2.7) and (2.12):

$$2Fe + 2H_2O + O_2 \rightarrow 2Fe^{2+} + 4OH^- \rightarrow 2Fe(OH)_2\downarrow \qquad (2.15)$$

Ferrous hydroxide precipitates from solution. However, this compound is unstable in oxygenated solutions and is oxidized to the ferric salt:

$$2Fe(OH)_2 + H_2O + \tfrac{1}{2}O_2 \rightarrow 2Fe(OH)_3 \qquad (2.16)$$

The final product is the familiar rust.

The classic example of a replacement reaction, the interaction of zinc with copper sulfate solution, illustrates metal deposition:

$$Zn + Cu^{2+} \rightarrow Zn^{2+} + Cu \qquad (2.17)$$

or, viewed as partial reactions:

$$Zn \rightarrow Zn^{2+} + 2e \qquad (2.3)$$

$$Cu^{2+} + 2e \rightarrow Cu \qquad (2.18)$$

The zinc initially becomes plated with copper and eventually the products are copper sponge and zinc sulfate solution.

During corrosion, more than one oxidation and one reduction reaction may occur. When an alloy is corroded, its component metals go into solution as their respective ions. More importantly, more than one reduction reaction can occur during corrosion. Consider the corrosion of zinc in aerated hydrochloric acid. Two cathodic reactions are possible: the evolution of hydrogen and the reduction of oxygen. This is illustrated schematically in Fig. 2-4. On the surface of the zinc there are two electron-consuming

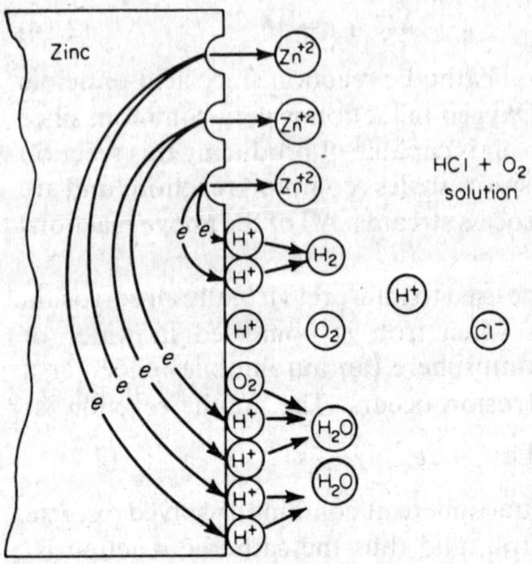

Figure 2-4 Electrochemical reactions occurring during corrosion of zinc in aerated hydrochloric acid.

reactions. Since the rates of oxidation and reduction must be equal, increasing the total reduction rate increases the rate of zinc solution. Hence, acid solutions containing dissolved oxygen will be more corrosive than air-free acids. Oxygen reduction simply provides a new means of "electron disposal." The same effect is observed if any oxidizer is present in acid solutions. A frequent impurity in commercial hydrochloric acid is ferric ion, present as ferric chloride. Metals corrode much more rapidly in such impure acid because there are two cathodic reactions, hydrogen evolution and ferric ion reduction:

$$Fe^{3+} + e \rightarrow Fe^{2+} \tag{2.19}$$

Since the anodic and cathodic reactions occuring during corrosion are mutually dependent, it is possible to reduce corrosion by reducing the rates of either reaction. In the above case of impure hydrochloric acid, it can be made less corrosive by removing the ferric ions and consequently reducing the total rate of cathodic reduction. Oxygen reduction is eliminated by preventing air from contacting the aqueous solution or by removing air that has been dissolved. Iron will not corrode in air-free water or seawater because there is no cathodic reaction possible.

If the surface of the metal is coated with paint or other nonconducting film, the rates of both anodic and cathodic reactions will be greatly reduced and corrosion will be retarded. A corrosion inhibitor is a substance that when added in small amounts to a corrosive reduces its corrosivity. Corrosion inhibitors function by interfering with either the anodic or cathodic reactions or both. Many of these inhibitors are organic compounds; they function by forming an impervious film on the metal surface or by interfering with either the anodic or cathodic reactions. High-molecular-weight amines retard the hydrogen evolution reaction and subsequently reduce corrosion rate. It is obvious that good conductivity must be maintained in both the metal and the electrolyte during the corrosion reaction. Of course, it is not practical to increase the electrical resistance of the metal, since the sites of the anodic and cathodic reactions are not known, nor are they predictable. However, it is possible to increase the electrical resistance of the electrolyte or corrosive and thereby reduce corrosion. Very pure water is much less corrosive than impure or natural waters. The low corrosivity of high-purity water is primarily due to its high electrical resistance. These methods for increasing corrosion resistance are described in greater detail in following chapters.

2-4 Polarization

The concept of polarization is briefly discussed here because of its importance in understanding corrosion behavior and corrosion reactions. The following discussion is simplified, and readers desiring a more comprehensive and quantitative discussion of this topic are referred to Chaps. 9 and 10.

The rate of an electrochemical reaction is limited by various physical and chemical factors. Hence, an electrochemical reaction is said to be polarized or retarded by these environmental factors. Polarization can be conveniently divided into two different types, activation polarization and concentration polarization.

Activation polarization refers to an electrochemical process that is controlled by the reaction sequence at the metal-electrolyte interface. This is easily illustrated by considering hydrogen-evolution reaction on zinc during corrosion in acid solution. Figure 2-5 schematically shows some of the possible steps in hydrogen reduction on a zinc surface. These steps can also be applied to the reduction of any species on a metal surface. The species must first be adsorbed or attached to the surface before the reaction can proceed according to step 1. Following this, electron transfer (step 2) must occur, resulting in a reduction of the species. As shown in step 3, two hydrogen atoms then combine to form a bubble of hydrogen gas (step 4). The speed of reduction of the hydrogen ions will be controlled by the slowest of these steps. This is a highly simplified picture of the reduction of hydrogen; numerous mechanisms have been proposed, most of which are much more complex than that shown in Fig. 2-5.

Concentration polarization refers to electrochemical reactions that are controlled by the diffusion in the electrolyte. This is illustrated in Fig. 2-6 for the case of hydrogen evolution. Here, the number of hydrogen ions in solution is quite small, and the reduction rate is controlled by the diffusion of hydrogen ions to the metal surface. Note that in this case the reduction rate is

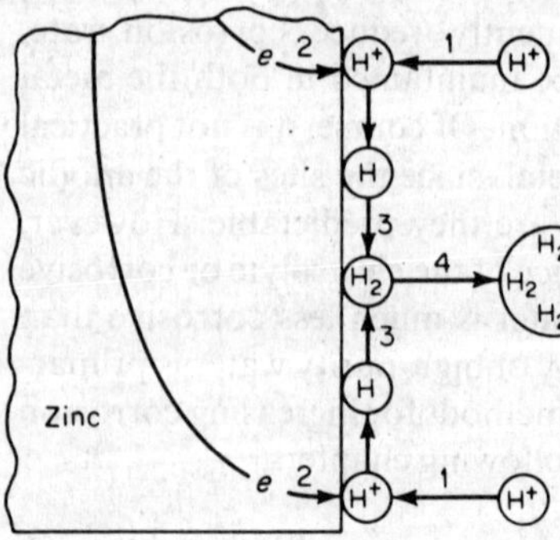

Figure 2-5 Hydrogen-reduction reaction under activation control (simplified).

controlled by processes occurring within the bulk solution rather than at the metal surface. Activation polarization usually is the controlling factor during corrosion in media containing a high concentration of active species (e.g., concentrated acids). Concentration polarization generally predominates when the concentration of the reducible species is small (e.g., dilute acids, aerated salt solutions). In most instances concentration polarization during metal dissolution is usually small and can be neglected; it is only important during reduction reactions.

The importance of distinguishing between activation and concentration polarization cannot be overemphasized. Depending on what kind of polarization is controlling the reduction reaction, environmental variables produce different effects. For example, any changes in the system that increase the diffusion rate will decrease the effects of concentration polarization and hence increase reaction rate. Thus, increasing the velocity or agitation of the corrosive medium will increase rate *only* if the cathodic process is controlled

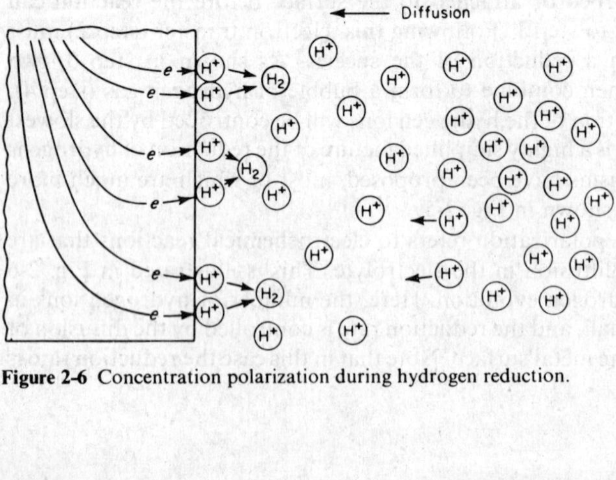

Figure 2-6 Concentration polarization during hydrogen reduction.

by concentration polarization. If both the anodic and cathodic reactions are controlled by activation polarization, agitation will have no influence on corrosion rate. These and other differences are discussed in detail below and also in Chaps. 9 and 10.

2-5 Passivity

The phenomenon of metallic passivity has fascinated scientists and engineers for over 120 years, since the days of Faraday. The phenomenon itself is rather difficult to define because of its complex nature and the specific conditions under which it occurs. Essentially, passivity refers to the loss of chemical reactivity experienced by certain metals and alloys under particular environmental conditions. That is, certain metals and alloys become essentially inert and act as if they were noble metals such as platinum and gold. Fortunately, from an engineering standpoint, the metals most susceptible to this kind of behavior are the common engineering and structural materials, including iron, nickel, silicon, chromium, titanium, and alloys containing these metals. Also, under limited conditions other metals such as zinc, cadmium, tin, uranium, and thorium have also been observed to exhibit passivity effects.

Passivity, although difficult to define, can be quantitatively described by characterizing the behavior of metals which show this unusual effect. First, consider the behavior of what can be called a normal metal, that is, a metal which does not show passivity effects. In Fig. 2-7 the behavior of such a metal is illustrated. Let us assume that we have a metal immersed in an air-free acid solution with an oxidizing power corresponding to point *A* and a corrosion rate corresponding to this point. If the oxidizing power of this solution is increased, say, by adding oxygen or ferric ions, the corrosion rate of the metal will increase rapidly. Note that for such a metal the corrosion rate increases as the oxidizing power of the solution increases. This increase in rate is exponential and yields a straight line when plotted on a semilogarithmic scale as in Fig. 2-7. The oxidizing power of the solution is controlled by both the specific oxidizing power of the reagents and the

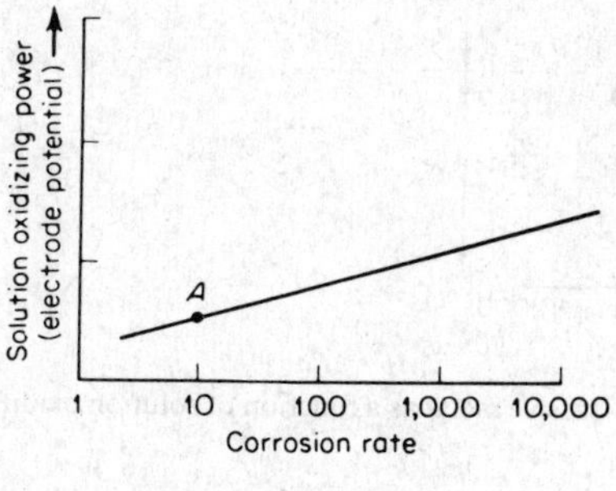

Figure 2-7 Corrosion rate of a metal as a function of solution oxidizing power (electrode potential).

concentration of these reagents. As will be described in Chaps. 9 and 10, oxidizing power can be precisely defined by electrode potential, but this is beyond our present discussion.

Figure 2-8 illustrates the typical behavior of a metal that demonstrates passivity effects. The behavior of this metal or alloy can be conveniently divided into three regions: active, passive, and transpassive. In the active region the behavior of this material is identical to that of a normal metal. Slight increases in the oxidizing power of the solution cause a corresponding rapid increase in the corrosion rate. If more oxidizing agent is added, the corrosion rate shows a sudden decrease. This corresponds to the beginning of the passive region. Further increases in oxidizing agents produce little, if any, change in the corrosion rate of the material. Finally, at very high concentrations of oxidizers or in the presence of very powerful oxidizers the corrosion rate again increases with increasing oxidizer power. This region is termed the transpassive region.

It is important to note that during the transition from the active to the passive region, a 10^3 to 10^6 reduction in corrosion rate is usually observed. The precise cause for this unusual active-passive-transpassive transition is not completely understood. It is a special case of activation polarization due to the formation of a surface film or protective barrier that is stable over a considerable range of oxidizing power and is eventually destroyed in strong oxidizing solutions. The exact nature of this barrier is not understood. However, for the purposes of engineering application, it is not necessary to understand the mechanism of this unusual effect completely since it can be readily characterized by data such as are shown in Fig. 2-8.

To summarize, metals that possess an active-passive transition become passive or very corrosion resistant in moderately to strongly oxidizing environments. Under extremely strong oxidizing conditions, these materials

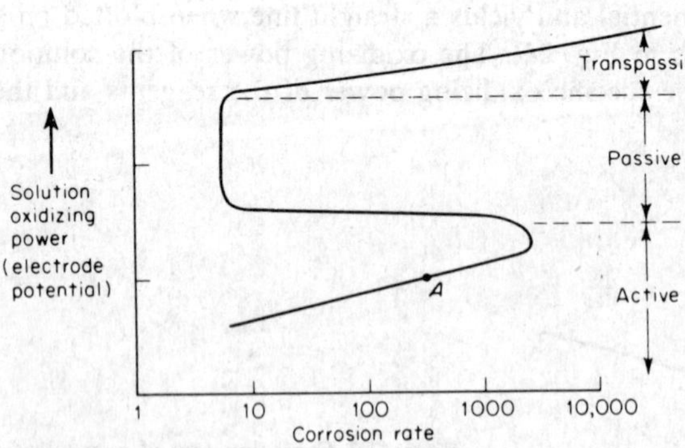

Figure 2-8 Corrosion characteristics of an active-passive metal as a function of solution oxidizing power (electrode potential).

lose their corrosion-resistance properties. These characteristics have been successfully used to develop new methods of preventing corrosion and to predict corrosion resistance. These applications are described in detail in succeeding chapters.

ENVIRONMENTAL EFFECTS

Frequently in the process industries, it is desirable to change process variables. One of the most frequent questions is: What effect will this change have on corrosion rates? In the following section, some of the more common environmental variables are considered on the basis of the concepts developed above.

2-6 Effect of Oxygen and Oxidizers

The effect of oxidizers and oxidizing power was discussed above in connection with the behavior of active-passive metals. The effect of oxidizers on corrosion rate can be represented by the graph shown in Fig. 2-9. Note that the shape of this graph is similar to that of Fig. 2-8 and that this figure is divided into three different sections. Behavior corresponding to section 1

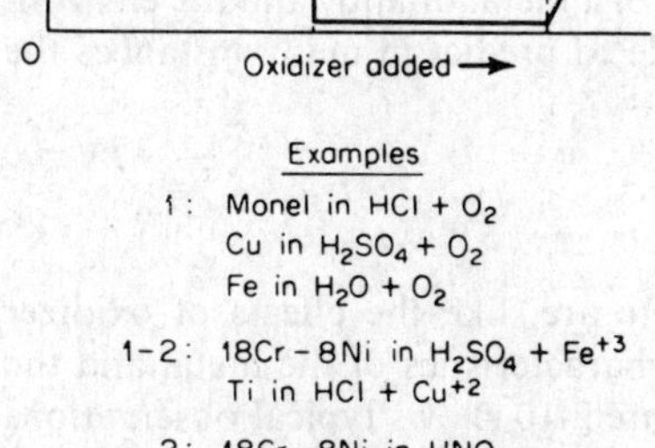

Examples

1: Monel in $HCl + O_2$
Cu in $H_2SO_4 + O_2$
Fe in $H_2O + O_2$

1–2: 18Cr – 8Ni in $H_2SO_4 + Fe^{+3}$
Ti in $HCl + Cu^{+2}$

2: 18Cr – 8Ni in HNO_3
Hastelloy C in $FeCl_3$

2–3: 18Cr – 8Ni in $HNO_3 + Cr_2O_3$

1–2–3: 18Cr – 8Ni in concentrated $H_2SO_4 + HNO_3$ mixtures at elevated temperatures

Figure 2-9 Effect of oxidizers and aeration on corrosion rate.

is characteristic of normal metals and also of active-passive metals when they exist only in the active state. For metals that demonstrate active-passive transition, passivity is achieved only if a sufficient quantity of oxidizer or a sufficiently powerful oxidizer is added to the medium. Increasing corrosion rate with increasing oxidizer concentrations as shown in section 1 is characteristic of Monel and copper in acid solutions containing oxygen. Both of these materials do not passivate. Although iron can be made to passivate in water, the solubility of oxygen is limited, and in most cases it is insufficient to produce a passive state as shown in Fig. 2-8.

An increase in corrosion rate, followed by a rapid decrease, and then a corrosion rate that is essentially independent of oxidizer concentration is characteristic of such active-passive metals and alloys as 18Cr-8Ni stainless steel and titanium.

If an active-passive metal is initially passive in a corrosive medium, the addition of further oxidizing agents has only a negligible effect on corrosion rate. This condition frequently occurs when an active-passive metal is immersed in an oxidizing medium such as nitric acid or ferric chloride. The behavior represented by sections 2 and 3 results when a metal, initially in the passive state, is exposed to very powerful oxidizers and makes a transition into the transpassive region. This kind of behavior is frequently observed with stainless steel when very powerful oxidizing agents such as chromates are added to the corrosive medium. In hot nitrating mixtures containing concentrated sulfuric and nitric acids, the entire active-passive-transpassive transition can be observed with the increased ratios of nitric to sulfuric acid.

It is readily seen that the effect of oxidizer additions or the presence of oxygen on corrosion rate depends on both the medium and the metals involved. The corrosion rate may be increased by the addition of oxidizers, oxidizers may have no effect on the corrosion rate, or a very complex behavior may be observed.

By knowing the basic characteristics of a metal or alloy and the environment to which it is exposed it is possible to predict in many instances the effect of oxidizer additions.

2-7 Effects of Velocity

The effects of velocity on corrosion rate are, like the effects of oxidizer additions, complex and depend on the characteristics of the metal and the environment to which it is exposed. Figure 2-10 shows typical observations when agitation or solution velocity are increased. For corrosion processes that are controlled by activation polarization, agitation and velocity have no effect on the corrosion rate as illustrated in curve *B*. If the corrosion process is under cathodic diffusion control, then agitation increases the corrosion rate as shown in curve *A*, section 1. This effect generally occurs

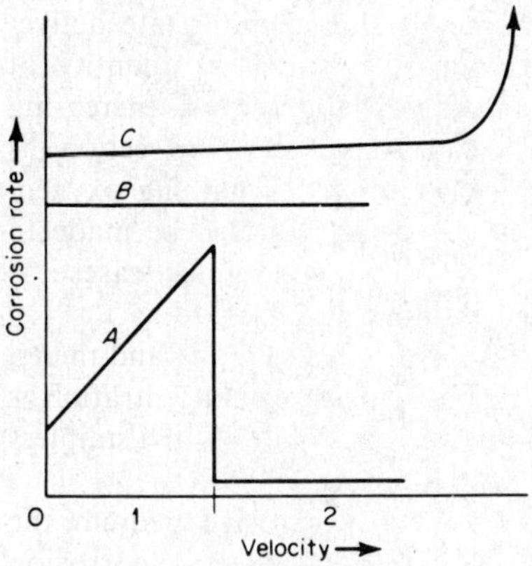

Examples

Curve A:
1: Fe in $H_2O + O_2$
Cu in $H_2O + O_2$
1–2: 18Cr–8Ni in $H_2SO_4 + Fe^{+3}$
Ti in $HCl + Cu^{+2}$

Curve B: Fe in dilute HCl
18Cr–8Ni in H_2SO_4

Curve C: Pb in dilute H_2SO_4
Fe in concentrated H_2SO_4

Figure 2-10 Effect of velocity on corrosion rate.

when an oxidizer is present in very small amounts, as is the case for dissolved oxygen in acids or water.

If the process is under diffusion control and the metal is readily passivated, then the behavior corresponding to curve *A*, sections 1 and 2, will be observed. That is, with increasing agitation, the metal will undergo an active-to-passive transition. Easily passivated materials such as stainless steel and titanium frequently are more corrosion resistant when the velocity of the corrosion medium is high.

Some metals owe their corrosion resistance in certain mediums to the formation of massive bulk protective films on their surfaces. These films differ from the usual passivating films in that they are readily visible and much less tenacious. It is believed that both lead and steel are protected from attack in sulfuric acid by insoluble sulfate films. When materials such as these are exposed to extremely high corrosive velocities, mechanical damage or removal of these films can occur, resulting in accelerated attack as shown in curve *C*. This is called erosion corrosion and is discussed in Chap. 3. In the case of curve *C*, note that until mechanical damage actually occurs, the effect of agitation or velocity is virtually negligible.

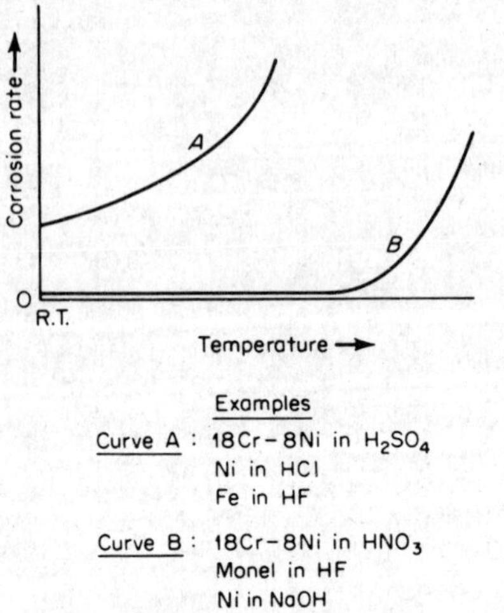

Figure 2-11 Effect of temperature on corrosion rate.

2-8 Effect of Temperature

Temperature increases the rate of almost all chemical reactions. Figure 2-11 illustrates two common observations on the effect of temperature on the corrosion rates of metals. Curve *A* represents the behavior noted above, a very rapid or exponential rise in corrosion rate with increasing temperature. Behavior such as noted in curve *B* is also quite frequently observed. That is, an almost negligible temperature effect is followed by a very rapid rise in corrosion rate at higher temperatures. In the case of 18-8 stainless steel in nitric acid, this effect is readily explained. Increasing the temperature of nitric acid greatly increases its oxidizing power. At low or moderate temperatures, stainless steels exposed to nitric acid are in the passive state very close to the transpassive region. Hence, an increase in oxidizing power causes a very rapid increase in the corrosion rate of these materials. A similar sort of mechanism may explain the behavior of Monel and nickel, as noted in Fig. 2-11. However, it is possible that curves such as *B* in many instances erroneously represent actual behavior. If the corrosion rate at low temperature is very low, and increases exponentially, linear plots will appear as curve *B*. That is, corrosion rate increases rapidly with temperature; this is not evident in the usual plots of corrosion rate versus temperature because of the choice of scales.

2-9 Effects of Corrosive Concentration

Figure 2-12 shows schematically the effects of corrosive concentration on corrosion rate. Note that curve *A* has two sections, 1 and 2. Many materials that exhibit passivity effects are only negligibly affected by wide changes in

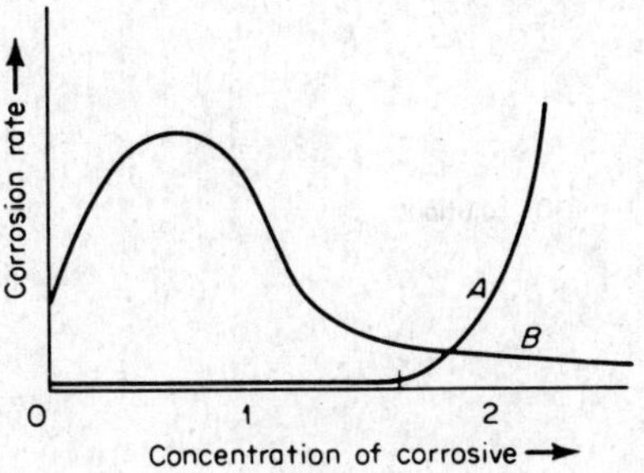

Examples

Curve A:

1: Ni in NaOH
$18Cr-8Ni$ in HNO_3
Hastelloy B in HCl
Ta in HCl

1–2: Monel in HCl
Pb in H_2SO_4

Curve B:

Al in acetic acid and HNO_3
$18Cr-8Ni$ in H_2SO_4
Fe in H_2SO_4

Figure 2-12 Effect of corrosive concentration on corrosion rate.

corrosive concentration, as shown in curve *A*, section 1. Other materials show similar behavior except at very high corrosive concentrations, when the corrosion rate increases rapidly as shown in curve *A*, sections 1 and 2. Lead is a material that shows this effect, and it is believed to be due to the fact that lead sulfate, which forms a protective film in low concentrations of sulfuric acid, is soluble in concentrated sulfuric acid. The behavior of acids that are soluble in all concentrations of water often yield curves similar to curve *B* in Fig. 2-12. Initially, as the concentration of corrosive is increased, the corrosion rate is likewise increased. This is primarily due to the fact that the amount of hydrogen ions, which are the active species, are increased as acid concentration is increased. However, as acid concentration is increased further, corrosion rate reaches a maximum and then decreases. This is undoubtedly due to the fact that at very high concentrations of acids ionization is reduced. Because of this, many of the common acids—such as sulfuric, acetic, hydrofluoric, and others—are virtually inert when in the pure state, or 100% concentration, and at moderate temperatures.

2-10 Effect of Galvanic Coupling

In many practical applications the contact of dissimilar materials is unavoidable. In complex process streams and piping arrangements, different metals and alloys are frequently in contact with each other and the corrosive medium. The effects of galvanic coupling will be considered in detail later

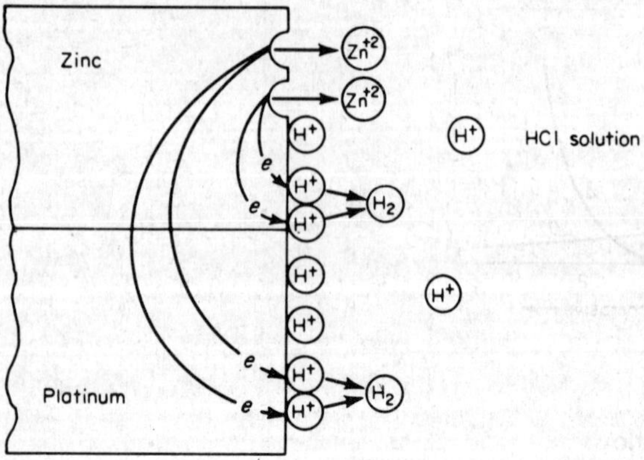

Figure 2-13 Electrochemical reactions occurring on galvanic couple of zinc and platinum.

and are only briefly mentioned here. Consider a piece of zinc immersed in a hydrochloric acid solution and contacted to a noble metal such as platinum (Fig. 2-13). Since platinum is inert in this medium, it tends to increase the surface at which hydrogen evolution can occur. Further, hydrogen evolution occurs much more readily on the surface of platinum than on zinc. These two factors increase the rate of the cathodic reaction and consequently increase the corrosion rate of the zinc. Note that the effect of galvanic coupling in this instance is virtually identical to that of adding an oxidizer to a corrosive solution. In both instances, the rate of electron consumption is increased and hence the rate of metal dissolution increases. It is important to recognize that galvanic coupling does not always increase the corrosion rate of a given metal; in some cases it decreases the corrosion rate. These specialized cases will be discussed in later chapters.

METALLURGICAL AND OTHER ASPECTS

2-11 Metallic Properties

Metals and alloys are crystalline solids. That is, the atoms of a metal are arranged in a regular, repeating array. The three most common crystalline arrangements of metals are illustrated in Fig. 2-14. Iron and steel have a body-centered cubic structure, the austenitic stainless steels are face-centered cubic, and magnesium possesses a hexagonal, close-packed lattice structure. Metallic properties differ from those of other crystalline solids such as ceramics and chemical salts. They are ductile (can be deformed plastically without fracturing) and are good conductors of electricity and heat. These

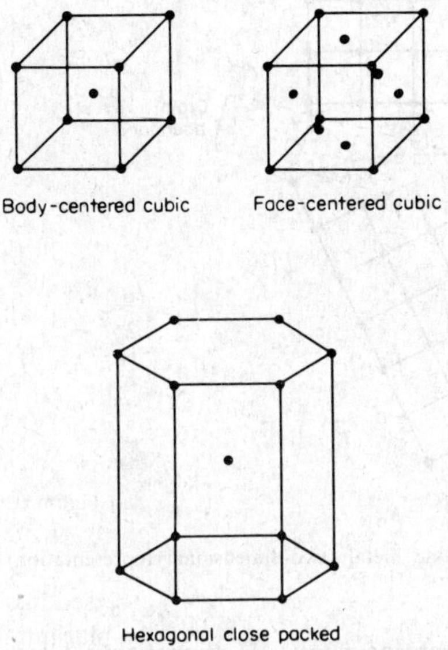

Figure 2-14 Metallic crystal structures.

properties result from the nondirectional bonding of metals; that is, each atom is bonded to many of its neighbors. Hence, the crystal structures are simple and closely packed as shown in Fig. 2-14. Ductility is probably the most important property of metals. Their ductility permits almost unlimited fabrication. Further, when highly stressed, metals usually yield plastically before fracturing. This property is, of course, invaluable in engineering applications.

When a metal solidifies during casting, the atoms, which are randomly distributed in the liquid state, arrange themselves in a crystalline array. However, this ordering usually begins at many points in the liquid, and as these blocks of crystals or grains meet, there is a mismatch at their boundary. When the metal has solidified and cooled, there will be numerous regions of mismatch between each grain. These regions are called grain boundaries. Figure 2-15 shows this using a two-dimensional representation of a grain boundary. Since the most stable configuration of the metal is its particular crystal lattice, grain boundaries are high-energy areas and are more active chemically. Hence, grain boundaries are usually attacked slightly more rapidly than grain faces when exposed to a corrosive. Metallographic etching, in many cases, depends on this difference in chemical reactivity to develop contrast between grains. Figure 2-16 shows a magnified view of 18-8 stainless steel that has been etched in acid solution. The grain boundaries appear dark because they have been more severely attacked than the grains.

Alloys are mixtures of two or more metals or elements. There are two

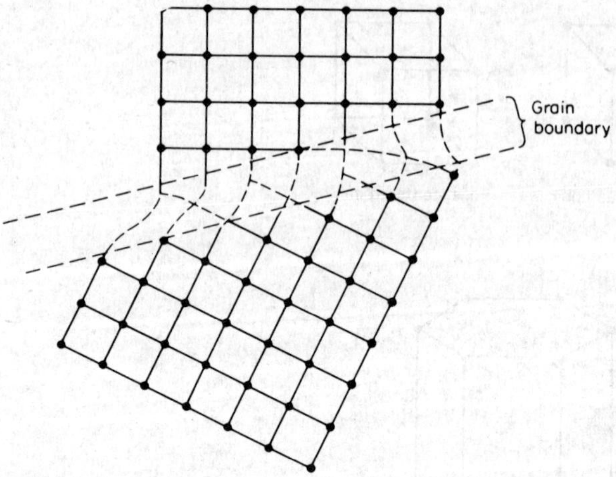

Figure 2-15 Grain boundary in a polycrystalline metal (two-dimensional representation).

kinds of alloys—homogeneous and heterogeneous. Homogeneous alloys are solid solutions. That is, the components are completely soluble in one another, and the material has only one phase. An example of a homogeneous or solid-solution alloy is 18-8 stainless steel (Fig. 2-16). The iron, nickel, chromium, and carbon are dissolved completely, and the alloy has a uniform composition. Heterogeneous alloys are mixtures of two or more separate phases. The components of such alloys are not completely soluble and exist as separate phases. The composition and structure of these alloys are not uniform. Figure 2-17 shows a photomicrograph of low-carbon steel. The carbon combines with some of the iron to form iron carbide, which usually appears in a lamellar form. Each type of alloy has advantages and disadvantages. Solid-solution alloys are generally more ductile and have lower strength than heterogeneous alloys. The choice between these two

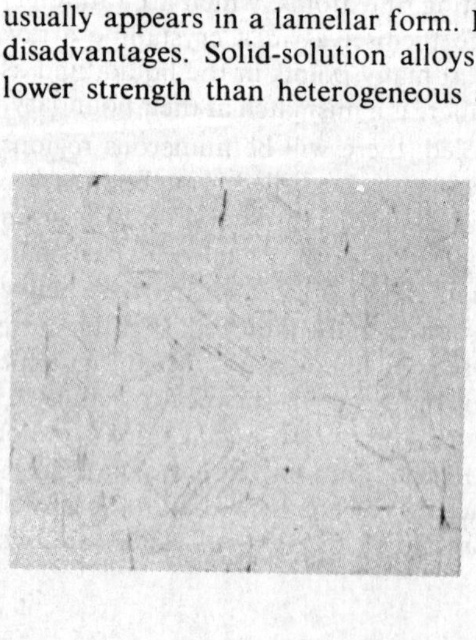

Figure 2-16 Photomicrograph of 18Cr-8Ni stainless steel etched to reveal grain boundaries (100×).

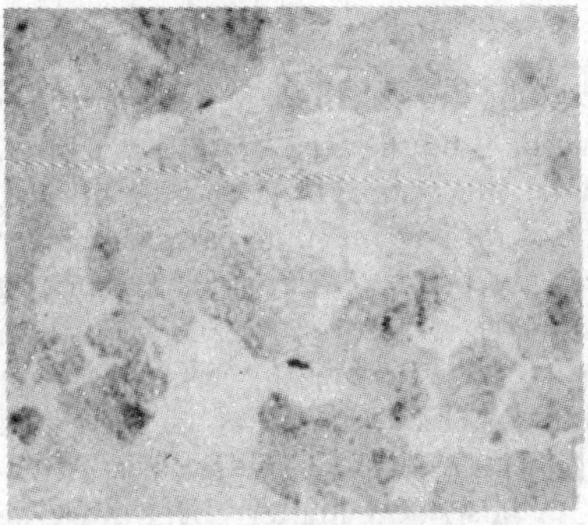

Figure 2-17 Photomicrograph of carbon steel etched to reveal iron carbide platelets (600 ×).

types depends on the mechanical properties desired. Solid-solution alloys are usually more corrosion resistant than alloys with two (or more) phases, since galvanic coupling effects are not present. However, there are important exceptions to this generalization and they are described in the following chapters.

Alloys are quite similar to aqueous solutions. Some substances can be dissolved, whereas others are insoluble. Solubility usually increases rapidly with increasing temperature. For example, iron carbide is completely soluble in iron at high temperatures; hence steel becomes a solid solution when heated to a high temperature. Precipitation of a phase can occur from supersaturated solid solutions as it does in the case of liquid solutions. As noted above, grain boundaries are high-energy areas, so precipitation frequently begins at the grain interfaces.

Other differences in the metal can be chemical, metallurgical, or mechanical in nature. Examples are impurities such as oxides and other inclusions, mill scale, orientation of grains, dislocation arrays, differences in composition of the microstructure, precipitated phases, localized stresses, scratches, and nicks. Highly polished surfaces are used in only special cases. Very pure metals are more corrosion resistant than commercial materials. For example, very pure and smooth zinc will not corrode in very pure hydrochloric acid, yet their commercial counterparts react rapidly. However, pure metals are expensive, and they are usually weak—one would not build a bridge of pure iron.

The following shows the effect of purity of aluminum on corrosion by hydrochloric acid:

% aluminum	Relative corrosion rate
99.998	1
99.97	1,000
99.2	30,000

Differences in the environment will be discussed in Chap. 3.

2-12 Economic Considerations

Control of corrosion is primarily an economic problem. Whether or not to apply a control method is usually determined by the cost savings involved. The method or methods utilized must be the optimum economic choice. Reduction of plant investment means less money must be earned. Lower maintenance or operating cost increases the profit. Companies are not in business primarily to make steel, chemicals, or automobiles-they are in business to make a profit. Percent return on investment (before or after taxes) is a common criterion. If a less expensive material is used and equivalent performance obtained (the rare case), the choice is easy. Alternative corrosion control systems vary in cost, and higher costs must be justified. Different companies use a variety of criteria. A chemical company requires a shorter time for payoff than an electric power plant because the process of the former is more likely to become obsolete in a shorter time.

Some plants are actually designed for as short a time as 1 year—others for 50 years or longer. A bridge is designed for a 100-year life, an auto for 5 to 10, and a rocket for a minute or less. Corrosion engineers must be familiar with an organization's practices so they can properly and effectively present cases to management (it approves expenditures). If the return is only 3%, more profit would be made by keeping the money in a bank instead of using it to change a process or build a new plant.

The reader is referred, for more details, to an excellent paper by C. P. Dillon.* This paper discusses factors in economic appraisal such as costs, equipment life, interest rate, tax rate, depreciation, and also methods of economic evaluation. A simple case from this paper involves a steel heat exchanger costing \$10,000, with a 2-year life, and a type 316 stainless steel exchanger costing \$20,000 and lasting 8 years. Return on investment is:

$$\mathrm{ROI} = 100\,\frac{(10{,}000/2) - (20{,}000/8)}{20{,}000 - 10{,}000} = 25\%$$

The general and more comprehensive case involves

$$\mathrm{ROI} = \frac{(O_a + I_a/n_a) - (O_b + I_b/n_b)}{I_b - I_a}\,100$$

where O = annual costs including maintenance, production losses, etc.; I/n = linear depreciation, with I = investment or installed cost and n = anticipated life in years; and subscripts a and b refer to present and proposed (or alternate) installations, respectively.

With regard to costs of metals and alloys, composition is the first guideline. Type 430 costs more than ordinary steel because of the added 17%

*C. P. Dillon, Economic Evaluation of Corrosion Control Measures, *Materials Protection*, **4**:38–45 (May 1965).

chromium; 304 costs more because of the nickel content; and 316 costs even more because molybdenum is an expensive alloying element. Copper costs more than iron. However, other factors strongly influence the price to the customer. For example, a given weight bar of steel may be worth $5 as a sash weight, but it may be worth $5,000 as sewing machine needles or $200,000 as balance wheels for watches. A tiny electric motor for a missile costs $300, whereas a 1-hp motor costs $50. Steel castings are much more expensive than cast iron because the former are more difficult to cast. Small castings cost more per pound than large ones in the same material because more labor is involved. Type 403 (aircraft quality 410) costs more than type 410 because 403 calls for better inspection. High-alloy materials such as Hastelloy C are expensive not only because of alloy content but also because they require high rolling temperatures. A more expensive material such as titanium may be more economical than steel for seawater heat exchangers because of less fouling and better heat transfer. In fact this is one of the bases for the use of Teflon tubing in heat exchangers. A fabrication plant producing mainly 316 equipment would charge more for 304 because it is a "special." Low-production items are generally more expensive than those in high production. Intricate shapes cost more per pound. Scarcity also determines price. In times of crisis when nickel is scarce, the corrosion resistance of type 430 increases! In a specialty chemical plant where a variety of products are made intermittently, type 316 vessels are preferred over 304 because they are more versatile from the corrosion standpoint.

The first nylon plant contained many type 304 parts (later plants used steel and cast iron) because it was desirable to keep the "bugs" in a new process plant at a minimum. Appearance, plant shutdowns, contamination of product, safety, and reliability are discussed in Chap. 1.

Corrosion is not a necessary evil. Large savings can be obtained by controlling corrosion. In one case costs were reduced from $2,000,000 to $53,000 per year through proper and intensive effort. Complete cost and maintenance data are helpful to delineate the high cost items and to determine return on investment. Satisfactory performance and desired life at a minimum total cost per year are all important.

The economics of corrosion control have become more complex because of the rising costs of labor, materials, and energy, coupled with rapid variations in interest rates and taxes. The net present value (*NPV*) provides the most accurate basis for analyzing business costs and can be directly applied to the economics of corrosion control. Although *NPV* involved extensive computations, these can be easily performed with pocket calculators, especially those with programmable functions.*

The concept of present value is relatively simple; it is based on the

*For an excellent discussion of this topic see Jon M. Smith, *Financial Analysis & Business Decisions on the Pocket Calculator*, John Wiley & Sons, Inc., New York, 1976.

growth of an investment receiving compound interest:

$$FV = PV(1+i)^n \tag{2.20}$$

where PV is the present value of the investment; FV is the future value; i is the interest rate per compounding period; and n is the number of periods. In the simplest case, the compounding period is one year. Thus i is the annual interest and n equals the number of years. For example, \$1000 invested for 10 years at 6% interest yields:

$$FV = 1000(1+0.06)^{10} = \$1790.85 \tag{2.21}$$

The above calculations can be restated in terms of present value: the present value of \$1790.85 in 10 years from now at 6% interest is \$1000.00. This can be calculated by rearranging Eq. (2.20) as follows:

$$PV = \frac{FV}{(1+i)^n} \tag{2.22}$$

Therefore, any future profit (or loss) can be related to present value. This concept can be expanded to include any number of future cash receipts or payments by the following general formula:

$$NPV = -I + \frac{C_1}{(1+i)^1} + \frac{C_2}{(1+i)^2} + \cdots + \frac{C_n}{(1+i)^n} \tag{2.23}$$

where NPV is the net present value of a number of future cash flows C_1, C_2, C_n, which follow an initial investment I. The net present value can be positive or negative corresponding to a gain or loss respectively. The interest i is the expected or actual rate of return.

The use of Eq. (2.23) is best illustrated by example. Consider three alternate selections for a heat exchanger: (1) steel at a cost of \$8000 with a lifetime of 2 years; (2) anodically protected steel, which lasts 8 or more years and requires a \$7000 potentiostat and annual labor costs of \$1100 to monitor the control system; and (3) stainless steel, which lasts 8 years and costs \$20,000. Basing the calculations on an 8-year period and neglecting maintenance costs, the annual costs for the three processes are:

	Costs, dollars		
Year	Steel	Steel (anodic. prot.)	Stainless steel
0	8000	15000	20,000
1		1100	—
2	8000	1100	—
3		1100	—
4	8000	1100	—
5		1100	—
6	8000	1100	—
7		1100	—
8	—	1100	—

The steel unit is replaced at the end of each 2-year period. At 10% interest, the *NPV* of each alternative is calculated by substitution into Eq. (2.23). The cash flows C_n are negative since they are costs and not profits.

$$\text{Steel} \qquad NPV = -8000 - \frac{8000}{(1.10)^2} - \frac{8000}{(1.10)^4} - \frac{8000}{(1.10)^6} = -\$24{,}592 \qquad (2.24)$$

$$\text{Steel (anodic. prot.)} \qquad NPV = -15000 - \frac{1100}{(1.10)^1} - \frac{1100}{(1.10)^2} + \frac{1100}{(1.10)^3} - \cdots - \frac{(1100)}{(1.10)^8} = -\$20{,}869 \qquad (2.25)$$

$$\text{Stainless steel} \qquad NPV = -20{,}000 - 0 = -\$20{,}000 \qquad (2.26)$$

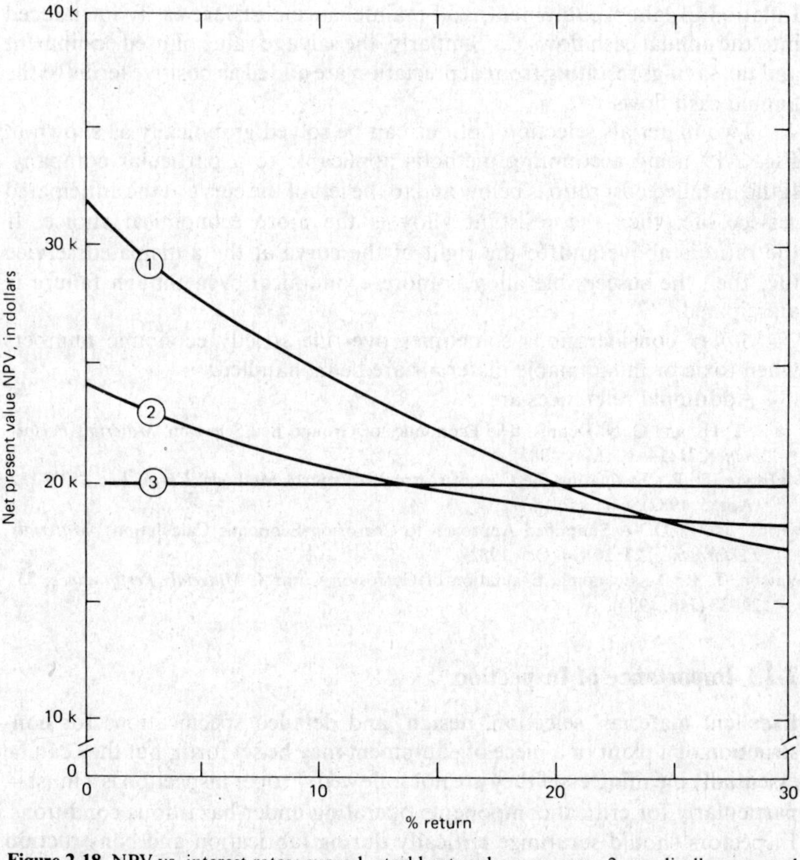

Figure 2-18 NPV vs. interest rates: curve 1, steel heat exchanger; curve 2, anodically protected steel heat exchanger; curve 3, stainless steel heat exchanger.

The stainless steel exchanger is the most economical at 10% interest. However, NPV is a function of interest rate. Substituting various interest rates into the above three equations permits a comparison of *NPV* vs. interest as shown in Fig. 2-18. Note that at interests above 14.7% the anodically protected steel exchanger is the most economical. The steel unit is the least expensive at interests above 24.1%.

Although the above interest rates are high compared to conventional savings accounts, they are not uncommon in business enterprises. Presently, tax-free municipal bonds yield 16% or more at corporate tax levels. Thus, investments in new plant equipment should equal or exceed this yield. If the process is a high risk venture funded by borrowed capital, returns of 25% or more may be required. It follows that the most economical choice often depends on the expected rate of return. The best choice for one company may differ from that of another.

Equation (2.23) is general and can be used for a variety of cost analyses. Inflation of labor, equipment, and maintenance costs are easily introduced into the annual cash flows C_n. Similarly, the salvage value of used equipment and tax savings resulting from depreciation are added as positive terms to the annual cash flows.

Two materials selection options can be solved graphically as shown in Fig. 2-19 using accounting methods applicable to a particular company. If the installed cost ratio is below and to the left of the curve at the anticipated service life, then the resistant alloy is the more economical choice. If the ratio is above and to the right of the curve at the anticipated service life, then the susceptible alloy is more economical even though failure is anticipated.

Safety considerations sometimes override strictly economic numbers when toxic or inflammable materials are being handled.

Additional references are:

Lewis, T. H., and C. N. Dennis: The Economics of Ground Bed Selection, *Materials Performance*, **21**:14–17 (May 1982).

McIntyre, D. R.: Evaluating the Cost of Corrosion-Control Methods, *Chem. Eng.*, 127–132 (Apr. 5, 1982).

Verink, Jr., E. D.: A Simplified Approach to Corrosion Economic Calculations, *Materials Performance*, **21**:26–34 (Oct. 1982).

Watson, T. R. B.: Economic Evaluation of Corrosion Control, *Materials Performance*, **23**: 29–33 (Jan. 1984).

2-13 Importance of Inspection

Excellent materials selection, design, and detailed specifications for construction of a plant or a piece of equipment may be set forth, but they can be essentially meaningless if they are not followed. Proper inspection is a must—particularly for critical components operating under hazardous conditions. Inspectors should scrutinize critically during fabrication and construction

—not limiting inspection to the final product only. In addition to being capable and well qualified, inspectors should have substantial authority. Inspectional aspects are as important as design and materials selection.

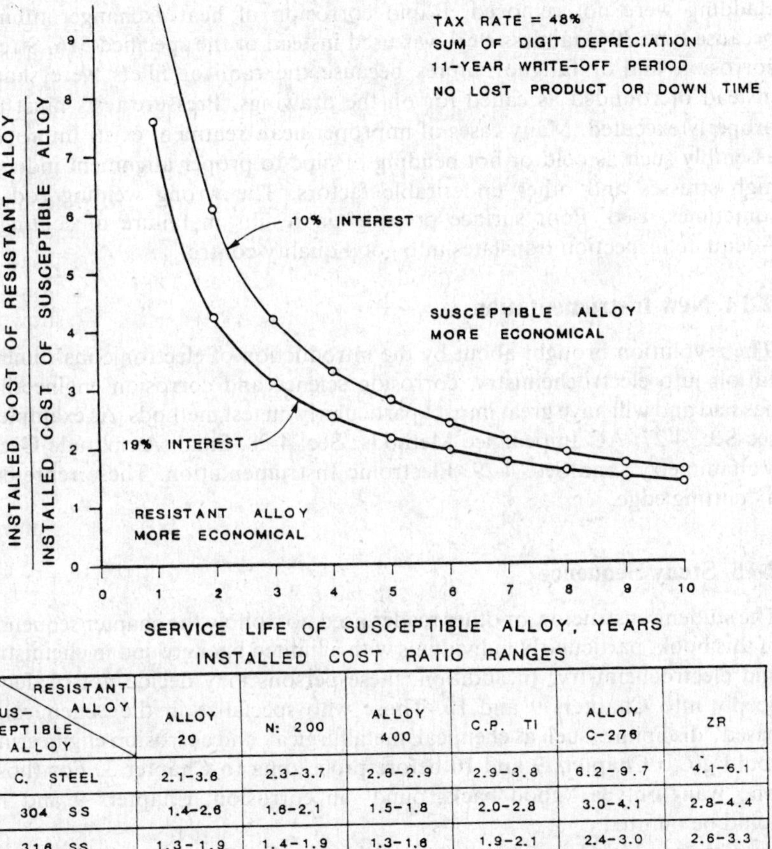

INSTALLED COST RATIO RANGES

RESISTANT ALLOY / SUSCEPTIBLE ALLOY	ALLOY 20	N: 200	ALLOY 400	C.P. TI	ALLOY C-276	ZR
C. STEEL	2.1–3.6	2.3–3.7	2.6–2.9	2.9–3.8	6.2–9.7	4.1–6.1
304 SS	1.4–2.6	1.6–2.1	1.5–1.8	2.0–2.8	3.0–4.1	2.8–4.4
316 SS	1.3–1.9	1.4–1.9	1.3–1.6	1.9–2.1	2.4–3.0	2.6–3.3

Figure 2-19 Economics of materials selection for SCC service. Based on annual cost analysis per NACE RP-02-72. *(NOTE: Installed Cost ratio ranges are from NACE Corrosion Engineer's Reference Book, R. S. Treseder, ed., p. 110, (1980).)*

Many examples of premature and sometimes catastrophic failures are known: A section of welded 10-in. pipe failed because the weld penetration at the joint was only $\frac{1}{16}$ in. (merely an overlay). Incomplete weld penetration is not uncommon. Tube hangers in an oil refinery furnace failed because these castings were extremely porous (over 50% of the cross section at the point of fracture consisted of voids). Unsatisfactory performances obtained because cleaning procedures were not followed. Cladding metal did not bond to the substrate steel because paper labels on the inner surface of the cladding were not removed. Rapid corrosion of heat exchanger tubing because type 304 stainless steel was used instead of the specified 316. Stress corrosion and/or fatigue failures because the radii at fillets were sharp instead of rounded as called for on the drawings. Pressure tests must be properly executed. Many cases of improper heat treatment exist. Improper assembly such as cold or hot bending of pipe to proper alignment induces high stresses and other undesirable factors. The wrong welding rod is sometimes used. Poor surface preparation results in failure of coatings. Adequate inspection translates into good quality control.

2-14 New Instrumentation

The revolution brought about by the introduction of electronic instrumentation into electrochemistry, corrosion science and corrosion engineering has had and will have great impact particularly on test methods. As examples, see Sec. 4-27, AC Impedance Methods; Sec. 4-28, Small Amplitude Cycle Voltammetry; and Sec. 4-29, Electronic Instrumentation. These represent a "cutting edge."

2-15 Study Sequence

The student, instructor, or other reader need not follow the chapter sequence of this book, particularly individuals with a limited background in chemistry and electrochemistry. In addition, these persons may decide not to delve deeply into Chapters 9 and 10. Those who specialize in the "chemically based" disciplines such as chemical, metallurgical, and corrosion engineering could go to Chapters 9 and 10 before proceeding to Chapter 3. For those who want only a "good background" in corrosion, Chapters 9 and 10 could be omitted.

Chapter 3 is designed to "stand on its own feet." Some instructors may wish to present additional methods for corrosion prevention (Chapter 6) at the same time, whether or not they apply to a particular form of corrosion.

In the case of a college course, the material covered could be affected by the time involved (i.e., quarter vs. semester). Some of the flavor of the course would be influenced by the instructor's expertise in corrosion.

What this all means is that considerable flexibility exists for material to be covered or presented.

CHAPTER

THREE

EIGHT FORMS OF CORROSION

It is convenient to classify corrosion by the forms in which it manifests itself, the basis for this classification being the appearance of the corroded metal. Each form can be identified by mere visual observation. In most cases the naked eye is sufficient, but sometimes magnification is helpful or required. Valuable information for the solution of a corrosion problem can often be obtained through careful observation of the corroded test specimens or failed equipment. Examination *before* cleaning is particularly desirable.

Some of the eight forms of corrosion are unique, but all of them are more or less interrelated. The eight forms are: (1) *uniform*, or general attack; (2) *galvanic*, or two-metal corrosion; (3) *crevice corrosion*; (4) *pitting*; (5) *intergranular corrosion*; (6) *selective leaching*, or parting; (7) *erosion corrosion*; and (8) *stress corrosion*. This listing is arbitrary but covers practically all corrosion failures and problems. The forms are not listed in any particular order of importance.

Below, the eight forms of corrosion are discussed in terms of their characteristics, mechanisms, and preventive measures. Hydrogen damage, though not a form of corrosion, often occurs indirectly as a result of corrosive attack and is therefore included in this chapter.

UNIFORM ATTACK

Uniform attack is the most common form of corrosion. It is normally characterized by a chemical or electrochemical reaction that proceeds uniformly over the entire exposed surface or over a large area. The metal becomes thinner and eventually fails. For example, a piece of steel or zinc immersed in dilute sulfuric acid will normally dissolve at a uniform rate over its entire surface. A sheet iron roof will show essentially the same degree of rusting over its entire outside surface. Figure 3-1 shows a steel tank in an abandoned gold-smelting plant. The circular section near the center of the photograph was thicker than the rest of the tank. This section is now supported by a "lace curtain" of tank bottom metal.

Uniform attack, or general overall corrosion, represents the greatest destruction of metal on a tonnage basis. This form of corrosion, however, is not of too great concern from the technical standpoint, because the life of equipment can be accurately estimated on the basis of comparatively simple tests. Merely immersing specimens in the fluid involved is often sufficient. Uniform attack can be prevented or reduced by (1) proper materials, including coatings, (2) inhibitors, or (3) cathodic protection. These expedients, which can be used singly or in combination, are described further in Chap. 6.

Most of the other forms of corrosion are insidious in nature and are considerably more difficult to predict. They are also localized; attack is limited to specific areas or parts of a structure. As a result, they tend to cause unexpected or premature failures of plants, machines, or tools.

Figure 3-1 Rusting of abandoned steel tank.

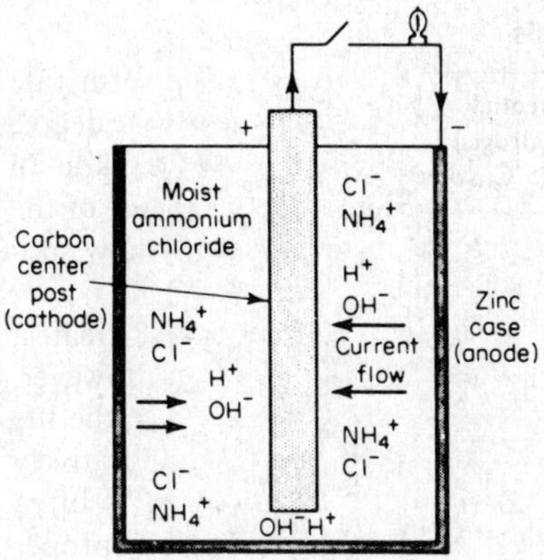

Figure 3-2 Section of dry-cell battery.

GALVANIC OR TWO-METAL CORROSION

A potential difference usually exists between two dissimilar metals when they are immersed in a corrosive or conductive solution. If these metals are placed in contact (or otherwise electrically connected), this potential difference produces electron flow between them. Corrosion of the less corrosion-resistant metal is usually increased and attack of the more resistant material is decreased, as compared with the behavior of these metals when they are not in contact. The less resistant metal becomes *anodic* and the more resistant metal *cathodic*. Usually the cathode or cathodic metal corrodes very little or not at all in this type of couple. Because of the electric currents and dissimilar metals involved, this form of corrosion is called galvanic, or two-metal, corrosion. It is electrochemical corrosion, but we shall restrict the term *galvanic* to dissimilar-metal effects for purposes of clarity.

The driving force for current and corrosion is the potential developed between the two metals. The so-called dry-cell battery depicted in Fig. 3-2 is a good example of this point. The carbon electrode acts as a noble or corrosion-resistant metal—the cathode—and the zinc as the anode, which corrodes. The moist paste between the electrodes is the conductive (and corrosive) environment that carries the current. Magnesium may also be used as the anodic material or outer case.

3-1 EMF and Galvanic Series

The potential differences between metals under reversible, or noncorroding, conditions form the basis for predicting corrosion tendencies as described in Chap. 9. Briefly, the potential between metals exposed to solutions containing approximately one gram atomic weight of their respective ions

Table 3-1 Standard emf series of metals

	Metal-metal ion equilibrium (unit activity)	Electrode potential vs. normal hydrogen electrode at 25°C, volts
↑	$Au\text{-}Au^{+3}$	+1.498
	$Pt\text{-}Pt^{+2}$	+1.2
Noble or	$Pd\text{-}Pd^{+2}$	+0.987
cathodic	$Ag\text{-}Ag^{+}$	+0.799
	$Hg\text{-}Hg_2^{+2}$	+0.788
	$Cu\text{-}Cu^{+2}$	+0.337
	$H_2\text{-}H^{+}$	0.000
	$Pb\text{-}Pb^{+2}$	−0.126
	$Sn\text{-}Sn^{+2}$	−0.136
	$Ni\text{-}Ni^{+2}$	−0.250
	$Co\text{-}Co^{+2}$	−0.277
	$Cd\text{-}Cd^{+2}$	−0.403
	$Fe\text{-}Fe^{+2}$	−0.440
	$Cr\text{-}Cr^{+3}$	−0.744
	$Zn\text{-}Zn^{+2}$	−0.763
Active or	$Al\text{-}Al^{+3}$	−1.662
anodic	$Mg\text{-}Mg^{+2}$	−2.363
	$Na\text{-}Na^{+}$	−2.714
↓	$K\text{-}K^{+}$	−2.925

Source: A. J. de Bethune and N. A. S. Loud, "Standard Aqueous Electrode Potentials and Temperature Coefficients at 25°C," Clifford A. Hampel, Skokie, Ill., 1964. See also Table 9-1. These potentials are listed in accordance with the Stockholm Convention. See J. O'M. Bockris and A. K. N. Reddy, *Modern Electrochemistry*, Plenum Press, New York, 1970.

(unit activity) are precisely measured at a constant temperature. Table 3-1 presents such a tabulation, often termed the electromotive force or emf series. For simplicity, all potentials are referenced against the hydrogen electrode (H_2/H^+) which is arbitrarily defined as zero. Potentials between metals are determined by taking the absolute differences between their standard emf potentials. For example, there is a potential of 0.462 volt between reversible copper and silver electrodes and 1.1 volt between copper and zinc. It is not possible to establish a reversible potential for alloys containing two or more reactive components, so only pure metals are listed in Table 3-1.

In actual corrosion problems, galvanic coupling between metals in equilibrium with their ions rarely occurs. As noted above, most galvanic corrosion effects result from the electrical connection of two *corroding* metals. Also, since most engineering materials are alloys, galvanic couples usually include one (or two) metallic alloys. Under these conditions, the

Table 3-2 Galvanic series of some commercial metals and alloys in seawater

↑	Platinum
	Gold
Noble or	Graphite
cathodic	Titanium
	Silver
	Chlorimet 3 (62 Ni, 18 Cr, 18 Mo
	Hastelloy C (62 Ni, 17 Cr, 15 Mo)
	18-8 Mo stainless steel (passive)
	18-8 stainless steel (passive)
	Chromium stainless steel 11-30% Cr (passive)
	Inconel (passive) (80 Ni, 13 Cr, 7 Fe)
	Nickel (passive)
	Silver solder
	Monel (70 Ni, 30 Cu)
	Cupronickels (60-90 Cu, 40-10 Ni)
	Bronzes (Cu-Sn)
	Copper
	Brasses (Cu-Zn)
	Chlorimet 2 (66 Ni, 32 Mo, 1 Fe)
	Hastelloy B (60 Ni, 30 Mo, 6 Fe, 1 Mn)
	Inconel (active)
	Nickel (active)
	Tin
	Lead
	Lead-tin solders
	18-8 Mo stainless steel (active)
	18-8 stainless steel (active)
	Ni-Resist (high Ni cast iron)
	Chromium stainless steel, 13% Cr (active)
	Cast iron
	Steel or iron
	2024 aluminum (4.5 Cu, 1.5 Mg, 0.6 Mn)
Active or	Cadmium
anodic	Commercially pure aluminum (1100)
	Zinc
↓	Magnesium and magnesium alloys

galvanic series listed in Table 3-2 yields a more accurate prediction of galvanic relationships than the emf series. Table 3-2 is based on potential measurements and galvanic corrosion tests in unpolluted seawater conducted by The International Nickel Company at Harbor Island, N.C. Because of variations between tests, the relative positions of metals, rather than their potentials, are indicated. Ideally, similar series for metals and alloys in all environments at various temperatures are needed, but this would require an almost infinite number of tests.

In general, the positions of metals and alloys in the galvanic series agree closely with their constituent elements in the emf series. Passivity influences galvanic corrosion behavior. Note in Table 3-2 the more noble position assumed by the stainless steels in the passive state as compared with the lower position of these materials when in the active condition. Similar behavior is exhibited by Inconel, which can be considered as a stainless nickel.

Another interesting feature of the galvanic series is the brackets shown in Table 3-2. The alloys grouped in these brackets are somewhat similar in base composition—for example, copper and copper alloys. The bracket indicates that in most practical applications there is little danger of galvanic corrosion if metals in a given bracket are coupled or in contact with each other. This is because these materials are close together in the series and the potential generated by these couples is not great. The farther apart in the series, the greater the potential generated.

In the absence of actual tests in a given environment, the galvanic series gives us a good indication of possible galvanic effects. Consider some actual failures in view of the data shown in Table 3-2. A yacht with a Monel hull and steel rivets became unseaworthy because of rapid corrosion of the rivets. Severe attack occured on aluminum tubing connected to brass return bends. Domestic hot-water tanks made of steel fail where copper tubing is connected to the tank. Pump shafts and valve stems made of steel or more corrosion-resistant materials fail because of contact with graphite packing.

Galvanic corrosion sometimes occurs in unexpected places. For example, corrosion was noted on the leading edges of inlet cowlings on jet engines. This attack was caused by the fabric used on the engine inlet duct plugs. This was a canvas fabric treated with a copper salt to prevent mildew. Treatment of fabric is common practice for preventing mildew, for flameproofing, and for other reasons. The copper salt deposited copper on the alloy steel, resulting in galvanic attack of the steel. This problem was solved by using a vinyl-coated nylon containing no metal.

These examples emphasize the fact that design engineers should be particularly aware of the possibilities of galvanic corrosion, since they specify the detailed materials to be used in equipment. It is sometimes economical to use dissimilar materials in contact—for example, water heaters with copper tubes and cast iron or steel tube sheets. If galvanic corrosion occurs, it accelerates attack on the heavy tube sheet (instead of the thin copper tubes), and long life is obtained because of the thickness of the tube sheets. Accordingly, expensive bronze tube sheets are not required. For more severe corrosion conditions, such as dilute acidic solutions, bronze tube sheets would be necessary.

The potential generated by a galvanic cell consisting of dissimilar metals can change with time. The potential generated causes a flow of current and corrosion to occur at the anodic electrode. As corrosion progresses, reaction products or corrosion products may accumulate at either the anode or cathode, or both. This reduces the speed at which corrosion proceeds.

In galvanic corrosion, polarization of the reduction reaction (cathodic polarization) usually predominates. Since the degree of cathodic polarization and its effectiveness varies with different metals and alloys, it is necessary to know something about their polarization characteristics before predicting the extent or degree of galvanic corrosion for a given couple. For example, titanium is very noble (shows excellent resistance) in seawater, yet galvanic corrosion on less resistant metals when coupled to titanium is usually not accelerated very much or is much less than would be anticipated. The reason is that titanium cathodically polarizes readily in seawater.

Summarizing, the galvanic series is a more accurate representation of actual galvanic corrosion characteristics than the emf series. However, there are exceptions to the galvanic series, as will be discussed later, so corrosion tests should be performed whenever possible.

3-2 Environmental Effects

The nature and agressiveness of the environment determine to a large extent the degree of two-metal corrosion. Usually the metal with lesser resistance to the *given* environment becomes the anodic member of the couple. Sometimes the potential reverses for a given couple in different environments. Table 3-3 shows the more or less typical behavior of steel and zinc in aqueous environments. Usually both steel and zinc corrode by themselves, but when they are coupled, the zinc corrodes and the steel is protected. In the exceptional case, such as some domestic waters at temperatures over 180°F, the couple reverses and the steel becomes anodic. Apparently the corrosion products on the zinc, in this case, make it act as a surface noble to steel.

Haney* shows that zinc becomes less active and potentials may reverse in the presence of inhibiting ions such as nitrates, bicarbonates and/or carbonates in water.

Tantalum is a very corrosion-resistant metal. It is anodic to platinum and carbon, but the cell is active only at high temperatures. For example, in

Table 3-3 Change in weight of coupled and uncoupled steel and zinc, g

	Uncoupled		Coupled	
Environment	Zinc	Steel	Zinc	Steel
0.05 *M* $MgSO_4$	0.00	−0.04	−0.05	+0.02
0.05 *M* Na_2SO_4	−0.17	−0.15	−0.48	+0.01
0.05 *M* NaCl	−0.15	−0.15	−0.44	+0.01
0.005 *M* NaCl	−0.06	−0.10	−0.13	+0.02

*E. G. Haney, The Zinc-Steel Potential Reversal in Cathodic Protection, *Materials Performance*, **21**:44–50 (Apr. 1982).

the tantalum-platinum couple, current does not begin to flow until 110°C is reached and 100 mA/ft^2 flows at 265°C. Tantalum is cathodic to *clean* high-silicon iron in strong sulfuric acid, but the current drops rapidly to zero. Above 145°C the polarity of the cell is reversed. Tantalum should not be used in contact with anodic metals because it absorbs cathodic hydrogen and becomes brittle.

Galvanic corrosion also occurs in the atmopshere. The severity depends largely on the type and amount of moisture present. For example, corrosion is greater near the seashore than in a dry rural atmosphere. Condensate near a seashore contains salt and therefore is more conductive (and corrosive) and a better electrolyte than condensate in an inland location, even under equal humidity and temperature conditions. Atmospheric exposure tests in different parts of the country have shown zinc to be anodic to steel in all cases, aluminum varied, and tin and nickel always cathodic. Galvanic corrosion does not occur when the metals are completely dry since there is no electrolyte to carry the current between the two electrode areas.

3-3 Distance Effect

Accelerated corrosion due to galvanic effects is usually greatest near the junction, with attack decreasing with increasing distance from that point. The distance affected depends on the conductivity of the solution. This becomes obvious when the path of the current flow and the resistance of the circuits are considered. In high-resistance, or quite pure, water the attack may be a sharp groove. Two-metal corrosion is readily recognized by the localized attack near the junction.

3-4 Area Effect

Another important factor in galvanic corrosion is the area effect, or the ratio of the cathodic to anodic areas. An unfavorable area ratio consists of a *large* cathode and a *small* anode. For a given current flow in the cell, the current density is greater for a small electrode than for a larger one. The greater the current density at an anodic area the greater the corrosion rate. Corrosion of the anodic area may be 100 or 1000 times greatest than if the anodic and cathodic areas were equal in size. Figure 3-3 shows two good examples of the area effect. The specimens are riveted plates of copper and steel both exposed in the ocean for 15 months at the same time. On the left are steel plates with copper rivets; on the right, copper plates with steel rivets. Copper is the more noble, or more resistant, material to seawater. The steel plates in the left specimen are somewhat corroded, but a strong joint still exists. The specimen on the right has an unfavorable area ratio, and the steel rivets are completely corroded. The *rate* or intensity of attack is obviously much greater on the specimen (the steel rivets) coupled to the large copper cathodic area.

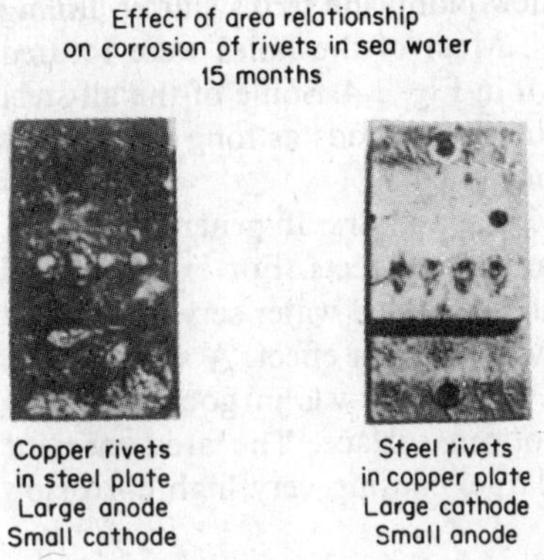

Figure 3-3 Area effect on steel-copper couple. (*International Nickel Company*.)

Violation of the above simple principle often results in costly failures. For example, a plant installed several hundred large tanks in a major expansion program. Most of the older tanks were made of ordinary steel and completely coated on the inside with a baked phenolic paint. The solutions handled were only mildly corrosive to steel, but contamination of the product was a major consideration. The coating on the floor was damaged also because of mechanical abuse, and some maintenance was required. To overcome this situation the bottoms of the new tanks were made of mild steel clad with 18-8 stainless steel. The tops and sides were of steel, with the sides welded to the stainless clad bottoms as illustrated by Fig. 3-4. The steel was coated with the same phenolic paint, with the coating covering only a small portion of the stainless steel below the weld.

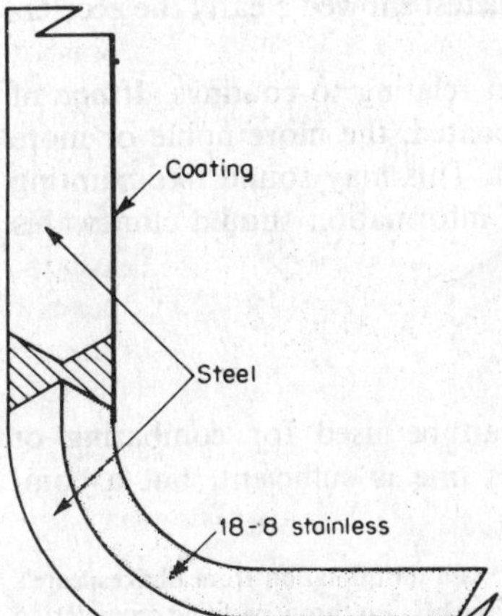

Figure 3-4 Detail of welded steel and stainless clad tank construction.

A few months after start-up of the new plant, the tanks started failing because of perforation of the side walls. Most of the holes were located within a 2-in. band above the weld shown in Fig. 3-4. Some of the all-steel tanks had given essentially trouble-free life for periods as long as 10 to 20 years as far as side-wall corrosion was concerned.

The explanation for the above failure is as follows. In general, all paint coatings are permeable and may contain some defects. For example, this baked phenolic coating would fail in double-distilled water service. Failure of the new tanks resulted from the unfavorable area effect. A small anode developed on the mild steel side plates. This area was in good electrical contact with the large stainless steel bottom surface. The area ratio of cathode to anode was almost infinitely large, causing very high corrosion rates in the order of 1000 mpy.

An interesting sidelight was the plant's claim that the tanks failed because of a poor coating job near the welds. They demanded recoating by the applicator; this would have cost more than the original job because of the need for sandblasting to remove the adherent phenolic coating instead of sandblasting a rusted surface. But failure would still occur at a rapid rate.

The plant "proved" that galvanic corrosion was not an important factor by conducting corrosion tests on specimens of equal area in boiling solutions. The solutions were boiled to accelerate the test, but boiling removed dissolved gases and actually decreased the aggressiveness of the environment. This problem was solved by coating the stainless steel tank bottoms, which reduced the exposed cathode area.

In another plant using similar solutions, failure of the coating was accelerated because of uncoated bronze manhole doors. Bronze doors had been substituted for cast steel ones because delivery time for the former was better! In this plant, comparative tests were made on two large tanks side by side in actual service, with the only known variable consisting of bronze doors—one coated and one not coated. This test showed clearly the acceleration of failure because of the bronze.

These examples demonstrate an axiom relating to coatings. If *one* of two dissimilar metals in contact is to be coated, the more noble or more corrosion-resistant metal should be coated. This may sound like painting the lily* to the uninitiated, but the above information should clarify this point.

3-5 Prevention

A number of procedures or practices can be used for combating or minimizing galvanic corrosion. Sometimes one is sufficient, but a com-

*The more popular expression "gilding the lily" is a misquotation from Shakespeare's *King John*, which states "to gild refined gold, to paint the lily—to throw perfume on violet, to smooth the ice, or add another hue unto the rainbow—is wasteful and ridiculous excess."

bination of one or more may be required. These practices are as follows:

1. Select combinations of metals as close together as possible in the galvanic series.
2. Avoid the unfavorable area effect of a small anode and large cathode. Small parts such as fasteners sometimes work well for holding less resistant materials.
3. Insulate dissimilar metals wherever practicable. It is important to insulate *completly* if possible. A common error in this regard concerns bolted joints such as two flanges, like a pipe to a valve, where the pipe might be steel or lead and the valve a different material. Bakelite washers under the bolt heads and nuts are assumed to insulate the two parts, yet the shank of the bolt touches both flanges! This problem is solved by putting plastic tubes over the bolt shanks, plus the washers, so the bolts are isolated completely from the flanges. Figure 3-5 shows proper insulation for a bolted joint. Tape and paint to increase resistance of the circuit are alternatives.
4. Apply coatings with caution. Avoid situations similar to one described in connection with Fig. 3-4. Keep the coatings in good repair, particularly the one on the anodic member.
5. Add inhibitors, if possible, to decrease the aggressiveness of the environment.
6. Avoid threaded joints for materials far apart in the series. As shown in Fig. 3-5, much of the effective wall thickness of the metal is cut away during the threading operation. In addition, spilled liquid or condensed moisture can collect and remain in the thread grooves. Brazed joints are preferred, using a brazing alloy more noble than at least one of the metals to be joined. Welded joints using welds of the same alloy are even better.

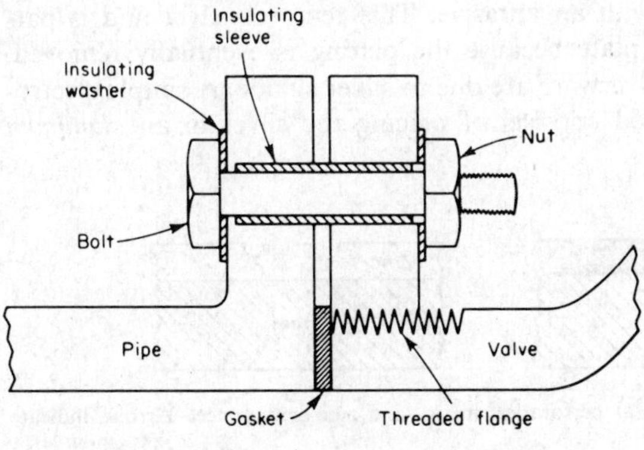

Figure 3-5 Proper insulation of a flanged joint.

7. Design for the use of readily replaceable anodic parts or make them thicker for longer life.
8. Install a third metal that is anodic to *both* metals in the galvanic contact.

3-6 Beneficial Applications

Galvanic corrosion has several beneficial or desirable applications. As noted before, dry cells and other primary batteries derive their electric power by galvanic corrosion of an electrode. It is interesting to note that if such a battery is used to the point where the zinc case is perforated and leakage of the corrosive electrolyte occurs, it becomes a galvanic corrosion problem! Some other beneficial applications are briefly described below:

Cathodic protection The concept of cathodic protection is introduced at this point because it often utilizes the principles of galvanic corrosion. This subject is discussed in more detail in Chap. 6. Cathodic protection is simply the protection of a metal structure by making it the cathode of a galvanic cell. Galvanized (zinc-coated) steel is the classic example of cathodic protection of steel. The zinc coating is put on the steel, not because it is corrosion resistant but because it is not. The zinc corrodes preferentially and protects the steel, as shown by Table 3-3 and Fig. 3-6. Zinc acts as a sacrificial anode. In contrast, tin, which is more corrosion resistant than zinc, is sometimes undesirable as a coating because it is usually cathodic to steel. At perforations in the tin coating, the corrosion of the steel is accelerated by galvanic action. Magnesium is often connected to underground steel pipes to suppress their corrosion (the magnesium preferentially corrodes). Cathodic protection is also obtained by impressing a current from an external power source through an inert anode (see Chap. 6).

Cleaning silver Another useful application concerns the use of galvanic corrosion for cleaning silverware in the home. Most household silver is cleaned by rubbing with an abrasive. This removes silver and is particularly bad for silver plate because the plating is eventually removed. Many of the stains on silverware are due to silver sulfide. A simple electrochemical cleaning method consists of placing the silver in an *aluminum*

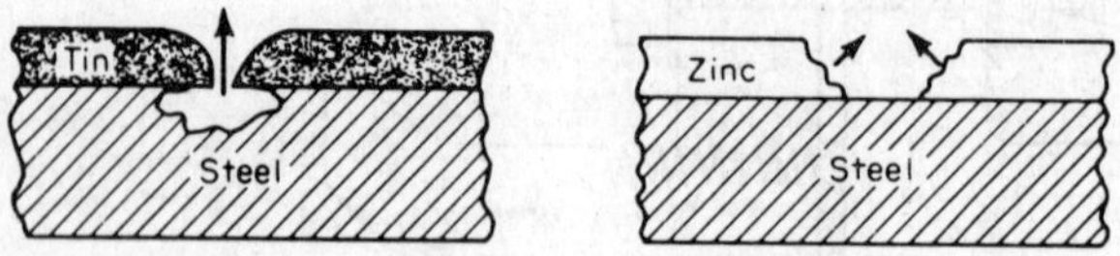

Figure 3-6 Galvanic corrosion at perforation in tin- and zinc-coated steel. Arrows indicate corrosive attack.

pan containing water and baking soda (do not use sodium chloride). The current generated by the contact between silver and aluminum causes the silver sulfide to be reduced back to silver. No silver is actually removed. The silver is then rinsed and washed in warm soapy water. It does not look quite as nice as a polished surface but it saves wear and tear on the silver and also on the individual who has to do the job. Simultaneous use of ultrasonic cleaning is faster and better, but this equipment is not generally available.

One will sometimes see for sale a piece of "magic metal" that will do the same thing. The directions call for placing it in an *enameled* pan. The so-called magic metal is usually a piece of magnesium or aluminum.

CREVICE CORROSION

Intensive localized corrosion frequently occurs within crevices and other shielded areas on metal surfaces exposed to corrosives. This type of attack is usually associated with small volumes of stagnant solution caused by holes, gasket surfaces, lap joints, surface deposits, and crevices under bolt and rivet heads. As a result, this form of corrosion is called *crevice corrosion* or, sometimes, *deposit* or *gasket corrosion*.

3-7 Environmental Factors

Examples of deposits that may produce crevice corrosion (or deposit attack) are sand, dirt, corrosion products, and other solids. The deposit acts as a shield and creates a stagnant condition thereunder. The deposit could also be a permeable corrosion product. Figure 3-7 shows crevice corrosion of a pure-silver heating coil after a few hours of operation. Solids in suspension or solution tend to deposit on a heating surface. This happened in this case, causing the corrosion shown. The silver lining in the tank containing this coil showed no attack because no deposit formed there.

Contact between metal and nonmetallic surfaces can cause crevice corrosion as in the case of a gasket. Wood, plastics, rubber, glass, concrete,

Figure 3-7 Crevice corrosion of a silver heating coil.

Figure 3-8 Gasket (crevice) corrosion on a large stainless steel pipe flange (*E. V. Kunkel.*)

asbestos, wax, and fabrics are examples of materials that can cause this type of corrosion. Figure 3-8 is a good example of crevice corrosion at a gasket–stainless steel interface. The inside of the pipe is negligibly corroded. Stainless steels are particularly susceptible to crevice attack. For example, a sheet of 18-8 stainless steel can be cut by placing a stretched rubber band around it and then immersing it in seawater. Crevice attack begins and progresses in the area where the metal and rubber are in contact.

To function as a corrosion site, a crevice must be wide enough to permit liquid entry but sufficiently narrow to maintain a stagnant zone. For this reason, crevice corrosion usually occurs at openings a few thousandths of an inch or less in width. It rarely occurs within wide (e.g., $\frac{1}{8}$-in.) grooves or slots. Fibrous gaskets, which have a wick action, form a completely stagnant solution in contact with the flange face; this condition forms an almost ideal crevice corrosion site.

3-8 Mechanism

Until recently it was believed that crevice corrosion resulted simply from differences in metal ion or oxygen concentration between the crevice and its surroundings. Consequently, the term *concentration cell corrosion* has been used to describe this form of attack. More recent studies* have shown that although metal-ion and oxygen concentration differences do exist during crevice corrosion, these are not its basic causes.

To illustrate the basic mechanism of crevice corrosion, consider a riveted plate section of metal M (e.g., iron or steel) immersed in aerated seawater (pH 7) as shown in Fig. 3-9. The overall reaction involves the dissolution of metal M and the reduction of oxygen to hydroxide ions as discussed in Chap. 2. Thus:

$$\text{Oxidation} \qquad M \rightarrow M^+ + e \tag{3.1}$$

$$\text{Reduction} \qquad O_2 + 2H_2O + 4e \rightarrow 4OH^- \tag{3.2}$$

Initially, these reactions occur uniformly over the entire surface, including the interior of the crevice. Charge conservation is maintained in both the metal and solution. Every electron produced during the formation of a metal ion is immediately consumed by the oxygen reduction reaction. Also, one hydroxyl ion is produced for every metal ion in the solution. After a short interval, the oxygen within the crevice is depleted because of the restricted convection, so oxygen reduction ceases in this area. This, by itself, does not cause any change in corrosion behavior. Since the area within a crevice is usually very small compared with the external area, the overall rate of oxygen reduction remains almost unchanged. Therefore, the rate of corrosion within and without the crevice remains equal.

Oxygen depletion has an important indirect influence, which becomes more pronounced with increasing exposure. After oxygen is depleted, no further oxygen reduction occurs, although the dissolution of metal M continues as shown in Fig. 3-10. This tends to produce an excess of positive charge in the solution (M^+), which is necessarily balanced by the migration of chloride ions into the crevice.† This results in an increased concentration of metal chloride within the crevice. Except for the alkali metals (e.g., sodium and potassium), metal salts, including chlorides and sulfates, hydrolize in water:

$$M^+Cl^- + H_2O = MOH\downarrow + H^+Cl^- \tag{3.3}$$

*G. J. Schafer and P. K. Foster, *J. Electrochem. Soc.*, **106**:468 (1959); G. J. Schafer, J. R. Gabriel, and P. K. Foster, *ibid.*, **107**:1002 (1960); L. Rosenfeld and I. K. Marshakov, *Corrosion*, **20**:115*t* (1964).

†Hydroxide ions also migrate from the outside, but they are less mobile than chloride and, consequently, migrate more slowly.

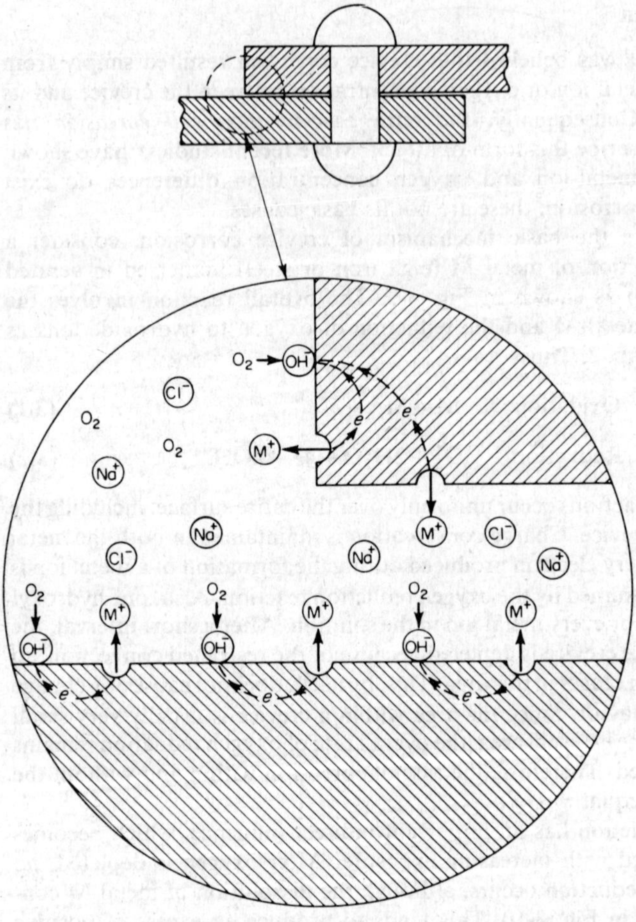

Figure 3-9 Crevice corrosion—initial stage.

Equation (3.3) shows that an aqueous solution of a typical metal chloride dissociates into an insoluble hydroxide and a free acid. For reasons that are not yet understood, both chloride and hydrogen ions accelerate the dissolution rates [Eq. (3.1)] of most metals and alloys. These are both present in the crevice as a result of migration and hydrolysis, and consequently the dissolution rate of M is increased, as indicated in Fig. 3-10. This increase in dissolution increases migration, and the result is a rapidly accelerating, or autocatalytic, process. The fluid within crevices exposed to neutral dilute sodium chloride solutions has been observed to contain 3 to 10 times as much chloride as the bulk solution and to possess a pH of 2 to 3.

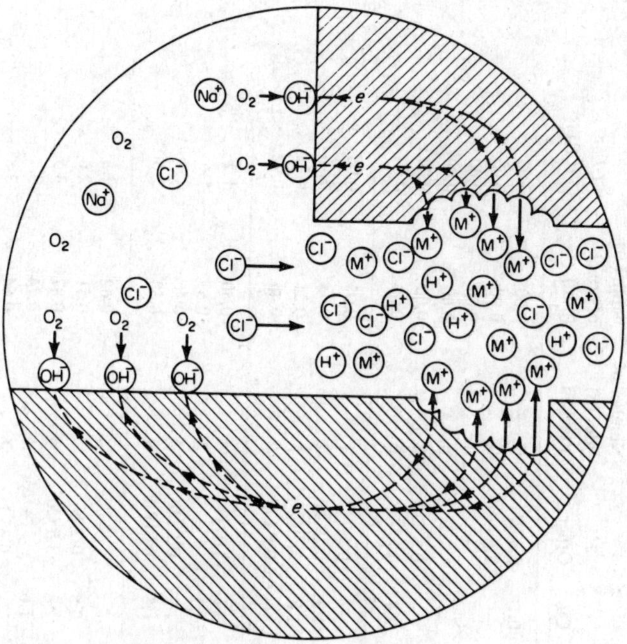

Figure 3-10 Crevice corrosion—later stage.

Table 3-4 Effect of geometric and electrochemical parameters on crevice corrosion resistance*

Parameter†	Increasing parameter causes crevice corrosion resistance to:
Critical anodic current density, i_c	Decrease
Crevice width, w	Increase
Passive potential range, E_p	Increase
Active potential range, E_a	Decrease
Solution specific resistance, ρ	Decrease

*B. J. Fitzgerald, Thesis, University of Connecticut, 1976. See also: C. Edeleanu and J. G. Gibson, *Chem. & Ind.*, 301 (1961); M. N. Folkin and V. A. Timonin, *Dokl. Akad. Nauk. SSSR*, **164**:150 (1965); W. D. France and N. D. Greene, *Corrosion*, **24**:247 (1968).

†Electrochemical parameters must be determined under actual or simulated crevice conditions.

Table 3-5 Ranking of alloys in U.S. Navy tests for resistance to crevice corrosion in filtered seawater at 30° C

30 day tests on 3 panels with 120 gritfinish and torque of 75 in.-lb

Rank	Alloy	Composition (wt%) Cr	Ni	Mo	Mn	Cu	Other	Number of sides (*S*) attacked	Maximum depth (*D*) of attack (mm)	*CCI* ($S \times D$)
	Hastelloy C-276	15.5	54.7	15.5	0.5	0.1	3.8 W	0	0.00	0.00
	Inconel 625	22.3	61.0	8.5	0.1	—	3.6 Nb	0	0.00	0.00
1	29-4 (A.L.)	29.6	0.1	4.0	—	—	—	0	0.00	0.00
	29-4-2 (A.L.)	29.5	2.2	4.0	—	—	—	0	0.00	0.00
	29-4C* (A.L.)	28.8	0.8	3.8	0.2	—	0.6 Ti	0	0.00	0.00
	Monit (Uddeholm)	25.3	4.1	3.8	0.4	0.4	—	0	0.00	0.00
2	Crucible SC-1	25.6	2.1	2.9	0.2	—	0.5 Ti	1	0.05	0.05
3	Ferralium 255 (Cabot)	26.2	5.6	3.2	0.8	1.8	0.19 N	2	0.08	0.16
4	Hastelloy G-3 (Cabot)	22.8	43.7	7.0	0.8	1.8	3.5 Co	1	0.21	0.21
5	Haynes 20 Mod (Cabot)	21.6	25.5	5.0	0.9	—	0.5 Co	2	0.46	0.92
6	26-1S	25.0	0.2	1.0	0.2	—	1.1 Ti	4	0.30	1.2
7	20Mo-6	23.9	33.4	5.6	0.4	3.3	—	3	0.53	1.6
8	EB 26-1 (A.L.)	25.9	0.1	1.0	—	—	0.1 Nb	4	0.46	1.8
9	A.L. 4X	20.2	24.4	4.4	1.4	1.5	0.019 P	4	0.50	2.0
10	A.L. 6X	20.4	24.6	6.4	1.4	—	—	4	0.62	2.5
11	254 SMO (Avesta)	20.0	17.9	6.1	0.5	0.8	0.2 N	5	0.51	2.6
12	Hastelloy G (Cabot)	22.2	46.8	5.8	1.5	1.8	3.5 Co	4	0.87	3.5
13	904L (Uddeholm)	20.5	24.7	4.7	1.5	1.6	—	5	0.74	3.7
14	AISI 216	20.0	6.0	2.5	8.0	—	0.35 N	6	0.64	3.8

15	254SFER (Avesta)	29.4	22.2	2.1	1.7	0.1	0.15 N	5	0.90	4.5
16	254 SLX (Avesta)	19.9	25.0	4.7	1.6	1.7	0.04 N	6	0.92	5.5
17	Rex 734	21.3	9.4	2.7	3.8	—	0.42 N	6	1.00	6.0
18	Type 317 LM	19.5	14.5	4.1	1.3	0.2	0.056 N	6	1.07	6.4
19	Nitronic 50	21.1	13.7	2.3	4.8	—	0.26 N	6	1.10	6.6
20	Jessop 700	20.7	25.2	4.4	1.6	0.2	0.28 Nb	5	2.00	10
21	Type 316	17.5	10.7	2.4	1.6	0.3	—	6	1.93	12
22	Carpenter 20 Cb-3	19.4	33.2	2.2	0.4	3.2	0.51 Nb	5	3.10	16
23	Jessop 777	20.8	25.6	4.5	1.4	2.2	0.24 Nb	6	2.90	17
24	44 LN	25.0	5.9	1.5	1.8	0.1	0.2 N	6	3.35	20
			Perforated							
	AISI 444	18.9	0.1	2.0	0.4	—	0.4 Nb	6	1.21	7.2
	AISI 329	27.0	4.2	1.4	0.3	0.1	—	6	1.29	7.7
	34 LN	16.8	13.8	4.2	1.6	—	0.14 N	6	1.04	6.2
			Attack outside crevice areas							
	AISI 439	17.7	0.3	—	0.3	—	0.4 Ti	6	0.72	4.3
	AISI 317L	18.9	12.2	3.6	1.7	—	0.056 N	6	1.92	12
	AISI 317L+	18.3	15.8	4.2	1.5	0.2	0.16 Co	6	1.09	6.5
	Incoloy 825	22.0	44.0	2.7	0.4	1.7	0.7 Ti	6	2.42	15

*Three additional panels were tested for 82 days with the same results.

As the corrosion within the crevice increases, the rate of oxygen reduction on adjacent surfaces also increases, as shown in Fig. 3-10. This cathodically protects the external surfaces. Thus during crevice corrosion the attack is localized within shielded areas, while the remaining surface suffers little or no damage.

The above mechanism is consistent with the observed characteristics of crevice corrosion. This type of attack occurs in many mediums, although it is usually most intense in ones containing chloride. There is often a long incubation period associated with crevice attack. Six months to a year or more is sometimes required before attack commences. However, once started, it proceeds at an ever-increasing rate.

Metals or alloys that depend on oxide films or passive layers for corrosion resistance are particularly susceptible to crevice corrosion. These films are destroyed by high concentrations of chloride or hydrogen ions (see Chap. 9), and dissolution rate markedly increases. A striking example of this has been reported concerning a hot saline water solution in a stainless steel (18-8) tank in a dyeing plant. A stainless steel bolt had fallen into the bottom of the stainless tank. Rapid attack with red rust developed under the bolt after a brief period. Aluminum is also susceptible because of the Al_2O_3 film required for corrosion protection.

Section 4-17 describes tests for crevice corrosion (also called occluded cell corrosion), and Fig. 4-19 illustrates a multiple crevice corrosion test technique that is very popular. This form of corrosion is difficult to study since the area of corrosion is hidden and test results are often scattered because of variations in the incubation period preceding the start of the attack. Theoretical and experimental studies offer promise in evaluating this type of attack. The effects of various geometric and electrochemical parameters on crevice corrosion resistance are now more clearly understood and are summarized in Table 3-4. The critical anodic current density and the active and passive potential ranges have been described in Chap. 9. Examination of this table shows that optimum crevice corrosion resistance will be achieved with an active-passive metal possessing:

1. A narrow active-passive transition
2. A small critical current density
3. An extended passive region

Titanium is an example of such a material, as are the high-nickel alloys (e.g., Hastelloy C). Type 430 stainless steel with a large critical current density, a wide active-passive transition, and a limited passive region is extremely susceptible to crevice corrosion. The crevice width is also an important variable. All materials are susceptible to crevice corrosion provided the crevice width is sufficiently narrow (e.g., 1 micrometer or less). This information provides a basis for estimating the probable crevice corrosion resistance of a given alloy.

Streicher* developed a Crevice Corrosion Index (CCI) to assist in the selection of materials of construction. This is based on the product of the number of sides attacked, S, and the maximum depth of attack, D; that is, $CCI = S \times D$. Table 3-5 ranks stainless alloys with the best in the top group. This excellent paper includes discussion of initiation, growth, effects of alloying elements, materials selection, and also a good bibliography.

Materials Technology Institute published MTI Technical Report No. 8, *A Sensor for Monitoring Crevice Corrosion—An Analysis and Evaluation—Phase 2*, (March 1983). A device is described that could be used for exposure to plant and other solutions for determining whether or not susceptibility to crevice corrosion exists.

3-9 Combating Crevice Corrosion

Methods and procedures for combating or minimizing crevice corrosion are as follows:

1. Use welded butt joints instead of riveted or bolted joints in new equipment. Sound welds and complete penetration are necessary to avoid porosity and crevices on the inside (if welded only from one side).
2. Close crevices in existing lap joints by continuous welding, caulking, or soldering.
3. Design vessels for complete drainage; avoid sharp corners and stagnant areas. Complete draining facilitates washing and cleaning and tends to prevent solids from settling on the bottom of the vessel.
4. Inspect equipment and remove deposits frequently.
5. Remove solids in suspension early in the process or plant flow sheet, if possible.
6. Remove wet packing materials during long shutdowns.
7. Provide uniform environments, if possible, as in the case of backfilling a pipeline trench.
8. Use "solid," nonabsorbent gaskets, such as Teflon, wherever possible.
9. Weld instead of rolling in tubes in tube sheets.

3-10 Filiform Corrosion

Although not immediately apparent, *filiform corrosion* (filamentary corrosion occuring on metal surfaces) is a special type of crevice corrosion. In most instances it occurs under protective films, and for this reason it is often referred to as *underfilm corrosion*. This type of corrosion is quite common; the most frequent example is the attack of enameled or lacquered surfaces of food and beverage cans that have been exposed to the atmosphere. The red-brown corrosion filaments are readily visible.

*M. A. Streicher, Analysis of Crevice Corrosion Data From Two Sea Water Exposure Tests on Stainless Alloys, *Materials Performance*, **22**:37–50 (May 1983).

Filiform corrosion has been observed on steel, magnesium, and aluminum surfaces covered by tin, silver, gold, phosphate, enamel, and lacquer coatings. It has also been observed on paper-backed aluminum foil, corrosion occuring at the paper-aluminum interface.

Filiform corrosion is an unusual type of attack, since it does not weaken or destroy metallic components but only affects surface appearance. Appearance is very important in food packaging, and this peculiar form of corrosion is a major problem in the canning industry. Although filiform attack on the exterior of a food can does not affect its contents, it does affect the sale of such cans.

Under transparent surface films, the attack appears as a network of corrosion product trails. The filaments consist of an active head and a red-brown corrosion product tail as illustrated in Fig. 3-11. The filaments are $\frac{1}{10}$ in. or less wide, and corrosion occurs only in the filament head. The blue-green color of the active head is the characteristic color of ferrous ions, and the red-brown coloration of the inactive tail is due to the presence of ferric oxide or hydrated ferric oxide.

Interaction between corrosion filaments is most interesting (see Fig. 3-12). Corrosion filaments are initiated at edges and tend to move in straight lines. Filaments do not cross inactive tails of other filaments. As is illustrated in (*a*), a corrosion filament upon striking the inactive tail of another filament is reflected. The angle of incidence is usually equal to the angle of reflection. If an actively growing filament strikes the inactive tail of another filament at a 90° angle, it may become inactive or, more frequently, it splits into two new filaments, each being reflected at an angle of approximately 45 degrees as shown in (*b*). The active heads of two filaments may join, forming a single new filament if they approach each other obliquely (*c*). Perhaps the most interesting interaction is the "death trap" illustrated in (*d*). Since growing filaments cannot cross inactive tails, they frequently become trapped and "die" as available space is decreased. Examples of "death traps" are easily found on the surface of discarded can lids, which have been exposed to moist atmospheres.

Environmental factors The most important environmental variable in filiform corrosion is the relative humidity of the atmosphere. Table 3-6

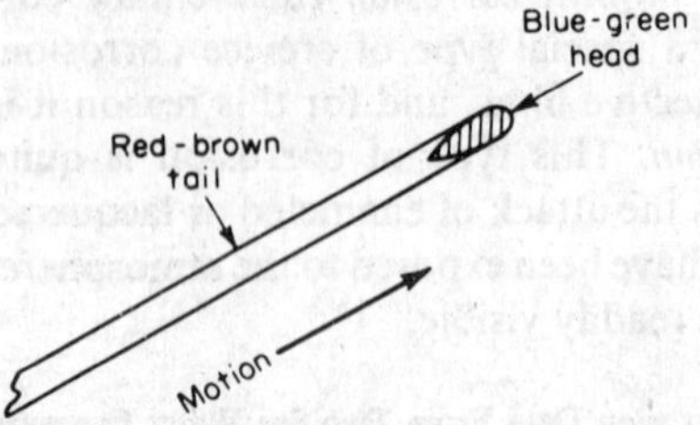

Figure 3-11 Schematic diagram of a corrosion filament growing on an iron surface (magnified).

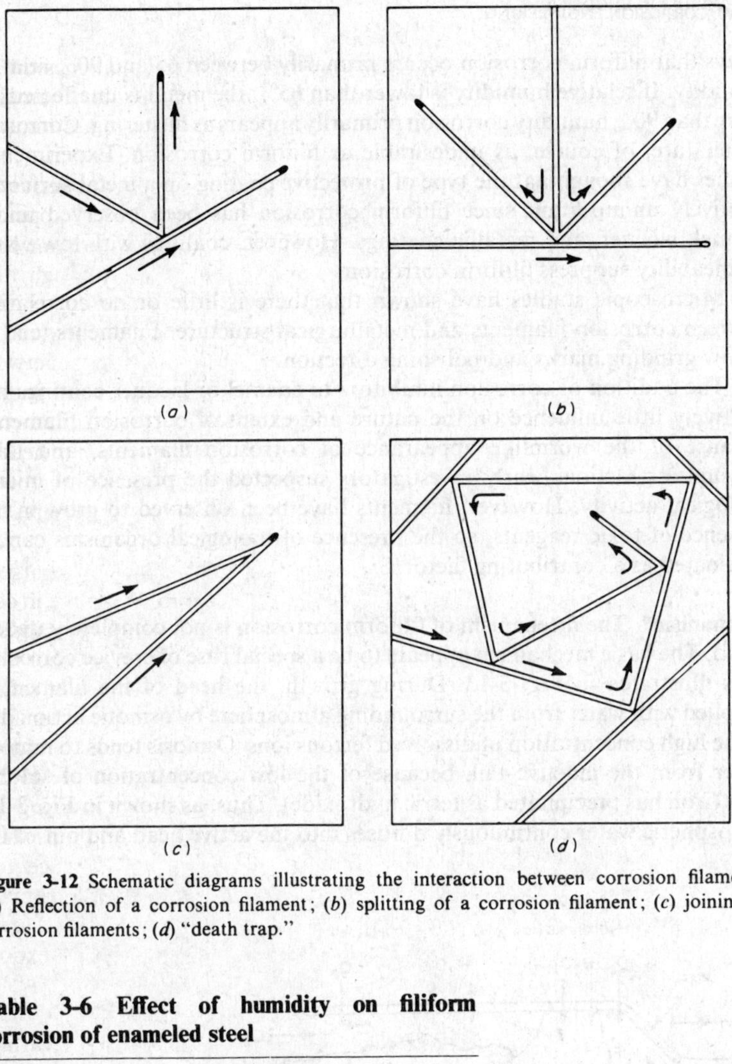

Figure 3-12 Schematic diagrams illustrating the interaction between corrosion filaments. (*a*) Reflection of a corrosion filament; (*b*) splitting of a corrosion filament; (*c*) joining of corrosion filaments; (*d*) "death trap."

Table 3-6 Effect of humidity on filiform corrosion of enameled steel

Relative humidity, %	Appearance
0–65	No corrosion
65–80	Very thin filaments
80–90	Wide corrosion filaments
93	Very wide filaments
95	Mostly blisters, scattered filiform
100	Blisters

Source: M. Van Loo, D. D. Laiderman, and R. R. Bruhn, *Corrosion*, **9**:2 (1953).

shows that filiform corrosion occurs primarily between 65 and 90% relative humidity. If relative humidity is lower than 65%, the metal is unaffected; at more than 90% humidity corrosion primarily appears as blistering. Corrosion blisters are, of course, as undesirable as filiform corrosion. Experimental studies have shown that the type of protective coating on a metal surface is relatively unimportant since filiform corrosion has been observed under enamel, lacquer, and metallic coatings. However, coatings with low water permeability suppress filiform corrosion.

Microscopic studies have shown that there is little or no correlation between corrosion filaments and metallurgical structure. Filaments tend to follow grinding marks and polishing direction.

The addition of corrosion inhibitors to enamel or lacquer coatings has relatively little influence on the nature and extent of corrosion filaments. Because of the wormlike appearance of corrosion filaments, and their unusual interactions, early investigators suspected the presence of microbiological activity. However, filaments have been observed to grow in the presence of toxic reagents, so the presence of biological organisms can be eliminated as a contributing factor.

Mechanism* The mechanism of filiform corrosion is not completely understood. The basic mechanism appears to be a special case of crevice corrosion as is illustrated in Fig. 3-13. During growth, the head of the filament is supplied with water from the surrounding atmosphere by osmotic action due to the high concentration of dissolved ferrous ions. Osmosis tends to remove water from the inactive tail, because of the low concentration of soluble salts (iron has precipitated as ferric hydroxide). Thus, as shown in Fig. 3-13, atmospheric water continuously diffuses into the active head and out of the

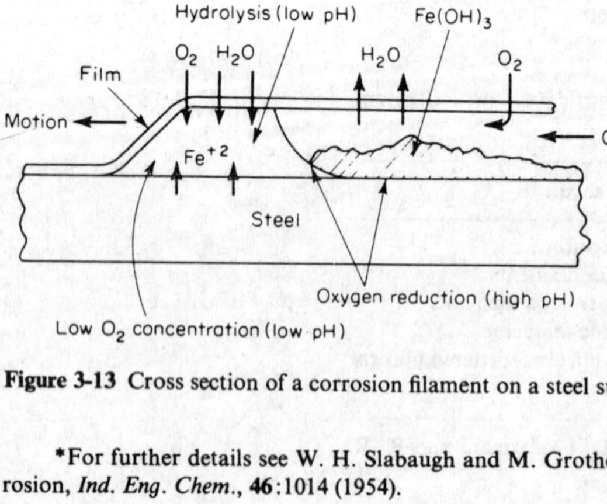

Figure 3-13 Cross section of a corrosion filament on a steel surface.

*For further details see W. H. Slabaugh and M. Grotheer, Mechanism of Filiform Corrosion, *Ind. Eng. Chem.*, **46**:1014 (1954).

inactive tail. Although oxygen diffuses through the film at all points, the concentration of oxygen at the interface between the tail and the head is high because of lateral diffusion. Corrosion is restricted to the head, where hydrolysis of the corrosion products produces an acidic environment. Thus, filiform corrosion can be viewed as a self-propagating crevice. Although Fig. 3-13 adequately explains the basic corrosion mechanism, the unusual growth characteristics (i.e., lack of spreading) and interactions between filaments are not understood.

Prevention There is no completely satisfactory way to prevent filiform corrosion. An obvious method is to store coated metal surfaces in low-humidity environments. Although this technique can be used in some instances, it is not always practical for long-time storage. Another preventive measure that has been employed consists of coating with brittle films. If a corrosion filament begins growing under a brittle coating, the film cracks at the growing head. Oxygen is then admitted to the head, and the differential oxygen concentration originally present is removed and corrosion ceases. However, as noted above, corrosion filaments usually start at edges. Hence, a new corrosion filament begins at the point of rupture. Although brittle films suppress the growth rate of corrosion filaments, they do not offer much advantage since articles coated with such film must be handled very carefully to prevent damage. Recent developments with films of very low water permeability hold some promise in preventing filiform corrosion.

PITTING

Pitting is a form of extremely localized attack that results in holes in the metal. These holes may be small or large in diameter, but in most cases they are relatively small. Pits are sometimes isolated or so close together that they look like a rough surface. Generally a pit may be described as a cavity or hole with the surface diameter about the same as or less than the depth.

Pitting is one of the most destructive and insidious forms of corrosion. It causes equipment to fail because of perforation with only a small percent weight loss of the entire structure. It is often difficult to detect pits because of their small size and because the pits are often covered with corrosion products. In addition, it is difficult to measure quantitatively and compare the extent of pitting because of the varying depths and numbers of pits that may occur under identical conditions. Pitting is also difficult to predict by laboratory tests. Sometimes the pits require a long time—several months or a year—to show up in actual service. Pitting is particularly vicious because it is a localized and intense form of corrosion, and failures often occur with extreme suddenness.

3-11 Pit Shape and Growth

Figure 3-14 is an example of pitting of 18-8 stainless steel by sulfuric acid containing ferric chloride. Note the sharply defined holes and the lack of attack on most of the metal surface. This attack developed in a few days. However, this is an extreme example, since pitting usually requires months or years to perforate a metal section. Figure 3-15 shows a copper pipe that handled potable water and failed after several years' service. Numerous pits are visible, together with a surface deposit.

Pits usually grow in the direction of gravity. Most pits develop and grow downward from horizontal surfaces. Lesser numbers start on vertical surfaces, and only rarely do pits grow upward from the bottom of horizontal surfaces.

Pitting usually requires an extended initiation period before visible pits appear. This period ranges from months to years, depending on both the specific metal and the corrosive. Once started, however, a pit penetrates the

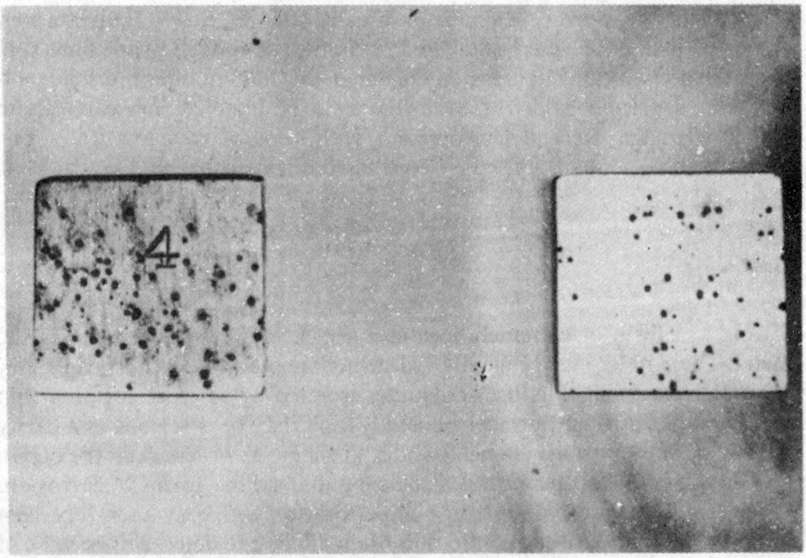

Figure 3-14 Pitting of 18-8 stainless steel by acid-chloride solution.

Figure 3-15 Pitting of a copper pipe used for drinking water.

metal at an ever-increasing rate. In addition, pits tend to undermine or undercut the surface as they grow. This aspect, illustrated in Fig. 3-16, shows a magnified section of a 16% Cr stainless steel (Type 430) tube which failed because of small pinhole leaks. The tube contained circulating water for cooling nitric acid in a plant making this acid. The outside of the tube (bottom) was exposed to the process side, or nitric acid side, and no measurable corrosion occurred on this surface. The cooling water contained a small amount of chlorides. Pitting started on the inside (upper) surface and progressed outwards. The hole in the bottom surface is the actual leak. The tendency of pits to undercut the surface makes their detection much more difficult. Subsurface damage is usually much more severe than is indicated by surface appearance.

Pitting may be considered as the intermediate stage between general overall corrosion and complete corrosion resistance. This is shown diagrammatically in Fig. 3-17. Specimen *A* shows no attack whatsoever. Specimen *C* has metal removed or dissolved uniformly over the entire exposed surface. Intense pitting occurred on specimen *B* at the points of breakthrough. This situation can be readily demonstrated by exposing three identical specimens of 18-8 stainless steel to ferric chloride and increasing the concentration and/or the temperature as we move to the right in Fig. 3-17. Very dilute, cold, ferric chloride produces no attack (in a short time) on *A*, but strong hot ferric chloride dissolves specimen *C*. Riggs, Sudbury, and Hutchinson* observed a striking example of this during a study of the effects

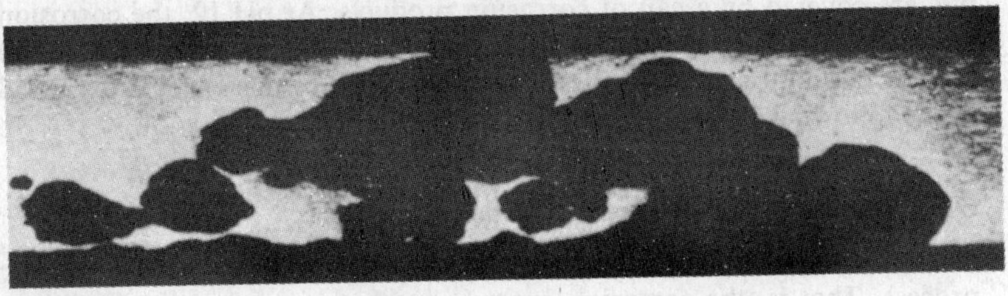

Figure 3-16 Pitting of stainless steel condenser tube.

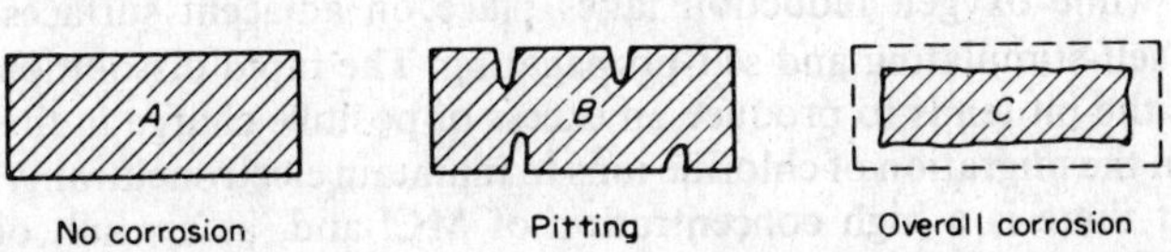

Figure 3-17 Diagrammatic representation of pitting corrosion as an intermediate stage.

*O. L. Riggs, J. D. Sudbury, and M. Hutchinson, *Corrosion*, **16**:94–98 (June 1960).

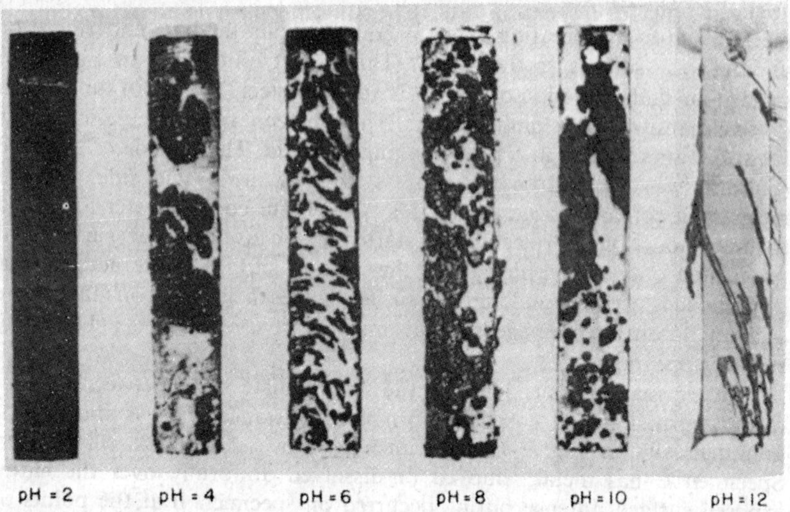

Figure 3-18 Corrosion of steel after 24 hours in 5% NaCl and 500-lb/in.^{2} oxygen pressure (*Continental Oil Co.*)

of high oxygen pressure and pH on the corrosion of steel by a 5% NaCl brine. Figure 3-18 shows that as pH is increased, the corrosion progresses from general corrosion to highly localized pitting. Beginning at pH 4, the pits are covered by a cap of corrosion products. At pH 12, the corrosion products assume an unusual tubular shape and corrosion rates are 17,000 mpy at the bottom of the tubes! The mechanism of this effect is discussed in the following section.

3-12 Autocatalytic Nature of Pitting

A corrosion pit is a unique type of anodic reaction. It is an autocatalytic process. That is, the corrosion processes within a pit produce conditions which are both stimulating and necessary for the continuing activity of the pit. This is illustrated schematically in Fig. 3-19. Here a metal M is being pitted by an aerated sodium chloride solution. Rapid dissolution occurs within the pit, while oxygen reduction takes place on adjacent surfaces. This process is self-stimulating and self-propagating. The rapid dissolution of metal within the pit tends to produce an excess of positive charge in this area, resulting in the migration of chloride ions to maintain electroneutrality. Thus, in the pit there is a high concentration of MCl and, as a result of hydrolysis [see Eq. (3.3)], a high concentration of hydrogen ions. Both hydrogen and chloride ions stimulate the dissolution of most metals and alloys, and the entire process accelerates with time. Since the solubility of oxygen is virtually zero in concentrated solutions, no oxygen reduction occurs within a pit. The cathodic oxygen reduction on the surfaces adjacent to pits tends to suppress corrosion. In a sense, pits cathodically protect the rest of the metal surface.

Although Fig. 3-19 indicates how a pit grows through self-stimulation, it does not immediately suggest how this process is initiated. Evans* has indicated how it could lead to the start of pitting. Consider a piece of metal M devoid of holes or pits, immersed in aerated sodium chloride solution. If, for any reason, the rate of metal dissolution is momentarily high at one particular point, chloride ions will migrate to this point. Since chloride stimulates metal dissolution, this change tends to produce conditions that

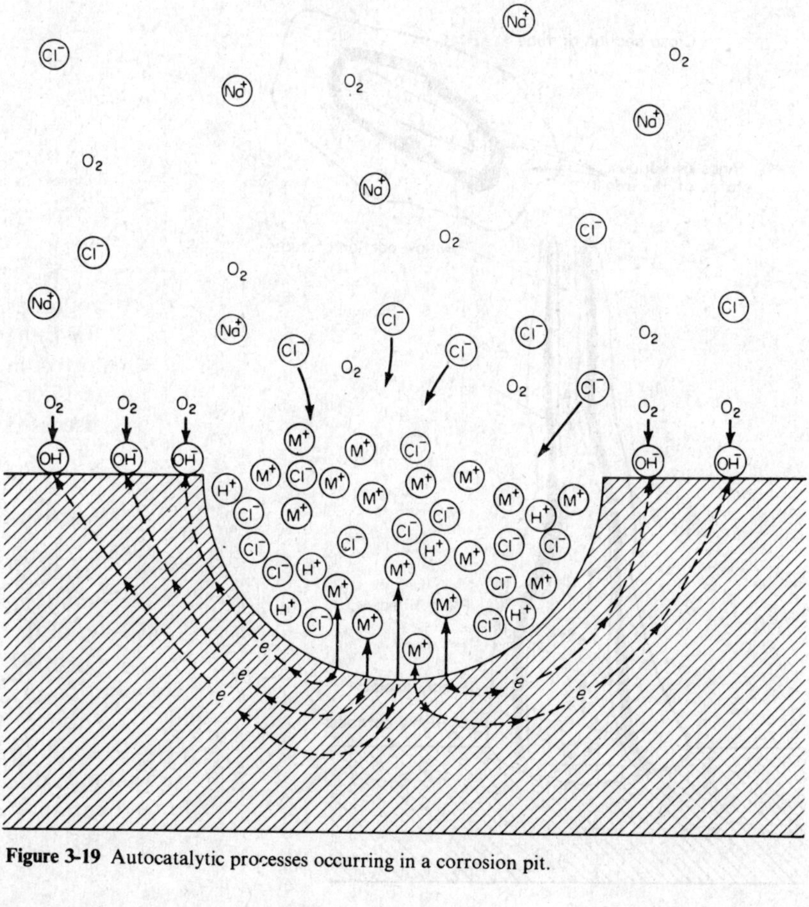

Figure 3-19 Autocatalytic processes occurring in a corrosion pit.

*U. R. Evans, *Corrosion*, 7:238 (1951).

are favorable to further rapid dissolution at this point. Locally, dissolution may be momentarily high because of a surface scratch, an emerging dislocation or other defect, or random variations in solution composition. It is apparent that during the initiation or early growth stages of a pit, conditions are rather unstable. The locally high concentration of chloride and hydrogen ions may be swept away by stray convection currents in the solution since a protective pit cavity does not exist. The author has observed that new pits are indeed unstable—many become inactive after a few minutes' growth.

The gravity effect mentioned before is a direct result of the autocatalytic nature of pitting. Since the dense, concentrated solution within a pit is

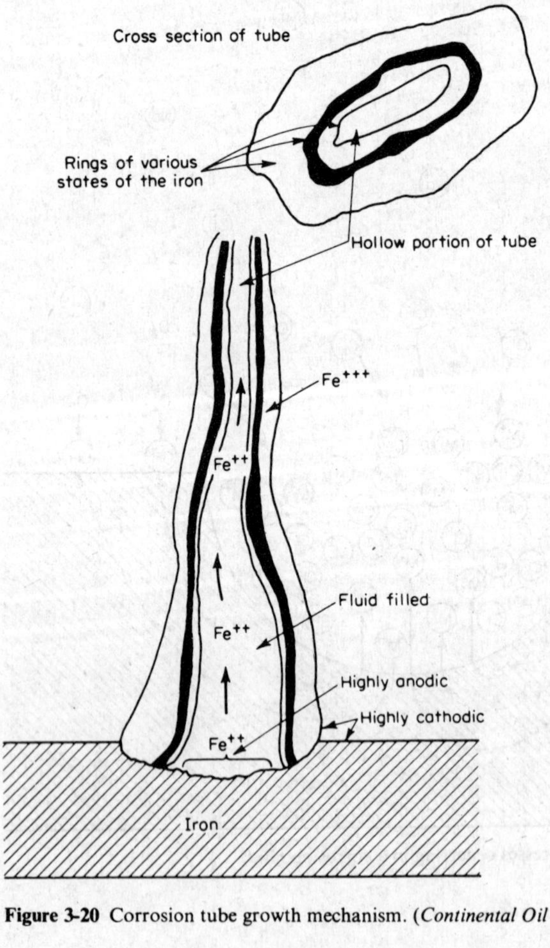

Figure 3-20 Corrosion tube growth mechanism. (*Continental Oil Co.*)

necessary for its continuing activity, pits are most stable when-growing in the direction of gravity. Also, pits are generally initiated on the upper surfaces of specimens because chloride ions are more easily retained under these conditions.

The pits with tubular corrosion products shown in Fig. 3-18 grow by a mechanism similar to that described above. Figure 3-20 indicates the mechanism proposed by Riggs, Sudbury, and Hutchinson. At the interface between the pit and the adjacent surface, iron hydroxide forms due to interaction between the OH^- produced by the cathodic reaction and the pit corrosion product. This is further oxidized by the dissolved oxygen in the solution to $Fe(OH)_3$, Fe_3O_4, Fe_2O_3, and other oxides. This "rust" rim grows in the form of a tube as shown in Fig. 3-20. The oxides forming the tube were identified by x-ray diffraction.

Comparison of Figs. 3-20, 3-19, and 3-10 shows that the mechanism of pit growth is virtually identical to that of crevice corrosion. This similarity has prompted some investigators to conclude that pitting is in reality only a special case of crevice corrosion. This view has some merit, since all systems that show pitting attack are particularly susceptible to crevice corrosion (e.g., stainless steels in seawater or ferric chloride). However, the reverse is not always correct—many systems that show crevice attack do not suffer pitting on freely exposed surfaces. It appears that pitting, though quite similar to crevice corrosion, deserves special consideration since it is a self-initiating form of crevice corrosion. Simply, it does not require a crevice—it creates its own.

3-13 Solution Composition

From a practical standpoint, most pitting failures are caused by chloride and chlorine-containing ions. Chlorides are present in varying degrees in most waters and solutions made with water. Much equipment operates in seawater and brackish waters. Hypochlorites (bleaches) are difficult to handle because of their strong pitting tendencies. Mechanisms for pitting by chlorides are controversial and not well established. Perhaps the best explanation is the acid-forming tendency of chloride salts and the high strength of its free acid (HCl). Most pitting is associated with halide ions, with chlorides, bromides, and hypochlorites being the most prevalent. Fluorides and iodides have comparatively little pitting tendencies.

Oxidizing metal ions with chlorides are aggressive pitters. Cupric, ferric, and mercuric halides are extremely aggressive. Even our most corrosion-resistant alloys can be pitted by $CuCl_2$ and $FeCl_3$. Halides of the nonoxidizing metal ions (e.g., NaCl, $CaCl_2$) cause pitting but to a much lesser degree of aggressiveness.

Cupric and ferric chlorides do not require the presence of oxygen to promote attack because their cations can be cathodically reduced. These ions are reducible as follows:

$$Cu^{2+} + 2e \rightarrow Cu \tag{3.4}$$

$$Fe^{3+} + e \rightarrow Fe^{2+} \tag{3.5}$$

In other words, they are electron acceptors. This is one reason ferric chloride is widely used in pitting studies. The reactions are not appreciably affected by the presence or absence of oxygen.

Pitting can be prevented or reduced in many instances by the presence of hydroxide, chromate, or silicate salts. However, these substances tend to accelerate pitting when present in small concentrations.

3-14 Velocity

Pitting is usually associated with stagnant conditions such as a liquid in a tank or liquid trapped in a low part of an inactive pipe system. Velocity, or increasing velocity, often decreases pitting attack. For example, a stainless steel pump would give good service handling seawater if it were run continuously, but would pit if it were shut down for extended periods. Figure 3-21 demonstrates this point. The material is type 316 stainless steel and the environment an acid-ferric chloride mixture at elevated temperature. This

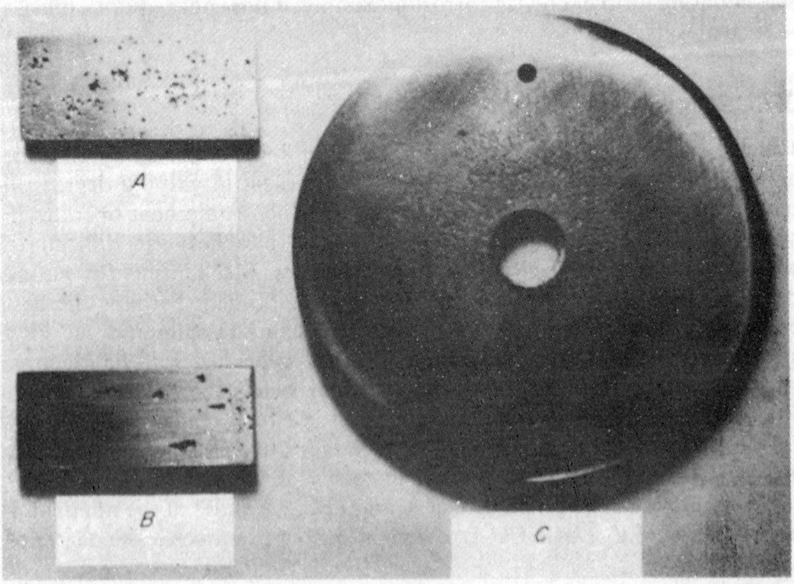

Figure 3-21 Effect of velocity on pitting of stainless steel.

test was run for 18 hours at the same time and in the same solution. Specimen *C* was exposed to high-velocity flow (about 40 ft/sec) and specimen *A* to a few feet per second, while specimen *B* was in a quiet or completely static solution. All specimens show pitting, but the depth of penetration in *C* is relatively small. Pitting is more intense on *A*, and *B* has deep and large "worm holes."

3-15 Metallurgical Variables

As a class, the stainless steel alloys are more susceptible to damage by pitting corrosion than are any other group of metals or alloys. As a result, numerous alloy studies have been devoted to improving the pitting resistance of stainless steels. The results are summarized in Table 3-7.

Holding types 304 and 316 stainless steel in the sensitizing temperature range (950 to 1450°F) decreases their pitting resistance. Austenitic stainless steels exhibit the greatest pitting resistance when solution-quenched above 1800° F.

Severe cold-working increases the pitting attack of 18-8 stainless steels in ferric chloride. Preferential edge pitting is usually observed on most wrought stainless products.

Surface finish often has a marked effect on pitting resistance. Pitting and localized corrosion are less likely to occur on polished than on etched or ground surfaces. Generally, the pits that form on a polished surface are larger and penetrate more rapidly than those on rough surfaces.

Ordinary steel is more resistant to pitting than stainless steel alloys. For example, the pitting of stainless steel condenser tubing exposed to brackish water or seawater often can be alleviated by the substitution of steel tubes.

Table 3-7 Effects of alloying on pitting resistance of stainless steel alloys

Element	Effect on pitting resistance
Chromium	Increases
Nickel	Increases
Molybdenum	Increases
Silicon	Decreases; increases when present with molybdenum
Titanium and columbium	Decreases resistance in $FeCl_3$; other mediums no effect
Sulfur and selenium	Decreases
Carbon	Decreases, especially in sensitized condition
Nitrogen	Increases

Source: N. D. Greene and M. G. Fontana, *Corrosion* **15**:25*t* (1959).

Although the general corrosion of steel is much greater than that of stainless steel, rapid perforation due to pitting does not occur.

3-16 Evaluation of Pitting Damage

Since pitting is a localized form of corrosion, conventional weight loss tests cannot be used for evaluation or comparison purposes. Metal loss is very small and does not indicate the depth of penetration. Measurements of pit depth are complicated by the fact that there is a statistical variation in the depths of pits on an exposed specimen as shown in Fig. 3-22. Note that the average pit depth is a poor way to estimate pit damage, since it is the deepest pit that causes failure.

Examination of Fig. 3-22 suggests that a measurement of maximum pit depth would be a more reliable way of expressing pitting corrosion. This is correct, but such measurements should never be used to predict equipment life since pit depth is also a function of sample size. This is shown in Fig. 3-23, which shows the relative probability of finding a pit of a given depth as a function of exposed area. For example, there is a probability of 0.2 (20%) of a pit with a depth of *d* occuring on a sample with an area of 1. On a specimen four times larger, it is a virtual certainty (probability = 1.0) that a pit of this depth will occur, and a 90% chance that a pit twice as deep will also occur. This clearly indicates that attempts to predict the life of a large

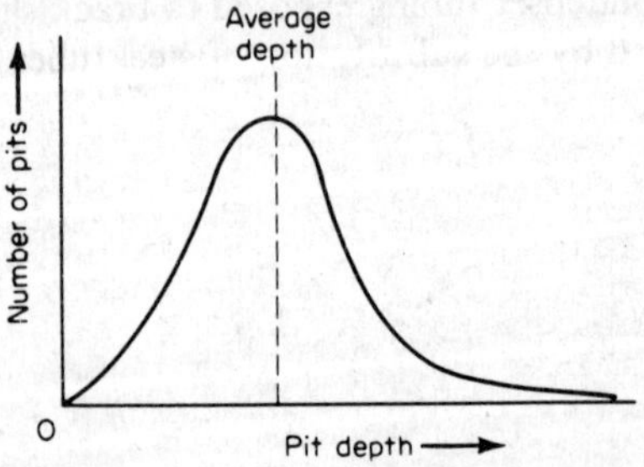

Figure 3-22 Relationship between pit depth and the number of pits appearing on a corroded surface.

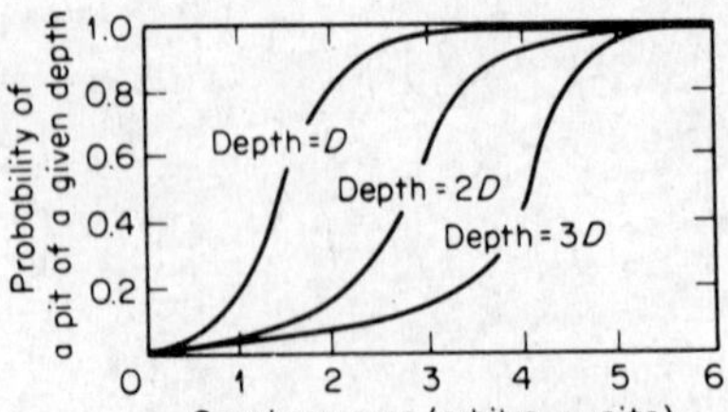

Figure 3-23 Pit depth as a function of exposed area.

plant on the basis of tests conducted on small laboratory specimens would be unwise. However, for laboratory comparisons of pitting resistance, maximum-pit-depth measurements are reasonably accurate.

3-17 Prevention

The methods suggested for combating crevice corrosion generally apply also for pitting. Materials that show pitting, or tendencies to pit, during corrosion tests should not be used to build the plant or equipment under consideration. Some materials are more resistant to pitting than others. For example, the addition of 2% molybdenum to 18-8S (type 304) to produce 18-8SMo (type 316) results in a very large increase in resistance to pitting. The addition apparently results in a more protective or more stable passive surface. These two materials behave so differently that one is considered unsuitable for seawater service but the other is sometimes recommended. The best procedure is to use materials that are known not to pit in the environment under consideration. As a general guide, the following list of metals and alloys may be used as a qualitative guide to suitable materials. However, tests should be conducted before final selection is made.

↑	Type 304 stainless steel
Increasing	Type 316 stainless steel
pitting	Hastelloy F, Nionel, or Durimet 20
resistance	Hastelloy C or Chlorimet 3
↓	Titanium

Adding inhibitors is sometimes helpful, but this may be a dangerous procedure unless attack is stopped *completely*. If it is not, the intensity of the pitting may be increased.*

INTERGRANULAR CORROSION

The more reactive nature of grain boundaries was discussed in Chap. 2. Grain boundary effects are of little or no consequence in most applications or uses of metals. If a metal corrodes, uniform attack results since grain boundaries are usually only slightly more reactive than the matrix. However, under certain conditions, grain interfaces are very reactive and intergranular corrosion results. Localized attack at and adjacent to grain boundaries,

*For additional reading on pitting, refer to N. D. Greene and M. G. Fontana, A Critical Analysis of Pitting Corrosion, *Corrosion*, **15**:41–47 (Jan. 1959). Also, An Electrochemical Study of Pitting Corrosion in Stainless Steels—Part 1, Pit Growth; Part 2, Polarization Measurements, pp. 48–60 in this same volume by the same authors.

with relatively little corrosion of the grains, is *intergranular corrosion*. The alloy disintegrates (grains fall out) and/or loses its strength.

Intergranular corrosion can be caused by impurities at the grain boundaries, enrichment of one of the alloying elements, or depletion of one of these elements in the grain-boundary areas. Small amounts of iron in aluminum, wherein the solubility of iron is low, have been shown to segregate in the grain boundaries and cause intergranular corrosion. It has been shown that based on surface tension considerations the zinc content of a brass is higher at the grain boundaries. Depletion of chromium in the grain-boundary regions results in intergranular corrosion of stainless steels.

3-18 Austenitic Stainless Steels

Numerous failures of 18-8 stainless steels have occurred because of intergranular corrosion. These happen in environments where the alloy should exhibit excellent corrosion resistance. When these steels are heated in approximately the temperature range 950 to 1450°F, they become *sensitized* or susceptible to intergranular corrosion. For example, a procedure to sensitize intentionally is to heat at 1200°F for 1 hr.

The almost universally accepted theory for intergranular corrosion is based on impoverishment or depletion of chromium in the grain-boundary areas. The addition of chromium to ordinary steel imparts corrosion resistance to the steel in many environments. Generally more than 10% chromium is needed to make a stainless steel. If the chromium is effectively lowered, the relatively poor corrosion resistance of ordinary steel is approached.

In the temperature range indicated, $Cr_{23}C_6$ (and carbon) is virtually insoluble and precipitates out of solid solution if carbon content is about 0.02% or higher. The chromium is thereby removed from solid solution, and the result is metal with lowered chromium content in the area adjacent to the grain boundaries. The chromium carbide in the grain boundary is not attacked. The chromium-depleted zone near the grain boundary is corroded because it does not contain sufficient corrosion resistance to resist attack in many corrosive environments. The common 18-8 stainless steel, type 304, usually contains from 0.06 to 0.08% carbon, so excess carbon is available for combining with the chromium to precipitate the carbide. This situation is shown schematically in Fig. 3-24. Carbon diffuses towards the grain boundary quite readily at sensitizing temperatures, but chromium is much less mobile. The surface already available at the grain boundary facilitates the formation of a new surface, namely that of the chromium carbide.

There is some evidence to indicate that the chromium content at the boundary may be reduced to a very low level or zero. Assume that the chromium content is reduced to 2%. Corrosion resistance is lowered, two dissimilar metal compositions are in contact, and a large unfavorable area ratio is present. The depleted area protects the grains. The net effect is rapid attack in the impoverished area, with little or no attack on the grains.

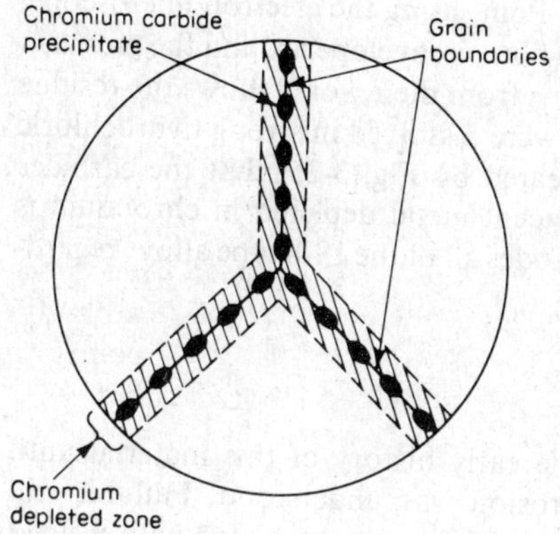

Figure 3-24 Diagrammatic representation of a grain boundary in sensitized type 304 stainless steel.

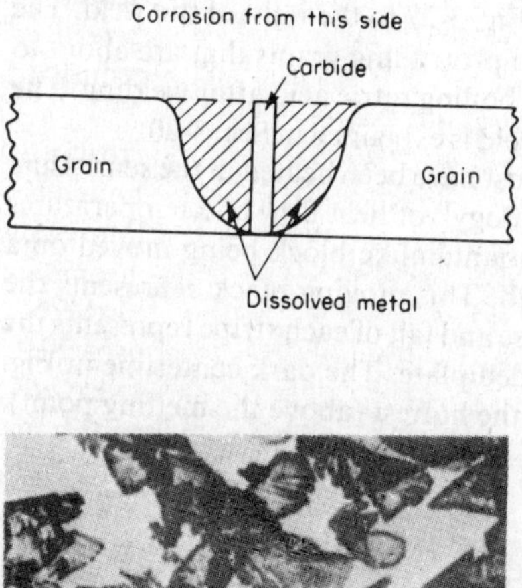

Figure 3-25 Cross section of area shown in Fig. 3-24.

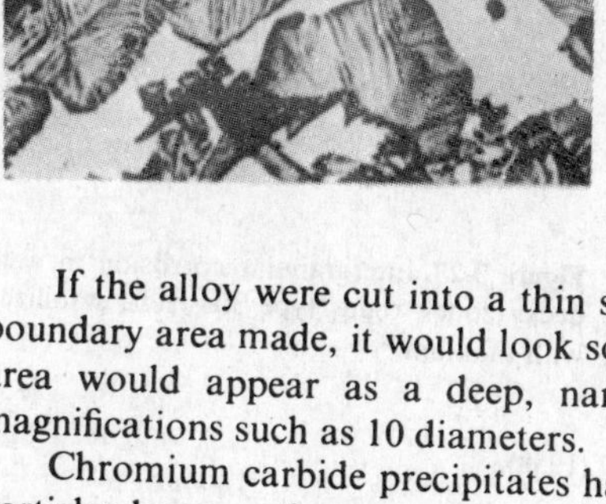

Figure 3-26 Electron photomicrograph of carbides isolated from sensitized type 304 stainless steel (13,000 ×).

If the alloy were cut into a thin sheet and a cross section of the grain-boundary area made, it would look something like Fig. 3-25. The corroded area would appear as a deep, narrow trench when observed at low magnifications such as 10 diameters.

Chromium carbide precipitates have been described for many years as particles because they are too small for detailed examination by the light microscope. Mahla and Nielsen of Du Pont, using the electron microscope, have shown that the carbide forms as a film or envelope around the grains in a leaflike structure. Figure 3-26, which is from their work, shows the residue after the metallic portions of the alloy were dissolved in strong hydrochloric acid. This emphasizes the point, indicated by Fig. 3-25, that the carbides themselves are not attacked—the adjacent metal depleted in chromium is dissolved. In fact, this acid rapidly corrodes all of the 18-8 type alloys regardless of heat treatment.

3-19 Weld Decay

Many failures of 18-8 occurred in the early history of this material until the mechanism of intergranular corrosion was understood. Failures still occur when this effect is not considered. These are associated with welded structures, and the material attacked intergranularly is called *weld decay*. The weld decay zone is usually a band in the parent plate somewhat removed from the weld. Such a zone is shown in Fig. 3-27 to the right of the weld. The "sugary" appearance is due to the small protruding grains that are about to drop off. This specimen was exposed to boiling nitric acid after welding. The absence of weld decay to the left of the weld is explained in Sec. 3-20.

The metal in the weld decay zone must have been heated in the sensitizing range. Figure 3-28* is a "tablecloth analogy" of heat flow and temperatures associated with welding. Visualize a mountainlike block being moved on a table under an elastic striped tablecloth. This moving block represents the weld being made along the plate. The rise and fall of each stripe represents the rise and fall of temperature in the welded plate. The dark centerline in Fig. 3-28 is the center of the weld, which is the hottest (above the melting point).

Figure 3-27 Intergranular corrosion in weld decay zone—right, type 304; left, stabilized with titanium.

*L. R. Honnaker, *Chem. Eng. Progr.*, **54**:79–82 (1958).

The lines with the crosses represent temperatures in the sensitizing zone. These lines correspond to the weld decay zone in Fig. 3-28.

Figure 3-29* depicts in different form essentially the same picture. Thermocouples were placed at points *A*, *B*, *C*, and *D*, and temperatures and times recorded during welding. The metal at points *B* and *C* (and between these points) is in the sensitizing temperature range for some time. Time and temperature relationships vary with the size or thickness of the material welded, the time to make the weld, and the type of welding. For example, thin sheet is rapidly welded, whereas heavy plate may take several weld passes. For sheet $\frac{1}{32}$ in. thick or less, time in the sensitizing range is sufficiently short so as not to cause intergranular corrosion in environments not particularly selective or aggressive to stainless steels. Cross welds would essentially double the time in this range, and appreciable carbide precipitation may occur.

Time and temperature effects provide one reason why electric arc welding is used more than gas welding for stainless steels. The former produces higher and more intense heating in shorter times. The latter would keep a wider zone of metal in the sensitizing range for a longer time, which means greater carbide precipitation.

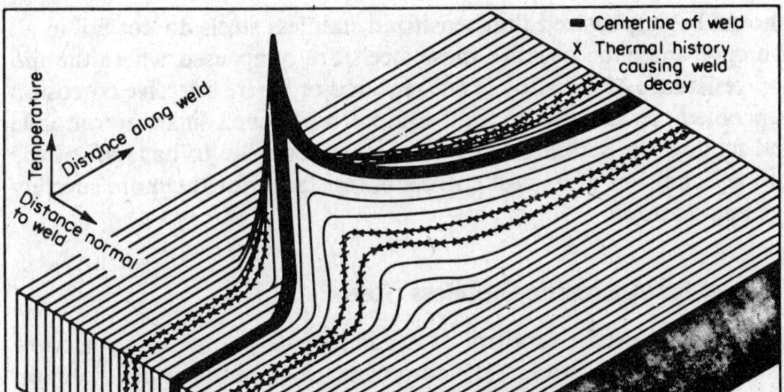

Figure 3-28 Tablecloth analogy of heat flow and temperatures during welding. Visualize a mountainlike block being moved beneath an elastic striped tablecloth. The rise and fall of each stripe represents the rise and fall of temperature in a welded plate. (*Du Pont Company*)

*L. R. Honnaker, *Chem. Eng. Progr.*, **54**:79–82 (1958).

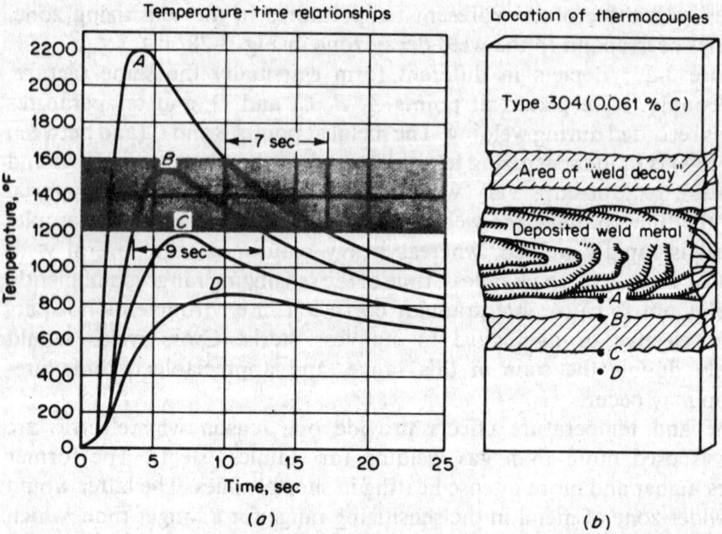

Figure 3-29 Temperatures during electric-arc welding of type 304 stainless steel. (*Du Pont Company*)

It should be emphasized that sensitized stainless steels do not fail in all corrosive environments, because these steels are often used where the *full* corrosion resistance of the alloy is not required or where selective corrosion is not a problem. Examples are food equipment, kitchen sinks, automobile trim, and facings on buildings. However, it is desirable to have all of the metal in the condition of its best corrosion resistance for the more severely corrosive applications.

3-20 Control for Austenitic Stainless Steels

Three methods are used to control or minimize intergranular corrosion of the austenitic stainless steels: (1) employing high-temperature solution heat treatment, commonly termed *quench-annealing* or *solution-quenching*, (2) adding elements that are strong carbide formers (called stabilizers), and (3) lowering the carbon content to below 0.03%

Commercial solution-quenching treatments consist of heating to 1950 to 2050°F followed by water quenching. Chromium carbide is dissolved at these temperatures, and a more homogeneous alloy is obtained. Most of the austenitic stainless steels are supplied in this condition. If welding is used during fabrication, the equipment must be quench-annealed to eliminate susceptibility to weld decay. This poses an expensive problem for large equipment and, in fact, furnaces are not available for heat-treating very large vessels. In addition, welding is sometimes necessary in the customer's plant to make repairs or, for example, to attach a nozzle to a vessel.

Quenching, or rapid cooling from the solution temperature, is very important. If cooling is slow, the *entire* structure will be susceptible to intergranular corrosion.

The strong carbide formers or stabilizing elements, columbium (or columbium plus tantalum) and titanium, are used to produce types 347 and 321 stainless steels, respectively. These elements have a much greater affinity for carbon than does chromium and are added in sufficient quantity to combine with all of the carbon in the steel. The stabilized steels eliminate the economic and other objections of solution-quenching the unstabilized steels after fabrication or weld repair. The left plate in Fig. 3-27 does not show weld decay, because it is type 321. The same picture would obtain if it were type 347.

Lowering the carbon to below 0.03% (type 304L) does not permit sufficient carbide to form to cause intergranular attack in most applications. One producer calls these the *extra-low-carbon* (ELC) steels. Figure 3-30 shows a situation similar to Fig. 3-27, except that here weld decay is absent in the low-carbon plate. The vertical trenches are due to a weld bead deposited on the back surface of the specimen.

The original 18-8 steels contained around 0.20% carbon, but this was quickly reduced to 0.08% because of rapid and serious weld decay failures. Lowering the carbon content much below 0.08% was not possible until it was discovered that it was possible to blow oxygen through the melt to burn out carbon and until low-carbon ferrochrome was developed.

These stainless steels have a high solubility for carbon when in the molten state and therefore have a tremendous propensity for picking up carbon. For example, the intent of the low-carbon grades is obviated when

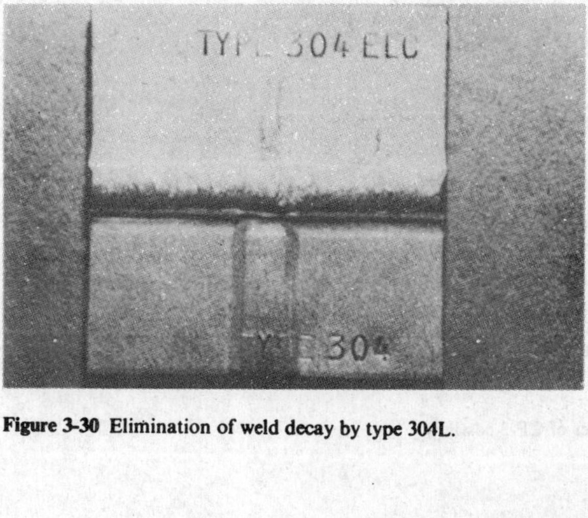

Figure 3-30 Elimination of weld decay by type 304L.

the welder carefully cleans the beveled plate with an oily or greasy rag before welding!

A few isolated carbides that may appear in type 304L are not destructive for many applications in which a continuous network of carbides would be catastrophic. In fact, the susceptibility to intergranular corrosion of the austenitic stainless steels can be reduced by severely cold-working the alloy. Cold-working produces smaller grains and many slip lines, which provide a much larger surface for carbide precipitation. This is not, however, a recommended or practical procedure.

Carbon pickup (surface carburization) during production of austenitic stainless steels has caused premature failures. It occurs when these steels are cast into molds containing carbonaceous materials such as organic binders and washes or baked oil sand. The hot metal absorbs carbon from the carbon-containing environment.

Increased carbon content of the stainless steel can degrade corrosion

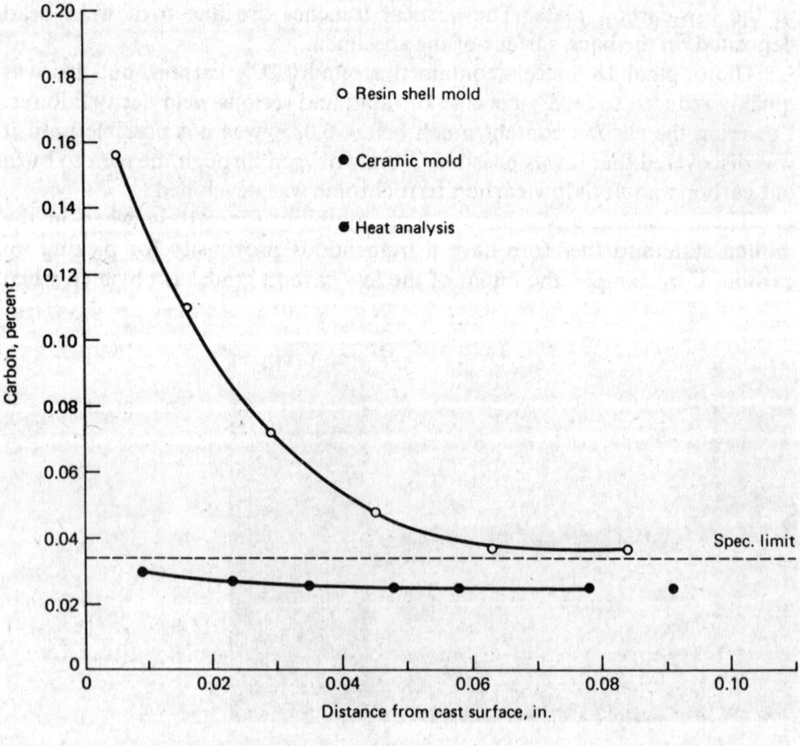

Figure 3-31 Carbon profiles of CF-3 castings.

resistance particularly to environments that are aggressive from the standpoint of intergranular attack. Resistance to pitting is also decreased.

Figure 3-31 shows carbon profiles for CF-3 (18-8, 0.03% C max) cast in a resin shell mold.* The carbon content near the surface is 0.16% as compared to 0.03% in the metal as poured. Higher carbon is in excess of the specification limit. Metal cast in a ceramic mold shows practically no carbon pickup. Similar situations occur in other austenitic stainless steels.

Figure 3-32 shows scanning electron microscope photographs of the surfaces after the Huey test (Sec. 4-19). Intergranular attack is evident on the resin shell casting but not on the ceramic mold casting.

Corrosion could continue into the metal beyond the carburized metal because intergranular corrosion, pitting, and stress corrosion could be initiated and propagated by crevice and/or notch effects. A number of case histories substantiate failure or reduced service life. Carbon pickup can be recognized if (1) castings are attacked more than wrought components, (2) intergranular attack occurs on a low-carbon (0.03%) material, (3) two castings of the same alloy show a substantial difference in attack, (4) machined surfaces show less attack than adjacent as-cast surfaces, and (5) carbon content near the surface is higher than in the main body of the casting.

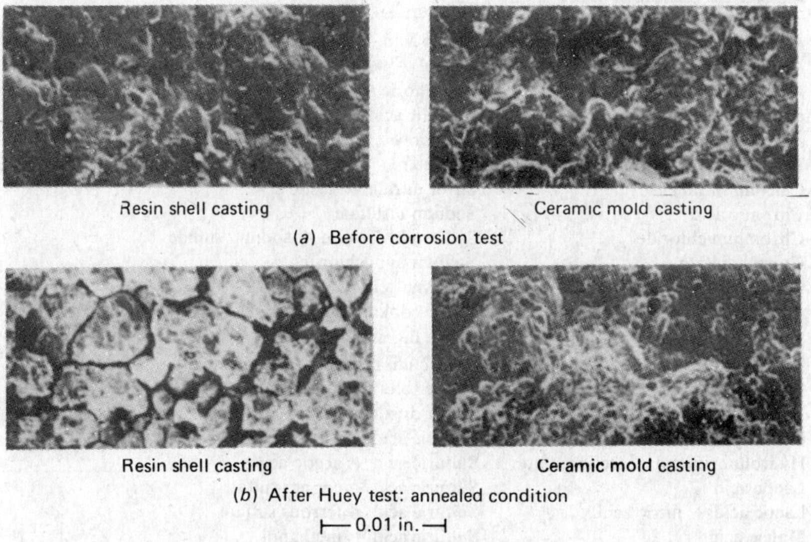

Figure 3-32 Scanning electron microscope (SEM) photographs of CF-3 casting surfaces.

*W. A. Luce, M. G. Fontana, and J. W. Cangi, *Corrosion*, **28**:115 (1972); also W. H. Herrnstein, J. W. Cangi, and M. G. Fontana, *Materials Performance*, **14**:21–27 (1975).

Table 3-8 Corrosives causing problems due to carbon pickup

Corrosives	Plants
Acetic acid	Adipodinitrile
Acetic + formic acid	Electroplating
Chromic acid	Metal refining
Fluoboric acid	Nitric acid
Metal plating solutions	Phosphoric acid
Nitric acid	Plutonium recovery
Nitric + adipic acid	Sulfuric acid
Nitric + hydrofluoric acid	Synthetic fiber
Phosphoric acid	Wet-process phosphoric acid
Phosphoric + sulfuric acid	Xylene
Sulfuric acid	
Sulfuric acid + copper sulfate	

Table 3-9 Corrosives that induce intergranular corrosion in austenitic stainless steel

Acetic acid	Oxalic acid
Acetic acid + salicylic acid	Phenol + naphthenic acid
Ammonium nitrate	Phosphoric acid
Ammonium sulfate	Phthalic acid
Ammonium sulfate + H_2SO_4	Salt spray
Beet juice	Seawater
Calcium nitrate	Silver nitrate + acetic acid
Chromic acid	Sodium bisulfate
Chromium chloride	Sodium hydroxide + sodium sulfide
Copper sulfate	Sodium hypochlorite
Crude oil	Sulfamic acid
Fatty acids	Sulfite cooking liquor
Ferric chloride	Sulfite digester acid
Ferric sulfate	(calcium bisulfite + sulfur dioxide)
Formic acid	Sulfite solution
Hydrocyanic acid	Sulfur dioxide (wet)
Hydrocyanic acid + sulfur dioxide	Sulfuric acid
Hydrofluoric acid + ferric sulfate	Sulfuric acid + acetic acid
Lactic acid	Sulfuric acid + copper sulfate
Lactic acid + nitric acid	Sulfuric acid + ferrous sulfate
Maleic acid	Sulfuric acid + methanol
Nitric acid	Sulfuric acid + nitric acid
Nitric acid + hydrochloric acid	Sulfurous acid
Nitric acid + hydrofluoric acid	Water + starch + sulfur dioxide
	Water + aluminum sulfate

Table 3-8 lists corrosive areas wherein corrosion due to carbon pickup was observed. Table 3-9 lists corrosives that can induce intergranular attack.*

If a certain specified carbon content is desired for a given corrosive environment, then the metal surface should meet the specification.

Carbon pickup is particularly critical for the very low carbon stainless steels (0.03% C max) and, of course, for the newer high-purity ferritic stainless steels. Sometimes the entire casting is "out of specification."

Future demands on stainless steel castings Consumers and users of high-alloy castings, such as the chemical process industries, will demand better quality castings in the future. There are indications that castings for "commercial" use may have to meet "nuclear" industry standards in many cases. Higher pressures and temperatures and increasing costs of maintenance and downtime call for better corrosion resistance and integrity of castings. Consumerism and product liability considerations call for predictable and reliable performance of equipment. Castings must be characterized by good quality, accurate dimensions, and easy reproducibility and will require documentation and certification. Inspection will be more critical. Premature failure because of defects or incorrect alloy composition must be avoided. Specifications must be carefully followed. Good liaison between the designer and producer is a must. Quality control and quality assurance are key words. These remarks also apply to wrought products.

3-21 Knife-Line Attack

The stabilized austenitic stainless steels are attacked intergranularly, under certain conditions, because of chromium carbide precipitation. Columbium or titanium fails to combine with the carbon. Figure 3-33 shows a section of a type 347 (18-8 + Cb) drum that contained fuming nitric acid. Severe intergranular attack occurred in a narrow band, a few grains wide, on both sides of the weld and immediately adjacent to it. Practically no corrosion is observable on the remainder of the container. This phenomenon was studied at Ohio State University and the basic mechanism for failure established.† It was christened knife-line attack because of its distinctive appearance.

Knife-line attack (KLA) is similar to weld decay in that they both result from intergranular corrosion and both are associated with welding. The two major differences are: (1) KLA occurs in a narrow band in the parent metal immediately adjacent to the weld, whereas weld decay develops at an appre-

*G. A. Nelson, "Corrosive Data Survey," NACE, 1967.

†M. L. Holzworth, F. H. Beck, and M. G. Fontana, *Corrosion*, 7:441–449 (1951).

Figure 3-33 Knife-line attack on type 347 stainless steel.

ciable distance from the weld; (2) KLA occurs in the stabilized steels; and (3) the thermal history of the metal is different.

The mechanism for the failure of this drum is based on the solubility of columbium in the stainless steel. Columbium and columbium carbides dissolve in the metal when it is heated to a very high temperature and they *remain* in solution when cooled rapidly from this temperature. The columbium stays in solution when the metal is then heated in the chromium carbide precipitation range; columbium carbide does not form, and the metal behaves (sensitizes) as though it were 18-8 without columbium.

The temperature of the weld metal is high enough to melt the alloy during welding—say, 3000°F. The metal adjacent to the weld is also at a high temperature because it is in contact with molten metal. The unmelted sheet is therefore just below the melting point, which is around 2600 to 2700°F. A sharp thermal gradient exists in the metal because of the relatively poor thermal conductivity of 18-8 and because the welding operation on this thin ($\frac{1}{16}$-in.) sheet is rapid (to avoid "burning through"). The thin sheet cools rapidly after welding.

This situation can be better explained by means of the chart shown as Fig. 3-34. The stainless steel as received from the steel mill contains columbium carbides and essentially no chromium carbides because it was heat-treated by water-quenching from 1950°F. Focus attention now on the narrow band of metal adjacent to the weld. This was heated to around

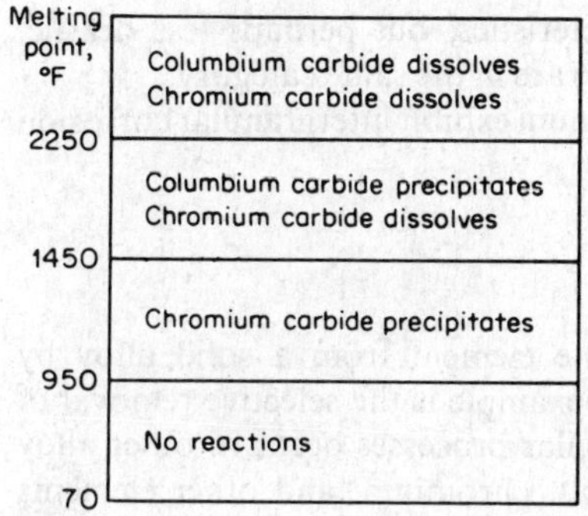

Figure 3-34 Schematic chart showing solution and precipitation reactions in types 304 and 347.

2600°F and cooled rapidly. According to the chart this band of metal has everything in solution (no precipitation of either carbide). If this metal is now heated in the sensitizing range of about 950 to 1400°F (as the drum was to relieve stress), only chromium carbide precipitates because the temperature is not high enough to form columbium carbide. If the drums were not heated after welding, failure would not have occurred because no carbides would have been present.

A simple experiment proves the mechanism. Take a sample of 18-8 + Cb, heat to 2300°F, and quench in water. Now heat it for $\frac{1}{2}$ hr at 1200°F and cool. The *entire sample* sensitizes essentially the same as 18-8 (no Cb).

The obvious remedy for avoiding knife-line attack is to heat the completed structure (after welding) to around 1950°F. According to the chart, chromium carbide dissolves and columbium carbide forms, which is the desired situation. The rate of cooling after the 1950°F treatment is not important.

Titanium-stabilized stainless steel (type 321) is also subject to knife-line attack under conditions similar to type 347. Type 304L steels have given superior performance in cases where the stabilized steels exhibited knife-line attack.

3-22 Intergranular Corrosion of Other Alloys

High-strength aluminum alloys depend on precipitated phases for strengthening and are susceptible to intergranular corrosion. For example, the Duraluminum-type alloys (Al-Cu) are strong because of precipitation of the compound $CuAl_2$. Substantial potential differences between the copper-depleted areas and adjacent material have been demonstrated. When these alloys are solution-quenched to keep the copper in solution, their susceptibility to intergranular corrosion is very small but they possess low strength. Other precipitates, such as $FeAl_3$, Mg_5Al_8, Mg_2Si, $MgZn_2$, and $MnAl_6$, along grain boundaries or slip lines in other aluminum alloy systems show somewhat similar characteristics, but perhaps less drastic. Some magnesium- and copper-base alloys are in the same category.

Die-cast zinc alloys containing aluminum exhibit intergranular corrosion by steam and marine atmospheres.

SELECTIVE LEACHING

Selective leaching is the removal of one element from a solid alloy by corrosion processes. The most common example is the selective removal of zinc in brass alloys (*dezincification*). Similar processes occur in other alloy systems in which aluminum, iron, cobalt, chromium, and other elements are removed. Selective leaching is the general term that describes these processes, and its use precludes the creation of terms such as *dealuminumification*, *decobaltification*, etc. *Parting* is a metallurgical term that is sometimes applied, but selective leaching is preferred.

3-23 Dezincification: Characteristics

Common yellow brass consists of approximately 30% zinc and 70% copper. Dezincification is readily observed with the naked eye because the alloy assumes a red or copper color that contrasts with the original yellow. There are two general types of dezincification, and both are readily recognizable. One is uniform, or layer-type, and the other is localized, or plug-type, dezincification. Figure 3-35 shows an example of uniform attack. The dark inner layer is the dezincified portion, and the outer layer is the unaffected yellow brass. Penetration of about 50% of the pipe wall occured after several years in potable-water service.

Figure 3-36 is a good example of plug-type dezincification. The dark areas are the dezincified plugs. The remainder of the tube is not corroded to any appreciable extent. This tube was removed from a powerhouse heat

Figure 3-35 Uniform dezincification of brass pipe.

Figure 3-36 Plug-type dezincification.

exchanger with boiler water on one side and fuel combustion gases on the outside. Figure 3-37 is a section through one of the plugs. Attack started on the water side of the tubing. Addition of zinc to copper lowers the corrosion resistance of the copper. If the dezincified area were good solid copper, the corrosion resistance of the brass would be improved. Unfortunately, the dezincified portion is weak, permeable, and porous as indicated in Fig. 3-37. The material is brittle and possesses little aggregate strength. This tube failed because of holes caused by some of the plugs being blown out by the water pressure (darkest areas in Fig. 3-36).

Overall dimensions do not change appreciably when dezincification occurs. If a piece of equipment is covered with dirt or deposits, or not inspected closely, sudden unexpected failure may occur because of the poor strength of dezincified material.

Uniform, or layer-type, dezincification seems to favor the high brasses (high zinc content) and definitely acid environments. The plug types seem to occur more often in the low brasses (lower zinc content) and neutral, alkaline, or slightly acidic conditions. These are general statements, and many exceptions occur. Stagnant conditions usually favor dezincification because of scale formation or foreign deposits settling on the metal surface. This can result in crevice corrosion and/or higher temperatures because of the insulating effect of the deposit (if a heat exchanger is involved).

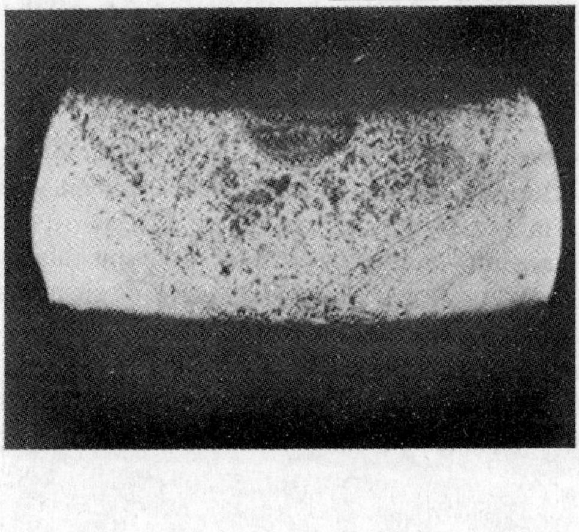

Figure 3-37 Section of one of the plugs shown in Fig. 3-36.

Metal structure and composition are important. Some brasses contain over 35% zinc. In these cases a zinc-rich phase forms (duplex structure) and localized corrosion may occur. Sometimes the beta phase is attacked first, and then dezincification spreads to the alpha matrix.

Failures of red brass (15% zinc) because of dezincification rarely occur. Tensile specimens of red brass, naval brass (35% zinc), and Muntz metal (40% zinc) exposed for several months to a chloride solution at 80°C showed losses in tensile strength of 5%, 30%, and 100%, respectively. Again, this shows the very low strength of dezincified alloy.

3-24 Dezincification: Mechanism

Two theories have been proposed for dezincification. One states that zinc is dissolved, leaving vacant sites in the brass lattice structure. This theory is not proven. A strong argument against it is that dezincification to appreciable depths would be impossible or extremely slow because of difficulty of diffusion of solution and ions through a labyrinth of small vacant sites.

The commonly accepted mechanism consists of three steps, as follows: (1) the brass dissolves, (2) the zinc ions stay in solution, and (3) the copper plates back on. Zinc is quite reactive, whereas copper is more noble. Zinc can corrode slowly in pure water by the cathodic ion reduction of H_2O into hydrogen gas and hydroxide ions. For this reason dezincification can proceed in the absence of oxygen. Oxygen also enters into the cathodic reaction and hence increases the rate of attack when it is present. Analyses of dezincified areas show 90 to 95% copper with some of it present as copper oxide. The amount of copper oxide is related to oxygen content of the environment. The porous nature of the deposit permits easy contact between the solution and the brass.

3-25 Dezincification: Prevention

Dezincification can be minimized by reducing the aggressiveness of the environment (i.e., oxygen removal) or by cathodic protection, but in most cases these methods are not economical. Usually a less susceptible alloy is used. For example, red brass (15% Zn) is almost immune.

One of the first steps in the development of better brasses was the addition of 1% tin to a 70-30 brass (Admiralty Metal). Further improvement was obtained by adding small amounts of arsenic, antimony, or phosphorus as "inhibitors." For example, arsenical Admiralty Metal contains about 70% Cu, 29% Zn, 1% Sn, and 0.04% As. Apparently these inhibiting elements are redoposited on the alloy as a film and thereby hinder deposition of copper. Arsenic is also added to aluminum (2% Al) brasses.

For severely corrosive environments where dezincification occurs, or for critical parts, cupronickels (70–90% Cu, 30–10% Ni) are utilized.

3-26 Graphitization

Gray cast iron sometimes shows the effects of selective leaching particularly in relatively mild environments. The cast iron appears to become "graphitized" in that the surface layer has the appearance of graphite and can be easily cut with a penknife. Based on this appearance and behavior, this phenomenon was christened "graphitization." This is a misnomer because the graphite is present in the gray iron before corrosion occurs. It is also called *graphitic corrosion*.

Figure 5-1 shows the microstructure of gray cast iron. What actually happens is selective leaching of the iron or steel matrix leaving the graphite network. The graphite is cathodic to iron, and an excellent galvanic cell exists. The iron is dissolved, leaving a porous mass consisting of graphite, voids, and rust. The cast iron loses strength and its metallic properties. Dimensional changes do not occur, and dangerous situations may develop without detection. The surface usually shows rusting that appears superficial, but the metal has lost its strength. The degree of loss depends on the depth of the attack. Graphitization is usually a slow process. If the cast iron is in an environment that corrodes this metal rapidly, all of the surface is usually removed and more-or-less uniform corrosion occurs.

Graphitization does not occur in nodular or malleable cast irons (see Chap. 5) because the graphite network is not present to hold together the residue. White cast iron has essentially no free carbon and is not subject to graphitization.

Graphitization of gray cast iron has reached the public eye because of failures of underground pipelines, particularly those handling hazardous materials. Graphitized pipe has cracked because of soil settlement or impact by excavating or earth-moving equipment. In several cases explosions, fires, and fatalities have occurred. This was emphasized to me during a three-year term as a member of the Technical Pipeline Safety Standards Committee of the U.S. Department of Transportation. Underground pipelines made of gray cast iron should be selected only after consideration of graphitization.

Bloom and Tuovinen* conducted dispersive analysis and metallurgical studies of graphitic residues from gray cast iron pipe that had been in service for about 40 years in Columbus, Ohio. Years ago I recommended use here of ductile (nodular) cast iron instead of gray cast iron (brittle). Ductile iron pipe with a cement mortar lining has been giving excellent performance.

3-27 Other Alloy Systems

Selective leaching by aqueous environments occurs in other alloy systems under appropriate conditions, especially in acids. Selective removal of

*P. R. Bloom and O. H. Tuovinen, Characterization of Graphitic Corrosion Residue of Cast Iron in a Water Distribution System, *Materials Performance*, 21–23 (Feb. 1984).

aluminum in aluminum bronzes has been observed in hydrofluoric and other acids. A two-phase or duplex structure is more susceptible. Massive effects were observed in crevices on aluminum bronze where the solution contained some chloride ions.

Selective leaching has been observed in connection with removal of silicon from silicon bronzes (Cu-Si) and also removal of cobalt from a Co-W-Cr alloy. It should be emphasized that these are rare cases and not as well known as dezincification.

Sometimes selective corrosion of one element in an alloy may be beneficial. Enrichment of silicon observed in the oxide film on stainless steels results in better passivity and resistance to pitting.

With reference to other alloy systems another constructive application of dealloying involves the preparation of Raney nickel catalyst by selectively removing aluminum from an aluminum-nickel alloy by action of caustic.

The term *dealloying* is frequently used and is preferred by some corrosionists. Dealloying is defined* as a corrosion process whereby one constituent of an alloy is preferentially removed from the alloy leaving an altered residual structure (like dezincification). Pryor† shows that dealloying occurs much slower in an aluminum bronze than in yellow brass.

3-28 High Temperatures

M. G. Fontana's early work on high-temperature oxidation of stainless steels showed selective oxidation of chromium when exposed to low-oxygen atmospheres at high temperatures (1800°F). When there is competition for oxygen, the elements with higher free energies for their oxide formation (higher affinity for oxygen) are oxidized to a greater degree. In the case of stainless steels, this results in a more protective scale. However, the remaining or substrate metal will be deficient in chromium. This phenomenon was clearly demonstrated by Trax and Holzwarth.‡ Pitting of type 430 (17% Cr) trim on automobiles was attributed to depletion of chromium during bright-annealing operations. Chromium contents as low as 11% were determined at and near the surface of the steel. Another unusual case¶ showed the selective corrosion of chromium and iron from Inconel (75% Ni, 15% Cr, 9% Fe) by potassium-sodium-fluoride-chloride salt baths at about 1475°F. The alloy was destroyed by conversion to a spongy mass.

*R. H. Heidersbach and E. D. Verink, *Corrosion*, **28**:397–418 (Nov. 1972).

†M. J. Pryor, The Dealloying of a Cu-8.9% Al Solid Solution, *J. Electrochem. Soc.*, **130**: 1625–1627 (July 1983).

‡R. V. Trax and J. C. Holzwarth, *Corrosion*, **16**:105–108 (1960).

¶R. Bakish and F. Kern, *Corrosion*, **16**:89–90 (1960).

EROSION CORROSION

Erosion corrosion is the acceleration or increase in rate of deterioration or attack on a metal because of relative movement between a corrosive fluid and the metal surface. Generally this movement is quite rapid, and mechanical wear effects or abrasion are involved. Metal is removed from the surface as dissolved ions, or it forms solid corrosion products that are mechanically swept from the metal surface. Sometimes movement of the environment decreases corrosion, particularly when localized attack occurs under stagnant conditions, but this is not erosion corrosion because deterioration is not increased.

Erosion corrosion is characterized in appearance by grooves, gullies, waves, rounded holes, and valleys and usually exhibits a directional pattern. Figure 3-38 shows a typical wavy appearance of an erosion-corrosion failure. This pump impeller was taken out of service after three weeks of operation. Figure 3-39 is a sketch representing erosion corrosion of a heat-exchanger tube handling water. In many cases, failures because of erosion corrosion occur in a relatively short time, and they are unexpected largely because evaluation corrosion tests were run under static conditions or because the erosion effects were not considered.

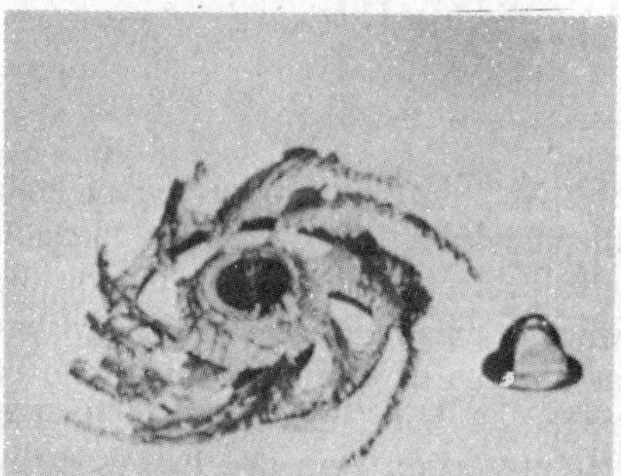

Figure 3-38 Erosion corrosion of stainless alloy pump impeller.

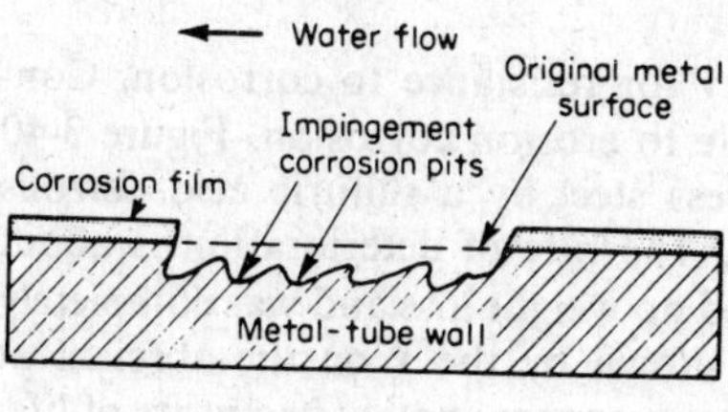

Figure 3-39 Erosion corrosion of condenser tube wall.

Most metals and alloys are susceptible to erosion-corrosion damage. Many depend upon the development of a surface film of some sort (passivity) for resistance to corrosion. Examples are aluminum, lead, and stainless steels. Erosion corrosion results when these protective surfaces are damaged or worn and the metal and alloy are attacked at a rapid rate. Metals that are soft and readily damaged or worn mechanically, such as copper and lead, are quite susceptible to erosion corrosion.

Many types of corrosive mediums could cause erosion corrosion. These include gases, aqueous solutions, organic systems, and liquid metals. For example, hot gases may oxidize a metal and then at high velocity blow off an otherwise protective scale. Solids in suspension in liquids (slurries) are particularly destructive from the standpoint of erosion corrosion.

All types of equipment exposed to moving fluids are subject to erosion corrosion. Some of these are piping systems, particularly bends, elbows, and tees; valves; pumps; blowers; centrifugals; propellers; impellers; agitators; agitated vessels; heat-exchanger tubing such as heaters and condensers; measuring devices such as an orifice; turbine blades; nozzles; ducts and vapor lines; scrapers; cutters; wear plates; grinders; mills; baffles; and equipment subject to spray.

Since corrosion is involved in the erosion-corrosion process all of the factors that affect corrosion should be considered. However, only the factors directly pertinent to erosion corrosion are discussed here.

3-29 Surface Films

The nature and properties of the protective films that form on some metals or alloys are very important from the standpoint of resistance to erosion corrosion. The ability of these films to protect the metal depends on the speed or ease with which they form when originally exposed to the environment, their resistance to mechanical damage or wear, and their rate of re-forming when destroyed or damaged. A hard, dense, adherent, and continuous film would provide better protection than one that is easily removed by mechanical means or worn off. A brittle film that cracks or spalls under stress may not be protective. Sometimes the nature of the protective film that forms on a given metal depends upon the specific environment to which it is exposed, and this determines its resistance to erosion corrosion by that fluid.

Stainless steels depend on passivity for resistance to corrosion. Consequently these materials are vulnerable to erosion corrosion. Figure 3-40 shows rapid attack of type 316 stainless steel by a sulfuric acid–ferrous sulfate slurry moving at high velocity. The rate of deterioration is about 4500 mpy at 55°C. This material showed no weight loss and was completely passive under stagnant conditions as shown by the x on the abscissa at 60°C. The impeller shown in Fig. 3-38 gave approximately two years of life,

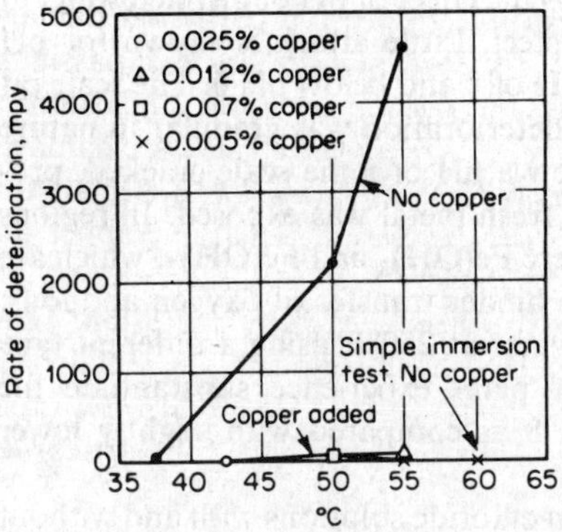

Figure 3-40 Effect of temperature and copper-ion addition on erosion corrosion of type 316 by sulfuric acid slurry (velocity, 39 ft/sec).

which was reduced to three weeks when the solution pumped was made more strongly reducing, thus destroying the passive film.

Lead depends on the formation of a lead sulfate–lead oxide protective surface for long life in sulfuric acid environments, and in many cases more than 20 years' service is obtained. Lead gains weight when exposed to sulfuric acid because of the surface coating or corrosion product formed except in strong acid wherein the lead sulfate is soluble and not protective. However, lead valves failed in less than one week and lead bends were rapidly attacked in a plant handling a 3% sulfuric acid solution at 90°C. As a result of these failures erosion-corrosion tests were made, and the results are plotted in Fig. 3-41. Under static conditions the lead showed no deterioration (slight gain in weight) as shown by the points on the abscissa. Under high-velocity conditions, attack increased with temperature as shown by the curve.

Variations in amount of attack on steel by water with different pH values but constant velocity are apparently due to the nature and composition of the

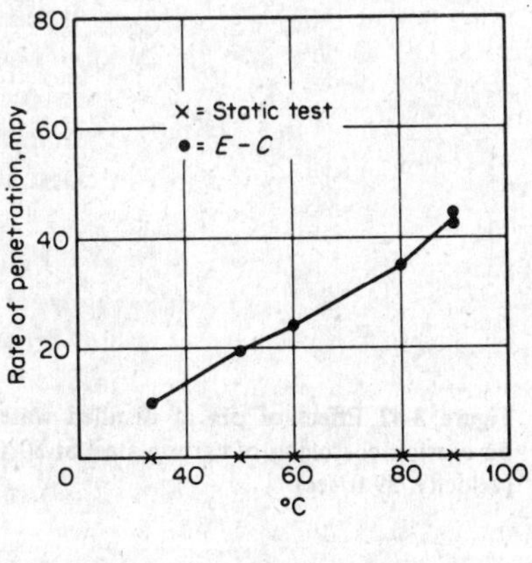

Figure 3-41 Erosion corrosion of hard lead by 10% sulfuric acid (velocity, 39 ft/sec).

surface scales formed. Figure 3-42 shows the effect of pH of distilled water at 50°C on erosion corrosion of carbon steel. Little attack is shown for pH values of 6 and 10 and high rates at a pH of 8 and below pH 6. The scale on the specimens exhibiting high rates of deterioration was granular in nature and consisted of magnetic Fe_3O_4. Below a pH of 5 the scale cracked, probably because of internal stresses, and fresh metal was exposed. In regions of low attack the corrosion products were $Fe(OH)_2$ and $Fe(OH)_3$, which are more protective probably because they hinder transfer of oxygen and ions. Erosion-corrosion tests in boiler feedwater at 250°F using a different type of testing equipment and also power plant experience substantiate the results indicating higher attack at pH 8 as compared with slightly lower values.

Tests on copper and brass in sodium chloride solutions with and without oxygen show that copper is attacked more than brass in the oxygen-saturated solutions. The copper was covered with a black and yellow-brown film ($CuCl_2$). The brass was covered with a dark gray film (CuO). The better resistance of the brass to attack was attributed to the greater stability or protectiveness of the dark gray film. Difficulty was encountered in obtaining reproducible results until a controlled alkali cleaning and drying procedure for the specimens was adopted. This indicates that surface films formed on copper and brass because of atmospheric exposure, abrading, or other reasons can have a definite effect on erosion-corrosion performance under some conditions.

Titanium is a reactive metal but is resistant to erosion corrosion in many environments because of the stability of the TiO_2 film formed. It shows excellent resistance to seawater and chloride solutions and also to fuming nitric acid.

The behavior of steel and low-alloy-steel tubes handling oils at high

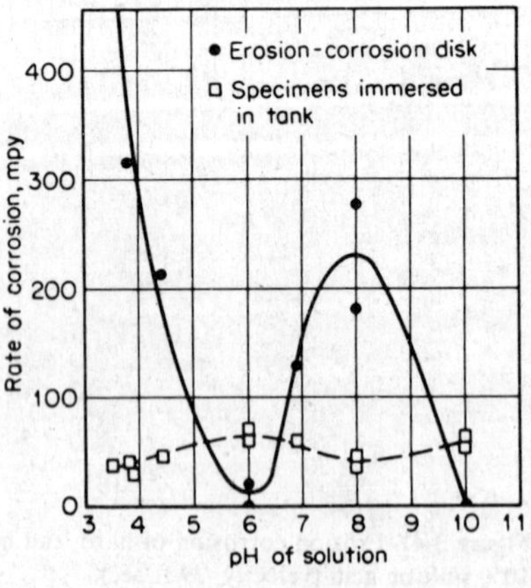

Figure 3-42 Effect of pH of distilled water on erosion corrosion of carbon steel at 50°C (velocity, 39 ft/sec).

temperatures in petroleum refineries depends somewhat on the sulfide films formed. When the film erodes, rapid attack occurs. For example, a normally tenacious sulfide film becomes porous and nonprotective when cyanides are present in these organic systems.

The effective use of inhibitors to decrease erosion corrosion depends, in many cases, on the nature and type of films formed on the metal as a result of reaction between the metal and the inhibitor.

3-30 Velocity

Velocity of the environment plays an important role in erosion corrosion. Velocity often strongly influences the mechanisms of the corrosion reactions. It exhibits mechanical wear effects at high values and particularly when the solution contains solids in suspension. Figure 3-40 and 3-41 show large increases in attack because of velocity. Figure 3-42 indicates that misleading results could be obtained when only static tests, or tests at very low velocities, are made. The specimens in the tank were subjected to only a mild swirling motion. Table 3-10 shows the effect of velocity on a variety of metals and

Table 3-10 Corrosion of metals by seawater moving at different velocities

	Typical corrosion rates, mdd*		
Material	1 ft/sec†	4 ft/sec‡	27 ft/sec¶
Carbon steel	34	72	254
Cast iron	45	—	270
Silicon bronze	1	2	343
Admiralty Brass	2	20	170
Hydraulic bronze	4	1	339
G bronze	7	2	280
Al bronze (10% Al)	5	—	236
Aluminum brass	2	—	105
90-10 Cu Ni (0.8% Fe)	5	—	99
70-30 Cu Ni (0.05% Fe)	2	—	199
70-30 Cu Ni (0.5% Fe)	<1	<1	39
Monel	<1	<1	4
Stainless steel type 316	1	0	<1
Hastelloy C	<1	—	3
Titanium	0	—	0

*Milligrams per square decimeter per day.
†Immersed in tidal current.
‡Immersed in seawater flume.
¶Attached to immersed rotating disk.
Source: International Nickel Co.

alloys exposed to seawater. These data show that the effect of velocity may be nil or extremely great.

Increases in velocity generally result in increased attack, particularly if substantial rates of flow are involved. The effect may be nil or increase slowly until a *critical velocity* is reached, and then the attack may increase at a rapid rate. Table 3-10 lists several examples exhibiting little effect when the velocity is increased from 1 to 4 ft/sec but destructive attack at 27 ft/sec. This high velocity is below the critical value for other materials listed at the bottom of the table.

Erosion corrosion can occur on metals and alloys that are completely resistant to a particular environment at low velocities. For example, hardened straight chromium stainless steel valve seats and plugs give excellent service in most steam applications, but grooving or so-called "wire drawing" occurs in high-pressure steam reducing or throttling valves.

Increased velocity may increase or reduce attack, depending on its effect on the corrosion mechanism involved. It may increase attack on steel by increasing the supply of oxygen, carbon dioxide, or hydrogen sulfide in contact with the metal surface.

Velocity can decrease attack and increase the effectiveness of inhibitors by supplying the chemical to the metal surface at a higher rate. It has been shown that less sodium nitrite is needed at high velocity to protect steel in tap water. Similar mechanisms have been postulated for other types of inhibitors.

Higher velocities may also decrease attack in some cases by preventing the deposition of silt or dirt which would cause crevice corrosion. On the other hand, solids in suspension moving at high velocity may have a scouring effect and thus destroy surface protection. This was the case in connection with Fig. 3-40, which involved rapid erosion corrosion of type 316 centrifugals handling a sulfuric acid slurry.

Erosion-corrosion studies of aluminum and stainless alloys in fuming nitric acid produced unusual and interesting results. Attack on aluminum increased and attack on type 347 stainless steel decreased as velocity was increased, because of the different corrosion mechanisms involved. Figure 3-43 shows increasing attack on aluminum with increasing velocity. Aluminum can form films of aluminum nitrates and aluminum oxide in fuming nitric acid. Little or no attack occurs at zero or very low velocities. At intermediate velocities of 1 to 4 ft/sec, the action of the solution is sufficient to remove the nitrate film but not strong enough to destroy the more adherent oxide film. Velocities above 4 ft/sec apparently remove much of the oxide, and erosion corrosion occurs at a faster rate.

Figure 3-44 shows *decrease* in attack on type 347 stainless steel as velocity is increased. Under stagnant conditions this steel in nitric acid is attacked autocatalytically because of formation of nitrous acid as a cathodic reaction product. Increasing velocity sweeps away the nitrous acid and thus removes one of the corrosive agents in the environment.

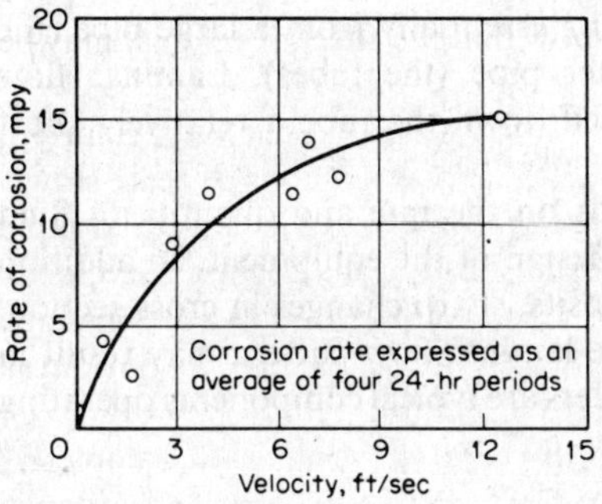

Figure 3-43 Erosion corrosion of 3003 aluminum by white fuming nitric acid at 108°F.

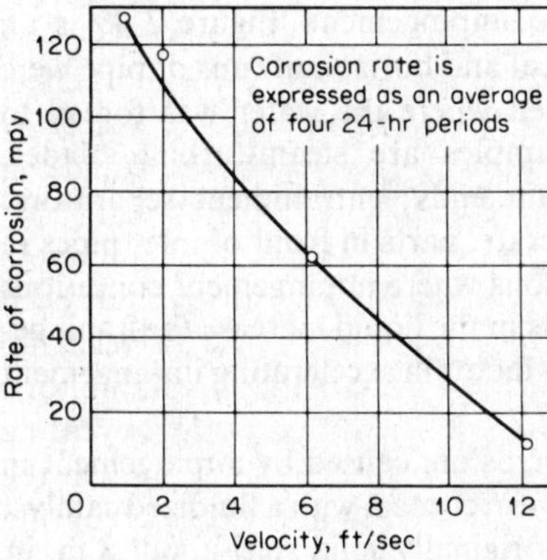

Figure 3-44 Erosion corrosion of type 347 stainless steel by white fuming nitric acid at 108°F.

Many stainless steels have a strong tendency to pit and suffer crevice corrosion in seawater and other chlorides. However, some of these materials are used successfully in seawater, provided the water is kept moving at a substantial velocity. This motion prevents formation of deposits and retards the initiation of pits.

3-31 Turbulence

Many erosion-corrosion failures occur because turbulence or turbulent flow conditions exist. Turbulence results in greater agitation of the liquid at the metal surface than is the case for laminar (straight line) flow. Turbulence results in more intimate contact between the environment and the metal. Perhaps the most frequently occuring example of this type of failure occurs in the inlet ends of tubing in condensers and similar shell-and-tube heat exchangers. It is designated as *inlet-tube corrosion.* The attack is usually confined to the first few inches of the tubing at the inlet end. Turbulence exists in this area because the liquid is flowing essentially from a large pipe (the exchanger head) into a small-diameter pipe (the tubes). Laminar flow develops after the liquid has progressed down the tube a relatively short distance.

The type of flow obtained depends on the rate and quantity of fluid handled and also on the geometry or design of the equipment. In addition to high velocities, ledges, crevices, deposits, sharp changes in cross section, and other obstructions that disturb the laminar flow pattern may result in erosion corrosion. Impellers and propellers are typical components operating under turbulent conditions.

3-32 Impingement

Many failures are directly attributed to impingement. Figure 3-45 is an example of this type of failure. The vertical and horizontal runs of pipe were relatively unaffected, but the metal failed where the water was forced to turn its direction of flow. Other examples are steam-turbine blades, particularly in the exhaust or wet-steam ends; entrainment separators; bends; tees; external components of aircraft; parts in front of inlet pipes in tanks; cyclones; and any other applications where impingement conditions exist. Solids and sometimes bubbles of gas in the liquid increase the impingement effect. Air bubbles are an important factor in accelerating impingement attack.

Figure 3-46 shows severe erosion corrosion, caused by impingement in less than one year of service, of a slide valve in contact with a fluidized catalyst and oil at 900°F in a refinery. This was originally solid steel about 3 in. in diameter. Figure 3-47 shows two types of erosion corrosion in a thermal cracking furnace for oil. The tube on the left contained superheated steam.

Figure 3-45 Impingement failure of elbow in steam condensate line.

Figure 3-46 Erosion corrosion of slide valve at 900°F in petroleum refinery.

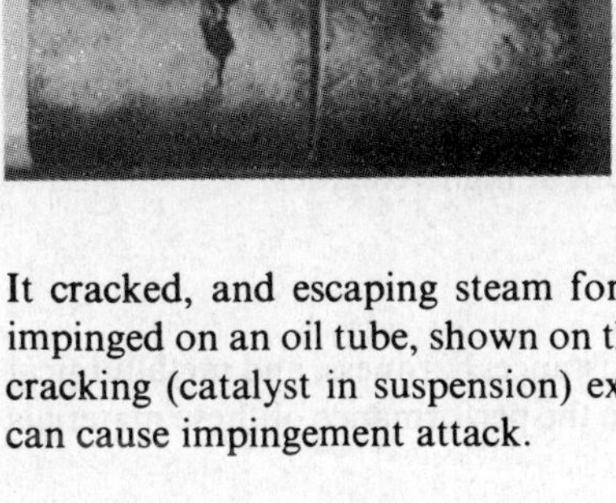

Figure 3-47 Impingement by escaping steam from cracked tube (left).

It cracked, and escaping steam formed the two holes shown. This steam impinged on an oil tube, shown on the right, and a leak developed. Catalytic cracking (catalyst in suspension) experience indicates that an angle of 25° can cause impingement attack.

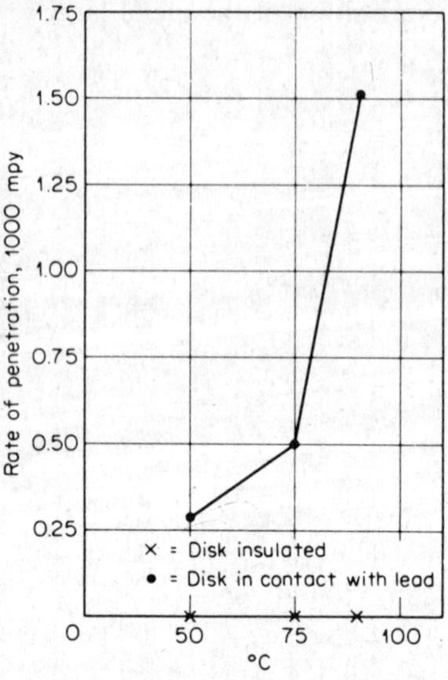

Figure 3-48 Effect of contact with lead on erosion corrosion of type 316 by 10% sulfuric acid (velocity, 39 ft/sec).

3-33 Galvanic Effect

Galvanic, or two-metal corrosion can influence erosion corrosion when dissimilar metals are in contact in a flowing system. The galvanic effect may be nil under static conditions but may be greatly increased when movement is present. Figure 3-48 shows that attack on type 316 by itself was nil in high-velocity sulfuric acid but increased to very high values when this alloy was in contact with lead. The passive film was destroyed by the combined forces of galvanic corrosion and erosion corrosion. Couples of lead and type 316 showed no corrosion under static conditions. Cracks in the Fe_3O_4 scale formed in the lower pH ranges of Fig. 3-42 doubtless contributed to increased attack because the scale is cathodic to the substrate steel by about 500 millivolts (mV).

Velocity changes can produce surprising galvanic effects. In seawater at low velocity the corrosion of steel is not appreciably affected by coupling with stainless steel, copper, nickel, or titanium. At high velocities the attack on steel is much less when coupled to stainless steel and titanium than when coupled to copper or nickel. This is attributed to the more effective cathodic polarization of stainless steel and titanium at high velocities.

3-34 Nature of Metal or Alloy

The chemical composition, corrosion resistance, hardness, and metallurgical history of metals and alloys can influence the performance of these materials under erosion-corrosion conditions. The composition of the metal largely determines its corrosion resistance. If it is an active metal, or an alloy composed of active elements, its corrosion resistance is due chiefly to its ability to form and maintain a protective film. If it is a more noble metal, it possesses good *inherent* corrosion resistance. A material with better inherent resistance would be expected to show better performance when all other factors are equal. For example, an 80% nickel–20% chromium alloy is superior to an 80% iron–20% chromium alloy because nickel has better inherent resistance than iron. For the same reason a nickel-copper alloy is better than one of zinc and copper.

The addition of a third element to an alloy often increases its resistance to erosion corrosion. The addition of iron to cupronickel produces a marked increase in resistance to erosion corrosion by seawater as shown in Table 3-10. The addition of molybdenum to 18-8 to make type 316 makes it more resistant to corrosion and erosion corrosion. In both cases the addition of the third element produces a more stable protective film. Aluminum brasses show better erosion-corrosion resistance than straight brass.

Resistance of steel and iron-chromium alloys to acid mine waters under erosion-corrosion conditions showed a straight-line increase in resistance with increasing chromium up to 13%. At this content and above, no attack occurred. Low-alloy chromium steels show better erosion-corrosion resistance than straight carbon steels in high-temperature boiler feedwater. Type 3 Ni-Resist (30% nickel, 3% chromium cast iron) showed practically no attack by seawater after 60 days under erosion-corrosion conditions, whereas ordinary cast iron was badly deteriorated.

Erosion-corrosion resistance of stainless steels and stainless alloys varies depending upon their compositions. Durimet 20 (30% Ni, 20% Cr, 3.5% Cu, 2% Mo) exhibits better performance than 18-8 steels in fuming nitric acid, seawater, and many other environments not only because of better inherent resistance but also because of the more protective films formed.

The soft metals are more susceptible to erosion corrosion because they are more subject to mechanical wear. Hardness is a fairly good criterion for resistance to mechanical erosion or abrasion but it is *not* necessarily a good criterion for predicting resistance to erosion corrosion. There are many methods for producing hard metals and alloys or for hardening them. One sure method for producing good erosion-corrosion resistance is solid-solution hardening. This involves adding one element to another to produce a solid solution that is corrosion resistant and is *inherently* hard. It cannot be softened or further hardened by heat treatment. The best and most familiar example is high-silicon (14.5% Si) iron. It is perhaps the most universally corrosion resistant of the nonprecious metals and the only alloy that can be used in many severe erosion-corrosion conditions.

Hardening by heat treatment results in changes in microstructure and heterogeneity and generally decreases resistance to corrosion, as noted in Chap. 2. For example, the precipitation-hardened stainless steels would not be expected to give as good performance as type 304 stainless steel under erosion-corrosion conditions.

A good example of poor performance by a high-hardness material concerned the centrifugals and conditions discussed in connection with Fig. 3-40. Both type 316 and type 329 stainless steels showed no measurable corrosion in the sulfuric acid slurry under static conditions, even when the type 329 was age-hardened to 450 Brinell hardness. Under the erosion-corrosion conditions in the centrifugal, however, the hard type 329 steel deteriorated more than 10 times faster than the soft (150 Brinell) type 316.

Cast iron sometimes shows better performance than steel under erosion-corrosion conditions, particularly in hot strong sulfuric acid. The iron in the cast iron is corroded, but the remaining graphitized layer consisting of the original graphite network and corrosion products provides some protection.

3-35 Combating Erosion Corrosion

Five methods for the prevention or minimization of damage due to erosion corrosion are used. In order of importance, or extent of use, they are: (1) materials with better resistance to erosion corrosion, (2) design, (3) alteration of the environment, (4) coatings, and (5) cathodic protection.

Better materials The reasons for using better materials that give improved performance are obvious. This method represents the economical solution to most erosion-corrosion problems.

Design This is an important method in that the life of presently used, or less costly, materials can be extended considerably or the attack practically eliminated. Design here involves change in shape, or geometry, and not selection of material. Erosion-corrosion damage can be reduced through better design as illustrated by the following examples. Increasing pipe diameter helps from the mechanical standpoint by decreasing velocity and also ensures laminar flow. Increasing the diameter and streamlining bends reduces impingement effects. Increasing the thickness of material strengthens vulnerable areas. In one instance of severe erosion corrosion of lead, maintenance costs were reduced to a satisfactory level by using a sweeping bend and doubling the thickness of the pipe. The design of other equipment, such as inlets and outlets, should be streamlined to remove obstructions for the same reasons. Readily replaceable impingement plates or baffles should be inserted. Inlet pipes should be directed toward the center of a tank instead of near its wall. Tubes should be designed to extend several inches beyond the tube sheet at the inlet end. In several cases, life of tubing was practically doubled by increasing the length 4 inches. The protruding tube ends were attacked, but operation was not affected.

Ferrules, or short lengths of flared tubing, can be inserted in the inlet ends. These could be made of the same material as the tubes or of material with better resistance. Bakelite and other plastic ferrules are readily available and widely used in condensers. The end of the ferrule should be "feathered" to blend the flow. If this is not done, erosion corrosion occurs on the tube just beyond the end of the ferrule because of the step present. Galvanic corrosion must be considered when using metallic inserts. The life of tubing in a vertical evaporator was doubled by turning the evaporator upside down when the inlet or bottom ends of the tubes became thin. The outlet ends, which were not appreciably attacked, became inlet ends. Equipment should be designed so that parts can be replaced readily. Tube bundles that can be readily removed and replaced by spares can be repaired at leisure. Buckets and conveyor flights that are easily replaced on centrifugals and other conveying equipment reduce costs of erosion corrosion. Use of pumps with interchangeable parts in different alloys helps reduce costs when an unsatisfactory alloy is originally selected. In one case, some of the blades of a steam turbine were out of line and the protruding blades suffered severe erosion-corrosion damage from water droplets in the steam. Misalignment from one pipe section to the next can cause erosion corrosion in both flanged and welded joints. Good design implies proper construction and workmanship.

Alteration of the environment Deaeration and the addition of inhibitors are effective methods, but in many cases they are not sufficiently economical for minimizing erosion-corrosion damage. Settling and filtration are helpful in removing solids. Whenever possible, the temperature of the environment should be reduced. This has been done in many cases without appreciably affecting the process. Temperature is our worst enemy in erosion corrosion, as it is in all types of corrosion.

Coatings Applied coatings of various kinds that produce a resilient barrier between the metal and its environment are sometimes utilized but are not always feasible for solving erosion-corrosion problems. Hard facings, or welded overlays, are sometimes helpful, provided the facing has good corrosion resistance. Repair of attacked areas by welding is often practical.

Cathodic protection This helps to reduce attack, but it has not found widespread use for erosion corrosion. One plant uses steel plates on condenser heads to provide cathodic protection of the inlet ends of tubes in heat exchangers handling seawater. Others use zinc plates. Zinc plugs are frequently used in water pumps.

Fortunately, all pumps, valves, lines, pipes, elbows, etc., do not fail because of erosion corrosion. However, serious trouble may develop if erosion corrosion is not considered.

3-36 Cavitation Damage

A special form of erosion corrosion, *cavitation damage*, is caused by the formation and collapse of vapor bubbles in a liquid near a metal surface. Cavitation damage occurs in hydraulic turbines, ship propellers, pump impellers, and other surfaces where high-velocity liquid flow and pressure changes are encountered. Before considering cavitation damage, let us examine the phenomenon of cavitation. If the pressure on a liquid such as water is reduced sufficiently, it boils at room temperature. Consider a cylinder full of water that is fitted with a tight piston in contact with the water. If the piston is raised away from the water, pressure is reduced and the water vaporizes, forming bubbles. If the piston is now pushed toward the water, pressure is increased and bubbles condense or collapse. Repeating this process at high speed such as in the case of an operating water pump, bubbles of water vapor form and collapse rapidly. Calculations have shown that rapidly collapsing vapor bubbles produce shock waves with pressures as high as 60,000 $lb/in.^2$ Forces this high can produce plastic deformation in many metals. Evidence of this is indicated by the presence of slip lines in pump parts and other equipment subjected to cavitation.

The appearance of cavitation damage is somewhat similar to pitting, except that the pitted areas are closely spaced and the surface is usually considerably roughened. Cavitation damage has been attributed to both corrosion and mechanical effects. In the former case, it is assumed that the collapsing vapor bubbles destroy protective surface films and thus increase corrosion. This mechanism is shown schematically in Fig. 3-49. The steps are as follows: (1) A cavitation bubble forms on the protective film. (2) The bubble collapses and destroys the film. (3) The newly exposed metal surface corrodes and the film is reformed. (4) A new cavitation bubble forms at the same spot. (5) The bubble collapses and destroys the film. (6) The exposed area corrodes and the film reforms. The repetition of this process results in deep holes.

Examination of Fig. 3-49 shows that it is not necessary to have a protective film for cavitation damage to occur. An imploding cavitation bubble has sufficient force to tear metal particles away from the surface. Once the surface has been roughened at a point this serves as a nucleus for new cavitation bubbles in a manner similar to that shown in Fig. 3-49. In actual practice, it appears that cavitation damage is the result of both mechanical and chemical action.

In general, cavitation damage can be prevented by the techniques used in preventing erosion corrosion outlined above. Also, there are some specific measures. Cavitation damage can be reduced by changing design to minimize hydrodynamic pressure differences in process flow streams. More corrosion-resistant materials may be substituted. Smooth finishes on pump impellers and propellers reduce damage since smooth surfaces do not provide sites

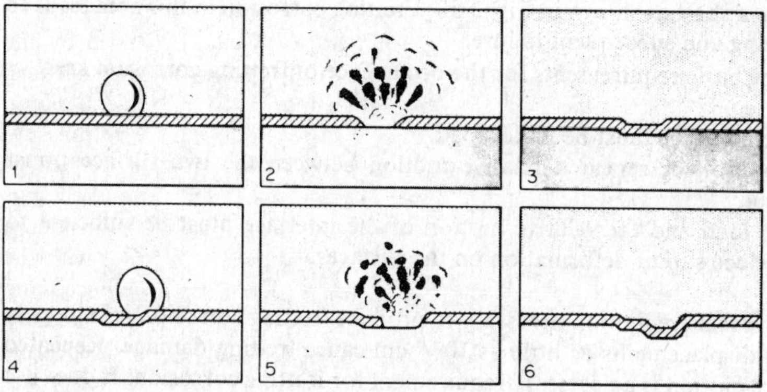

Figure 3-49 Schematic representation of steps in cavitation. (*R. W. Henke*)

for bubble nucleation. Coating metallic parts with resilient coatings such as rubber and plastic have also proved beneficial. It is important to use caution in applying such coatings, since bonding failures between the metal-coating interface frequently occur during operation. Cathodic protection also reduces cavitation damage, apparently because of the formation of hydrogen bubbles on the metal surface, which cushions the shock wave produced during cavitation.

3-37 Fretting Corrosion

Fretting describes corrosion occuring at contact areas between materials under load subjected to vibration and slip. It appears as pits or grooves in the metal surrounded by corrosion products. Fretting is also called *friction oxidation*, *wear oxidation*, *chafing*, and *false brinelling* (so named because the resulting pits are similar to the indentations made by a Brinell hardness test). It has been observed in engine components, automotive parts, bolted parts, and other machinery. Essentially, fretting is a special case of erosion corrosion that occurs in the atmosphere rather than under aqueous conditions.

Fretting corrosion is very detrimental because of the destruction of metallic components and the production of oxide debris. Seizing and galling often occur, together with loss of tolerances and loosening of mating parts. Further, fretting causes fatigue fracture since the loosening of components permits excessive strain, and the pits formed by fretting act as stress raisers.

A classic case of fretting occurs at bolted tie plates on railroad rails. Frequent tightening of these plates is required because the parts are not lubricated and fretting corrosion proceeds rapidly. Another common case of fretting corrosion occurs at the interface between a press-fitted ball-bearing race on a shaft as shown in Fig. 3-50. Fretting corrosion in this area leads to loosening and subsequent failure.

The basic requirements for the occurrence of fretting corrosion are:

1. The interface must be under load.
2. Vibration or repeated relative motion between the two surfaces must occur.
3. The load and the relative motion of the interface must be sufficient to produce slip or deformation on the surfaces.

The relative motion necessary to produce fretting corrosion is extremely small; displacements as little as 10^{-8} cm cause fretting damage. Repeated relative motion is a necessary requirement for fretting corrosion. It does not occur on surfaces in continuous motion, such as axle bearings or the ball bearings shown in Fig. 3-50, but rather on interfaces, which are subject to repeated small relative displacements. This point is best illustrated by considering fretting corrosion occuring on automobile axles during long-distance shipment by rail or boat. This is caused by the load on these surfaces and the continuous vibration or jiggling that occurs during shipment. Normal operation of an automobile does not show this difficulty because the relative motion between the axle bearing surfaces is very large (complete revolutions).

The two major mechanisms proposed for fretting corrosion are the wear-oxidation and oxidation-wear theories, which are schematically illustrated in Figs. 3-51 and 3-52, respectively. The wear-oxidation mechanism is based on the concept that cold welding or fusion occurs at the interface between metal surfaces under pressure, and during subsequent relative motion, these contact points are ruptured and fragments of metal are removed. These fragments, because of their small diameter and the heat due to friction, are immediately oxidized. This process is then repeated with the resulting loss of metal and accumulation of oxide residue. Thus,

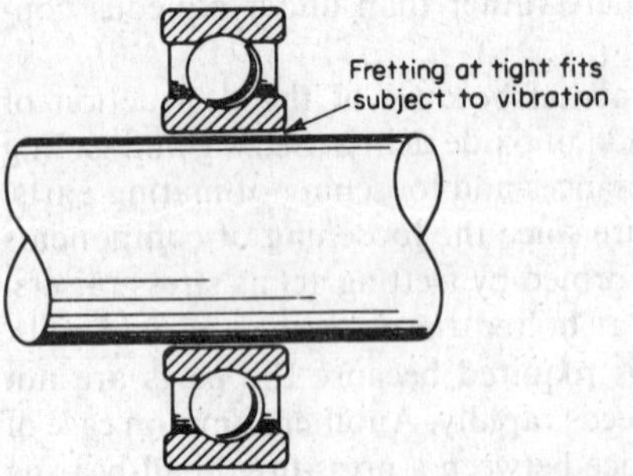

Figure 3-50 Example of typical fretting corrosion location.

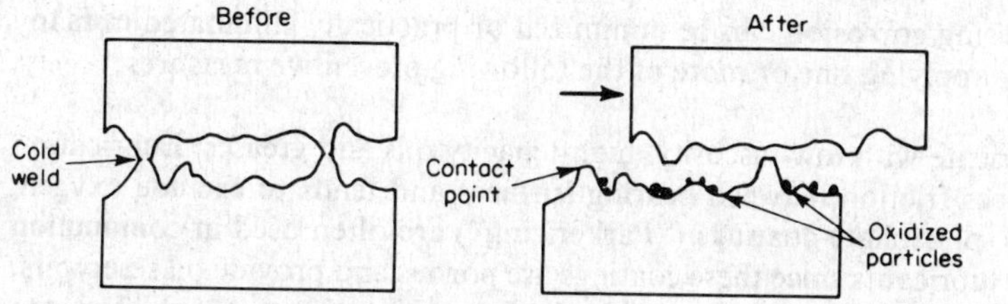

Figure 3-51 Schematic illustration of the wear-oxidation theory of fretting corrosion.

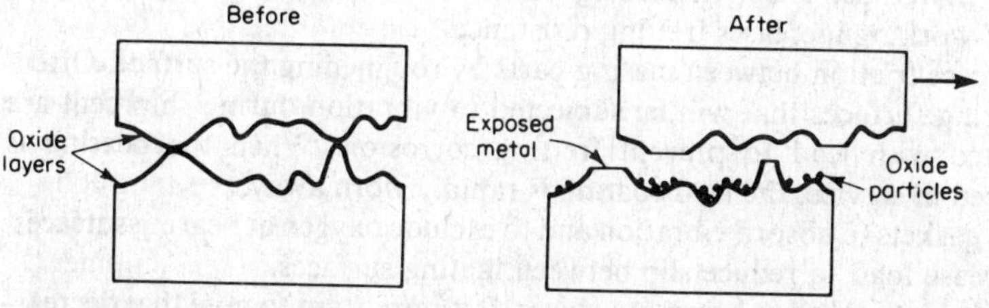

Figure 3-52. Schematic illustration of the oxidation-wear theory of fretting corrosion.

the wear-oxidation hypothesis is based on the concept that frictional wear causes the damage and subsequent oxidation is a secondary effect.

The oxidation-wear concept, illustrated in Fig. 3-52, is based on the hypothesis that most metal surfaces are protected from atmospheric oxidation by a thin, adherent oxide layer. When metals are placed in contact under load and subjected to repeated relative motion, the oxide layer is ruptured at high points and results in oxide debris, as shown schematically in Fig. 3-52. It is assumed that the exposed metal reoxidizes and the process is repeated. The oxidation-wear theory is essentially based on a concept of accelerated oxidation due to frictional effects.

Considering Figs. 3-51 and 3-52 and the two theories outlined above, it is obvious that both theories lead to the same conclusion—namely, the production of oxide debris and destruction of metal interfaces. Recent investigations suggest that both of the above mechanisms operate during fretting corrosion. The presence of an oxide layer does not appear to be necessary in every case, since fretting damage has been observed on almost every kind of surface including the noble metals, mica, glass, and ruby. Oxygen, however, does have an effect since its presence accelerates fretting attack of many materials, especially ferrous alloys. The actual mechanism of the fretting corrosion is probably a combination of the mechanisms illustrated in Figs. 3-51 and 3-52.

Fretting corrosion can be minimized or practically eliminated in many cases by applying one or more of the following preventive measures:

1. Lubricate with low-viscosity, high-tenacity oils and greases. Lubrication reduces friction between bearing surfaces and tends to exclude oxygen. Also, phosphate coatings ("Parkerizing") are often used in conjunction with lubricants since these coatings are porous and provide oil reservoirs.
2. Increase the hardness of one or both of the contacting materials. This can be accomplished by choosing a combination of hard materials or hard alloys. Table 3-11 lists the relative fretting corrosion resistance of various material combinations. As shown, hard materials are more resistant than soft materials. Also, increasing surface hardness by shot-peening or cold-working increases fretting resistance.
3. Increase friction between mating parts by roughening the surface. Often, bearing surfaces that will be subjected to vibration during shipment are coated with lead to prevent fretting corrosion. When the bearing is placed in service, the lead coating is rapidly worn away.
4. Use gaskets to absorb vibration and to exclude oxygen at bearing surfaces.
5. Increase load to reduce slip between mating surfaces.
6. Decrease the load at bearing surfaces. It is important to note that decreasing the load is not always successful, since very small loads are capable of producing damage.
7. If possible, increase the relative motion between parts to reduce attack.

Table 3-11 Fretting resistance of various materials

Poor	Average	Good
Aluminum on cast iron	Cast iron on cast iron	Laminated plastic on gold plate
Aluminum on stainless steel	Copper on cast iron	Hard tool steel on tool steel
Magnesium on cast iron	Brass on cast iron	Cold-rolled steel on cold-rolled steel
Cast iron on chrome plate	Zinc on cast iron	Cast iron on cast iron with phosphate coating
Laminated plastic on cast iron	Cast iron on silver plate	Cast iron on cast iron with coating of rubber cement
Bakelite on cast iron	Cast iron on copper plate	Cast iron on cast iron with coating of tungsten sulfide
Hard tool steel on stainless	Cast iron on amalgamated copper plate	Cast iron on cast iron with rubber gasket
Chrome plate on chrome plate	Cast iron on cast iron with rough surface	Cast iron on cast iron with Molykote lubricant
Cast iron on tin plate	Magnesium on copper plate	Cast iron on stainless with Molykote lubricant
Cast iron on cast iron with coating of shellac	Zirconium on zirconium	

Source: J. R. McDowell, *ASTM Special Technical Publication No. 144*, p. 24, American Society for Testing Materials, Philadelphia, 1952.

A comprehensive text on this subject, covering mechanisms, testing techniques, and case histories is recommended to readers desiring further information.* Two points are worth noting. First, Waterhouse suggests that fretting rather than fretting corrosion should be used to describe the phenomenon since corrosion products are not always present. Second, there appears to be a misconception concerning the cause of corrosion at screw-plate interfaces and other shielded sites on surgical implants. Cohen† was the first to suggest that the observed attack is due to fretting, and this idea has been amplified by Waterhouse (pp. 56–59) and others. This appears to be incorrect since the attack is characteristic of crevice corrosion rather than fretting. Relative motion between orthopedic plates and screws may initiate crevice attack, but it is not the primary cause of the observed corrosion since similar attack is observed at shielded sites on nonstressed components.

STRESS CORROSION

Stress-corrosion cracking (SCC) refers to cracking caused by the simultaneous presence of tensile stress and a specific corrosive medium. Many investigators have classified all cracking failures occurring in corrosive mediums as stress-corrosion cracking, including failures due to hydrogen embrittlement. However, these two types of cracking failures respond differently to environmental variables. To illustrate, cathodic protection is an effective method for preventing stress-corrosion cracking, whereas it rapidly accelerates hydrogen-embrittlement effects. Hence, the importance of considering stress-corrosion cracking and hydrogen embrittlement as separate phenomena is obvious. For this reason, the two cracking phenomena are discussed separately in this chapter.

During stress-corrosion cracking, the metal or alloy is virtually unattacked over most of its surface, while fine cracks progress through it. This is illustrated in Fig. 3-53. This cracking phenomenon has serious consequences since it can occur at stresses within the range of typical design stress. The stresses required for stress-corrosion cracking are compared with the total range of strength capabilities for type 304 stainless steel in Fig. 3-54. Exposure to boiling $MgCl_2$ at 310°F (154°C) is shown to reduce the strength capability to approximately that available at 1200°F.

The two classic cases of stress-corrosion cracking are "season cracking" of brass, and the "caustic embrittlement" of steel. Both of these obsolete terms describe the environmental conditions present that led to stress-corrosion cracking. Season cracking refers to the stress-corrosion cracking

*R. B. Waterhouse, *Fretting Corrosion*, Pergamon Press, New York, 1972.
†J. Cohen, *J. Bone Joint Surg.* **44A**:307 (1962).

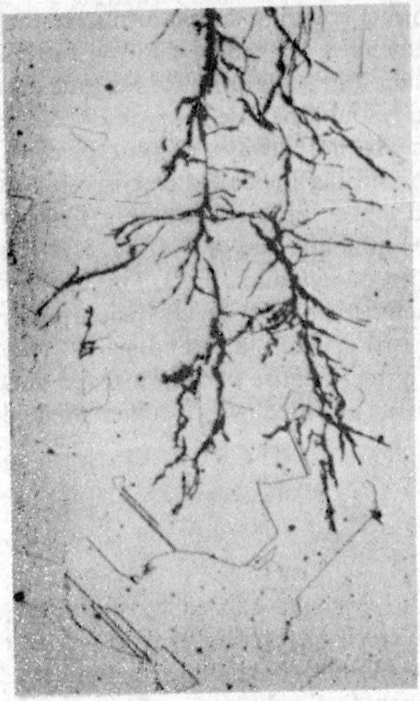

Figure 3-53 Cross section of stress-corrosion crack in stainless steel (500 ×).

failure of brass catridge cases. During periods of heavy rainfall, especially in the tropics, cracks were observed in the brass cartridge cases at the point where the case was crimped to the bullet. It was later found that the important environmental component in season cracking was ammonia, resulting from the decomposition of organic matter. An example of this is shown in Fig. 3-55.

Many explosions of riveted boilers occurred in early steam-driven locomotives. Examination of these failures showed cracks or brittle failures at the rivet holes. These areas were cold-worked during riveting operations, and analysis of the whitish deposits found in these areas showed caustic, or sodium hydroxide, to be the major component. Hence, brittle fracture in the presence of caustic resulted in the term caustic embrittlement. Figure 3.56 shows a plate that failed by caustic embrittlement. The cracks are numerous and very fine and have been revealed by application of a penetration dye solution. While stress alone will react in ways well known in mechanical metallurgy (i.e., creep, fatigue, tensile failure) and corrosion alone will react to produce characteristic dissolution reactions; the simultaneous action of both sometimes produces the disastrous result shown above.

Not all metal-environment combinations are susceptible to cracking. A good example is the comparison between brasses and austenitic stainless

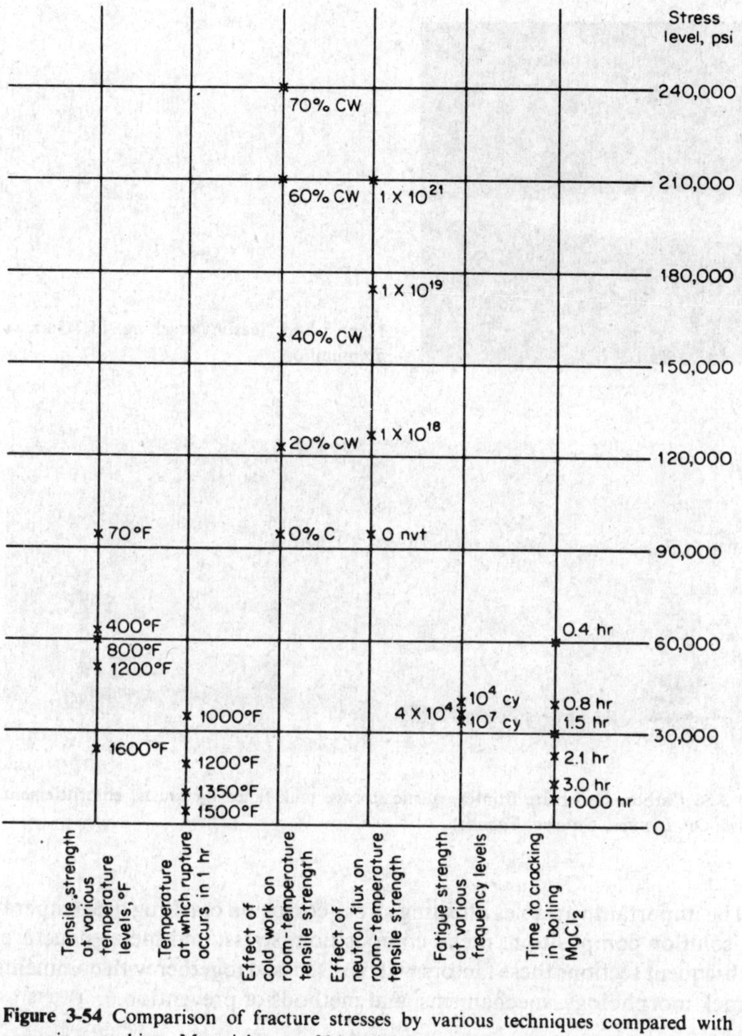

Figure 3-54 Comparison of fracture stresses by various techniques compared with stress-corrosion cracking. Material: type 304 stainless. (*Courtesy Dr. R. W. Staehle, Ohio State University*)

steels. Stainless steels crack in chloride environments but not in ammonia-containing environments, whereas brasses crack in ammonia-containing environments but not in chlorides. Further, the number of different environments in which a given alloy will crack is generally small. For example, stainless steels do not crack in sulfuric acid, nitric acid, acetic acid, or pure water, but they do crack in chloride and caustics.

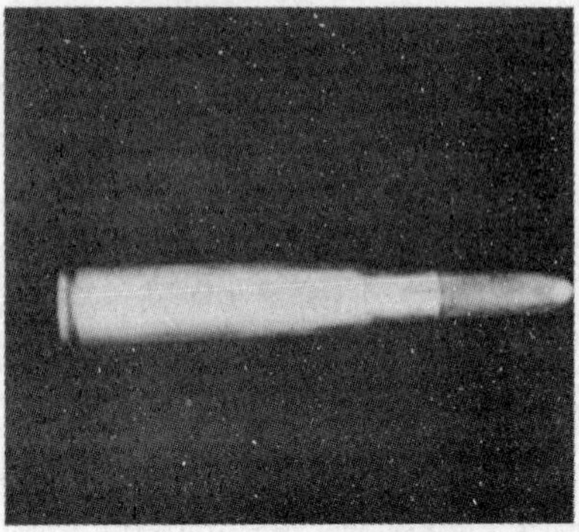

Figure 3-55 Season cracking of German ammunition.

Figure 3-56 Carbon steel plate from a caustic storage tank failed by caustic embrittlement. (*Imperial Oil, Limited, Ontario, Canada*)

The important variables affecting stress-corrosion cracking are temperature, solution composition, metal composition, stress, and metal structure. In subsequent sections these factors will be discussed together with comments on crack morphology, mechanisms, and methods of prevention.

3-38 Crack Morphology

Stress-corrosion cracks give the appearance of a brittle mechanical fracture, when, in fact, they are the result of local corrosion processes. However, even though stress-corrosion cracking is not strictly a mechanical process, it is still convenient to label the process and general features of Fig. 3-53 as a crack.

Both intergranular and transgranular stress-corrosion cracking are observed. Intergranular cracking proceeds along grain boundaries, while

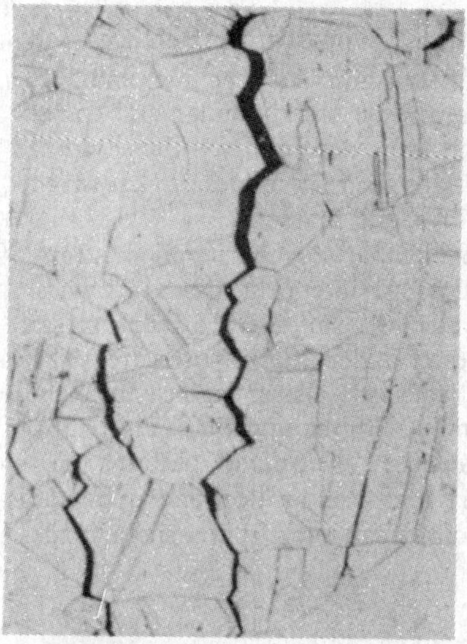

Figure 3-57 Intergranular stress corrosion cracking of brass. (*E. N. Pugh*)

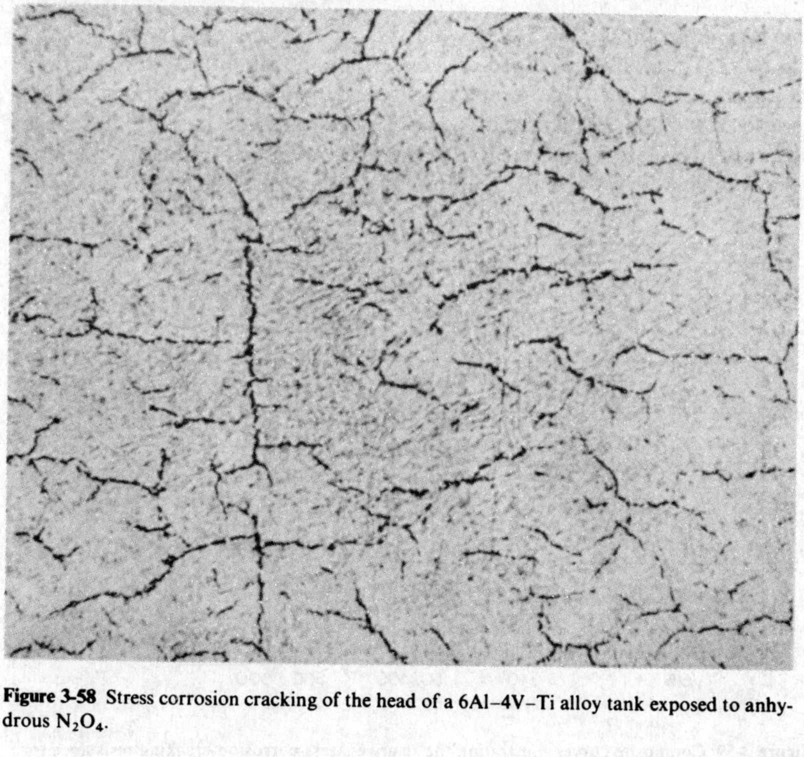

Figure 3-58 Stress corrosion cracking of the head of a 6Al–4V–Ti alloy tank exposed to anhydrous N_2O_4.

transgranular cracking advances without apparent preference for boundaries. Figure 3-53 is an example of transgranular cracking, and Fig. 3-57 shows the intergranular mode of cracking. Intergranular and transgranular cracking often occur in the same alloy, depending on the environment or the metal structure. Such transitions in crack modes are known in the high-nickel alloys, iron-chromium alloys, and brasses.

Cracking proceeds generally perpendicular to the applied stress. Cracking in Figs. 3-53 and 3-57 is of this type. An interesting case is shown in Fig. 3-58, in which the metal is subjected to uniform biaxial tensile stresses (the hemispherical head of a pressure vessel under internal pressure). The cracks appear to be randomly oriented. Cracks vary also in degree of branching. In some cases the cracks are virtually without branches (Fig. 3-56), and in other cases they exhibit multibranched "river delta" patterns (Fig. 3-53). Depending on the metal structure and composition and upon the environment composition, crack morphology can vary from a single crack to extreme branching.

3-39 Stress Effects

Increasing the stress decreases the time before cracking occurs, as shown in Fig. 3-59. There is some conjecture concerning the minimum stress required to prevent cracking. This minimum stress depends on temperature, alloy composition, and environment composition. In some cases it has been observed to be as low as about 10% of the yield stress. In other cases, cracking does not occur below about 70% of the yield stress. For each alloy-environ-

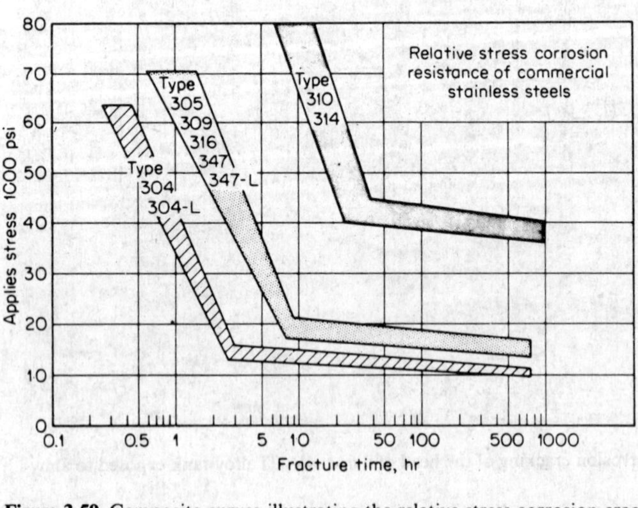

Figure 3-59 Composite curves illustrating the relative stress-corrosion-cracking resistance for commercial stainless steels in boiling 42% magnesium chloride.

ment combination there is probably an effective minimum, or threshold, stress. This threshold value must be used with considerable caution since environmental conditions may change during operation.

The criteria for the stresses are simply that they be tensile and of sufficient magnitude. These stresses may be due to any source: applied, residual, thermal, or welding. In fact, numerous cases of stress-corrosion cracking have been observed in which there is no externally applied stress. As-welded steels contain residual stresses near the yield point.

Corrosion products have been shown to be another source of stress. Stresses up to 10,000 lb/in.^{2} can be generated by corrosion products in constricted regions. A stress-corrosion crack that has been propagated by corrosion-product stresses is shown in Fig. 3-60. In this figure, the corrosion products appear to exert a wedging action.

This *wedging action* of 10,000 lb/in.^{2} (10 ksi) results in very high stresses at the crack tip because the tip is a sharp notch which is a great stress concentrator. Hudak and Page* show that highly localized stresses of about

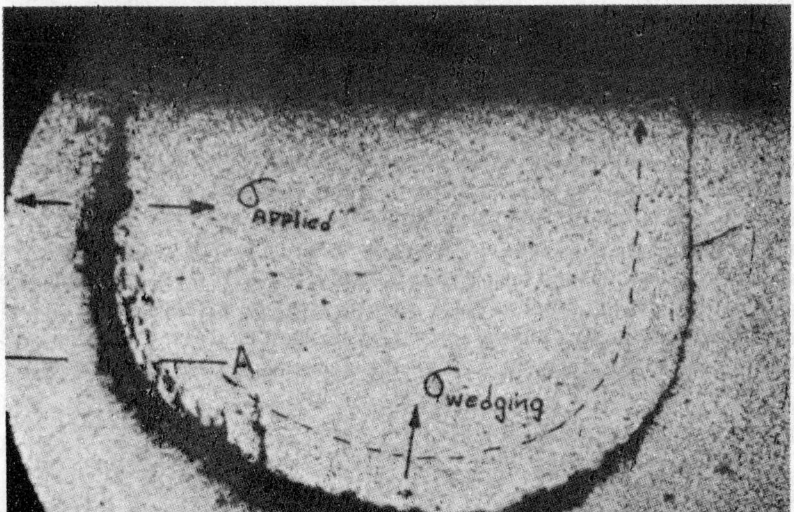

Figure 3-60 The wedging action of corrosion products. This crack in stainless steel has proceeded in its circular path under the influence of stresses produced only by corrosion products. [*H. W. Pickering, M. G. Fontana, and F. H. Beck, Corrosion, 18: 230t (June, 1962)*]

*S. J. Hudak and R. A. Page, Analysis of Oxide Wedging During Environment Assisted Crack Growth, *Corrosion*, 285–290 (July 1983).

2000 MPa (megapascals), or approximately 289 ksi, may be achieved. These authors approached the problem using fracture mechanics. High stresses result in *denting* of heat exchanger tubes as described in the next paragraph.

A phenomenon termed *denting* has been observed in nuclear steam generators. Inconel tubes are crushed (dented inward) where they pass through carbon steel tube supports and also at contact with steel tube sheets. These annular spaces become filled with steel corrosion products whose volume is greater than the metal consumed and the consequent pressure moves the Inconel tube wall inward. Obviously, stresses greater than the yield point of the alloy are produced. This situation is similar to the wedging action of corrosion products in the stress-corrosion crack shown in Fig. 3-60.

A similar situation involves high-silicon iron and ordinary gray cast iron exposed to oleum (fuming sulfuric acid). The acid penetrates along the graphitic flakes (see Fig. 5-1), corrodes the "steel" matrix, builds up pressure in these confined spaces, and cracks the iron, sometimes catastrophically (see Sec. 7-2). An experience of this type in the 1930s provided my first acquaintance with wedging action.

3-40 Time to Cracking

The parameter of time in stress-corrosion cracking phenomena is important since the major physical damage during stress-corrosion cracking occurs during the later stages. As stress-corrosion cracks penetrate the material, the cross-sectional area is reduced and the final cracking failure results entirely from mechanical action. This is illustrated in Figs. 3-61 and 3-62. Figure 3-61 illustrates the rate of cracking as a function of crack depth for a specimen under constant tensile load. Initially, the rate of crack movement is more or less constant, but as cracking progresses the cross-sectional area of the specimen decreases and the applied tensile stress increases. As a result, the rate of crack movement increases with crack depth until rupture occurs.

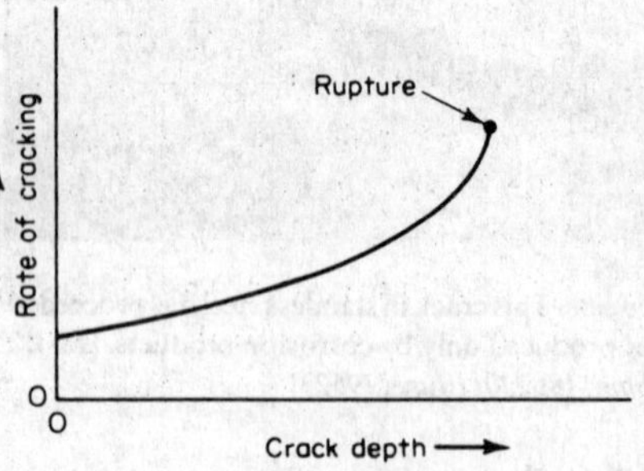

Figure 3-61 Rate of stress-corrosion crack propagation as a function of crack depth during tensile loading.

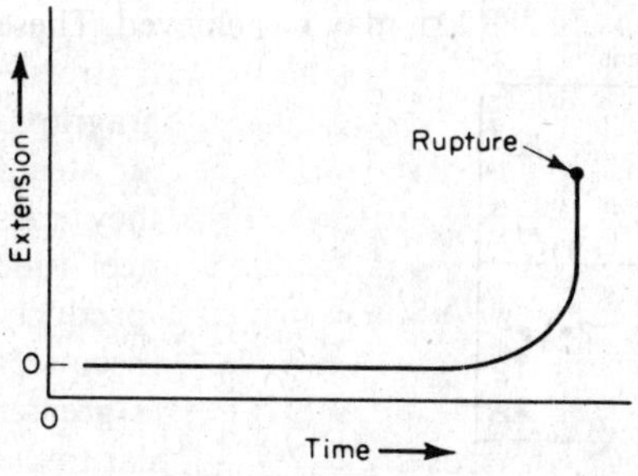

Figure 3-62 Specimen extension as a function of time during constant-load stress-corrosion cracking test.

Immediately preceding rupture, the cross section of the material is reduced to the point where the applied stress is equal to or greater than the ultimate strength of the metal, and failure occurs by mechanical rupture. Figure 3-62 illustrates the relationship between the time of exposure and the extension of a specimen during stress-corrosion cracking. The width of the crack is narrow during the early stages of cracking, and little change in extension is observed. During later stages, the crack widens. Prior to rupture, extensive plastic deformation occurs and a large change in extension is observed.

A common and important question frequently asked concerning stress corrosion cracking is: How long should a stress-corrosion cracking test be conducted? Figures 3-61 and 3-62 indicate that the test should be conducted until failure occurs. Short-term stress-corrosion cracking tests should be avoided since very little physical and mechanical evidence of cracking is apparent until after it has occurred.

3-41 Environmental Factors

At present there appears to be no general pattern to the environments that cause stress-corrosion cracking of various alloys. Stress-corrosion cracking is well known in various aqueous mediums, but it also occurs in certain liquid metals, fused salts, and nonaqueous inorganic liquids (see Figs. 3-58 and 8-7).

The presence of oxidizers often has a pronounced influence on cracking tendencies. Figure 3-63 shows the combined effects of chloride and dissolved oxygen on the stress-corrosion cracking of type 304 stainless steel. In fact, the presence of dissolved oxygen or other oxidizing species is critical to the cracking of austenitic stainless steels in chloride solutions, and if the oxygen is removed, cracking will not occur.

Table 3-12 lists a number of environment-alloy systems in which cracking occurs. New environments that cause stress-corrosion cracking in various alloys are constantly being found. Thus, it is always necessary to evaluate a given alloy in stress-corrosion tests when the environmental composition is

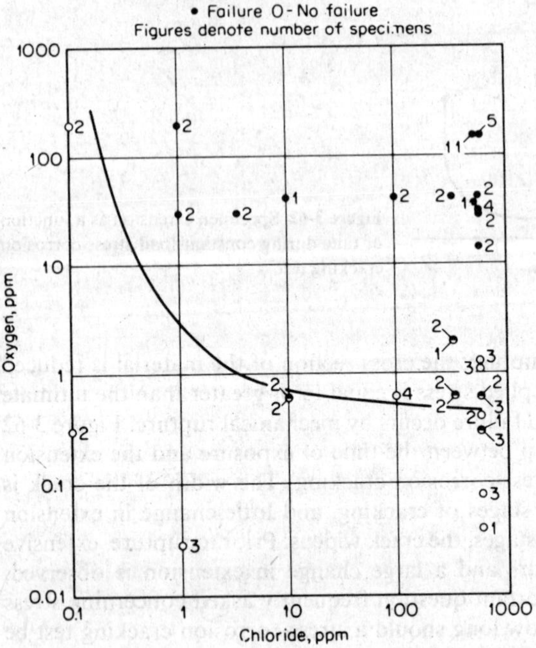

Figure 3-63 Proposed relationship between chloride and oxygen content of alkaline-phosphate-treated boiler water, and susceptibility to stress-corrosion cracking of austenitic stainless steel exposed to the steam phase with intermittent wetting. [*W. Lee Williams, Corrosion, 13*:*539t, (Aug. 1957)*]

changed. It is usually characteristic of crack-producing environments that the alloy is negligibly attacked in the nonstressed condition. Although stress-corrosion cracking of steel is frequently reported in hydrogen sulfide solutions and cyanide-containing solutions, as shown in Table 3-12, these failures are undoubtedly due to hydrogen embrittlement rather than stress-corrosion cracking. (See Secs. 3-47 to 3-50.)

As is the case with most chemical reactions, stress-corrosion cracking is accelerated by increasing temperature. In some systems, such as magnesium alloys, cracking occurs readily at room temperature. In other systems, boiling temperatures are required. Most alloys susceptible to cracking will begin cracking at least as low as 100°C. The effect of temperature on the cracking of austenitic stainless steels is shown in Fig. 3-64. Similar data for the caustic embrittlement of as-welded steel are presented in Fig. 3-65.

The physical state of the environment is also important. Alloys exposed to single-phase aqueous environments are sometimes less severely attacked than metals at the same temperature and stress when exposed to alternate wetting and drying conditions.

Table 3-12 Environments that may cause stress corrosion of metals and alloys

Material	Environment	Material	Environment
Aluminum alloys	$NaCl-H_2O_2$ solutions	Ordinary steels	NaOH solutions
	NaCl solutions		$NaOH-Na_2SiO_2$ solutions
	Seawater		Calcium, ammonium, and sodium nitrate solutions
	Air, water vapor		Mixed acids ($H_2SO_4-HNO_3$)
Copper alloys	Ammonia vapors and solutions		HCN solutions
	Amines		Acidic H_2S solutions
	Water, water vapor		Seawater
Gold alloys	$FeCl_3$ solutions		Molten Na-Pb alloys
	Acetic acid–salt solutions	Stainless steels	Acid chloride solutions such as $MgCl_2$ and $BaCl_2$
Inconel	Caustic soda solutions		$NaCl-H_2O_2$ solutions
Lead	Lead acetate solutions		Seawater
Magnesium alloys	$NaCl-K_2CrO_4$ solutions		H_2S
	Rural and coastal atmospheres		$NaOH-H_2S$ solutions
	Distilled water		Condensing steam from chloride waters
Monel	Fused caustic soda	Titanium alloys	Red fuming nitric acid, seawater, N_2O_4, methanol-HCl
	Hydrofluoric acid		
	Hydrofluosilicic acid		
Nickel	Fused caustic soda		

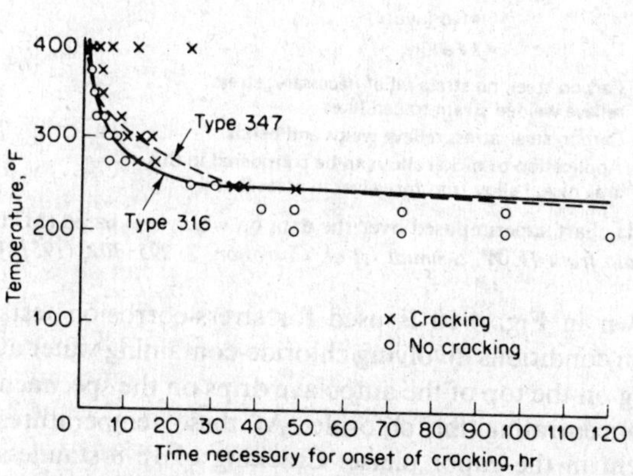

Figure 3-64 Effect of temperature on time for crack initiation in types 316 and 347 stainless steels in water containing 875 ppm NaCl. [*W. W. Kirk, F. H. Beck, M. G. Fontana, Stress Corrosion Cracking of Austenitic Stainless Steels in High Temperature Chloride Waters, in T. Rhodin (ed.), Physical Metallurgy of Stress Corrosion Fracture, Interscience Publishers, Inc., New York,* 1959.]

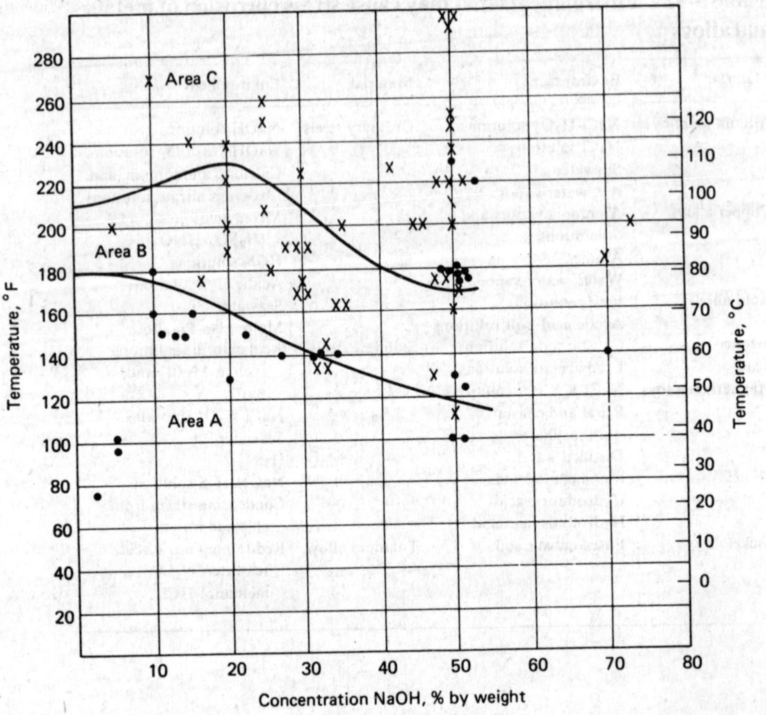

Legend:

● No failure

X Failure

Area A - Carbon steel, no stress relief necessary; stress relieve welded steam-traced lines

Area B - Carbon steel; stress relieve welds and bends

Area C - Application of nickel alloys to be considered in this area nickel alloy trim for valves in areas B and C

Figure 3-65 NACE caustic soda chart superimposed over the data on which it is based (MTI Publication No. 15, 1985). [*Data from H. W. Schmidt, et al, Corrosion, 7, 295–302, (1951).*]

The autoclave* shown in Fig. 4-10 is used for stress-corrosion tests under vapor condensation conditions involving chloride-containing water at 400°F. Liquid condensing on the top of the autoclave drips on the specimen and flash-dries, thus concentrating the chloride. At these temperatures sodium chloride is present in the vapor phase. Cracking of 18-8 stainless steels in two hours at applied stresses as low as 2000 lb/in.^{2} occurs under these conditions. The specimen immersed in the liquid requires high stresses and long times for cracking. Similar results are obtained when the specimen is alternately immersed in and removed from the water.

*R. W Staehle, F. H. Beck, and M. G. Fontana, Mechanism of Stress Corrosion of Austenitic Stainless Steels, *Corrosion*, **15**:51 (1959).

Good correlation is obtained between these tests and actual service failures. Figure 3-66 is an excellent example. This high-pressure autoclave was forged from 18-8 stainless steel with a 2-in. wall and cost $20,000. It was in operation for only a few batches with total times in hours. Dy-Chek penetrant was used to emphasize the appearance of the many cracks on the outside surface. This surface was cooled by a good grade of city water. The cooling-jacket system drained after each operation. The droplets of water clinging to the autoclave surface dried and the chloride concentrated.

Figure 3-67 shows cracking of an 18-8 tank from the outside surface. Cracks are accentuated by dye penetrant. This vessel handled warm distilled water. The outside was covered with an insulating material containing a few parts per million of chloride. Rain penetrated the insulation and leached out the chlorides, and then the solution dried and concentrated. This plant experienced many such cracks on insulated vessels and lines. Similar experiences are frequent and have been called external stress-corrosion cracking.

Figure 3-68 shows the location of cracks in a vertical stainless steel condenser. Splashing in the dead space caused alternate wetting and drying.

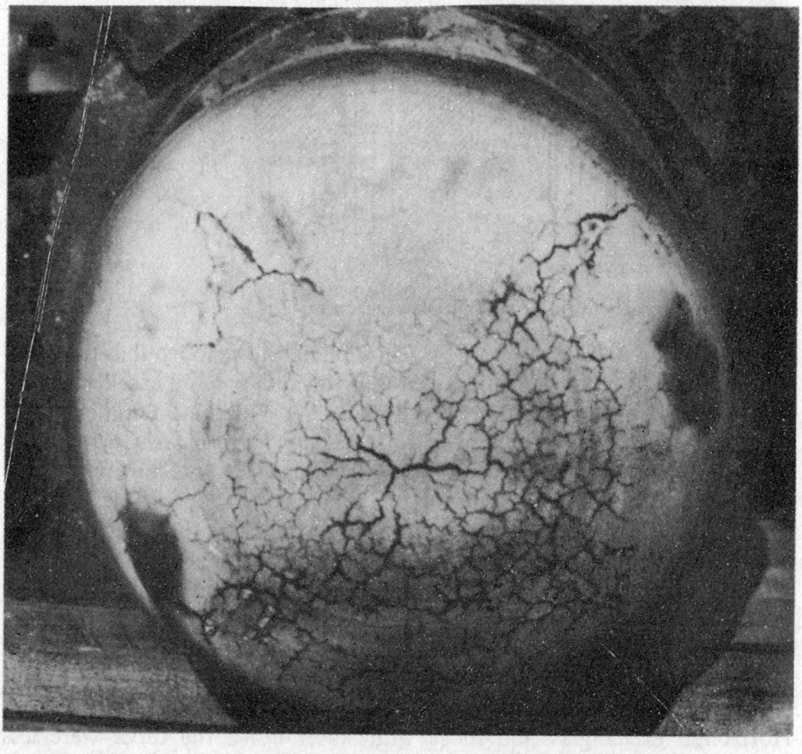

Figure 3-66 Stress corrosion of type 304 autoclave. (*Mallinckrodt Chemical Works*)

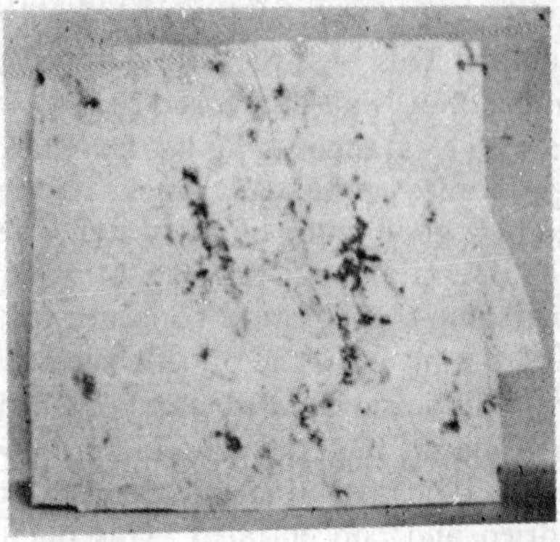

Figure 3-67 External stress corrosion of type 304 vessel.

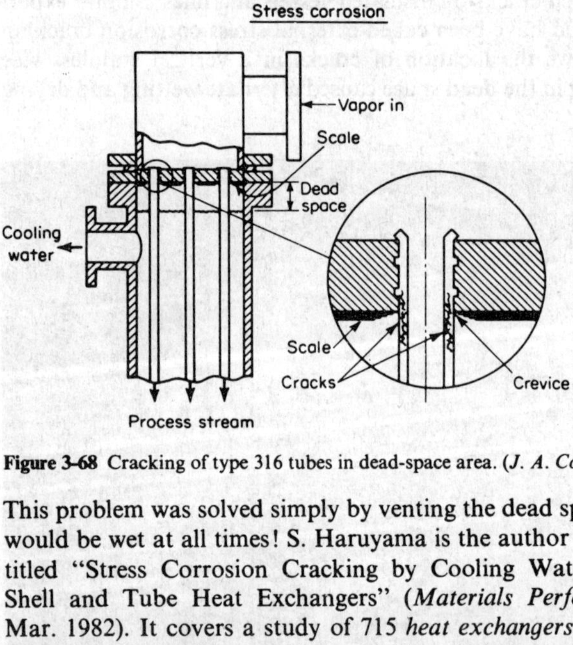

Figure 3-68 Cracking of type 316 tubes in dead-space area. (*J. A. Collins*)

This problem was solved simply by venting the dead space so that the tubes would be wet at all times! S. Haruyama is the author of an excellent paper titled "Stress Corrosion Cracking by Cooling Water of Stainless Steel Shell and Tube Heat Exchangers" (*Materials Performance*, pp. 14–19, Mar. 1982). It covers a study of 715 *heat exchangers* in commercial plant service for several years. Some of his conclusions follow. The mode of cracking (SCC) was transgranular in 47% of the cases followed by transgranular plus intergranular at 16% and intergranular at 8%. Causes for cracking were led by a wide margin by existence of *vapor space* in 34% of the cases. Replacement by another material at 31% was by far the favoured method for preventing SCC. Types 304, 316, and 321 showed similar performance but more failures occurred with the low carbon grades (see Chapter 5) for 304 and 316. No failures of ferritic or duplex alloys were reported. Horizontal exchangers and those with water *inside* the tubes were less susceptible. Cooling by seawater was not included in this study.

A survey by D. R. McIntyre (*Chem. Eng.*, p. 132, Apr. 5, 1982) of several

Gulf Coast chemical plants "revealed that each lost an average of one vertical condenser *per year* to improper venting practice." Two other pertinent papers are "Coping with an Improperly Vented Condenser" by K. J. Bell (*Chem. Eng. Progr*., pp. 54–55, July 1983) and "Troubleshooting Shell and Tube Heat Exchangers" by S. Yokell (*Chem. Engin*., pp. 57–80, July 25, 1983).

In connection with heat exchanger tubing, Smallwood* presents an excellent summary of tubing reliability. He describes corrosion-type defects as: (1) substitution of an inadequate alloy through error, (2) selective weld metal attack, (3) improper pickling, (4) corrosion during testing or handling, (5) residual stresses, (6) improper heat treatment, (7) imbedded tramp metal, (8) preferred-grain orientation, (9) surface roughness, (10) dents, and (11) high-temperature contamination. An example of the latter is carbon pickup because of local or general carburization of the metal. (Carbon pickup of castings is discussed in Sec. 3-20). Examples of all of these defects are discussed. Mechanical defects are also covered, and methods for defect detection, such as ultrasonic, are presented. A quality-assurance program is outlined.

3-42 Metallurgical Factors

The susceptibility to stress-corrosion cracking is affected by the average chemical composition, preferential orientation of grains, composition and distribution of precipitates, dislocation interactions, and progress of the phase transformation (or degree of metastability). These factors further interact with the environmental composition and stress to affect time to cracking, but these are secondary considerations.

Figures 3-69 (Ni added to 18 Cr-Fe base) and 3-70 show the effects of alloy composition in austenitic stainless steels and mild steels. In both cases there is a minimum in time to cracking as a function of composition. In fact, this observation of a minimum in time to cracking versus composition is a common (although not universal) observation in other alloy systems (e.g., Cu-Au).

In the past it has been a common generalization that pure metals do not crack. This has been challenged by observations of cracking in 99.999% pure copper exposed in ammoniacal solutions containing $Cu(NH_3)_5{}^{2+}$ complex ions.† While generally the use of pure metals is often an available avenue for preventing cracking, it should be pursued only with caution.

High-strength aluminum alloys exhibit a much greater susceptibility to stress-corrosion cracking in directions transverse to the rolling direction than in those parallel to the longitudinal direction. This effect is due to the distribution of precipitates which results from rolling.

Figure 3-71 shows the increase in resistance to stress/corrosion as the

*R. E. Smallwood, Heat Exchanger Tubing Reliability, *Materials Performance*, **16**:27–34 (Feb. 1977).

†E. N. Pugh, W. G. Montague, A. R. C. Westwood, *Corrosion Sci*., **6**:345 (1966).

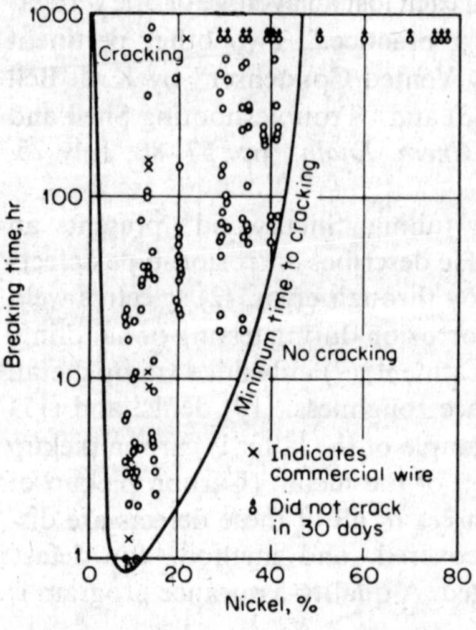

Figure 3-69 Stress-corrosion cracking of iron-chromium nickel wires in boiling 42% magnesium chloride. [*H. R. Copson, Effect of Composition on Stress Corrosion Cracking of Some Alloys Containing Nickel, in T. Rhodin (ed.), Physical Metallurgy of Stress Corrosion Fracture, Interscience Publishers, Inc., New York, 1959*]

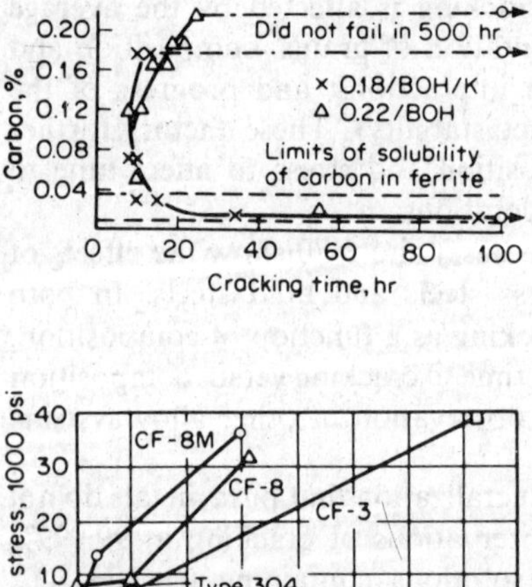

Figure 3-70 Effect of carbon content on the cracking time of mild steel exposed to boiling calcium ammonium nitrate. *(R. N. Parkins)*

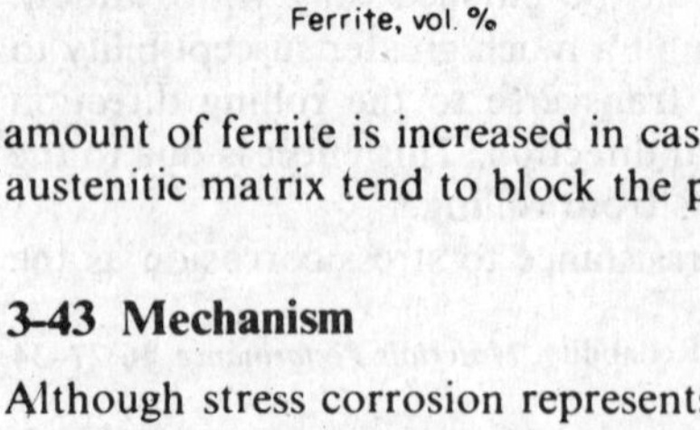

Figure 3-71 Effect of ferrite on stress required to induce stress-corrosion cracking in several cast stainless alloys. Type 304 and 316 with zero ferrite also plotted. Specimens exposed 8 hr in condensate from 875-ppm chloride water at 400°F. [*M. G. Fontana, F. H. Beck, J. W. Flowers, Metal Progr., 86:99 (Dec. 1961)*]

amount of ferrite is increased in cast stainless steels. Pools of ferrite in the austenitic matrix tend to block the progress of cracks.

3-43 Mechanism

Although stress corrosion represents one of the most important corrosion problems, the mechanism involved is not well understood. This is one of the big unsolved questions in corrosion research. The main reason for this situation is the complex interplay of metal, interface, and environment properties. Further, it is unlikely that a specific mechanism will be found that applies to all metal-environment systems. The most reliable and useful information has been obtained from empirical experiments. Some of the possible "operating steps" or processes involved are discussed immediately below.

Corrosion plays an important part in the initiation of cracks. A pit, trench, or other discontinuity on the surface of the metal acts as a *stress raiser*. Stress concentration at the tip of the "notch" increases tremendously as the radius of notch decreases. Stress-corrosion cracks are often observed to start at the base of a pit.

Once a crack has started, the tip of the advancing crack has a small radius and the attendant stress concentration is great. Using audio amplification methods, Pardue* showed that a mechanical step or jump can occur during crack propagation. In fact, "pings" were heard with the naked ear.

The *conjoint* action of stress and corrosion required for crack propagation was demonstrated by Priest.† An advancing crack was stopped when cathodic protection was applied (corrosion stopped—stress condition not changed). When cathodic protection was removed, the crack started moving again. This cycle was repeated several times. In this research, the progress of the crack was photographed and projected at the actual speed of propagation.

Plastic deformation of an alloy can occur in the region immediately preceding the crack tip because of high stresses. If the alloy is metastable, a phase transformation could occur (e.g., austenite to martensite in the nickel stainless steels). The newly formed phase could have different strength, susceptibility to hydrogen, or reactivity. If the alloy is not metastable, the cold-worked (plastically deformed) region might be less corrosion resistant than the matrix because of the continuous emergence of slip steps. This is a dynamic process and could explain why severely deformed metals (before exposure to a corrosive) do not exhibit sufficiently high corrosion rates to account for rapid penetration of cracks.

The role of tensile stress has been shown to be important in rupturing protective films during both initiation and propagation of cracks. These films could be tarnish films (as in the case of brasses), thin oxide films, layers richer in the more noble component (as in the case of copper-gold alloys and some of the stainless steels and alloys), or other passive films. Breaks in the passive film or enriched layer on stainless steel allows more rapid corrosion at various points on the surface and thereby initiates cracks. Breaking of films ahead of the advancing crack would not permit healing, and propa-

*W. M. Pardue, F. H. Beck, and M. G. Fontana, *Am. Soc. Metals Trans. Quart.*, **54**: 539–548 (1961).

†D. K. Priest, F. H. Beck, and M. G. Fontana, *Trans. Am. Soc. Metals*, **47**:473–492 (1955).

gation would continue. Rapid local dissolution without stifling is required for rapid propagation.

In the case of intergranular cracking, the grain-boundary regions could be more anodic, or less corrosion resistant, because of precipitated phases, depletion, enrichment, or adsorption, thus providing a susceptible path for the crack. Another example of local dissolution concerns mild steels that crack in nitrate solutions. In this case, iron carbide is cathodic to ferrite.

These examples indicate the complex interplay between metal and environment and account for the specificity of environmental cracking of metals and alloys.

A very large amount of research and development work has been done during the past decade or so on stress-corrosion cracking. The importance of this subject is emphasized by the fact that more effort and funds have been expended on stress corrosion than on all other forms of corrosion combined. Extensive study is continuing, particularly in the fields of nuclear energy and coal-conversion systems. The collapse of the Silver Bridge into the Ohio River with a loss of about two-score lives has focused public attention on this problem. Costly failures in industrial plants have prompted extensive investigation.

Many detailed steps and mechanisms for stress-corrosion cracking of specific metal-environment combinations have been postulated. Two basic "models" for a general mechanism are (1) the *dissolution* model wherein anodic dissolution (Fig. 3-79) occurs at the crack tip because strain ruptures the passive film at the tip, and (2) the *mechanical* model, wherein specific species adsorb and interact with strained metal bonds and reduce bond strength. The first seems more universal than the second. Many ramifications of these models have been postulated. Hydrogen embrittlement may be an operative factor particularly for high-strength alloys. For most engineering work past experience is the best guide, with reliable and valid testing second. In all cases chemistry, metallurgy, and mechanics (stress field) must be considered.

At this writing, a handbook on stress-corrosion cracking and corrosion fatigue is under preparation by Roger W. Staehle. This project is sponsored by the Advanced Research Projects Agency (ARPA). The primary objective of the handbook is to serve the engineering design community. Publication is scheduled for 1986. This book should provide the best and most useful engineering information on stress-corrosion cracking and its corollary, corrosion fatigue.

3-44 Multienvironment Charts

In addition to information presented elsewhere in this book, here are several charts (tables) showing cracking tendencies for metal and alloy systems in a *variety* of environments including *liquid* metals. These tables are from

Materials Technology Institute of the Chemical Process Industries, Inc. (MTI) Manual No. 15, titled *Guidelines for Preventing Stress Corrosion Cracking in the Chemical Process Industries*. It should be emphasized that these are only guidelines, but they do present a good picture.

Table 3-13 shows the situation for carbon steels. The A-designation means ASTM, Gr is grade, and HSLA is high-strength low alloy steel. Chapter 5 describes materials including metals and nonmetallics. *Low-alloy* or medium-strength steels exhibit yield strengths below 180,000 lb/in.^{2}. Low-alloy steels with *very high strengths* (i.e., AISI 4340) are much more susceptible to stress corrosion than weaker steels. In general, susceptibility to cracking increases with strength level. Values of K_{ISCC} are usually a smaller fraction of the tensile strength. K_{ICSS} is the critical plane strain intensity factor, which covers stress corrosion. Fracture mechanics is discussed later in this chapter.

Figure 3-72 shows potential ranges over which the SCC of carbon steels can occur in five solutions (from MTI manual). This means that changes in solution composition or temperature could shift the potential of

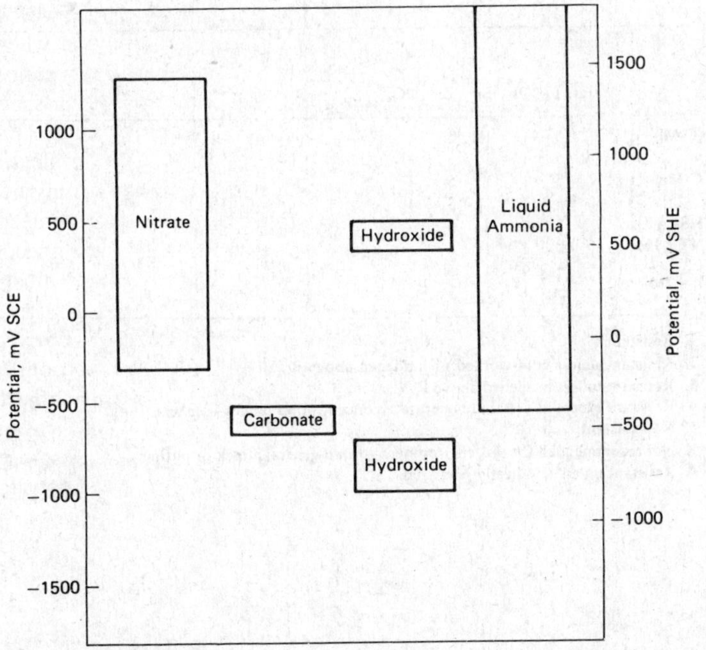

Figure 3-72 Schematic diagram showing potential ranges over which SCC of carbon steels occurs in various solutions.

Table 3-13 Environments vs. low-alloy steels

	Cl^-, Acid	Cl^-, Neutr.	Cl^-, Oxid.	Sea-water	High-Purity H_2O	OH^-	HCN	S^{2-}	S^{2-}/Cl^-	Carbon-ates
A516 Gr. 70 C-M	X	1	1	1	A		2	2	2	4
A106B C-Mn	X	1	1	1	A	5	2	2	2	4
A285 Gr. C C-Mn	X	1	1	1	A	5	2	2	2	4
A242 HSLA	X	1	1	1		5	2	2	2	4
A517 Gr. F HSLA	X	1	1	1		5	5	5	5	4
A387(22)2 1/4 Cr-1Mo	X	1	1	1		5	2	2	2	4
AISI 4140	X	3	3	3	3	5	5	5	5	4
AISI 4340	X	3	3	3	3	5	5	5	5	4

	SO_3^{2-}	SO_4^{2-}	CrO_4	NO_3^-	NH_3	$CO\text{-}CO_2\text{-}H_2O$	Amines	Hf	Cd	Al	Bi
A516 Gr. 70 C-M	1	A	1	5	4	4	4	1	5	X	1
A106B C-Mn	1	A	1	5	4	4	4	1	5	X	1
A285 Gr. C C-Mn	1	A	1	5	4	4	4	1	5	X	1
A242 HSLA	1	A	1		4	4	4	1	5	X	1
A517 Gr. F HSLA	1	A	1	5	4	4	4	1	5	X	1
A387(22)2 1/4 Cr-1Mo	4	A	1		4	4	4	5	5	X	1
AISI 4140	4	A	1	5	4	4	4	5	5	X	1
AISI 4340	4	A	1	5	4	4	4	5	5	X	1

	Hg	Li	Pb	Sn	Zn	PO_4^{3-}
A516 Gr. 70 C-M	1	5	1	1	5	4
A106B C-Mn	1	5	1	1	5	4
A285 Gr. C C-Mn	1	5	1	1	5	4
A242 HSLA	1	5	1	1	5	4
A517 Gr. F HSLA	1	5	1	1	5	
A387(22)2 1/4 Cr-1Mo	1	5	1	1	5	
AISI 4140	1	5	1	1	5	
AISI 4340	1	5	1	1	5	

Notes: 1 Resistant.
2 Resistant unless cold-worked or hardened above R_c22.
3 Resistant unless hardened above R_c33.
4 Resistant except at certain temperature-concentration ranges—see text.
5 Nonresistant.
X Not recommended for this environment—rapid general attack or pitting.
A Resistant unless anodically polarized.

Table 3-14 Environments vs. various wrought stainless steels

	Cl^-, Acid	Cl^-, Neutr.	Cl^-, Oxid.	F^-, SiF_6^{-2}	OH^-	$S^=$	S^{-2}/Cl^-	SO_4^{-2}	CrO_4^{-2}	NO_3^-	NH_3	O_2SCC	Sea-water	Cd	Al	Hg	Li	Pb	Zn
Ferritic																			
405	X	1	1		1	1	1		1	1	1	3	X	5	X	1	5	1	5
409	X	1	1		1	1	1		1	1	1	3	X	5	X	1	5	1	5
429	X	1	1		1	1	1		1	1	1	3	X	5	X	1	5	1	5
430	X	1	1		1	1	1		1	1	1	3	X	5	X	1	5	1	5
434	X	1	1		1	1	1		1	1	1	3	X	5	X	1	5	1	5
436	X	1	1		1	1	1		1	1	1	3	X	5	X	1	5	1	5
442	X	1	1		1	1	1		1	1	1	3	X	5	X	1	5	1	5
446	X	1	1		1	1	1		1	1	1	3	X	5	X	1	5	1	5
Martensitic																			
403	X	2	3	3	4	2	2	4	1	4			X	5	X	1	5	1	5
410	X	3	3	3	4	2	2	4	1	4			X	5	X	1	5	1	5
414	X	3	3	3	4	5	5	4	1	4			X	5	X	1	5	1	5
416	X	3	3	3	4	5	5	4	1	4			X	5	X	1	5	1	5
420	X	3	3	3	4	2	2	4	1	4			X	5	X	1	5	1	5
422	X	3	3	3	4	5	2	4	1	4			X	5	X	1	5	1	5
431	X	3	3	3	4	5	5	4	1	4			X	5	X	1	5	1	5
440A	X	3	3	3	4	5	5	4	1	4			X	5	X	1	5	1	5
440C	X	3	3	3	4	5	5	4	1	4			X	5	X	1	5	1	5
Duplex																			
SAF2205	4	4	4			2	2		1	1			X		X	1		1	5
Uranus 50	X	4	4		4	2	2		1	1			X		X	1		1	5
3RE60	X	4	4				5		1	1			X		X	1		1	5
Ferralium 255	4	4	4		4	2	2		1	1			X		X	1		1	5
AISI 329	X	4	4		4		5		1	1			X		X	1		1	5

Notes: 1 Resistant.

2 Resistant unless cold-worked or hardened above R_c22.

3 Resistant unless hardened above R_c33.

4 Resistant except at certain temperature-concentration ranges—see text.

5 Nonresistant.

X Not recommended for this environment—rapid general attack or pittting.

Table 3-15 Environments vs. copper alloys

CDA Number	Cl^-,	CrO_4^{-2}	HPO_4^{-2}	HCO_4^{-2}	OH^-	S^{-2}	S^{-2}/Cl^-	SO_2	SO_3^{-2}	SO_4^{-2}	NO_2^-	NO_3^-	NH_3	Steam	Sea-water	Bi	Hg	Li	Pb	Sn	Zn	Amines $+O_2$
110 ETP Copper	1				1	1							2		1	5	5	5	1	1		2
122 DHP Copper	1				1	1	X						2		1	5	5	5	1	1	1	2
220 Bronze	1				1	1							5		1	5	5	5		1	1	5
230 Red Brass	1				4	1		1		4	5	5	5		1	5	5	5	5	5	1	5
260 Cartridge Brass			4	4	4	1		5		4	5	5	5		1	5	5	5	5	5	1	5
270 Yellow Brass	4		4	4	4	1				4	5	5	5		2	5	5	5	5	5	1	5
280 Muntz Metal	4		4	4	4	1				4	5	5	5		2	5	5	5	5	5	1	5
443 Admiralty Brass	1	1	1	1	4	1		5		4	5	5	5		1	5	5	5	5	5	1	5
687 Inh. Al Brass	1				4	1		5		4	5	5	5		1	5	5	5	5	5	1	5
464 Naval Brass	1				4	1				4	5	5	5		1	5	5	5	5	5	1	5
510 Phosphor Bronze	1				1	1		1					5		1	5	5	5			1	5
613 Al Bronze	1				1	1							5		1	5	5	5			1	1
614 Al Bronze D	1				1	1							5	5	1	5	5	5			1	5
630 Ni-Al Bronze	1				1	1							5		1	5	5	5			1	5
655 High-si Bronze	1				1	1							2	5	1	5	5	5			1	2
675 Mn Bronze	3*				1	1							5		3*	5	5	5			1	5
70690-10 Cu-Ni	1				1	1	X	1		1	1	1	4		1	5	5	5			1	2
715 70-30 Cu-Ni					1	1	X	1		1	1	1	4		1	5	5	5			1	2
836 Ounce Metal	1				1	1							5		1	5	5	5			1	5
865 Mn Bronze	1				1	1							5		3*	5	5	5			1	5
905 G Bronze	1				1	1							5		1	5	5	5			1	5
922 M Bronze	1				1	1							5		1	5	5	5			1	5
957 Ni-Al-Mn Bronze	1				1	1							5		1	5	5	5			1	5
958 Ni-Al Bronze	1				1	1							5		1	5	5	5			1	5
964 Copper Nickel	1				1	1	X						4		1	5	5	5			1	2
770 Nickel Silver						1	1	5				5	5		5	5	5	5			1	5

Table 3-15

CDA Number	N_2H_4 $+O_2$	Pb-Sn Solder	Cl^-, Oxid.	Citrates	Tartrates
110 ETP Copper		1			
122 DHP Copper		1			
220 Bronze					
230 Red Brass	5	5	1		
260 Cartridge Brass	5	5	4	4	4
270 Yellow Brass	5	5	4	4	4
280 Muntz Metal	5	5	4		
443 Admiralty Brass	5	5	4	4	4
687 Inh. Al Brass	5	5	1		
464 Naval Brass	5	5			
510 Phosphor Bronze		1			

Notes: * Chlorinated solvents.
1 Resistant.
2 Resistant unless hardened or cold-worked.
3 Resistant unless sensitized.
4 Resistant except for special temperatures or concentrations—see text for discussion.
5 Nonresistant.
X Not recommended for this environment.

Table 3-16 Environments vs. aluminum alloys

	Cl⁻, Neutr.	OH^-	NH_3	NO_3^-	S^{-2}	S^{-2}/Cl^-	*	Cl⁻, Acid	Sea-water	SO_4^{-2}	Cd	Al	Bi	Hg	Li	Pb	Ga	Sn	**
1100	1	X	1	1	1	1	1	X	1		1	—	1	5	1	1	5	5	5
2024	5	X	1	1	1	1	5	X	5		1	—	1	5	1	1	5	5	5
3003	1	X	1	1	1	1	1	X	1		1	—	1	5	1	1	5	5	5
5052	1	X	1	1	1	1	1	X	1		1	—	1	5	1	1	5	5	5
5083	3	X	1	1	1	1	3	X	3		1	—	1	5	1	1	5	5	5
5154	1	X	1	1	1	1	1	X	1		1	—	1	5	1	1	5	5	5
6061	1	X	1	1	1	1	1	X	1		1	—	1	5	1	1	5	5	5
6063	1	X	1	1	1	1	1	X	1		1	—	1	5	1	1	5	5	5
7075	5	X	1		1	1	5	X	5		1	—	1	5	1	1	5	5	5
242	1	X	1	1	1	1	1	X	1		1	—	1	5	1	1	5	5	5
295	1	X	1	1	1	1	1	X	1		1	—	1	5	1	1	5	5	5
A332	1	X	1	1	1	1	1	X	1		1	—	1	5	1	1	5	5	5
A356	1	X	1	1	1	1	1	X	1		1	—	1	5	1	1	5	5	5
B443	1	X	1	1	1	1	1	X	1		1	—	1	5	1	1	5	5	5
514	1	X	1	1	1	1	1	X	1		1	—	1	5	1	1	5	5	5
920	1	X	1	1	1	1	1	X	1		1	—	1	5	1	1	5	5	5

Notes: * High purity H_2O.
** Alkali metals.
1 Resistant.
3 Resistant in some cases.
5 Nonresistant.
X Not recommended for this environment-rapid general attack or pitting.

Table 3-17 Environments vs. titanium and zirconium alloys

Type	UNS	Cl^-, Acid	Cl^-, Neutr.	Cl^-, Oxid.	OH^-	NH_3	NO_3^-	S^{-2}	S^{-2}/Cl^-
C.P. Gr. 1 Ti	R50250	1	1	1	X	1	4	1	1
C.P. Gr. 2 Ti	R50400	1	1	1	X	1	4	1	1
Ti-Pd Gr. 7	R52400	1	1	1	X	1	4	1	1
Ti-6Al-4V Gr. 5	R56400	2	2	2	X	1	4	1	1
Low Alloy Gr. 12		1	1	1	X	1	4	1	1
C.P. Zr	R70200	1	1	1	1	1	4	1	1
Zircaloy	R70400	A	A	A	1	1	4	1	1
Zircaloy	R70500	A	A	A	1	1	4	1	1

Type	O_2SCC	MeOH	Br_2	*	I	Hg	Ga	Cd	Sn	Ag	Ce
C.P. Gr. 1 Ti	1	5		1		5	5	5	1	1	
C.P. Gr. 2 Ti	1	5		1		5	5	5	1	1	
Ti-Pd Gr. 7	1	5		1		5	5	5	1	1	
Ti-6Al-4V Gr. 5	1	5		5		5	5	5	5	5	
Low Alloy Gr. 12	1	5		1		5	5	5	1	1	
C.P. Zr	1	5	5	5	5	5					5
Zircaloy	1	5	5	5	5	5					5
Zircaloy	1	5	5	5	5	5					5

Notes: * Chlorinated solvents.
1 Resistant.
2 Resistant unless hardened or cold-worked.
3 Resistant unless sensitized.
4 Resistant except for special temperatures or concentrations—see text for discussion.
5 Nonresistant.
X Not recommended for this environment.
A Resistant unless anodically polarized—see text for discussion.

Table 3-18 Stress-corrosion cracking in additional environment—alloys systems

This table represents a consensus of CFI experience and should be used as a screening guide only.

UNS	Cl^-, Acid	Cl^-, Neutr.	Cl^-, Oxid.	F^- Br^-	SiF_6^{2-}	I	OH^-	S^{-2}	S^{-2}/Cl	$S_4O_6^{2-}$	SO_3^{2-}	SO_4^{2-}	CrO_4^{2-}
S15700	5	4			1	1		5				5	1
S17400	5	4			1			5				5	1
S17700	5	5			1			5				5	1
S20200	5	5	5		1	1	4	2		3	3	1	1
S20910	5	5	5		1	1	4	2		3	3	1	1
S21600	5	5	5		1	1	4	2		3	3	1	1
S21603	5	5	5		1	1	4	2		3,4	3,4	1	1
S21800	5	4	5		1	1	4	2		3	3	1	1
S30400	5	5	5		1	1^h	4	2	5	3	3	1	1
S30403	5	5	5		1	1^h	4	2	5	3,4	3,4	1	1
S30908	5	5	5		1	1	4	2		3	3	1	1
S31008	5	5	5		1	1	4	2		3	3	1	1
S31600	5	5	5	4	1	1^j	4	2	1^i	3	3	1	1
S31603	5	5	5		k	k	4	2	1^i	3,4	3,4	1	1
S31700	5	5	5		1	1	4	2	1^i	3	3	1	1
S31703	5	5	5		1	1	4	2	1^i	3,4	3,4	1	1
S32100	5	5	5		1	1	4	2		1	1	1	1
S32900	4	5	5		2	2	4					1	1
S34700	5	5	5		1	1	4	2		1	1	1	1
S38100	4	4	5		1	1	4					1	1
S43100	5	5	5		2		4					2	1
S44625	4	1	4				4	2					1
S44626	4	1	4				4	2					1
S44700	1	1	1				4	2					1
S45000	4	1	4		2		4	2				2	1
S45500	4	1	4		2		4	2				2	1
N02200	4	1	X	1	2	1	1	X				1	1
N02201	4	1	X	1	2	2,4	1	X				1	1
N04400	a	1	X	1	5	1	4					1	4
N05500	1	1	X		5	1	4					1	1
N06007	4	1	1	1	1	1	4				1	1	1
N06455	4	1		1	1	1		2	2^k	1			1
N06600	1	1		1	2	1	2	2,3		3	3	4^b	1
N06601	1	1		1	2	1	4	2					1
N06625	4^g	1		1	1	1		2	2^k	3	3		1
N06690	1	1		1		1							1
N07718	5^c	1											1
N07750		1		5^c					5^c				1
N08020	4	4	4	1	1	1	4		1	1		1	1
N08700	4	1	4	1	1	1						1	1
	4	1	4	1	1	1						1	1
N08366	4	1	4	1	1	1						1	1
	4	1	4	1	1	1						1	1
	4	1	4	1	1	1						1	1
N08904	4	1		1	1	1						1	1
N08800	4	1		1	1	1	4		1	3	3		1
N08825	4	1		1	1	1	4			1	1	1	1
N10001	4^f	1			4		4			1		4	1
N10002	4^g	1	3		4	1		1	2	1		4	1
N10276	2,3,4	1			1	1	4	1	2,3	1			1
N10655	1	1			4					1	1		1

Notes: 1 Resistant. 2 Resistant unless cold-worked or hardened. 3. Resistant unless sensitized. 4 Resistant except special temperature-concentration conditions—see text. 5 Nonresistant. X Not recommended for this environment.

NO_3^-	NH_3	Azo Dyes	O_2SCC	Sea-water	Cd	Al	Bi	Hg	Li	Pb	S	Sn	Zn	Alloy
1	1	1		X									5	PH15-7Mo
1	1			X									5	17-4PH
1	1	1		X									5	17-7PH
1	1	1		X	5	5							5	202
1	1	1		4	5	5							5	Nitronic 50
1	1	1		X	5	5							5	216
1	1	1		X	5	5							5	216L
1	1	1		X	5	5							5	Nitronic 60
1	1	1	3	X	5	5							5	304
1	1	1	3,4	X	5	5		4					5	304L
1	1	1	3	X	5	5		4					5	309S
1	1	1	3	X	5	5		4					5	310S
1	1	1	3	X	5	5		4					5	316
1	1	1	3,4	X	5	5		4					5	316L
1	1	1	3	X	5	5		4					5	317
1	1	1	3	X	5	5		4					5	317L
1	1	1	1	X	5	5		4					5	321
1	1	1		X	5	5							5	329
1	1	1	1	X	5	5		4					5	347
1	1	1		X	5	5							5	18-18-2
1	1	1		X									5	431
1	1	1		X									5	26-1 (XM-27)
1	1	1		X									5	26-1S (XM-33)
1	1	1		4									5	29-4
1	1	1		X									5	Custom 450
1	1	1		X									5	Custom 455
1	1		1				5	4	5	5	5	4		Ni 200
1	1		1				5	4	5	5	5	4		Ni 201
4	4		1	4			5	5	5	5	5	4		Monel 400
							5	5	5		5	5		K-500
1	1				5	5							1	Alloy G
1	1	1			5	5					4		1	Alloy C-4
1	1	1	5[c]	4	5	5	1			4	4		4	Alloy 600
1	1	1	1		5	5					4			Alloy 601
1	1	1	5		5	5					4			Alloy 625
1	1	1	5		5	5					4			Alloy 690
1	1	1	5[c]								5			Alloy 718
1	1	1	5[c]	4							5	5		Alloy 750
1	1	1			5	5		4					5	Alloy 20 Cb-3
1	1	1			5	5		4					5	JS 700
1	1	1			5	5		4					5	2RN65
1	1	1			5	5		4					5	A-L 6X
1	1	1			5	5		4					5	JS-777
1	1	1			5	5		5						2RK65
1	1	1			5	5		4					5	904L
1	1	1	1		5	5		4					4	Alloy 800
1	1	1	5		5	5								Alloy 825
1	1	5	5		5	5			5		5		1	Alloy B
1	1	1	5	4	5	5					4		1	Alloy C
1	1	1	5		5	5					4		1	Alloy C-276
1	1				5	5					5		1	Alloy B-2

[a]10% $CuCl_2$ in HCl [b]pH 2.5, 289°C (552°F) [c]Aged 1 hour at 955°C (1750°F)/air cooled plus 8 hours at 718°C (1325°F)/furnace cooled to 621°C (1150°F)/air cooled [d]Aged 2 hours at 1149°C (2100°F)/air-cooled plus 24 hours at 843°C (1550°F)/air cooled plus 20 hours at 704°F (1300°F)/air cooled [e]Threshold stress yield point [f]Zinc chloride [g]Aluminum chloride in HCl at 450°C (842°F) [h]One suspected case [i]Under anaerobic conditions [j]Iodine in nuclear fuel elements [k]As cathode coupled to carbon steel in galvanic cell

the system into or out of the danger zone. Cathodic protection and/or inhibitor additions can inhibit cracking.

Table 3-14 is for some wrought stainless steels, Table 3-15 for copper alloys, Table 3-16 for aluminum alloys, Table 3-17 for titanium and zirconium alloys, and Table 3-18 for other stainless steels and high-nickel alloys. For additional information on titanium, the reader is referred to an excellent and lengthy review by Blackburn, Feeney, and Beck.*

In addition to the information in Table 3-17, *zirconium* and its alloys are resistant to stress corrosion in pure water, moist air, steam, and many solutions of sulfates and nitrates. It could crack in $FeCl_3$ and $CuCl_2$ solutions halogens in water, halogen vapors, and organic liquids such as carbon tetrachloride, and fused salts at high temperatures.

For an excellent review of *uranium* and its alloys, see N. J. Mangani, "Hydrogen Embrittlement and Stress Corrosion Cracking of Uranium and Uranium Alloys," in M. G. Fontana and R. W. Staehle, eds., *Advances in Corrosion Science and Technology*, vol. 6, pp. 89–161, Plenum Press, New York, 1976.

Columbium (niobium) and *tantalum* are not subject to usual stress corrosion. They can be embrittled by hydrogen. This embrittlement of tantalum in hot acids can be inhibited by contact with platinum.

Magnesium alloys are being used in many cases where light weight is an important factor, contrary to the general impression of the poor corrosion resistance of magnesium. Bare alloys have shown good resistances to water. Reliable protection systems such as coatings have been developed. Alloys containing manganese have good resistance, but those with high aluminum or zinc content are quite susceptible to stress corrosion.

3-45 Classification of Mechanisms

As described above, the complexity of the interactions between various environments, nature of the alloy, metallurgical structure, etc., indicates the impossibility of one unified mechanism for stress corrosion of all metal-environment systems.

M. A. Streicher (in a private communication to be published in 1985) classified some SCC mechanisms that may be operative in different systems as follows:

1. *Metallurgical Mechanisms*
   a. *Dislocation coplanarity*. Resistance to cracking corresponds to the dislocation pattern. The pattern in susceptible stainless steels tends

*M. J. Blackburn, J. A. Feeney, and T. R. Beck, "Stress Corrosion Cracking of Titanium Alloys," in M. G. Fontana and R. W. Staehle, eds., *Advances in Corrosion Science and Technology*, vol. 3, pp. 67–292, Plenum Press, New York, 1973.

to form planar arrays, whereas in resistant alloys the dislocation patterns are cellular or tangled.

b. *Stress-aging and microsegregation*. In stress aging of austenitic stainless steels, jerky plastic flow occurs. This phenomenon is associated with microsegregation of solutc atoms to dynamic defects in the crystal structure. This type of segregation may account for transgranular stress-corrosion cracking behavior. The cracking rate is limited by the solute diffusion rate as well as electrochemical polarization.

c. *Adsorption*. Surface active species adsorb and interact with strained bonds at the crack tip, causing reduction in bond strength and leading to cracking propagation.

2. *Dissolution Mechanisms*

a. *Stress-accelerated dissolution*. Crack propagates by localized anodic dissolution. Principal role of plastic deformation is to accelerate the dissolution process.

b. *Film Formation at Cracking Wall*. Based on the coplanarity mechanism, cracks initiate at the place of slip step emergence. Propagation is a result of the dissolution of yielding metal. As the crack progresses, the film on the crack walls is repaired and serves as a cathodic site.

c. *Noble element enrichment*. The composition of the slip step has a lower nickel concentration than that of the enriched surface; and the slip step dissolves until the nickel is enriched to the same composition as the preexisting surface.

d. *Film rupture*. Stress-corrosion cracking proceeds by successively breaking a passive film. At the point of rupture, dissolution proceeds until repassivation occurs.

e. *Chloride ion migration*. The chloride ion migrates through the cracked film toward the region of highest stress. The chloride ion then acts to break down the film, thereby allowing metal dissolution.

3. *Hydrogen Mechanisms*

a. *Hydride formation*. Hydrogen enters type 304 stainless steel to form martensite, which will diffuse into stringers normal to the direction of stress and then cause cracking.

b. *Hydrogen embrittlement*. Hydrogen accumulates within the metal at the crack tip, leading to localized weakening, either by void formation or lowering the cohesive strength. Cracks propagate by mechanical fracture of the weakened region.

4. *Mechanical Mechanisms*

a. *Tunnel Pitting and Tearing*. Crack propagates by formation of deep pits or tunnels via dissolution followed by linking of these pits or tunnels by ductile rupture.

b. *Corrosion Product Wedging*. The corrosion products build up in existing cracks and then exert a wedging action.

3-46 Methods of Prevention

As metioned above, the mechanism of stress-corrosion cracking is imperfectly understood. As a consequence, methods of preventing this type of attack are either general or empirical in nature. Stress-corrosion cracking may be reduced or prevented by application of one or more of the following methods:

1. *Lowering the stress* below the threshold value if one exists. This may be done by annealing in the case of residual stresses, thickening the section, or reducing the load. Plain carbon steels may be stress-relief annealed at 1100 to 1200°F, and the austenitic stainless steels are frequently stress-relieved at temperatures ranging from 1500 to 1700°F.
2. *Eliminating the critical environmental species* by, for example, degasification, demineralization, or distillation.
3. *Changing the alloy* is one possible recourse if neither the environment nor stress can be changed. For example, it is common practice to use Inconel (raising the nickel content) when type 304 stainless steel is not satisfactory. Although carbon steel is less resistant to general corrosion, it is more resistant to stress-corrosion cracking than are the stainless steels. Thus, under conditions which tend to produce stress-corrosion cracking, carbon steels are often found to be more satisfactory than the stainless steels. For example, heat exchangers used in contact with seawater or brackish waters are often constructed of ordinary mild steel.
4. *Applying cathodic protection* to the structure with an external power supply or consumable anodes. Cathodic protection should only be used to protect installations where it is positively known that stress-corrosion cracking is the cause of fracture, since hydrogen embrittlement effects are accelerated by impressed cathodic currents.
5. *Adding inhibitors* to the system if feasible. Phosphates and other inorganic and organic corrosion inhibitors have been used successfully to reduce stress-corrosion cracking effects in mildly corrosive mediums. As in all inhibitor applications, sufficient inhibitor should be added to prevent the possibility of localized corrosion and pitting.
6. *Coatings* are sometimes used, and they depend on keeping the environment away from the metal—for example, coating vessels and pipes that are covered with insulation. In general, however, this procedure may be risky for bare metal.
7. *Shot-peening* (also known as shot-blasting) produces residual compressive stresses in the surface of the metal. Woelful and Mulhall* show very substantial improvement in resistance to stress corrosion as a result of peening with glass beads. Type 410 stainless was exposed to 3% NaCl

*M. Woelful and R. Mulhall, Glass Bead Impact Testing, *Metal Progr*. **57**–59 (Sept. 1982).

at room temperature; type 304 to 42% $MgCl_2$ at 150°C; and aluminum alloy 7075-T6 to a water solution of $K_2Cr_2O_7$-CrO_3-NaCl at room temperature. A paper on this subject by Daley* is also of interest.

All of the exposed surface of the *completed* equipment must be shot peened for good results. The surface layer under compressive stress is quite thin—usually a few thousandths of an inch. An example of a successful application involves a type 316 centrifuge handling organic chlorides at 60°C that exhibited extensive SCC after one year. A shot-peened *replacement* 316 centrifuge showed no cracking after 42 months. Peening of cracked surfaces is *not* recommended.

3-47 Corrosion Fatigue

Fatigue is defined as the tendency of a metal to fracture under repeated cyclic stressing. Usually, fatigue failures occur at stress levels below the yield point and after many cyclic applications of this stress. A schematic illustration of a typical fatigue fracture in a cylindrical bar is shown in Fig. 3-73. Characteristically, fatigue failures show a large smooth area and a smaller area which has a roughened and somewhat crystalline appearance. Studies have shown that during the propagation of a fatigue crack through a metal, the frequent cyclic stressing tends to hammer or pound the fractured surface smooth. A crack propagates until the cross-sectional area of the metal is reduced to the point where the ultimate strength is exceeded and rapid brittle fracture occurs. The surface of a brittle fracture usually has a roughened appearance. The unusual appearance of fatigue fractures has led to the common misstatement which attributes such failures to metal "crystallization." This is obviously incorrect, since all metals are crystalline, and the roughened surface which appears on the roughened fracture is the result of brittle fracture and not crystallization.

Fatigue tests are conducted by subjecting a metal to cyclic stresses of various magnitudes and measuring the time to fracture. Results of such tests

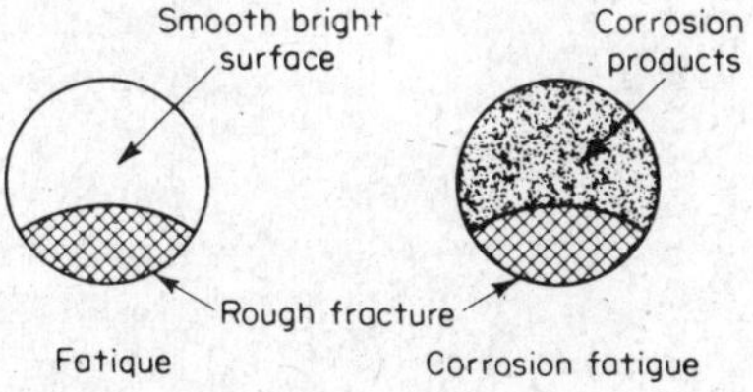

Figure 3-73 Schematic illustration of fatigue and corrosion-fatigue failures.

*J. J. Daley, Controlled Shot Peening Prevents Stress Corrosion Cracking, *Chem. Eng.*, 113–116 (Feb. 16, 1976).

are shown in Fig. 3-74. The fatigue life of steel and other ferrous materials usually becomes independent of stress at low stress levels. As shown in Fig. 3-74, this is called the *fatigue limit*. In general, it is assumed that if a metal is stressed below its fatigue limit, it will endure an infinite number of cycles without fracture. If the specimen used in the fatigue test is notched prior to testing, the fatigue resistance is reduced, as shown in Fig. 3-74. The fatigue resistance is directly related to the radius or the sharpness of the notch. As the notch radius is reduced, the fatigue resistance is likewise reduced. Nonferrous metals such as aluminum and magnesium do not possess a fatigue limit. Their fatigue resistance increases as the applied stress is reduced but does not become independent of stress level.

Corrosion fatigue is defined as the reduction of fatigue resistance due to the presence of a corrosive medium. Thus, corrosion fatigue is not defined in terms of the appearance of the failure, but in terms of mechanical properties. Figure 3-73 illustrates a typical corrosion-fatigue failure. There is usually a large area covered with corrosion products and a smaller roughened area resulting from the final brittle fracture. It is important to note that the presence of corrosion products at a fatigue-fracture point does not necessarily indicate corrosion fatigue. Superficial rusting can occur during ordinary fatigue fracture, and the presence of rust or other corrosion products does not necessarily mean that fatigue life has been affected. This can only be determined by a corrosion-fatigue test.

Corrosion fatigue is probably a special case of stress-corrosion cracking. However, the mode of fracture and the preventive measures differ and it is justifiable to consider it separately.

Renewed attention has been given to corrosion fatigue because of potential catastrophic failures in aerospace, nuclear, and marine (offshore platforms, submarines) structures. Extensive testing and detailed theoretical studies have been conducted. Although the mechanism (or mechanisms) of this type of corrosion remains unclear, it is known that crack initiation and crack growth respond differently to environmental factors.

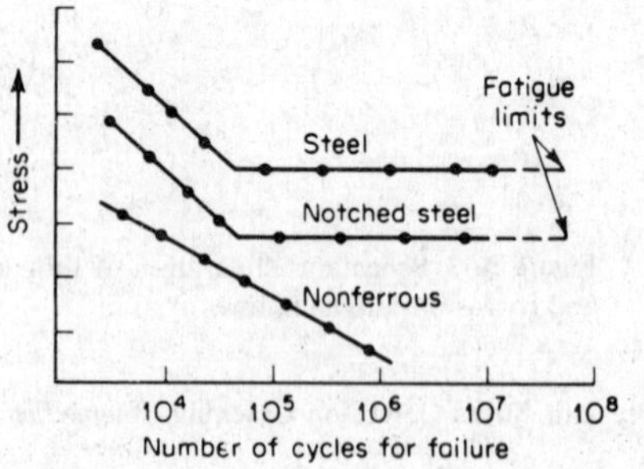

Figure 3-74 Schematic illustration of the fatigue behavior of ferrous and nonferrous alloys.

Environmental factors Environmental factors strongly influence corrosion-fatigue behavior. In ordinary fatigue the stress-cycle frequency has only a negligible influence on fatigue resistance. This factor is of great convenience in fatigue testing since tests can be conducted rapidly at high rates of cyclic stressing. However, corrosion-fatigue resistance is markedly affected by the stress-cycle frequency. Corrosion fatigue is most pronounced at low stress frequencies. This dependence is readily understood since low-frequency cycles result in greater contact time between metal and corrosive. Thus, in evaluating corrosion-fatigue resistance, it is important to conduct the test under conditions identical to those encountered in practice.

Corrosion fatigue is also influenced by the corrosive to which the metal is exposed. Oxygen content, temperature, pH, and solution composition influence corrosion fatigue. For example, iron, steel, stainless steels, and aluminum bronzes possess good corrosion fatigue resistance in water. In seawater, aluminum bronzes and austenitic stainless steels retain only about 70 to 80% of their normal fatigue resistance. High-chromium alloys retain only about 30 to 40% of their normal fatigue resistance in contact with seawater. It is apparent that corrosion fatigue must be defined in terms of the metal and its environment.

Bogar and Crooker* tested alloys of steel, aluminum, and chromium in natural seawater, ASTM seawater, and a 3% sodium chloride solution. They conclude "that solution composition seldom has a large or a consistent effect on marine corrosion fatigue test results." Davis, Vassilaros, and Gudas† tested metal matrix composites.

Mechanism The mechanism of corrosion fatigue has not been studied in detail, but the cause of this type of attack is qualitatively understood. Corrosion-fatigue tests of iron and ferrous-base materials show that their fatigue-life curves resemble those of nonferrous metals. Also, corrosion fatigue seems to be most prevalent in mediums that produce pitting attack. These two facts indicate that fatigue resistance is reduced in the presence of a corrosive because corrosion pits act as stress raisers and initiate cracks. It is most likely that the corrosion is most intense at the crack tip, and as a consequence there is no stable pit radius. Since the pit or radius continuously decreases due to simultaneous mechanical and electrochemical effects, the fatigue curve of a ferrous metal exposed to a corrosive resembles that of a nonferrous metal. A corrosion-fatigue failure is usually transgranular and does not show the branching that is characteristic of many stress-corrosion

*F. D. Bogar and T. W. Crooker, Fatigue Testing in Natural and Marine Corrosion Environments Substitute Ocean Waters, *Materials Performance*, 37 (Aug. 1983).

†D. A. Davis, M. G. Vassilaros, and J. P. Gudas, Corrosion Fatigue and Stress Corrosion Characteristics of Metal Matrix Composites in Seawater, *Materials Performance*, 38–42 (Mar. 1982).

cracks. The final stages of corrosion fatigue are identical to those occurring during ordinary fatigue; final fracture is purely mechanical and does not require the presence of a corrosive.

Wei, Shim, and Tanaka* emphasize the complex interactions of loading, environmental and metallurgical variables. To make test information applicable to proper design, a model must quantify crack growth in terms of processes, including transport of deleterious species to the crack tip, localized chemical reactions at the crack tip, entry and diffusion of hydrogen, and embrittlement effects. R. P. Gangloff† discusses small cracks and also models the many factors and their interplay in corrosion fatigue and SCC.

Prevention Corrosion fatigue can be prevented by a number of methods. Increasing the tensile strength of a metal or alloy improves ordinary fatigue but is detrimental to corrosion fatigue. In the case of ordinary fatigue resistance, alloys with high tensile strength resist the formation of nucleating cracks. It should be noted, however, that once a crack starts in a high-tensile-strength material, it usually progresses more rapidly than in a material with lower strength. During corrosion fatigue a crack is readily initiated by corrosive action; hence, the resistance of high-tensile material is quite low. Corrosion fatigue may be eliminated or reduced by reducing the stress on the component. This can be accomplished by altering the design, by stress-relieving heat treatments, or by shot-peening the surface to induce compressive stresses. Corrosion inhibitors are also effective in reducing or eliminating the effects of corrosion fatigue. Corrosion-fatigue resistance also can be improved by using coatings such as electrodeposited zinc, chromium, nickel, copper, and nitride coatings. When electrodeposited coatings are applied it is important to use plating techniques that do not produce tensile stresses in the coating or charge hydrogen into the metal.

Many engineers associate fatigue and corrosion fatigue with rotating parts, but other types of equipment (usually considered static) could fail. For example, I have investigated several expensive failures of heat-exchanger tubing because of vibration. The design of equipment should ensure the avoidance of structural vibrations.

Suggested Reading

Devereaux, O., A. J. McEvily, and R. W. Staehle, eds., *Corrosion Fatigue: Chemistry Mechanics and Microstructure*, National Association of Corrosion Engineers, Houston, Tex., 1972.

Corrosion Fatigue: Mechanics, Metallurgy, Electrochemistry, and Engineering, ASTM STP 801, Information Center, Battelle Laboratories, Columbus, Ohio, 1981.

Corrosion Fatigue: Mechanics, Metallurgy, Electrochemistry, and Engineering, ASTM STP 801, American Society for Testing and Materials, Philadelphia, May 1983.

*R. P. Wei, G. Shim, and K. Tanaka, "Corrosion Fatigue and Modeling," TMS-AIME Fall Meeting, 1983.

†R. P. Gandgloff, "Localized Chemistry Effects on the Growth Kinetics of Small Cracks," TMS-AIME Fall Meeting, 1983.

HYDROGEN DAMAGE

3-48 Characteristics

Hydrogen damage is a general term which refers to mechanical damage of a metal caused by the presence of, or interaction with, hydrogen. Hydrogen damage may be classified into four distinct types:

1. Hydrogen blistering
2. Hydrogen embrittlement
3. Decarburization
4. Hydrogen attack

Hydrogen blistering results from the penetration of hydrogen into a metal. An example of blistering is shown in Fig. 3-75. The result is local deformation and, in extreme cases, complete destruction of the vessel wall. Hydrogen embrittlement also is caused by penetration of hydrogen into a metal, which results in a loss of ductility and tensile strength. Decarburization, or the removal of carbon from steel, is often produced by moist hydrogen at high temperatures. Decarburization lowers the tensile strength of steel. Hydrogen attack refers to the interaction between hydrogen and a component of an alloy at high temperatures. A typical example of hydrogen attack is the disintegration of oxygen-containing copper in the presence of hydrogen. Decarburization and hydrogen attack are high-temperature processes; they are discussed in detail in Chap. 11.

Hydrogen blistering and hydrogen embrittlement may occur during exposure to petroleum, in chemical process streams, during pickling and welding operations, or as a result of corrosion. Since both of these effects

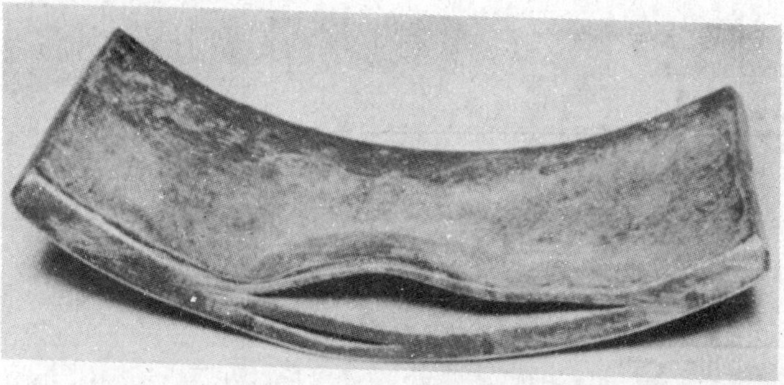

Figure 3-75 Cross section of a carbon steel plate removed from a petroleum process stream showing a large hydrogen blister. Exposure time: 2 years. (*Imperial Oil Limited, Ontario, Canada*)

produce mechanical damage, catastrophic failure may result if they are not prevented.

3-49 Environmental Factors

Atomic hydrogen (H) is the only species capable of diffusing through steel and other metals. The molecular form of hydrogen (H_2) does not diffuse through metals. Thus, hydrogen damage is produced only by the atomic form of hydrogen. There are various sources of nascent or atomic hydrogen—high-temperature moist atmospheres, corrosion processes, and electrolysis. The reduction of hydrogen ions involves the production of hydrogen atoms and the subsequent formation of hydrogen molecules. Hence, both corrosion and the application of cathodic protection, electroplating, and other processes are major sources of hydrogen in metals. Certain substances such as sulfide ions, phosphorous, and arsenic compounds reduce the rate of hydrogen-ion reduction. Apparently most of these function by decreasing the rate at which hydrogen combines to form molecules. In the presence of such substances there is a greater concentration of atomic hydrogen on the metal surface.

3-50 Hydrogen Blistering

A schematic illustration of the mechanism of hydrogen blistering is shown in Fig. 3-76. Here, the cross-sectional view of the wall of a tank is shown. The interior contains an acid electrolyte, and the exterior is exposed to the atmosphere. Hydrogen evolution occurs on the inner surface as a result of a corrosion reaction on cathodic protection. At any time there is a fixed concentration of hydrogen atoms on the metal surface, and some of these

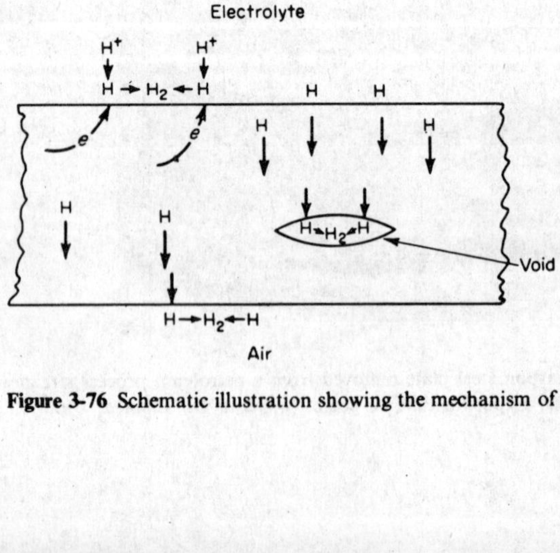

Figure 3-76 Schematic illustration showing the mechanism of hydrogen blistering.

diffuse into the metal rather than combining into molecules, as shown. Much of the hydrogen diffuses through the steel and combines to form hydrogen molecules on the exterior surface. If hydrogen atoms diffuse into a void, a common defect in rimmed steels, they combine into molecular hydrogen. Since molecular hydrogen cannot diffuse, the concentration and pressure of hydrogen gas within the void increases. The equilibrium pressure of molecular hydrogen in contact with atomic hydrogen is several hundred thousand atmospheres, which is sufficient to rupture any known engineering material.

Hydrogen blistering is most prevalent in the petroleum industry. It occurs in storage tanks and in refining processes. One method for control is to add an inhibitor, such as the polysulfide ion.

3-51 Hydrogen Embrittlement

The exact mechanism of hydrogen embrittlement is not as well known as that of hydrogen blistering. The initial cause is the same: penetration of atomic hydrogen into the metal structure. For titanium and other strong hydride-forming metals, dissolved hydrogen reacts to form brittle hydride compounds. In other materials, such as iron and steel, the interaction between dissolved hydrogen atoms and the metal is not completely known.

There are indications that a large fraction of all the environmentally activated cracking of ferritic and martensitic iron-base alloys and the titanium-base alloys is due in some way to the interaction of the advancing crack with hydrogen. The general characteristics of such cracking susceptibility are illustrated in Fig. 3-77 for the cracking of type 4340 steel (C-0.40, Mn-0.70, P-0.04, S-0.04, Si-0.30, Ni-1.8, Cr-0.8, Mo-0.25). This figure* shows that higher strength levels are more susceptible to cracking and that higher stresses cause cracking to occur more rapidly. These trends are in fact general for most alloys subject to hydrogen embrittlement; i.e., the alloys are most susceptible to cracking in their highest strength level. The tendency for embrittlement is also increased with hydrogen concentration in the metal as shown in Fig. 3-78. This figure† shows that after a given length of time, cracking occurs at successively higher stresses as the cathodically charged hydrogen is removed by baking treatments and the tremendous differences in stresses involved.

Most of the mechanisms that have been proposed for hydrogen embrittlement are based on slip interference by dissolved hydrogen. This slip interference may be due to the accumulation of hydrogen near dislocation sites or microvoids, but the precise mechanism is still in doubt.

Hydrogen embrittlement is distinguished from stress-corrosion cracking

*R. A. Davis, G. H. Dreyer, and W. C. Gallaugher, *Corrosion*, **20**:93*t* (1964).
†H. H. Johnson, E. J. Schneider, and A. R. Troiano, *Trans. AIME*, **212**:526–536 (1958).

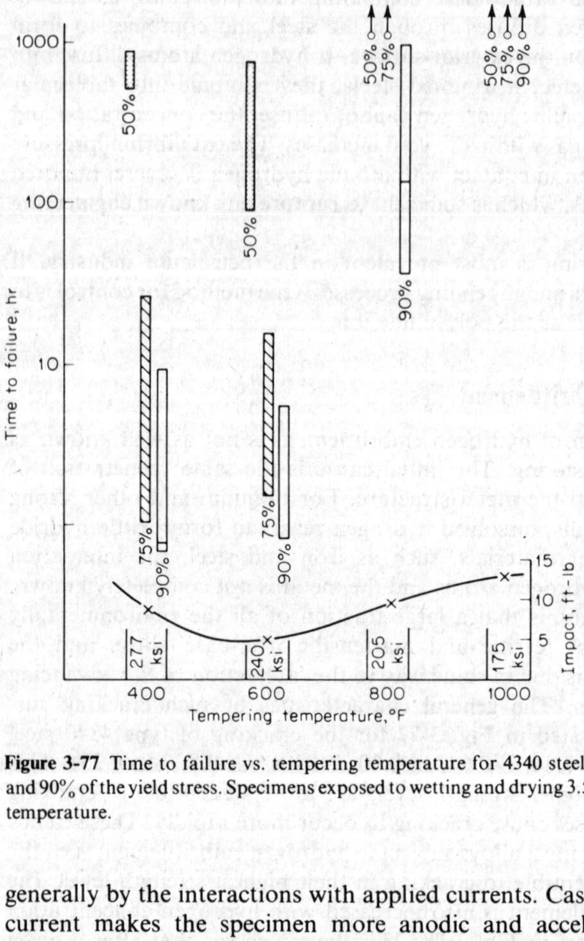

Figure 3-77 Time to failure vs. tempering temperature for 4340 steel at stress levels of 50, 75, and 90% of the yield stress. Specimens exposed to wetting and drying 3.5% NaCl solution at room temperature.

generally by the interactions with applied currents. Cases where the applied current makes the specimen more anodic and accelerates cracking are considered to be stress-corrosion cracking, with the anodic-dissolution process contributing to the progress of cracking. On the other hand, cases where cracking is accentuated by current in the opposite direction, which accelerates the hydrogen evolution reaction, are considered to be hydrogen embrittlement. These two phenomena are compared with regard to cracking mode and applied current in Fig. 3-79.

Although hydrogen embrittlement is more or less the "universal" description, other terms are used. If absorption is due to contact with hydrogen gas, it is often described as hydrogen stress cracking. If hydrogen is absorbed because of the corrosion reaction it is called SCC or sometimes hydrogen stress cracking. If corrosion is due to the presence of hydrogen sulfide, a common term is sulfide stress cracking. A few ppm of absorbed hydrogen can cause cracking.

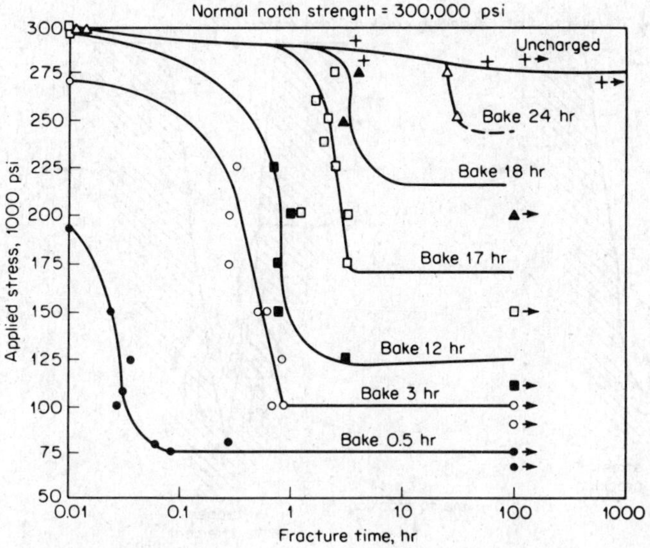

Figure 3-78 Static fatigue curves for various hydrogen concentrations obtained by baking 4340 steel different times at 300°F.

When hydrogen is initially present (before use) such as in electroplated articles, baking removes the hydrogen (Fig. 3-78). This procedure is also used during shutdowns of hydrogenation equipment involving high-strength steel components.*

Hydrogen cracking tendency decreases with increasing temperature, and significant change occurs above about 150°F (70°C). Except in corrosion reactions involving hydrofluoric acid or hydrogen sulfide, hydrogen stress cracking is usually not a problem with steels having yield strengths below 150 lb/in.^{2} (1,000 MPa). For these acids the limit drops to about 80 lb/in.^{2} (550 MPa).*

Stevens and Bernstein† state that crack growth results indicate that crack tip plasticity and branching plays a role in hydrogen-induced crack growth in HSLA steel. Pasco and associates‡ present an absorption model and also equations for calculating experimental results.

*R. S. Treseder, Guarding Against Hydrogen Embrittlement, *Chem. Eng.*, 105–108 (June 29, 1981).

†M. F. Stevens and I. M. Bernstein, "Microstructural Effects on Hydrogen Embrittlement of a Ti-Containing HSLA Steel."

‡R. W. Pasco, K. Sieradzki, and P. J. Ficalora, "An Absorption Model for Stage II Crack Growth Rates."

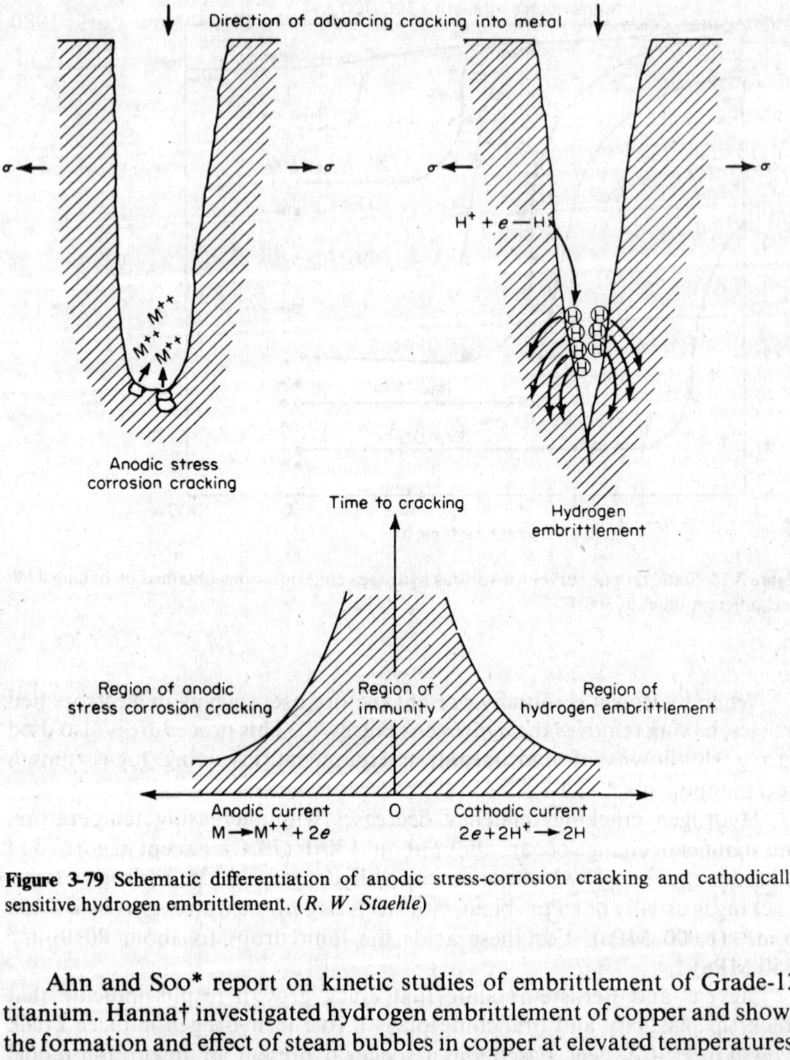

Figure 3-79 Schematic differentiation of anodic stress-corrosion cracking and cathodically sensitive hydrogen embrittlement. (*R. W. Staehle*)

Ahn and Soo* report on kinetic studies of embrittlement of Grade-12 titanium. Hanna† investigated hydrogen embrittlement of copper and shows the formation and effect of steam bubbles in copper at elevated temperatures.

For an in-depth study of metallurgical variables the reader is referred to "The Role of Metallurgical Variables in Hydrogen-Assisted Environmental Fracture," in M. G. Fontana and R. W. Staehle, eds., *Advances in Corrosion*

*T. M. Ahn and P. Soo, "The Kinetics of Hydrogen Embrittlement in ASTM Grade-12 Titanium Alloy."

†M. D. Hanna, Hydrogen Embrittlement of Copper: Formation and Growth of Intergranular H_2O Bubbles."

Note: The last four of these references are papers presented during the TMS Fall Meeting 1983.

Science and Technology, vol. 7, pp. 53–175, Plenum Press, New York, 1980. Most engineering alloy systems and anodic cracking are included. Practically all conceivable variables are discussed, and an excellent bibliography is included.

In his paper Treseder also discusses *sulfide stress cracking*, which occurs in the presence of water and hydrogen sulfide. This problem is of great importance in the petroleum industry all the way from production through refining. Apparently the iron sulfide formed on the metal surface has a catalytic effect in increasing the amount of hydrogen that enters the metal as opposed to other corrosion reactions. Here again the critical strength level of the steel is 80 lb/in.^{2} (550 MPa). A hydrogen sulfide partial pressure above 0.05 lb/in.^{2} (absolute) can cause cracking.

Section 4-24 describes NACE test method TM-01-77 for "sour" environments to determine acceptable alloys, and some results are presented there. For most carbon and low-alloy steels a maximum hardness of Rockwell C 22 is recommended. Where higher strength is required, some alloys are acceptable up to RC-35. Selection of suitable materials is the *major* method for controlling sulfide stress cracking.

Turn, Wilde, and Troianos* discuss cracking of pipe steels and present additional information.

Treseder also describes *monitoring* as follows: "In-plant monitoring of process conditions that might lead to hydrogen embrittlement effects is being done by use of hydrogen probes. These are of two types: one consists of a thin-walled, closed-end steel tube inserted into the process stream, the other is in the form of a patch on the outside of the pipe or vessel. In each case the permeation rate of hydrogen through the steel is measured. These data can be correlated with plant conditions, and used to monitor mitigation measures such as inhibition."

3-52 Prevention

Hydrogen blistering may be prevented by application of one or more of the following preventative measures:

1. *Using "clean" steel*. Rimmed steels tend to have numerous voids, and the substitution of killed steel greatly increases the resistance to hydrogen blistering because of the absence of voids in this material.
2. *Using coatings*. Metallic, inorganic, and organic coatings and liners are often used to prevent the hydrogen blistering of steel containers. To be successful, the coating or liner must be impervious to hydrogen penetration and be resistant to the mediums contained within the tank. Steel clad with austenitic stainless steel or nickel is often used for this

*J. C. Turn, B. E. Wilde, and C. A. Troianos, On the Sulfide Cracking of Line Pipe Steels, *Corrosion*, 364–370 (Sept. 1983).

purpose. Also, rubber and plastic coatings and brick linings are frequently employed.

3. *Using inhibitors.* Inhibitors can prevent blistering since they reduce corrosion rate and the rate of hydrogen-ion reduction. Inhibitors, however, are primarily used in closed systems and have limited use in once-through systems.
4. *Removing poisons.* Blistering usually occurs in corrosive mediums containing hydrogen-evolution poisons such as sulfides, arsenic compounds, cyanides, and phosphorous-containing ions and rarely occurs in pure acid corrosives. Many of these poisons are encountered in petroleum process streams, which explains why blistering is a major problem in the petroleum industry.
5. *Substituting alloys.* Nickel-containing steels and nickel-base alloys have very low hydrogen diffusion rates and are often used to prevent hydrogen blistering.

Although *hydrogen embrittlement*, like hydrogen blistering, results from the penetration of hydrogen into a metal or alloy, methods for preventing this form of damage are somewhat different. For example, the use of clean steels has relatively little influence on hydrogen embrittlement since the presence of voids is not involved. Hydrogen embrittlement may be prevented by application of one or more of the following preventive measures:

1. *Reducing corrosion rate.* Hydrogen embrittlement occurs frequently during pickling operations where corrosion of the base metal produces vigorous hydrogen evolution. By careful inhibitor additions, base-metal corrosion can largely be eliminated during pickling with a subsequent decrease in hydrogen pickup.
2. *Altering plating conditions.* Hydrogen pickup during plating can be controlled by the proper choice of plating baths and careful control of plating current. If electroplating is performed under conditions of hydrogen evolution, poor deposits and hydrogen embrittlement are the result.
3. *Baking.* Hydrogen embrittlement is an almost reversible process, especially in steels. That is, if the hydrogen is removed, the mechanical properties of the treated material are only slightly different from those of hydrogen-free steel. A common way of removing hydrogen in steels is by baking at relatively low temperatures (200 to 300°F). See Fig. 3-78.
4. *Substituting alloys.* The materials most susceptible to hydrogen embrittlement are the very-high-strength steels. Alloying with nickel or molybdenum reduces susceptibility.
5. *Practicing proper welding.* Low-hydrogen welding rods should be specified for welding if hydrogen embrittlement is a problem. Also, it is important to maintain dry conditions during welding since water and water vapor are major sources of hydrogen.

3-53 Fracture Mechanics

A thorough discussion of fracture mechanics is beyond the scope of this book, but a few remarks are appropriate. The reader is referred to MTI Manual No. 8, *Fracture Control for the Chemical Process Industries* (1983). This is a primer (simplified) manual intended for corrosion engineers and others who may not have a substantial background in engineering mechanics. Example of application are described. The manual states in a few words the basic premise of fracture mechanics; namely, fracture resistance of a material is expressed by the material's fracture toughness, termed K_{Ic}. Fracture occurs when the stress intensity K is equal to K_{Ic}.

In its simplest form, stress intensity K is expressed as $K = 1.77\sigma\sqrt{\alpha}$, where σ is the acting stress and α is the size of an existing crack. As stated above, fracture occurs when $K = 1.77\sigma\sqrt{\alpha} = K_{Ic}$. This applies for propagation of cracks due to mechanical stress. In the presence of a corrosive environment, the situation could be vastly different.

As described earlier in this chapter, SCC can greatly reduce the load-bearing capacity of a structural part. For SCC we replace K_{Ic} with K_{ISCC}, which is the threshold value for SCC and it is the highest-plane strain-stress intensity below which crack propagation does *not* occur. Figure 3-80 illustrates this point. For this case, stress-corrosion cracks should not propagate below a value of 15. In the area between the curve and the dotted line, crack growth may be slow, but eventually the rate would become fast and critical and catastrophic failure would result. In general, this applies to SCC and also corrosion fatigue.

The fracture mechanics approach is quite "clean" for high-strength low-alloy steels (i.e., 4340 in seawater), but the picture is somewhat confused

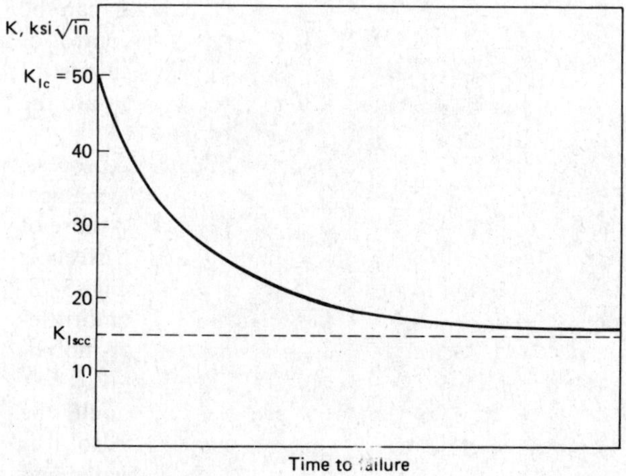

Figure 3-80 Stress corrosion test data.

for the austenitic stainless steels (i.e., 18-8) largely because of the branching nature of the cracks (Fig. 3-53). There are other limitations, for example, in thin-walled sheet and pipe. The best advice is not to fool with the fracture mechanics approach unless you have a good understanding of the subject.

I believe fracture mechanics entered the corrosion picture because times for crack initiation varied all over the place for many corrosion-resistant alloys. This variable is removed when you start with a precracked specimen and then follow its propagation.

CHAPTER

FOUR

CORROSION TESTING

4-1 Introduction

Thousands of corrosion tests are made every year. The value and reliability of the data obtained depend on details involved. Unfortunately, many tests are not conducted or reported properly, and the information obtained is misleading. Most tests are made with a specific objective in mind. This may vary from tests designed to teach a student the procedures involved to the loading of an airplane wing on the seashore for studying susceptibility to stress corrosion. Precise results or merely qualitative comparisons may be required. In any case, the reliability of the test is no better than the thinking and planning involved. Well-planned and executed tests usually result in *reproducibility* and *reliability*. These are two of the most important factors in corrosion testing. Corrosion tests, and application of the results, are considered to be a most important aspect of corrosion engineering.

Many corrosion tests are made to select materials of construction for equipment in the process industries. It is very important for the tests to duplicate the actual plant service conditions as closely as possible. The greater the deviation from plant conditions the less reliable the test will be.

4-2 Classification

Corrosion testing is divided into four types of classifications: (1) laboratory tests, including acceptance or qualifying tests; (2) pilot-plant or semiworks tests; (3) plant or actual service tests; and (4) field tests. The last two could be combined, but to avoid confusion in terminology the following distinction is made: the third involves tests in a particular service or a given plant, whereas the fourth involves field tests designed to obtain more general information. Examples of field tests are atmospheric exposure of a large number of specimens in racks at one or more geographical locations and similar tests in soils or seawater.

Laboratory tests are characterized by small specimens and small volumes of solutions, and actual conditions are simulated insofar as conveniently possible. The best that can be done in this regard is the use of actual plant solutions or environment. Laboratory tests serve a most useful function as screening tests to determine which materials warrant further investigation. Sometimes plants are built based primarily on laboratory tests, but results could be catastrophic and sometimes are.

Pilot-plant or semiworks tests are usually the best and most desirable. Here the tests are made in a small-scale plant that essentially duplicates the intended large-scale operation. Actual raw materials, concentrations, temperatures, velocities, and volume of liquor to area of metal exposed are involved. Pilot plants are usually run long enough to ensure good results. Specimens can be exposed in the pilot plant, and the equipment itself is studied from the corrosion standpoint. One possible disadvantage is that conditions of operation may be widely varied in attempting to determine optimum operation. This means careful "logging" and keeping of thorough records.

An important point to be emphasized here is cooperation between operating personnel and the corrosion engineer. Research and development people are primarily interested in proving the process with least cost and are not usually concerned with materials of construction. An example may serve to illustrate the point. Research chemists and chemical engineers worked out a process for attacking an ore with sulfuric acid and obtained a rapid reaction and good yields. The operation was successfully carried out in a pilot plant within a short time. A decision was then made to construct a production plant. The corrosion engineer was consulted with regard to selection of materials of construction. The first questions raised concerned the materials used in the pilot plant and the corrosion indications obtained. The pilot plant was constructed from steel and cast iron, and most of the equipment was badly attacked during the few runs made. It would have required little effort and expense to expose various materials in the pilot plant to obtain some corrosion data under operating conditions. This would have eliminated much guesswork on the part of the corrosion engineer and also reduced the requirements for later laboratory tests that did not simulate operating conditions.

Actual plant tests are made when an operating plant is available. Interest here is in evaluating better or more economical materials or in studying corrosion behavior of existing materials as process conditions are changed.

Perhaps the ideal and logical sequence of testing for a proposed new plant would be to use laboratory tests to determine which materials are definitely unsatisfactory and those warranting further consideration and then pilot-plant tests on specimens and actual parts such as a valve, pump, heat-exchanger tube, or pipe section made from materials that showed promise during the laboratory investigation. This procedure would provide a sound basis for building the production plant. Unfortunately this situation is the exception, particularly in the process industries. Management sometimes takes months or even years to make the decision with regard to whether or not to build a new plant because of the many factors to be thoroughly considered. Then when the funds are allocated, the plant must be built and production sold "tomorrow"! Therefore, it behooves the corrosion engineer to know exactly how and what he is doing, because false starts and unreliable results may be disastrous.

4-3 Purposes

Perhaps the main justifications for corrosion testing are:

1. Evaluation and selection of materials for a specific environment or a given definite application. This could be a new or modified plant or process where previous operating history is not available. It could involve an old plant or process that is to be replaced or expanded with more economical materials of construction or materials that would exhibit less contamination of product, improved safety, more convenient design and fabrication, or substitution of less strategic materials.
2. Evaluation of new or old metals or alloys to determine the environments in which they are suitable. Much of this type of work is done by producers and vendors of materials. The information obtained aids in the selection of materials to be tested for a specific application. Inclusion of tests on other materials which are known to be in commercial use in these environments permits helpful comparisons. In the case of new metals and alloys, the data obtained provide information concerning possible applications. This category could also include the effects of changes in environment—such as additions of inhibitors or deaeration—on the corrosion of metals and alloys.
3. Control of corrosion resistance of the material or corrosiveness of the environment. These are usually routine tests to check the quality of the material. The Huey test (boiling 65% nitric acid) is used to check the heat treatment of stainless steels. Another example is the salt-spray test, where specimens are exposed in a box or cabinet containing a spray or fog of seawater or salt water. This type of test is often used for checking or evaluating paints and electroplated parts. These tests may not be directly related to the intended services but are sometimes incorporated in specifications as acceptance tests. In some cases periodic testing is required to determine changes in the aggressiveness of the environment because of operating changes such as temperatures, process raw materials, changes in concentrations of solutions, or other changes that are often regarded as insignificant from the corrosion standpoint by operating personnel. Probe techniques, described later, are also very helpful here.

4. Study of the mechanisms of corrosion or other research and development purposes. These tests usually involve specialized techniques, precise measurements, and very close control.

4-4 Materials and Specimens

The first step in corrosion testing concerns the specimens themselves. This is an important step and could be compared to the foundation of a house. If complete information on the materials is not known, the data obtained may be practically useless. Chemical composition, fabrication history, metallurgical history, and positive identification of specimens are all required. There are many cases where the material tested was described as "stainless steel." In order to avoid confusion and to increase reliability of the tests, many laboratories and companies maintain stocks of material for corrosion testing only. Material representative of the metal or alloy involved is obtained in substantial quantity and specimens cut from it. The mill heat number, the chemical composition, and heat treatment are determined. Metallographic examination to ensure normal structure is also desirable. The stock and specimens are immediately identified by a reference number. Stamping numbers on the specimens represents common practice. If brittle materials are involved, notches can be ground on the edges. The identification should be such that it will not be obliterated during testing. Sufficient material is obtained at one time to satisfy the requirements for several years, and a variety of materials are stocked. The person who wishes to conduct corrosion tests obtains material from this specimen stock room and is sure that he or she is working with known materials.

Metals and alloys are available in wrought or cast form or both. Rolled strip or cast bars are available from producers. If the equipment is to be made from wrought material, it is desirable to test wrought specimens. If castings are involved, cast specimens should be tested. However, the corrosion resistance of cast and wrought metals and alloys is generally regarded as identical. This reduces the number of tests required in some cases. If particular shapes are involved, representative material should be tested. A good example of this is cold-drawn wire. If welded construction is involved, then specimens containing welds, or weld beads, should be tested. These welds should have the same heat treatment (or lack of heat treatment) as the process equipment in question.

Size and shape of specimens vary, and selection is often a matter of convenience. Squares, rectangles, disks, and cylinders are often used. Flat samples are usually preferred because of easier handling and surface preparation. Specimens $\frac{1}{16}$ to $\frac{1}{4}$ in. thick, 1 in. wide, and 2 in. long are commonly employed in laboratory tests. Plant- and field-test specimens could be of these sizes also, but are usually larger. For wrought specimens, a large ratio of rolled area to edge area is desirable because in equipment made from sheet the rolled surface is exposed to corrosion. This is one reason for using thin specimens. Experiments have shown that the cut edge might corrode twice as fast as the rolled surface and accordingly a misleading picture may be obtained if, for example, disks are cut from a rolled rod. This results in a low ratio of rolled surface to cut edge. Small specimens also permit more accurate weighing and measuring of dimensions, particularly for short-time tests or where corrosion rates are low. Larger specimens are desirable when studying pitting corrosion because of the probability factor involved. If the effect of corrosion on the strength of the metal or alloy is under consideration, tensile specimens are used for corrosion tests in order to avoid working or machining of the specimen after exposure.

4-5 Surface Preparation

Ideally the surface of the test specimen should be identical with the surface of the actual equipment to be used in the plant. However, this is usually impossible because the surfaces of commercial metals and alloys vary as produced and as fabricated. The degree of scaling or amounts of oxide on the equipment varies and also the conditions of other surface contaminants. Because of this situation and because the determination of the corrosion resistance of the metal or alloy itself is of primary importance in most cases, a clean metal surface is usually used. A standard surface condition is also desirable and necessary in order to facilitate comparison with results of others.

A common and widely used surface finish is produced by polishing with No. 120 abrasive cloth or paper or its approximate equivalent. This is not a smooth surface, but it is not rough, and it can be readily produced. Prior treatments such as machining, grinding, or polishing with a coarse abrasive may be necessary if the specimen surface is very rough or heavily scaled. All these operations should be made so that excessive heating of the specimen is avoided. A good general rule is that the specimen could at all times be held by the naked hand. The 120 finish usually removes sufficient metal to get below any variations (such as decarburization or carburization) in the original metal surface.

Clean polishing belts or papers should be used to avoid contamination of the metal surface, particularly when widely dissimilar metals are being polished. For example, a belt used to polish steel should not then be used to polish brass or vice versa. Particles of one metal would be imbedded in the other and erroneous results obtained.

A smoother finish may be required in certain cases such as actual equipment that requires a highly polished surface or sometimes where extremely low rates of corrosion are anticipated.

Quite often test specimens are made by shearing from a thin plate or sheet. The edges must be machined, filed, or ground to remove the severely cold-worked metal and subsequently finished similarly to the remainder of the specimen. The edges and corners of the specimens should be slightly beveled or rounded to facilitate polishing.

Soft metals such as lead would tend to smear if polished on an emery belt. Rubbing with a hard eraser until a bright surface is obtained is a recommended procedure for lead and lead alloys. A sharp blade is sometimes used to shave or prepare lead specimens. The soft metals also present the problem of the abrasive being imbedded in the surface. Scrubbing with pumice powder and other fine abrasives is sometimes used on magnesium, aluminum, and their alloys. Electrolytic polishing is occasionally used for research work but is not generally recommended for plant tests.

Chemical treatments or passivating pretreatments for stainless steels and alloys are sometimes used but are not recommended because false and misleading results might be obtained. A passivity treatment may result in good corrosion resistance during testing but may not be effective during actual service of the equipment. In other words, a material should not be used in service if its corrosion resistance depends upon an artificial passivation treatment. Natural passivity effects would show up during tests to determine the effect of time on corrosion. Chemical treatments are utilized to decontaminate metal surfaces and serve a useful purpose here.

4-6 Measuring and Weighing

After surface preparation the specimens should be carefully measured to permit calculation of the surface area. Since area enters in the formula for calculating the corrosion rate, the results can be no more accurate than the accuracy of measurement of the surface area. The original area is used to calculate the corrosion rate throughout the test. If the dimensions of the specimen change appreciably during the test, the error introduced is not important because the material is probably corroding at too fast a rate for its practical use.

After measuring, the specimen is degreased by washing in a suitable solvent such as acetone, dried, and weighted to nearest 0.1 mg (for small specimens). The specimen should be exposed to the corrosion environment immediately or stored in a desiccator, particularly if the material is not corrosion-resistant to the atmosphere. Direct handling of the specimens is undesirable.

4-7 Exposure Techniques

A variety of methods are utilized for supporting specimens for exposure in the laboratory or in the plant. The important considerations are: (1) The corrosive should have easy access to the specimen; (2) the supports should not fail during the test; (3) specimens should be insulated or isolated electrically from contact with another metal unless galvanic effects are intended; (4) the specimen should be properly positioned if effects of complete immersion, partial immersion, or vapor phase are being studied; and (5) for plant tests, the specimens should be as readily accessible as possible.

Figure 4-1 shows a widely used arrangement for testing in the laboratory under boiling, warm, or room-temperature conditions. This particular setup is for boiling tests. The specimen is held in a glass cradle to permit circulation of the corrosive. The use of a cradle avoids the expense of drilling a hole to hang the specimen. The flask is an ordinary 1000-ml wide-mouth Erlenmeyer. The condenser is called an acorn or finger-type condenser. The condenser fits loosely, so the flasks and condensers are easily interchangeable. These parts are readily available and much less expensive than the older-type condensers, which have a ground-glass joint with the flask and are not interchangeable. The latter are expensive, and joint freezing is a problem. The acorn condenser is hung on a convenient hook in the hood when it is removed. This arrangement is also suitable for temperature bath tests and is used for room-temperature tests when liquids with high vapor pressure are involved. A number of flasks connected in series for water cooling are run on one hot plate. Where liquid loss during testing is a problem, special flasks with long necks and elongated acorn condensers are used. An important consideration for boiling tests is to be sure that sufficient heat is available to cause boiling in all the flasks. The upper end of the stem of the cradle is in the form of a hook so that it can be easily lifted out of the flask. One specimen per flask is desirable, but duplicate specimens are often run in the same flask. Different materials run in the same container often produce erroneous results because the ions of one metal may affect the corrosiveness of the environment on a different metal.

Although the acorn condenser in Fig. 4-1 has been used for years, M. A. Streicher (private communication—to be published in 1985) has shown that the Allihn condenser gives more reliable results for several stainless alloys in acid-chloride environments. Corrosion rates could vary substantially. The Allihn consists of a flask connected with a ground-glass joint to a vertical

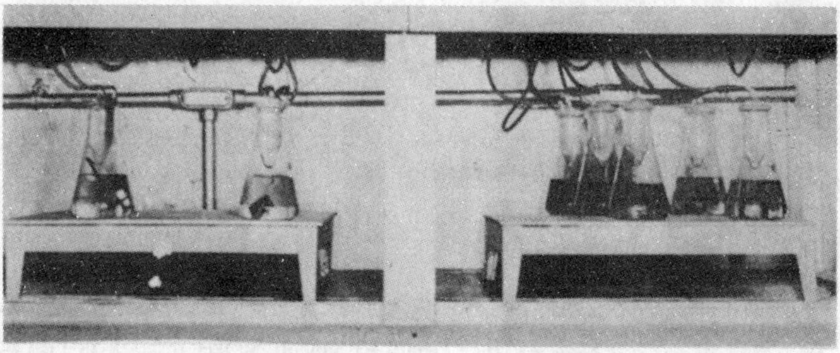

Figure 4-1 Laboratory tests in boiling solutions.

condenser. The main difference is that the acorn type introduces oxygen into the solution, whereas the Allihn does not.

Figure 4-2 shows an arrangement for a pilot-plant test. Glass tubing is used to cover the support rod and for the spacers. The sample in the glass cradle is hard, brittle, and could not be drilled. The specimens on the right are designed to determine the effects of contact with lead—the lead and stainless alloy specimens are held in contact by wrapping with lead wire. Figure 4-3 shows the bracket used to support a similar arrangement in a lead-lined tank in this pilot plant. The propeller on the end of the mixer rod, as shown in Fig. 4-3, is also a corrosion-test specimen. Figure 4-4 shows a specimen after a test in which it was badly attacked by the slurry involved. Figure 4-5 shows a spool-type specimen holder for tests in an actual operating plant

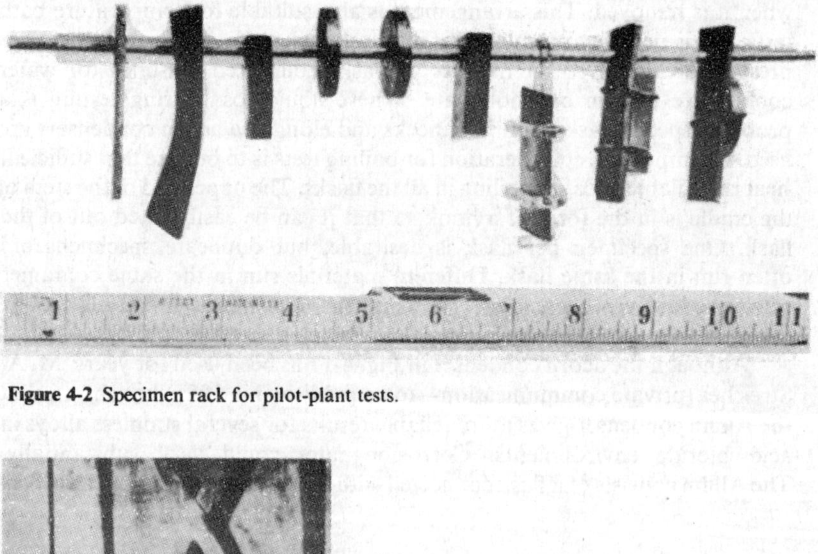

Figure 4-2 Specimen rack for pilot-plant tests.

Figure 4-3 Support bracket for rack in Fig. 4-2.

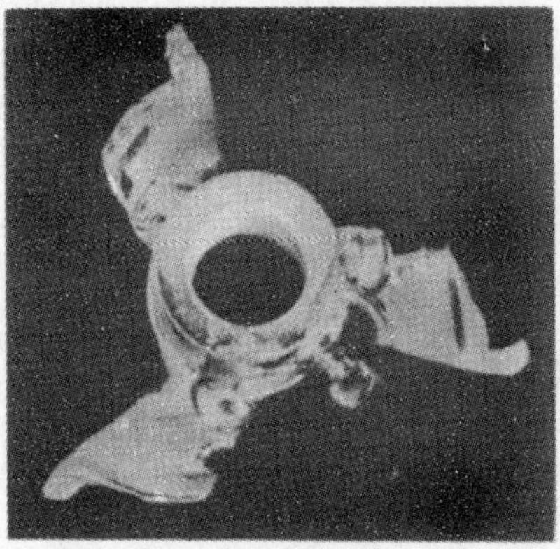

Figure 4-4 Propeller test specimen from mixer shown in Fig. 4-3.

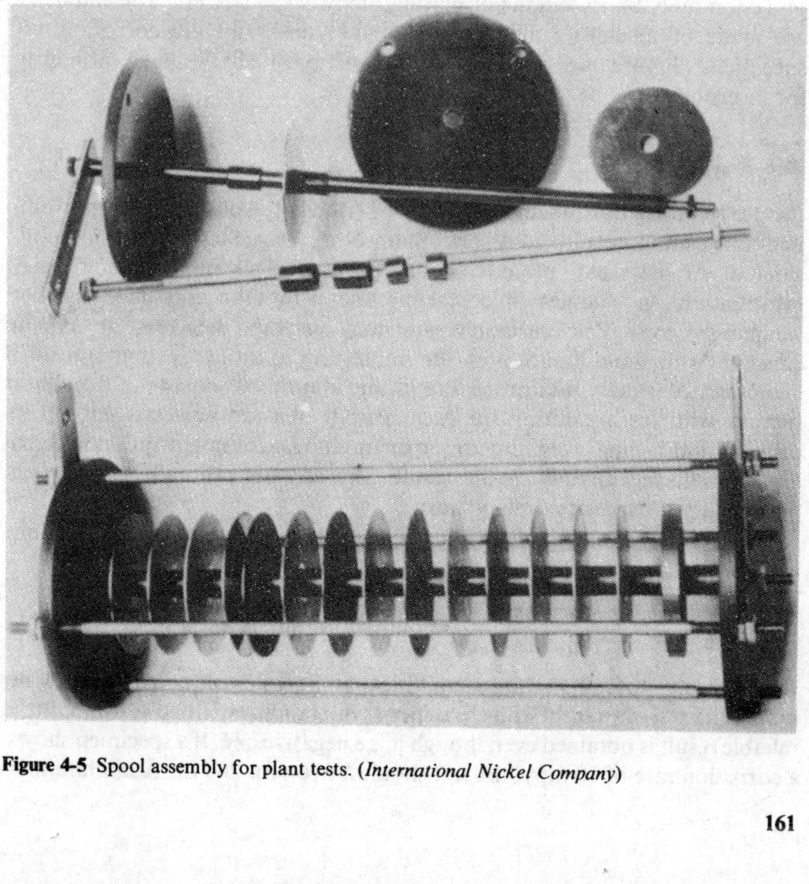

Figure 4-5 Spool assembly for plant tests. (*International Nickel Company*)

Figure 4-6 Test rack for insertion in pipeline.

or pilot plant. The metal specimen support rods are covered with a Bakelite or Teflon tube. Short lengths of plastic tubing act as spacers. The end disks are made of insulating material. The other disks are the corrosion-test specimens. Figure 4-6 shows a similar arrangement designed primarily for insertion in a pipe.

4-8 Duration

Proper selection of time and number of periods of exposure are important, and misleading results may be obtained if these factors are not considered. At least two periods should be used. This procedure provides information on changes in corrosion rate with time and may uncover weighing errors. The corrosion rate may increase, decrease, or remain constant with time. Quite often the initial rate of attack is high and then decreases. A widely used procedure in the laboratory consists of five 48-hr periods with fresh solution for each period. If a test consists only of an original and a final weighing, an error in either case might go undetected and be reflected directly in the result. The test time should be reported, particularly if exposure time is short.

A *very rough* rule for checking results with respect to minimum test time is the formula

$$\frac{2000}{\text{mils per year}} = \text{hours (duration of test)}$$

This formula is based on the general rule that the lower the corrosion rate the longer the test should be run. If a specimen completely dissolved in 2 hr, a reliable result is obtained even though it is a negative one. If a specimen shows a corrosion rate of 10 mpy, the test should be run for 200 hr. A minimum of two weeks and preferably one month is recommended for semiworks or plant tests.

Field tests such as exposure to atmospheric corrosion or soil corrosion usually involve very low rates of attack, and sometimes several years are needed to provide definitive results.

Wachter and Treseder* present an excellent procedure for evaluating the effect of time on corrosion of the metal and also on the corrosiveness of the

Table 4-1 Planned-interval test

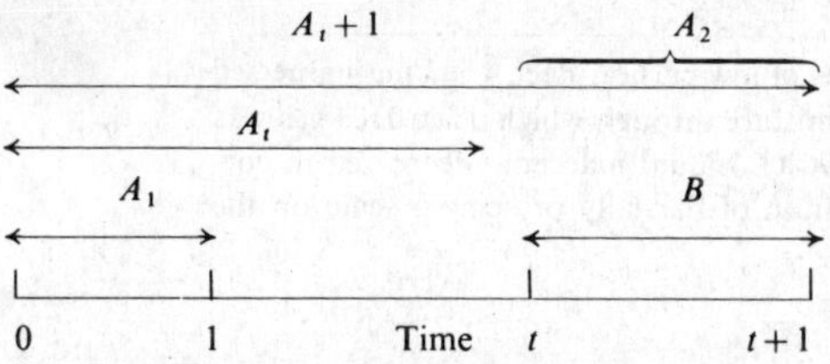

Identical specimens placed in same corrosive fluid; imposed conditions of test constant for entire time $(t+1)$; A_1, A_t, $A_t + 1$, B, represent corrosion damage experienced by each test specimen; A_2 is calculated by subtracting A_t from $A_t + 1$.

Occurrences during corrosion test

Liquid corrosiveness	Criteria	Metal corrodibility	Criteria
Unchanged	$A_1 = B$	Unchanged	$A_2 = B$
Decreased	$B < A_1$	Decreased	$A_2 < B$
Increased	$A_1 < B$	Increased	$B < A_2$

Combinations of situations

Liquid corrosiveness	Metal corrodibility	Criteria
1. Unchanged	Unchanged	$A_1 = A_2 = B$
2. Unchanged	Decreased	$A_2 < A_1 = B$
3. Unchanged	Increased	$A_1 = B < A_2$
4. Decreased	Unchanged	$A_2 = B < A_1$
5. Decreased	Decreased	$A_2 < B < A_1$
6. Decreased	Increased	$A_1 > B < A_2$
7. Increased	Unchanged	$A_1 < A_2 = B$
8. Increased	Decreased	$A_1 < B > A_2$
9. Increased	Increased	$A_1 < B < A_2$

*A. Wachter and R. S. Treseder, *Chem. Eng. Progr.*, **43**:315–326 (1947).

Table 4-2 Planned-interval corrosion test*

	Interval, days	Weight loss, mg	Penetration, mils	Apparent corrosion rate, mpy
A_1	0–1	1080	1.69	620
A_t	0–3	1430	2.24	270
A_t+1	0–4	1460	2.29	210
B	3–4	70	0.11	40
A_2	calc. 3–4	30	0.05	18
		$A_2 < B < A_1$		
		$0.05 < 0.11 < 1.69$		

*Conditions: Duplicate strips of low-carbon steel, $\frac{3}{4} \times 3$ in., immersed in 200 ml of 10% AlCl-90% SbCl misture through which dried HCl gas was slowly bubbled at atm pressure, 90°C. Liquid markedly decreased in corrosiveness during test, and formation of partially protective scale on the steel was indicated.

environment in laboratory tests. This plan is called the planned-interval test. Tables 4-1 and 4-2 and the following discussion are reproduced from the Wachter and Treseder paper.

4-9 Planned-Interval Tests

These tests involve not only the accumulated effects of corrosion at several times under a given set of conditions but also the initial rate of corrosion of fresh metal, the more or less instantaneous corrosion rate of metal after long exposure, and the initial corrosion rate of fresh metal during the same period of time as the latter. The rates, or damage in unit time interval, are referred to in the diagram of Table 4-1 as A_1, A_2, and B, respectively. Unit time interval may often be taken conveniently as one day in a planned-interval test extended over a total period of several days. It would be desirable to have duplicate specimens for each interval, and further time extensions of test could be made with similar added specimens and interval spacing.

Comparison for corrosion damage A_1 for the unit time interval from 0 to 1 with corrosion damage B for the unit time interval from t to $t+1$ shows the magnitude and direction of change in corrosiveness of the medium that may have occurred during the total time of the test. Comparison of A_2 with B, where A_2 is the corrosion damage calculated by subtracting A_t from A_t+1, correspondingly shows the magnitude and direction of change in corrodibility of the metal specimen during the test. These comparisons may be taken as criteria for the changes and are tabulated in Table 4-1. Also given in Table 4-1 are the criteria for all possible combinations of changes in corrosiveness of the medium and corrodibility of the metal. Additional information thus obtained on occurrences in the course of the test justifies the extra effort involved. An example of the data obtained from a planned-interval test is given in Table 4-2.

Causes for the changes in corrosion rate as a function of time are not given by the planned-interval test criteria. Corrosiveness of the liquid may decrease as a result of corrosion during the course of a test owing to reduction in concentration of the corrosive agent, to depletion of a corrosive contaminant, to formation of inhibiting products, or to other metal-catalyzed changes in the liquid. Corrosiveness of the liquid may increase owing to formation of autocatalytic products or to destruction of corrosion inhibiting substances, or to other catalyzed changes in the liquid. Changes in corrosiveness of the medium may arise also from changes in composition that would occur under the test conditions even in the absence of metal. To determine whether the latter effect occurs, an identical test is run without test strips for the total time t; then test strips are added and the test continued for unit time interval. Comparison with A_1 of corrosion damage from this test will show whether the corrosive character of the liquid changes significantly in the absence of metal.

Corrodibility of the metal in a test may decrease as a function of time owing to formation of protective scale or to removal of the less resistant surface layer of metal. Metal corrodibility may increase owing to formation of corrosion-accelerating scale or to removal of the more resistant surface layer of metal. Indications of the causes of changes in corrosion rate often may be obtained from close observation of tests and corroded specimens and from special supplementary tests designed to reveal effects that may be involved.

Changes in solution corrodibility are not a factor in most plant tests that consist of once-through runs or where large ratios of solution volume to specimen area are involved.

If the effect of corrosion on the mechanical properties of the metal or alloy is under consideration, a set of unexposed specimens is needed for comparison.

4-10 Aeration

Aeration, or the presence of dissolved oxygen in a liquid environment, may have a profound influence on the corrosion rates. Accordingly, this factor should be carefully considered in a corrosion-test program. Generally speaking, some metals and alloys are more rapidly attacked in the presence of oxygen, whereas others may show better corrosion resistance. One of the first methods of corrosion control consisted of deaeration of boiler water because of the marked effect of oxygen on the corrosion of steel and cast iron. Copper, brasses, bronzes, and other copper alloys such as Monel and nickel are also subject to increased attack in the presence of oxygen, particularly in acid solutions. These materials often show excellent resistance under neutral and reducing conditions. Aluminum and the stainless steels, on the other hand, sometimes show better corrosion resistance in the presence of air.

An excellent example concerns pumps and tubing made of an expensive alloy consisting of approximately two-thirds nickel and one-third molybdenum. These parts were in hydrochloric acid service where good corrosion resistance was expected. However, the parts failed rapidly because of an air sparger in the system.

Aeration effects sometimes are observed in connection with the water line or liquid line in a vessel. For example, corrosion may occur at this point if the liquid level is quite constant and the atmosphere over the liquid is air. If steel or copper specimens are completely immersed in water or a dilute acid, respectively, corrosion may be low. Under the same conditions except for semi-immersion of the specimens, substantial attack may occur.

It is quite difficult to exclude air completely from a plant operation. Pumps often introduce air into the liquid if the packing is not tight. Blanketing vessels with nitrogen is sometimes used to minimize solution of oxygen by the liquid involved.

The presence of dissolved oxygen in a process liquor may also result in corrosion at crevices, under deposits on the metal, or in other stagnant areas. In any case it is important to have some information on the degree of aeration expected and also on the effect of the solution on any oxygen that may be present. Many organic compounds, for example, may react with oxygen and thereby effectively remove it from solution as far as corrosion is concerned.

Perhaps the simplest and most widely used aeration test consists of merely bubbling air through the solution. The solution is then assumed to be saturated with air. In the same category except for deaeration, the test consists of bubbling nitrogen through solution. If saturation with respect to oxygen is desired, then pure oxygen gas should be used. If rather complete deaeration is required, then purified nitrogen or argon should be used. In most practical applications, air is involved so air is used in the test. If a gas is not bubbled through the solution, then the aeration effect depends on the air present in the solution at the start of the test and also on its rate of removal or escape, if any, or absorption from the atmosphere.

Common procedure is to introduce the gas through a porous Alundum thimble or a piece of porous brick. Cementing a glass tube to, say, a 1-in. cube of porous brick with silicate cement is a convenient device for most solutions. Direct impingement of the bubbles on the specimen is to be avoided. Sometimes a glass tube or chimney is placed around the porous brick to avoid impingement and also to cause some stirring or movement of the solution. A separate mechanical stirrer can be used also. A large flask or container is often used as a manifold for tubes leading to a number of containers in which the corrosion-test specimens are located. This procedure is generally suitable, but for precise work and better reproducibility, the gas to each test should be metered.

4-11 Cleaning Specimens After Exposure

This is one of the most important steps in corrosion testing, and proper procedures must be used. Before describing cleaning methods, the importance of examining the specimens prior to cleaning should be emphasized. In many cases, visual observation of the specimens on removal from test provides valuable information concerning the causes or mechanism of the corrosion involved. For example, deposits or encrustations may be the cause of pitting of the metal.

Change in weight of the specimen is most often used for calculation of the corrosion rate. Accordingly, complete or incomplete removal of corrosion products is directly reflected in the corrosion rate. Corrosion products can be classified as loose, or readily removable, and tight, or adherent, and also as protective and nonprotective. The protectiveness of the surface products can usually be determined by varying the length of exposure as described previously.

One common cleaning procedure consists of holding the specimen under a stream of tap water and vigorously scrubbing the surface with a rubber stopper. The assumption here is that corrosion products removed by this method would be removed in actual operation and those that did not come off would stay on during normal operation of most equipment such as tanks, lines, and valves. The rubber-stopper cleaning method has been found satisfactory for most corrosion tests in practical applications involving aqueous solutions and also for many other tests. If appreciable quantities of corrosion product adhere to the specimen, this method does not determine the true corrosion rate because all of the converted or corroded metal is not removed. If it is desirable to get down to bare or unaffected metal, then more drastic cleaning methods must be used. The danger is that uncorroded metal may also be removed and these methods should not be used when the corrosion rate is very low.

Cleaning methods may be classified as (1) mechanical, such as scraping, brushing, scrubbing with abrasives, sandblasting, and the rubber-stopper method described; (2) chemical, such as the use of chemicals* and solvents; and (3) electrolytic,* which involves making the specimen the cathode under an impressed current in a variety of chemical reagents with or without inhibitors added. Brushing or scraping is sometimes used to remove the products loosened by chemical or electrolytic methods. Blank determinations should be made to ascertain the amount of metal removed by the cleaning method itself. In any case the cleaning method should be stated when reporting results of the corrosion tests. In general, a chemical or electrolytic method is often specific for the metal or alloy under test. The main exception here involves solvents used to remove grease, oil, tar, and other organic matter.

*These methods are not commonly used but are helpful in (1) examining metal surface for pitting and other localized attack, (2) determining total metal "consumed" (i.e., weight loss in high-temperature oxidation), and (3) cleaning surfaces (i.e., restoration of art objects).

It should be emphasized that further cleaning is not necessary if the rubber-stopper treatment leaves fairly clean metal.

An electrolytic method consists of making the specimen the cathode under the following conditions:

Solution	Sulfuric acid, 5% by weight
Anode	Carbon
Cathode	Test specimen
Current density for cathode	20 A/dm^2 (1.3 A/in.^{2})
Inhibitor	Organic inhibitor (i.e., Rhodine), 2ml/liter of solution
Temperature	165°F
Exposure period	3 min

The blanks or weight losses for clean specimens as a result of this treatment are given in Table 4-3. This procedure is followed by an alkaline rinse and rubbing with a rubber stopper. Attack on metals such as zinc and magnesium is too high for the use of this procedure. A similar electrolytic method is used to descale steel specimens subjected to oxidation at high temperatures.

Another method for removing scale from ferrous alloys consists of making the specimen the cathode in a fused bath of 60% sodium hydroxide and 40% sodium carbonate at 400°C. A Monel beaker is satisfactory for the container and anode. A cathode current density of 1 A/in.^{2} and 5 min. are used. This specimen is then quenched in water and given the electrolytic

Table 4-3 Weight loss due to electrolytic cleaning

Material	Loss mg/in.^{2}
Aluminum 2S	0.10
Admiralty brass	0.013
Red brass	0.00
Yellow brass	0.026
5% tin bronze	0.00
Copper	0.013
Monel	0.00
Steel	0.051
18-8 stainless steel	0.00
Chemical lead	0.39
Nickel	0.14
Tin	0.014
Zinc	Too high

sulfuric acid treatment described above. Steel and stainless steel lose about 1 mg/in.^{2}.

Aluminum and its alloys Clean by dipping in concentrated (about 70%) nitric acid at room temperature for several minutes. Keep the time of immersion to a minimum because aluminum shows some attack by this acid. Scrub lightly in a stream of water with a rubber stopper or a bristle brush so as not to mechanically abrade these soft materials. Alternate dipping and scrubbing is recommended to hasten the cleaning.

Copper and its alloys Clean in 5 to 10% sulfuric acid or 15 to 20% hydrochloric acid at room temperature for several minutes and scrub with rubber stopper or bristle brush.

Iron and steel The electrolytic method described is often used. Chemical methods consist of treatment in warm 20% hydrochloric or sulfuric acid containing organic inhibitors. Another method consists of exposure to boiling 20% sodium hydroxide containing about 10% zinc dust. Intermittent brushing is usually helpful in most cases for all of these methods. In connection with specimens scaled at high temperatures, it is helpful to quench from temperature in order to crack or spall the scale. This is followed by electrolytic cleaning and brushing.

Stainless steels and alloys Electrolytic cleaning in aqueous solutions and fused salts has been described. The sodium hydroxide–zinc-dust treatment is also used. Hot nitric acid in concentrations to 70% are also utilized.

Lead and its alloys Clean by immersion in boiling 1% acetic acid or a hot solution of concentrated ammonium acetate for a few minutes; scrub very lightly.

Magnesium and its alloys Dip for 15 min in a boiling solution of 1% silver chromate and pure 15% chromic acid (CrO_3).

Nickel and its alloys Immerse in 15 to 20% hydrochloric acid or 10% sulfuric acid at room temperature.

Zinc and its alloys Dip in saturated ammonium acetate solution at room temperature and scrub lightly.

4-12 Temperature

Perhaps the most important single factor in corrosion is the effect of temperature. Accordingly, it is very important that the temperature of the surface of the specimen (which is the corroding temperature) be identified and known. Sometimes corrosion decreases with temperature—i.e. removal of dissolved oxygen in connection with copper alloys. In many cases corrosion increases rapidly with increase in temperature. An excellent example is the corrosion of 18-8S stainless steel by nitric acid, as shown in Table 4-4. Corrosion increases from a low rate to several thousand mils per year as a result of increasing the temperature about 100°F. Numerous cases show that increasing from room temperature to 150°F results in a two-, five-, or even tenfold increase in corrosion.

Laboratory tests are often made in controlled-temperature water or oil baths. Temperatures should be controlled to ±2°F or better. The bath should be large enough and the heaters properly spaced so that an even temperature distribution is obtained. Forced circulation of the bath liquid is usually not necessary. Commercially available temperature baths are fairly expensive. Inexpensive baths can be readily made using wooden tanks lined with sheet copper and fitted with heaters and temperature controllers.

Accelerated corrosion tests at temperatures above proposed operating temperatures are often made to decrease the time of testing. This is a dangerous procedure because the effect of temperature may be great, thus needlessly eliminating more economical materials.

A common error involves the assumption that the environment temperature is the corroding temperature. This is particularly true in the case of materials for heating surfaces. The average temperature of the liquid in a tank may be 150°F, but the corroding or surface temperature of the steam heating coil may be considerably higher. Tests at 150°F may, therefore, provide erroneous results. Surface temperatures can often be estimated by considering heat-transfer coefficients.

Figure 4-7 shows a recommended method of testing for heating tubes. The U-tube is made of the material under consideration and is heated by steam at the proposed operating pressure. This tube was made by welding

Table 4-4 Effect of temperature on corrosion of 18-8S by 65% nitric acid

Temperature, °F	Corrosion, mpy
Up to 250	Less than 20
260	100
280	200
320	500
330	1000
370	5000

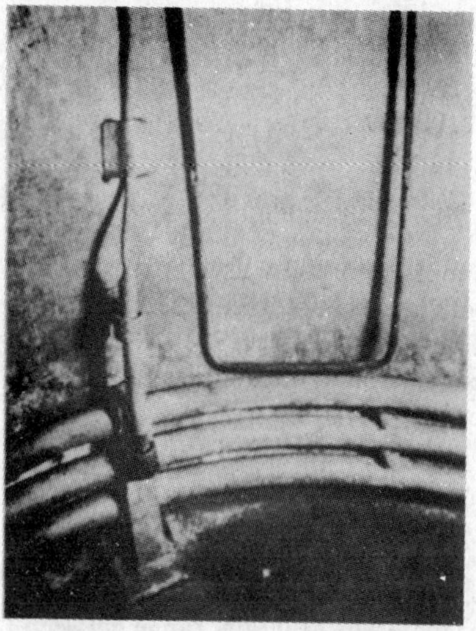

Figure 4-7 Heating-tube test in pilot plant.

several lengths of tubing because the actual installation would be similarly constructed. This test illustrated the importance of this factor also, because although the parent material showed good resistance, the welds were rapidly attacked. Another advantage of this type of test is that sometimes solids deposit on heating surfaces and cause rapid attack. Clipping a sheet specimen (like a clamp) on an existing heating coil or tube is another method for simulating actual temperatures.

4-13 Standard Expressions for Corrosion Rate

In most cases, aside from contamination problems, the primary concern is the life (usually life in years) of the equipment involved. A good corrosion rate expression should involve (1) familiar units, (2) easy calculation with minimum opportunity for error, (3) ready conversion to life in years, (4) penetration, and (5) whole numbers without cumbersome decimals.

I began promoting the expression mils per year in 1945, and it is now widely used. The formula for calculating this rate is

$$\text{Mils per year} = \frac{534W}{DAT}$$

as discussed in Chap. 2.

Conversion from other units to obtain mils per year is

Multiply	*By*
in./yr	1000
in./month	12,100
$mg/dm^2/day$ (mdd)	1.44/specific gravity

Mils penetration per year (mpy) is the most commonly used corrosion rate expression in the United States. It is popular because it expresses corrosion rate in terms of penetration using small integers (see Sec. 2-2). A substitute expression is required to facilitate the conversion to the metric system. Some equivalent metric penetration rates are:

$$1 \text{ mpy} = 0.0254 \frac{\text{mm}}{\text{yr}} = 25.4 \frac{\mu\text{m}}{\text{yr}} = 2.90 \frac{\text{nm}}{\text{hr}} = 0.805 \frac{\text{pm}}{\text{sec}}$$

where millimeter (mm) is 10^{-3} meter; micrometer or micron (μm) is 10^{-6} meter; nanometer (nm) is 10^{-9} meter; and picometer (pm) is 10^{-12} meter. The corrosion rates of usefully resistant materials generally range between 1 and 200 mpy. Table 4-5 compares mpy with the above metric expressions. Millimeters/year yields fractional numbers, while μm/yr generally produces large integers. The equivalent rates in terms of nm/hr are smaller integers. The most interesting metric expression is pm/sec, which yields numbers very close to mpy rates.

Before a substitute metric corrosion rate expression is selected, it is

Table 4-5 Comparison of mils penetration per year (mpy) with equivalent metric-rate expressions

		Approximate metric equivalent†			
Relative corrosion resistance*	mpy	$\frac{mm}{yr}$	$\frac{\mu m}{yr}$	$\frac{nm}{hr}$	$\frac{pm}{sec}$
Outstanding	<1	<0.02	<25	<2	<1
Excellent	1–5	0.02–0.1	25–100	2–10	1–5
Good	5–20	0.1–0.5	100–500	10–50	5–20
Fair	20–50	0.5–1	500–1000	50–150	20–50
Poor	50–200	1–5	1000–5000	150–500	50–200
Unacceptable	200+	5+	5000+	500+	200+

*Based on typical ferrous- and nickel-based alloys. For more expensive alloys, rates greater than 5 to 20 mpy are usually excessive. Rates above 200 mpy are sometimes acceptable for cheap materials with thick cross sections (e.g., cast-iron pump body).

†Approximate values to simplify ranges.

important to consider how metric measurements are used to describe dimensions. Centimeters, millimeters, and micrometers (microns) are employed for small lengths. Generally, the units are chosen to avoid decimals. For example, a 6-in. pipe with a $\frac{1}{4}$-in. wall thickness is described as a 10-cm pipe with a 6-mm wall. The changeover usually occurs at the 0.1 level. Thus, dimensions less than 0.1 mm are converted to micrometers (e.g., 100 μm = 0.1 mm).

Although nm/hr and especially pm/sec (Table 4-5) provide conveniently small whole numbers, they require additional conversion before they can be related to the typical dimensions of engineering structures. Micrometers/yr and mm/yr appear to be the most useful expressions; the former for low corrosion rates and the latter for rapid rates. After some experience with the metric system, engineers will probably use both rates interchangeably without difficulty (e.g., "steel corrodes at 2 millimeters per year as compared to 75 micrometers for type 316 stainless"). Possibly, in time, mpy will refer to mm penetration per year and μmpy to micrometers per year.

The above two expressions are easily calculated from weight-loss data by equations similar to that described in Sec. 2-2.

$$\frac{\mu\text{m}}{\text{yr}} = 87{,}600 \frac{W}{DAT}$$

$$\frac{\text{mm}}{\text{yr}} = 87.6 \frac{W}{DAT}$$

where W is the weight loss in milligrams, D is the density in g/cm^3, A is the area in cm^2, and T is the time of exposure in hours.

Increasingly, corrosion rates determined by linear polarization and other electrochemical techniques (Secs. 10-8, and 10-9) are expressed in terms of current density. These expressions can be converted into penetration rates by the following expression based on Faraday's law:

$$\text{Corrosion penetration rate} = K \frac{ai}{nD}$$

where a is the atomic weight of the metal; i is the current density in μa/cm^2; n is the number of electrons lost (valence change); D is the density in g/cm^3; and K is a constant depending on the penetration rate desired. For mpy, μm/yr, and mm/yr, the values of K are 0.129, 3.27, and 0.00327, respectively. The parameter n is determined by analysis of the solution or by measuring the potential and pH and referring to appropriate thermodynamic data.* The atomic weight a and n values for alloys can be calculated from weighted averages of their components. Consider iron corroding in air-free acid at an electrochemical corrosion rate of 1 μA/Cm^2. It dissolves as ferrous ions

*Potential-pH diagrams.

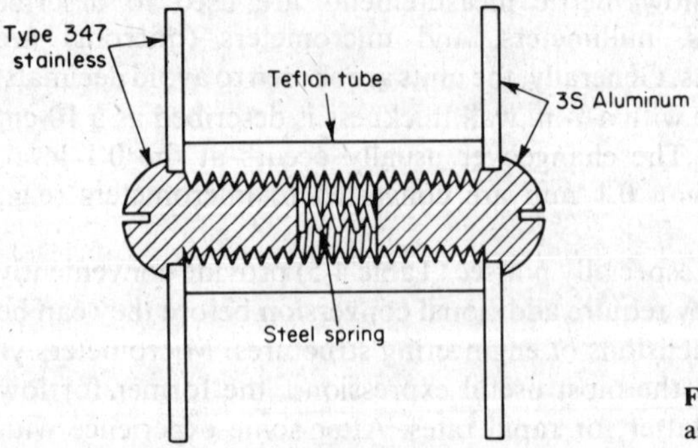

Figure 4-8 Galvanic test couple assembly.

(Fe^{2+}), and thus, $n = 2$. The corrosion rate in mpy is obtained by substitution:

$$\text{Penetration rate} = 0.129 \frac{(55.8)(1)}{(2)(7.86)} = 0.46 \text{ mpy}$$

4-14 Galvanic Corrosion

A method for conducting *galvanic* or *two-metal corrosion* tests is shown in Fig. 4-2. Figure 4-8 illustrates another method that minimizes crevice effects.

In connection with the "area" effect in two-metal corrosion, a simple and inexpensive procedure consisting of a large brass sheet with a steel nut bolted to the center of the sheet was used in a study of brass piping and a small steel valve in cooling water.

4-15 High Temperatures and Pressures

Autoclaves are often used for testing at high temperatures and pressures. Figure 4-9 shows a bracket and stress-corrosion specimen used in the autoclave illustrated in Fig. 4-10. Chloride-containing waters are placed in the autoclave. Condensation from the top of the autoclave drips on to the specimen to give what is termed "vapor-phase condensation conditions." This is an excellent test for stress corrosion, correlates well with actual service problems, and is used to evaluate materials.

Figure 4-11 shows a Teflon-lined autoclave used at Ohio State University in work for the Alloy Casting Institute in connection with corrosion by various chemicals at temperatures above the boiling point. Teflon protects the autoclave from the severely aggressive environments involved.

Figure 4-12 shows an arrangement for inserting and removing a plant test specimen, in this case an oil refinery unit, at high temperatures and pressures. The corrosion test specimen is the short length of tubing near

Figure 4-9 Stress-corrosion specimen and supporting bracket for autoclave test (see Fig. 4-10).

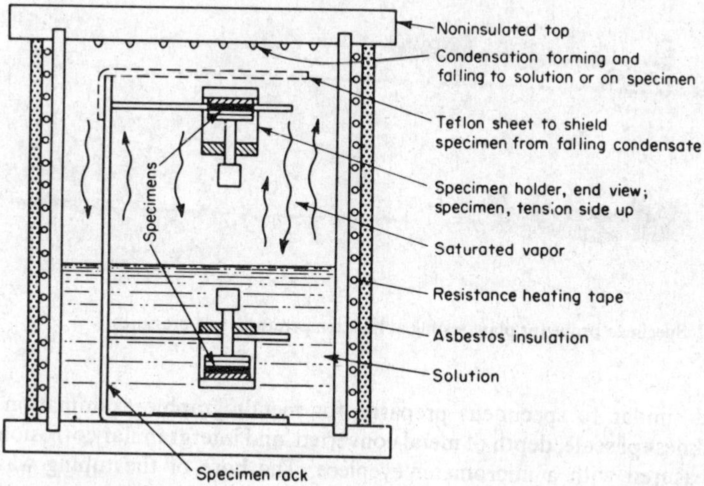

Figure 4-10 Schematic cross section of autoclave for stress-corrosion tests in chloride-containing waters at high temperatures.

the end of the rod. A valve connected to the main system is not shown. The assembly shown is connected to the valve. For insertion, the valve is opened and the rod pushed in. For removal, the valve is closed and the small amount of liquid in the assembly shown is slowly bled out.

In connection with pilot-plant testing of a refinery hydroforming process, lengths of $\frac{1}{8}$-in.-ID high-pressure tubing were welded in place. After exposure, the tubing was cut (destructive test) and the specimens

Figure 4-11 Teflon-lined autoclave for testing at high temperatures.

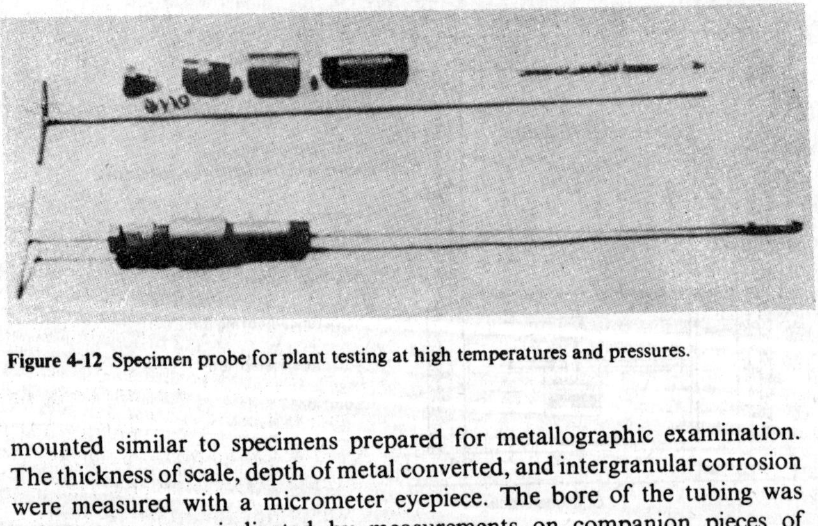

Figure 4-12 Specimen probe for plant testing at high temperatures and pressures.

mounted similar to specimens prepared for metallographic examination. The thickness of scale, depth of metal converted, and intergranular corrosion were measured with a micrometer eyepiece. The bore of the tubing was quite accurate as indicated by measurements on companion pieces of tubing.

Circulating loops are used for testing in fluids at high temperatures and pressures. For example, at Ohio State, a loop capable of 700°F and 7000 lb/in.^{2} is utilized. See Sec. 8-12 for testing in liquid metals and fused salts.

4-16 Erosion Corrosion

Figure 4-13 is a photograph of equipment used in laboratory tests. Details are drawn in Fig. 4-14. Figure 4-15 is an exploded view of the nonmetallic housing that contains the spinning-disk-test specimen. The liquid or slurry

Figure 4-13 Equipment for laboratory erosion-corrosion tests.

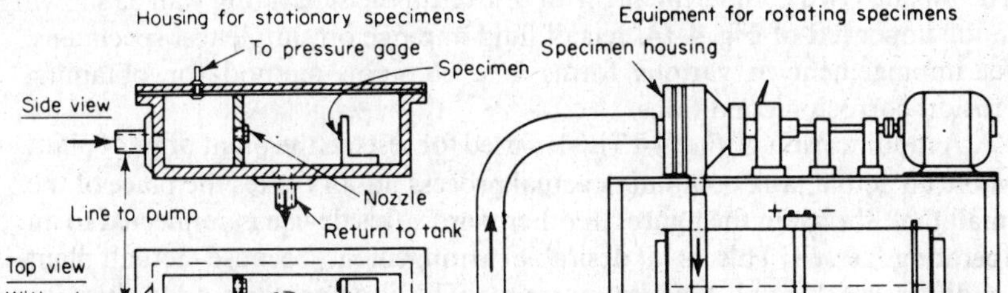

Figure 4-14 Details of equipment shown in Fig. 4-13.

Figure 4-15 Exploded view of housing in Fig. 4-14 showing spinning disk and renewable wear plate.

in the 30-gal glass-lined tank is recirculated and pumped directly into the face of the rotating-disk specimen. Specimens are also hung in the tank for comparison between low velocity and the high velocity to which the disk is subjected. This is an open tank, and therefore it is not suited for liquids involving toxic vapors or highly volatile liquids. This apparatus yielded much of the data presented in Chap. 3.

A box interchangeable with the rotating unit above the tank is also used. The box has two compartments with orifices in the separating wall as shown in the upper left of Fig. 4-14. Jets of fluid impinge on simple test specimens. Jet impingement in various forms is a common method for obtaining erosion-corrosion conditions.

A modification of Fig. 4-13 is also used for tests in the plant or pilot plant where an actual tank containing actual process liquors takes the place of the small tank shown in the figure. In other words, this device is connected to an operating vessel. This is a desirable arrangement because actual plant conditions exist for all practical purposes. This is an excellent procedure for evaluating pumps and valves at little cost.

Figure 4-16 shows a closed all-metal system that was developed primarily for testing in fuming nitric acid but is suitable for most other liquids. Test specimens are hung in the tank and also placed in a bracket so that the return from the pump impinges on the specimen in this bracket. Other test "specimens" in this system are the pump itself, lines, valves, elbows, tees, and an orifice plate that is also used to control the rate of flow. Galvanic or two-metal corrosion can also be tested here by joining pipes or flanges of different metals. The data on fuming nitric acid presented in Figs. 3-43 and 3-44 were produced by this equipment.

A number of other recirculating systems are used to study erosion corrosion under a great variety of conditions including high temperatures

Figure 4-16 All-metal closed system for erosion-corrosion testing.

and high pressures. Some of these are called *loop tests*, where the loop system operates under conditions duplicating closely those anticipated in actual operations. All-glass closed systems that permit atmosphere control and control of gas content of solutions are used also at Ohio State University.

The LaQue Center for Corrosion Technology, Inc., has an extensive testing station at Wrightsville Beach, N.C., where actual seawater is pumped for long periods of time through a variety of actual heat exchangers, valves, piping, and other equipment. (See Sec. 4-32 for photographs and description.) The effects of velocity are studied at various rates of flow. LCCT also uses flat 5-in. disks attached to the ends of rotating spindles. The disks are rotated at 1140 rpm in a wooden tank so arranged that fresh seawater enters one side and flows out another through an overflow pipe at a rate of approximately 3 gal/min. The specimens are weighed and also measured with a micrometer to determine depth of penetration. Many other means of rotating specimens are often used.

The Detroit Edison Co. uses units illustrated in Fig. 4-17 for determining the erosion-corrosion resistance of metals and alloys to high-velocity boiler feed-water at temperatures up to about 400°F.* The water impinges on the face of the plain disk specimen and escapes at right angles through the slot in the face of the lower disk. A typical test uses a differential pressure of 300 lb/in.^{2} across the specimen and an exposure time of 500 hr. The disk specimen can be made by welding different materials together, and thus galvanic corrosion effects may be studied. Interestingly, tests in this unit check the results obtained by the apparatus shown in Fig. 4-13 in that both show greater corrosion by water when the pH is raised from 7 to 8 (Fig. 3-42).

The Materials Technology Institute sponsored a project in the Fontana Corrosion Center at The Ohio State University. The work was done by D. D. Macdonald, A. Moccari, and T. Y. Chen and their report is titled "Development of Controlled Hydrodynamic Techniques for Corrosion

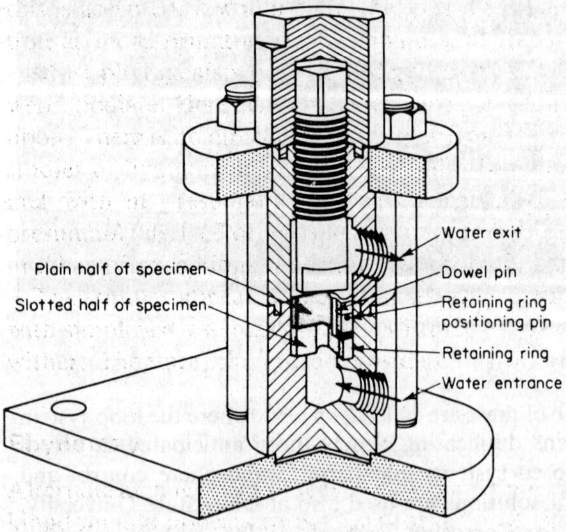

Figure 4-17 Sectional view of test unit for boiler feedwater.

*H. A. Wagner, J. M. Decker, and J. C. Marsh, Corrosion-Erosion of Boiler Feed Pumps and Regulating Valves, *Trans. ASME*, **69**:389–97 (1946).

Testing." This detailed study was on the effect of *hydrodynamic* variables (flow rate and geometry), oxygen concentration, and solution pH on the corrosion of 90-10 Cu-Ni alloy in 1 *M* sodium chloride solution at 25°C. Three hydrodynamic systems were investigated: a flow-through pipe system, an annular flow system, and a rotating-cylinder system. Flow rates spanned the range from laminar flow to highly turbulent conditions, and all hydrodynamic states were carefully characterized in terms of the Reynolds (Re) and Schmidt (Sc) numbers. Corrosion rates were measured using small-amplitude cyclic voltammetry (SACV), ac impedance techniques (see Sec. 4-27 and 4-28 in this chapter), and direct weight loss determinations. Corrosion potentials were also measured as a function of time, flow velocity, and solution composition (oxygen concentration, pH).

The three hydrodynamic systems developed in this work have been shown to be effective test systems for the evaluation of corrosion phenomena under carefully controlled hydrodynamic conditions. However, because of its compactness and simplicity, the rotating-cylinder system is believed to be the most effective and convenient test method for use in the laboratory and is recommended for routine corrosion testing and materials evaluation. The work reported provides the important link between the hydrodynamic and corrosion characteristics of the rotating-cylinder system and geometries frequently encountered in the field, most notably circular pipes.

Comparison of the performances of the ac impedance and SACV techniques for monitoring corrosion rates *in situ* has shown that the latter provides more consistent and accurate values for the polarization resistance. It is interesting to note that theoretically both small perturbation techniques should yield identical data. However, the SACV technique has proved to be superior to the ac impedance technique for the measurement of polarization resistance, and it is recommended that this technique be used in the in-plant environment, at least for monitoring the rate of corrosion of copper-nickel alloys in sodium chloride–containing solutions.

A through discussion of this study would be too lengthy and complicated for inclusion here. The interested reader is referred to the MTI report to be published in 1985 and also Poulson* and Syrett and Macdonald.† These relatively new corrosion-testing techniques will doubtless be used more extensively in the future.

4-17 Crevice Corrosion

Many methods are used for studying crevice, or "gasket," corrosion.

Figure 4-18 shows a plant test rack for studying gasket corrosion.‡ The gasket test materials are held between bolted strips of type 316 stainless steel.

*B. Poulson, Electrochemical Measurements in Flowing Solutions, *Corrosion Sc.*, **27**: 391–430 (1983).

†B. C. Syrett and D. D. Macdonald, *Corrosion*, **35**:505 (1979).

‡E. V. Kunkel, *Corrosion*, **10**:260–266 (1954).

Figure 4-18 Rack for evaluating gasket materials.

Duplicate gasket specimens are used, and they are separated by a third strip between the two outer strips. The gasket specimens are $2\frac{1}{2}$ in. long and extend about $\frac{1}{4}$ in. on each side. These tests were evaluated by estimating the percentage of the area attacked, measuring the depth of the pits, and by noting the condition of the gasket and its ease of removal after test.

Another test method uses two disks held together by a bolt and nut of the same material. The facing surfaces of the two disks are machined with a slight taper starting from a center flat about $\frac{1}{2}$ in. in diameter. This results in a tight center joint and a fine crevice near the center that increases in width as the periphery of the disk is approached. This arrangement is sometimes modified by use of flat strips instead of disks.

Another method consists of laying on a horizontal metal specimen a small pile of sand, some sludge, a piece of asbestos or gasket material, a piece of rubber, or any other material to be studied. This combination is placed in the corrosive under study. If sludges or slurries are involved, a specimen is sometimes placed vertically in a container with sludge settled on the bottom and clear liquid over it. One end of the specimen is thereby buried in the sludge. Another method consists of merely wrapping the specimen with string, cord, or rubber bands.

Another type of test specimen consists of two rods, one containing a male and the other a female thread. These are screwed together with or without gaskets or spacers. A different arrangement consists of stringing specimens on a rod and using spacers of different materials or sizes.

A method developed at Ohio State involves a test specimen that is readily prepared and gives good results. A $\frac{1}{2} \times 4$ in. strip is bent downwards on a large radius—just enough to obtain a slight curvature. It is then bent 180° back on itself in the opposite direction over a $\frac{1}{2}$- to 1-in.-diameter rod. The resulting "hairpin" is then squeezed together so the two ends are in firm contact. The original downward bend causes the tips to spring back slightly, thus forming a crevice.

Loosely rolling a tube into a hole in a metal plate produces a good test specimen because this arrangement duplicates cases wherein crevice corrosion often occurs—in shell and tube heat exchangers, for example.

A new multiple-crevice corrosion test technique that permits statistical

Figure 4-19 Grooved Delrin washer.

comparisons of alloys has been developed by D. B. Anderson.* Two grooved, Delrin washers (Fig. 4-19) are pressed on each side of a sheet specimen by compressing a threaded rod. This design offers 40 uniform crevices (20 per washer), which can be used to provide statistical data (Fig. 14-20). As shown in this graph, there is a 100% probability that crevice attack will occur on CA-15 alloy and a 52% chance the attacked region will exceed 50 mils (1.27 mm) in depth. In contrast, the probability of crevice attack of CF-8M is 25%, with only 1% of the attacked region being greater than 50 mils deep. This simple test method offers a quantitative way to compare the crevice corrosion tendencies of alloys. It is important to remember

*D. B. Anderson "Statistical Aspects of Crevice Corrosion in Seawater," *Galvanic and Pitting Corrosion—Field and Laboratory Studies*, Spec. Tech. Publ. 576, pp. 231–242, American Society for Testing and Materials, Philadelphia, 1976.

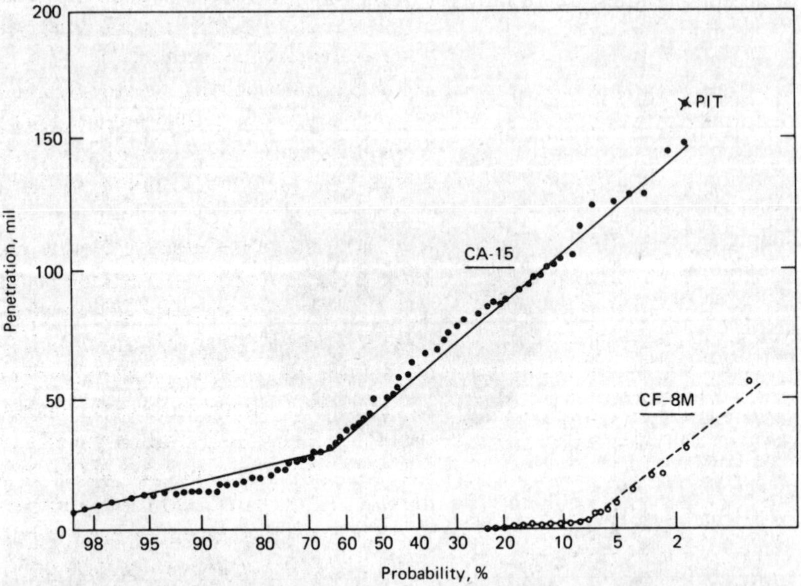

Figure 4-20 Probability of crevice corrosion for cast stainless steels. Exposure conditions: 2 ft/sec, 32 days, 95°F (15°C), 75:1 exposed/shielded area ratio. [*D. B. Anderson, Statistical Aspects of Crevice Corrosion in Seawater*, "*Galvanic and Pitting Corrosion—Field and Laboratory Studies,*" *Spec. Tech. Publ. 576:231–242, American Society for Testing and Materials, Philadelphia*, (1976)].

that these data are time-dependent; longer exposures increase both the probability and depth of attack.

4-18 Intergranular Corrosion

Any corrosion test can be considered as a test for intergranular corrosion because the specimens should be examined for localized attack such as intergranular and pitting. However, there are several tests designed primarily to show susceptibility to intergranular corrosion. The classic example is the nitric acid test for stainless steels. This test is used primarily to check heat treatment of these steels.

Tests on tensile specimens and bend tests are made to determine loss of mechanical properties because of intergranular attack.

4-19 Huey Test for Stainless Steels

This test (ASTM A-262) consists of exposure to boiling 65% nitric acid for five 48-hr periods. Sensitized material exhibits high corrosion rates. Examples of acceptable rates are 18 mpy for quench-annealed type 304, 30 mpy for CF-8, and 24 mpy for type 304L (after heating for 1 hour at 1250°F). Material is rejected if the corrosion rate is increasing rapidly in the later periods.

Controversy exists concerning the use of Huey tests to predict corrosion performance in other environments. However, this acceptance test serves a useful purpose in that it attempts to ensure the use of material for construction of actual equipment that is similar to the material tested in the given environment—whatever it may be. In addition, one who pays for stainless steel should have equipment constructed of steel that is in the condition of optimum corrosion resistance.

Maleic acid, lactic acid, copper sulfate–sulfuric acid, and ferric sulfate–sulfuric acid are also used to determine susceptibility to intergranular corrosion of stainless steels.

4-20 Streicher Test for Stainless Steels

The Huey test is time-consuming and relatively expensive. M. A. Streicher of the Du Pont Co. developed an oxalic acid screening test (ASTM A-262-55T) that quickly indicates sensitized material and thereby reduces by some 90% the number of Huey tests required. Borderline cases are tested in nitric acid.

The Streicher test consists of polishing a small specimen, or part of a specimen, through No. 000 emery paper, etching in 10% oxalic acid for 1.5 min under an applied current density of 1 A/cm^2, and then examining the surface at 250 to 500 magnification. The specimen is the anode and a stainless steel beaker is used as a cathode. A "step" structure indicates properly heat-treated material, while a "ditch" structure indicates intergranular susceptibility. Figure 4-21 shows a step structure and Fig. 4-22 a ditch structure for type 304 stainless steel. This test is applicable to both cast and wrought alloys.

In *addition* to the oxalic test, Streicher presents a detailed discussion of many other tests for stainless steels in a special technical publication of ASTM (STP-656, 84 pages, 1978). The title is *Theory and Application of Evaluation Tests for Detecting Susceptibility of Intergranular Attack in*

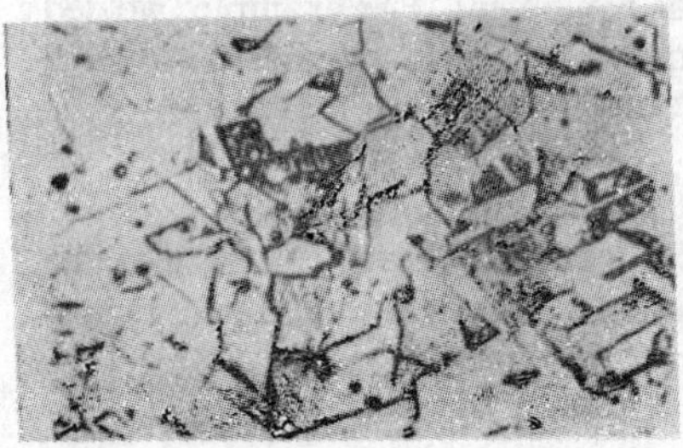

Figure 4-21 Step structure from electrolytic-oxalic acid etch. Quench-annealed type 304. Nitric acid rate, 7 mpy (400 ×).

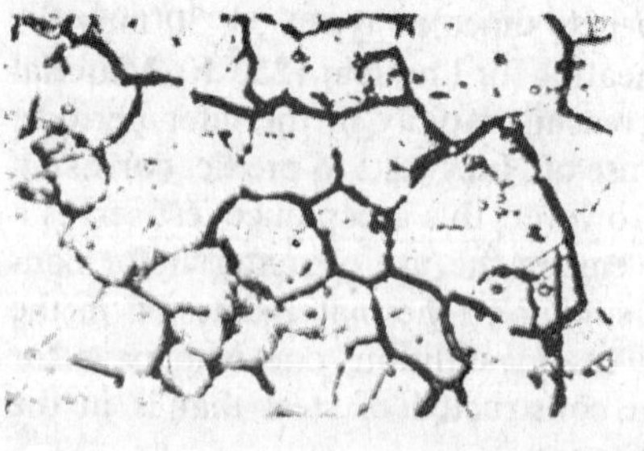

Figure 4-22 Ditch structure of sensitized type 304. Nitric acid rate, 60 mpy (400 ×).

Stainless Steels and Related Alloys—Problems and Opportunities. It includes discussion of the oxalic test, ferric sulfate–sulfuric acid test, nitric-hydrofluoric test, copper sulfate–16% sulfuric acid test, copper sulfate–50% sulfuric acid test, and new standard test methods for ferritic stainless steels. Also discussed are the effect of sensitization on SCC, pitting, and crevice corrosion.

4-21 Warren Test

Type 316L stainless steel poses a special problem in connection with the use of the Huey test for acceptance purposes. When sensitized for 1 hr at 1250°F, type 316L may form a sigma phase, particularly if ferrite is present in the quench-annealed structure. Sigma-containing material shows high corrosion rates and fails the Huey test even though chromium carbide precipitation is not present.

Warren* of Du Pont proposes the use of a 10% nitric acid–3% hydrofluoric acid mixture at 70 or 80°C to evaluate type 316L. Two 2-hr periods or five $\frac{1}{2}$-hr periods are used. This environment severely attacks material with precipitated chromium carbides but not with sigma phase. If the ratio of the corrosion rate of "sensitized" material (1 hr at 1250°F) to that of quench-annealed is less than 1.5, the steel is acceptable as far as chromium carbide precipitation is concerned. Sigma phase is not considered deleterious with regard to corrosion in most environments. However, the 316L situation illustrates the point that test specimens, for evaluation in a plant process, should be in the condition of the actual equipment. For example, welded specimens and also sensitized specimens of 316L should be tested, particularly if heavy welded sections are involved.

4-22 Pitting

The random (probability) and extremely localized character and also the long incubation times often required result in a difficult situation with regard

*D. Warren, *Corrosion*, **15**:213–220*t* (Apr. 1959).

to testing for pit resistance. Weight loss determinations are practically worthless. Loss-in-strength tests are of no help. Edge effects on specimens tend to confuse the issue. The best procedure is to expose specimens as large as possible (at least a few square inches) for as long a time as possible (at least a month) for the given problem. The area effect (ratio of cathodic to anodic area) is not very important for stainless steels based on our work, which shows that even small cathodic areas can support an active pit.

Several methods are used to report pitting results including maximum pit depth, number of pits per unit area, average depth, and also a complete statistical analysis. From the practical standpoint, maximum pit depth is important, regardless of number, because "time for the first leak" is most important. Many failures are due to one single hole in the entire structure. The picture is further confused by the fact that occasionally a pit stops growing, contrary to the usual situation.

The best approach is to avoid use of materials that show *any* tendency towards pitting during the corrosion tests.

4-23 Stress Corrosion

Figure 4-9 shows a simple loaded-beam specimen widely used for stress corrosion. Figure 4-10 shows an autoclave for high-temperature tests. Stress corrosion test specimens are stressed in a variety of ways, such as constant load, constant strain (constant-strain-rate test, or CSRT), and slow strain rate (slow-strain-rate test, or SSRT). The latter is probably the best test method for evaluating materials by means of laboratory tests. The U-bend represents a constant-strain test.

Specimens are usually stressed by two methods: (1) constant strain or deformation (stretched or bent to a fixed position at the start of the test) and (2) constant load. One criticism of the former is that the stress may be reduced during the test because of plastic deformation or relaxation and also because stress is relieved when a crack forms on the surface.

A wide variety of sizes and shapes of specimens and methods of stressing are used. Some are simple, but others require rather complicated apparatus. The most important factor is that the specimen must have a tensile stress, and it is desirable to know accurately the stress in the specimens. For simple beams and applied loads these stresses can be readily calculated from handbook formulas, but it is desirable to calibrate the setup through determination of the stresses by means of wire strain gages or other methods. In the case of stresses due to cold-working or welding, calculation or determination of the stress pattern is difficult, but this is not often necessary, particularly when a specific application is involved.

A common shape of specimen is the U-bend, hairpin, or horseshoe type. In this case, a bolt is placed through holes in the legs of the specimen and it is loaded by tightening a nut on the bolt, as illustrated in Fig. 4-23. The outer

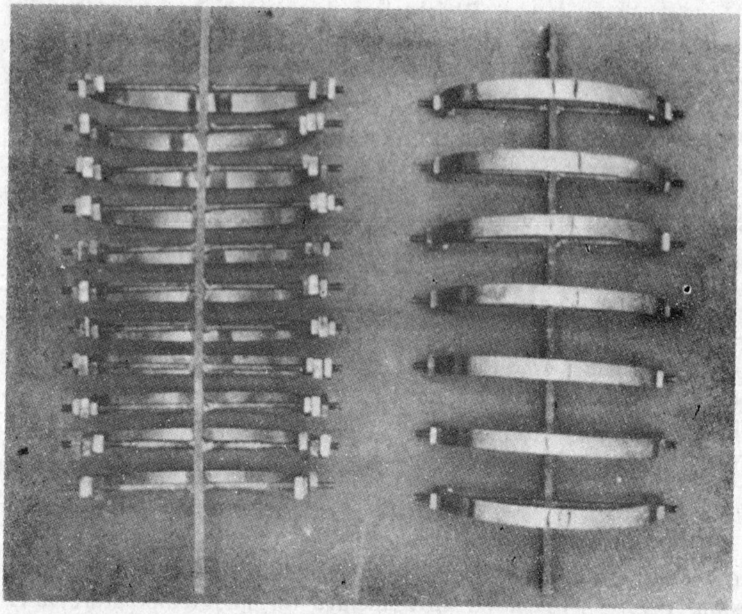

Figure 4-23 Stress-corrosion rack of welded specimens for plant test.

fibers (top of the bend) have the highest stress, with stresses decreasing as one moves away from the top point of the bend, where cracking usually occurs. The underside of the bend is in compression and the metal cannot crack or be pulled apart here. One disadvantage of this type of specimen is that the specimen should be stress-relieved after forming. A common error made in this type of test (and also others) is use of a different metal for the bolt. This often results in galvanic or two-metal corrosion, which may obscure the desired results. This objection can be overcome by insulating the bolt and nut from the specimen, using the same material for the bolt and nut, or immersing only the bent part of the specimen in the solution. Galvanic corrosion must be considered in any type of stress-corrosion test. The surface stress of flat strip-type specimens (Fig. 4-9) can be calculated from Euler's approximate long-column formula, which can be reduced to:

$$S = \frac{\pi^2 Ehy}{2L^2}$$

where S = extreme fiber stress, lb/in.^{2}
E = Young's modulus, lb/in.^{2}
L = length of specimen, in.
h = thickness of specimen, in.
y = deflection at center, in.

The ends and corners of the "snap-in" specimens should be rounded to prevent restraint.

Another type uses a strip specimen with a center fulcrum and the ends pulled down by bolts.

One of the criticisms of the specimens described above is that the highest stress is along a line at right angles to the specimen and at the top of the bend. In other words, only a very small part of the specimen is under the most severe condition of test, and probability becomes a consideration. This objection can be overcome by loading the specimen at two points to obtain uniform stress on the specimen between these two points.

Other devices for constant-load tests consist of suspended dead weights on the specimen, with or without levers, as in an ordinary tension test, and spring loading. In many cases, the specimen is identical or similar to specimens used to determine the tensile strength of metals, and they are stressed by pulling longitudinally (no bending). This could be done by a machine, springs, or jigs (frame where the specimen is loaded by tightening a nut—exactly like a bolt is tightened).

The stresses applied in stress-corrosion tests are usually high. Values from 50 to 100% of the yield strength are used. Sometimes precracked (by fatigue) specimens are used to avoid the crack-initiation or incubation period. When quantitative stress values are not required, plastically deformed specimens (i.e., bent strip or cupped sheet) are used. The specimen is exposed and then examined for cracks.

Corrosion-fatigue tests are conducted similarly to ordinary fatigue tests except that the specimen is in a corrosive environment. The specimen could be a rotating beam (eccentrically loaded) or a strip or round that is held fixed at one end and bent back and forth from the other end.

4-24 NACE Test Methods

Table 4-6 lists some of the standard corrosion test methods promulgated by the National Association of Corrosion Engineers. TM-01-77 is for metals and alloys for service in "sour" (hydrogen sulfide) environments. Figure 4-24 shows a test unit and Fig. 4-25 shows multiple units undergoing tests. One advantage of this test is the small space required as compared to other methods such as dead-weight loading.

Standard TM-01-77 considers 30 days of exposure sufficient to indicate susceptible alloys. If no failure occurs in this time, the applied stress is considered to be below the threshold stress. A threaded tensile specimen is attached within a steel proof ring. The ring is forced to deflect elastically a given amount. The ring tends to return to its original position, but the specimen prevents this and thus bears the load. The specimen is enclosed in a chamber containing a solution of 5% sodium chloride and 0.5% acetic acid. Hydrogen sulfide is continuously bubbled through the solution.

Table 4-6 NACE standard corrosion test methods*

TM-01-69	Laboratory Corrosion Testing of Metals for the Process Industries
TM-01-70	Visual Standard for Surfaces of New Steel Airblast Cleaned with Sand Abrasive
TM-02-70	Method of conducting Controlled Velocity Laboratory Corrosion Tests
TM-01-71	Autoclave Corrosion Testing of Metals in High Temperature Water
TM-01-72	Antirust Properties of Petroleum Products Pipeline Cargoes
TM-01-73	Methods for Determining Water Quality for Subsurface Injection Using Membrane Filters
TM-01-74	Laboratory Methods for the Evaluation of Protective Coatings Used as Lining Materials in Immersion Service
TM-02-74	Dynamic Corrosion Testing of Metals in High Temperature Water
TM-03-74	Laboratory Screening Tests to Determine the Ability of Scale Inhibitors tc Prevent the Precipitation of Calcium Sulfate and Calcium Carbonate from Solution
TM-01-75	Visual Standard for Surfaces of New Steel Centrifugally Blast Cleaned with Steel Grit and Shot
TM-02-75	Performance Testing of Sucker Rods by the Mixed String, Alternate Rod Method
TM-03-75	Abrasion Resistance Testing of Thin Film Baked Coatings and Linings Using the Falling Sand Method
TM-01-77	Testing of Metals for Resistance to Sulfide Stress Cracking at Ambient Temperatures
TM-01-84	Accelerated Test Procedures for Screening Atmospheric Surface Coating Systems for Offshore Platforms and Equipment

*Request current information on price from National Association of Corrosion Engineers, P.O. Box 218340, Houston, Tex. 77218.

Figure 4-24 Test cell according to TM-01-77. (*Courtesy Cortest, Inc.*)

Figure 4-25 Multiple units undergoing tests. (*Courtesy The Duriron Co. Inc.*)

Tests by the Duriron Company show that alloys CD4MCu (see Table 5-2) and Durcomet 101 (modified CD4MCu) are recommended for sour services according to TM-01-77.

4-25 Slow-Strain-Rate Test

In the SSRT a tensile specimen is subjected to a steadily increasing stress in a given environment. This procedure results in rupture of surface films and thus tends to eliminate initiation time required for a surface crack to form (which may be a long time in some cases). Figure 4-26 shows equipment for a SSRT.

A criticism of the CSRT is that the stress may be reduced during the test because of plastic deformation or relaxation and also because stress is relieved when a crack forms on the surface. In a constant-load test, when a crack grows, stress is increased because of the decreased cross-sectional area of the specimen.

MTI Manual No. 15, *Guidelines for Preventing Stress Corrosion Cracking in the Chemical Process Industry*, describes in great detail the above and a multitude of other tests, simple and complicated, such as a straining electrode under potentiostatic control.

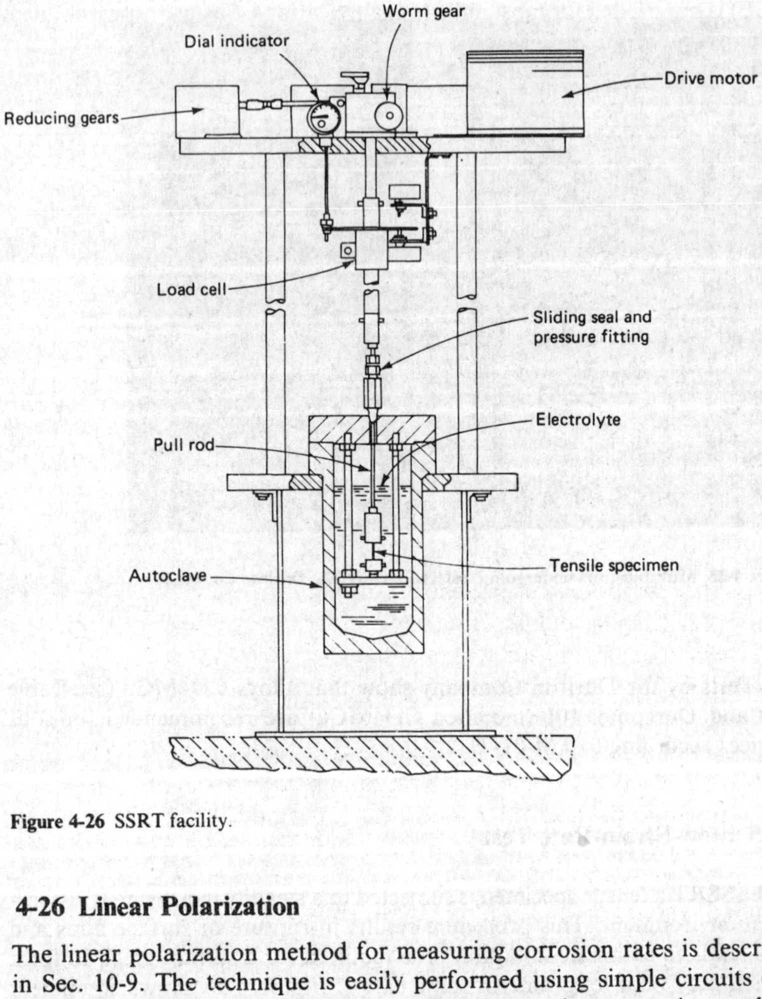

Figure 4-26 SSRT facility.

4-26 Linear Polarization

The linear polarization method for measuring corrosion rates is described in Sec. 10-9. The technique is easily performed using simple circuits (viz. Fig. 10-20) and can be used to continuously monitor corrosion under a variety of environmental conditions. As a result, there have been numerous studies of this method and its applications since the late 1960s. There are now a variety of commercial instruments available ranging from portable field-test units to automatic units capable of monitoring and recording corrosion rates at 10 or more points in a process stream. Numerous successful applications have been reported in petroleum production, chemical process plants, seawater, and selection of materials for surgical implants (see Sec. 4-30).

The technique has been simplified. In early studies it was not recognized that a conventional reference electrode such as a saturated calomel electrode

(SCE) is not needed for most industrial applications. Since the measurement requires only a few minutes to perform, the potential of a corroding metal is sufficiently stable during this period to act as a reference. Also, the counter or auxiliary electrode does not have to be a noble metal as usually employed in precision laboratory studies. Consequently, it is possible to fabricate the three electrodes (working, counter, and reference) from the same material, and assemble these into a compact test probe. Figure 4-27 shows a versatile plug probe, containing three concentric ring electrodes of steel that can be simply screwed into pipelines or process vessels. Another suggested simplification is the use of a two-rather than three-electrode assembly.* In this modification the reference electrode is eliminated and polarization is performed in both anodic and cathodic directions and corrosion rates are calculated by averaging methods. Although the two-electrode method has proved useful in some applications, it tends to be less accurate than the three-electrode technique. Since this modification is not simpler and is less accurate than the three-electrode method, it is not recommended for general use.

A review by Mansfeld† provides an excellent survey of the theory, historical evolution, and applications of linear-polarization methods.

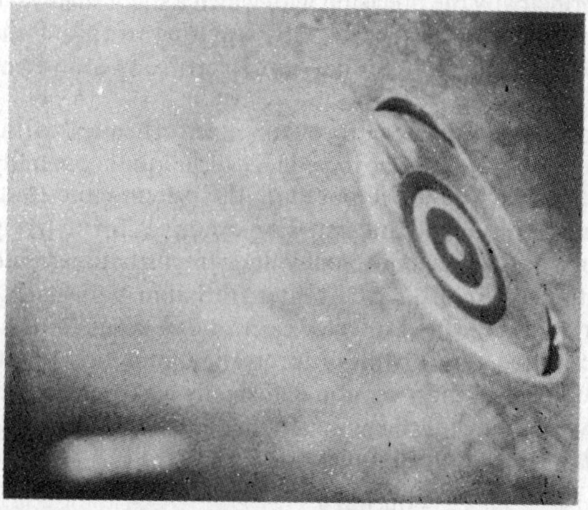

Figure 4-27 Flush-mounted three-electrode threaded probe for linear-polarization measurements within pipelines and vessels. *(Courtesy, E. C. French, Petrolite Corporation, Strafford, Tex.)*

*See R. Bandy and D. A. Jones, *Corrosion*, **32**:125 (1976).

†F. Mansfeld, The Polarization Resistance Technique for Measuring Corrosion Currents, in M. G. Fontana and R. W. Staehle, eds., *Advances in Corrosion Science and Technology*, vol. 6, pp. 163–262, Plenum Press, New York, 1976.

Mansfeld and co-workers have stimulated renewed interest in linear polarization by their emphasis on the inherent errors in the technique. Linear-polarization calculations involve approximation that can cause errors of up to 50%. However, these large errors only occur under special environmental conditions. Also, such errors are difficult to detect since the reproducibility of weight loss corrosion tests are rarely better than 20 to 50% even under ideal conditions.

Linear-polarization measurements should be performed by measuring a number of data points within 5 millivolts of the corrosion potential to ensure maximum accuracy. Tests performed in low-conductivity electrolytes can yield very erroneous results if the electrolyte resistance is not considered. *Linear-polarization results should always be compared with weight loss or other corrosion rate measurements* to ensure the accuracy of the technique and its suitability for a particular medium.*

4-27 AC Impedance Methods

These methods have frequently been applied to corroding systems in an effort to determine the mechanisms of the corrosion process, specifically the nature of the rate determining step, or to measure the corrosion resistance, R_{corr}, which is inversely related to the corrosion rate. The advantages of the ac impedance method are principally the precision with which a small sine-wave perturbation can be measured in an electrically noisy environment,† and the powerful methods that are available for the analysis of electrically equivalent circuits.

The sequence of operations necessary to obtain electrochemical parameters of a corroding electrode using the ac impedance technique is generally as follows. A sinusoidal perturbation is imposed on the system under test, centered around the dc condition of interest. For example, for a freely corroding electrode we may impose a sinusoidal voltage perturbation about the free corrosion potential, or a sinusoidal current perturbation with a mean of zero amperes, and for a cathodically protected electrode we would impose a voltage perturbation centered on the cathodic protection potential. Whatever perturbation is employed, it must be of a sufficiently small amplitude that the response of the system under test is a linear function of the amplitude of the input function. That is, the transfer function

$$T[j\omega] = \frac{\text{output}}{\text{input}}$$

must be independent of input function amplitude.

*See R. LeRoy, *Corrosion*, **29**:272 (1973) and **31**:173 (1975).

†For example, using an analog or digital phase-sensitive detector, under optimum conditions a 10-mV ac signal can be recovered from 100 V (80 dB) of asynchronous noise with 1% precision.

The next step is to determine the transfer function as a function of frequency, over as wide a range of frequencies as is practical. In general, 10^{-3} to 10^4 Hz is a suitable range, but for a rapidly corroding system a higher low-frequency limit may be necessary. If the input function is a sinusoidal current perturbation, then the transfer function is the system impedance

$$Z[j\omega]=\frac{V[j\omega]}{I[j\omega]}=R+jX$$

Conversely, if the input is a voltage perturbation, then the transfer function is the admittance

$$Y[j\omega]=\frac{I[j\omega]}{V[j\omega]}=A+jB$$

The complex operator, $j=\sqrt{-1}$, indicates that these are complex quantities, which may be expressed as a real and imaginary component, or as magnitude ($|Z|$) and phase (ϕ), at each frequency. For the impedance

$$|Z|=(R^2+X^2)^{1/2}$$

the phase angle is given by

$$\phi=\tan^{-1}(X/R)$$

The measured impedance or admittance spectra are next used in an attempt to postulate an electric circuit that has the same spectrum as the system under test. This "equivalent circuit" may contain any or all of the following impedance elements.

A *series resistance*, R_u, associated with the resistance of the electrolyte path between the tip of the reference electrode or Lugin probe and the corroding electrode,. and the resistance of any electrical connections to the corroding electrode. At the limit of high frequency, only this term remains in the impedance spectrum.

The *double-layer capacitance*, C_{dl}. This term is voltage-dependent and thus instrinsically nonlinear. In aqueous media this capacitance is typically a few tens of microfarads per cm^2, and the measured value can be used to estimate the effective area of a corroding electrode.

A *diffusional impedance* associated with the diffusion of the cathodic reactant (e.g., O_2 or H^+) to the cathodic sites of the corroding electrode. For diffusion across a stratified layer of thickness δ,

$$Z_d=\frac{RT}{nF(i_L-i)}\frac{\tanh(\sqrt{ju})}{\sqrt{ju}}$$

where i_L is the limiting current for diffusion, i is the steady-state current density and u is defined as

$$u=\frac{\omega\delta^2}{D}$$

For high frequencies or large values of δ, the diffusional impedance reduces to the form

$$Z_d=\sigma(1-j)\omega^{-1/2}$$

where σ is defined as

$$\sigma=\frac{RT}{n^2F^2A'c\sqrt{2D}}$$

The unknown parameters in this expression are the effective area A', the concentration c of diffusing species, and its diffusion coefficient D. If any two of these are known, then the third can be estimated from a measured value of Z_d as a function of frequency. At the limit of low frequency the diffusional impedance reduces to the form

$$Z_d^{\omega\to 0}=R_d=\sigma\delta\sqrt{2/D}$$

Thus, from a knowledge of σ (determined at higher frequencies) one can estimate the thickness of the diffusion layer. More importantly, from the relative magnitudes of R_d and the corrosion resistance one can determine if the corrosion process occurs under diffusion control. The *corrosion resistance*. This term is reserved to account for charge transfer in the anodic and cathodic reactions. Since *each* of these processes may have an anodic and cathodic component, the faradic current at a corroding electrode is the sum of four exponential (Tafel) terms. If the equilibrium potentials for the two processes are well separated, however, it is conventional to neglect the smaller terms (e.g., the partial current due to redeposition of dissolved metal), and define the faradic corrosion current as

$$I_f=I_a+I_c=I_{\text{corr}}\left[\exp\left(\frac{V_{\text{app}}}{\beta_a}\right)-\exp\left(\frac{V_{\text{app}}}{\beta_c}\right)\right]$$

where the applied voltage, $V_{\text{app}}=V-V_{\text{fc}}-V_{\text{ac}}$, and β_a and β_c are the Tafel constants for the anodic and cathodic processes, respectively. This expression is clearly nonlinear with respect to the amplitude of a sinusoidal perturbation, V_{ac}. Expressing the exponential terms as a Taylor series expression in V_{app}, we obtain

$$I_p=2.303I_{\text{corr}}\sum_{n-1}^{\infty}V_{\text{app}}^n\frac{1/\beta_a^n-1/(-\beta_c)^n}{n!}$$

Linearization is achieved only when the second and subsequent terms are negligible compared with the first term (in V_{app}^1). Provided this condition is met,* then

$$I_f = 2.303 I_{corr} V_{app} \left[\frac{1}{\beta_a} + \frac{1}{\beta_c} \right]$$

and we can define the corrosion resistance as follows:

$$R_{corr} = \frac{V_{app}}{I_f} = \frac{\beta_a \beta_c}{2.303 I_{corr}(\beta_a + \beta_c)}$$

The corrosion resistance appears in series with the diffusional impedance, and both are in parallel with the double-layer capacitance. Provided an accurate data set is obtained over a wide range of frequencies, it is generally possible to deconvolve R_{corr} from the remainder of the equivalent circuit by taking account of the frequency ranges in which the various terms contribute most to the measured impedance and mathematically extracting impedance elements from the measured data set. For example, at the highest frequencies ($f \geqslant 3$ kHz) only the uncompensated electrolyte resistance contributes to the measured impedance, and this can be measured and extracted. Intermediate frequencies $10 \leqslant f \leqslant 10^2$ Hz are dominated by the parallel combination of R_{corr} and C_{dl}, and a region of linear variation of the imaginary component of the impedance with $1/f$ can be used to determine C_{dl}. At low frequencies the impedance is dominated by diffusion processes.

In this manner, R_{corr} can be determined with some precision and can be used qualitatively to determine the general trend of corrosion rates. If quantitative information is desired, it is necessary to measure β_a and β_c by a separate method, for example by a slow-sweep-rate experiment in the vicinity of the free corrosion potential. This imposes a severe limitation on the impedance method when used to determine corrosion rates, since the Tafel coefficients may change during the course of an experiment, and it may not be convenient or possible to use the polarization method repetitively to monitor β_a and β_c. One method of obviating this limitation is to use the intrinsic nonlinearity of R_{corr} to determine theTafel coefficients.† By applying a somewhat larger sinusoidal perturbation in the range 20 to 50 mV; and measuring the harmonic components‡ in addition to the fundamental, sufficient information is obtained to measure β_a, β_c, and V_{fc}, in addition to R_{corr}.

*D. D. Macdonald and M. C. H. McKubre, "Electrochemical Impedance Techniques in Corrosion," in Florian Mansfeld and Ugo Bertocci, eds., *Electrochemical Corrosion Testing*, ASTM STP 727. American Society for Testing and Materials, Philadelphia, pp. 110–149, 1981.

†M. C. H. McKubre, "The Electrochemical Measurements of Corrosion Rates in Cathodically Protected Systems," Report CS-2858 to the Electric Power Research Institute, Palo Alto, Calif., 1983.

‡For an input perturbation at frequency f, the output is measured at frequency f (the fundamental), and at $2f$, $3f$, $4f$, etc.

The reader is referred to recent reviews for a more extensive discussion of ac impedance methods.*

4-28 Small-Amplitude Cycle Voltammetry

An alternate means of measuring the corrosion or polarization resistance is to use small-amplitude cyclic voltammetry (SACV). In this technique a small-amplitude (e.g., 10-mv) triangular voltage (Fig. 4-28) is imposed across the corroding interface using a potentiostat and function generator. The resulting current is then plotted against the voltage using an X-Y recorder to yield a Lissajous figure of the form shown in Fig. 4-29. The

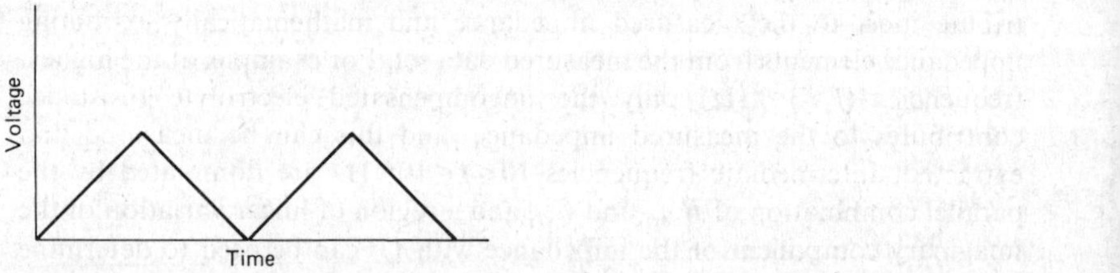

Figure 4-28 Voltage vs. time input used in small-amplitude cyclic voltammetry.

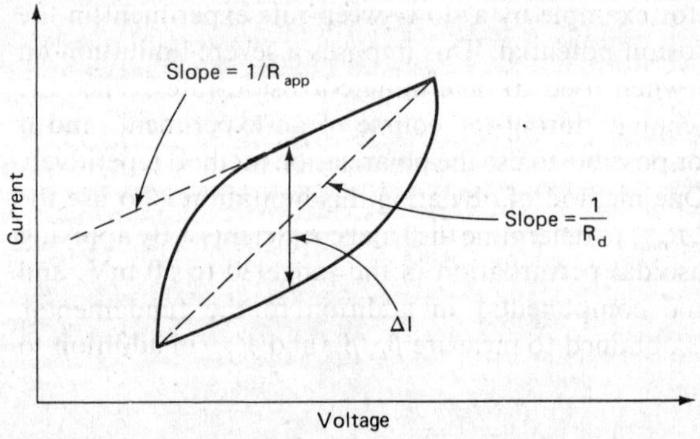

Figure 4-29 Schematic representation of Lissajous figure generated by SACV illustrating the diagonal and apparent resistances.

*C. Garbrielli, "Identification of Electrochemical Processes by Frequency Response Analysis," monograph commissioned by the Solartorn Research Group, Hampshire, England, 1980. See also D. D. Macdonald and M. C. H. McKubre, "Impedance Measurements in Electrochemical Systems," in J. O'M Bockris, B. E. Conway, and R. E. White, eds., *Modern Aspects of Electrochemistry*, vol. 14, Plenum Press, New York, 1982.

measurement is repeated at different voltage sweep rates and the apparent (R_{app}) and diagonal (R_d) resistances are plotted against the voltage sweep rate. These plots are then extrapolated to zero sweep rate to determine the true polarization resistance.

A mathematical analysis of the SACV technique has been published by Macdonald.* Assuming that the electrical properties of a corroding interface can be represented by the electrical equivalent circuit shown in Fig. 4-30, Macdonald was able to calculate the current response to a triangular voltage excitation using Laplace transforms. This analysis yields the following expressions for the apparent and diagonal resistances:

$$\frac{1}{R_{\text{app}}}=\frac{1}{R_p+R_s}+\frac{2R_p}{R_s(R_s+R_p)(1+e^{aE_m/vb})}$$

$$\frac{1}{R_d}=\frac{1}{R_p+R_s}+\frac{2R_p^2C(e^{aE_m/vb}-1)v}{E_m(R_p+R_s)^2(e^{aE_m/vb}+1)}$$

In these equations, $a=R_p+R_s$ and $b=R_sR_pC$, where R_p is the polarization resistance, R_s is the solution resistance between the corroding interface and the reference electrode, and C is the interfacial capacitance. These equations show that both R_{app} and R_d are functions of the voltage sweep rate, but that at $v=0$ both quantities are equal to R_p+R_s. The numerical simulations shown in Fig. 4-31 demonstrate that $1/R_{\text{app}}$ and $1/R_d$ are highly sensitive to voltage sweep rate at low values of v. This is precisely the region over which SACV is carried out, and the data plotted in Fig. 4-31 indicate that experimental values for $1/R_{\text{app}}$ and $1/R_d$ are required at very low sweep rates if accurate polarization resistances are to be obtained.

The question as to how low the voltage sweep rate must be in order that the apparent and diagonal resistance be negligibly different from R_s+R_p is best answered by noting that as $v \to 0$, $I \to 0$, and $R_{app} \to R_d \to R_s+R_p$ (Fig. 4-29). Thus, a suitable criterion is that the sweep rate should be sufficiently small that the ratio $\Delta IR_d/E_m$ is less than some critical value (e.g., 0.01). In practice, sweep rates as low as 0.1 mV/sec frequently are required to obtain reliable values for the polarization resistance (see example Fig. 4-31).

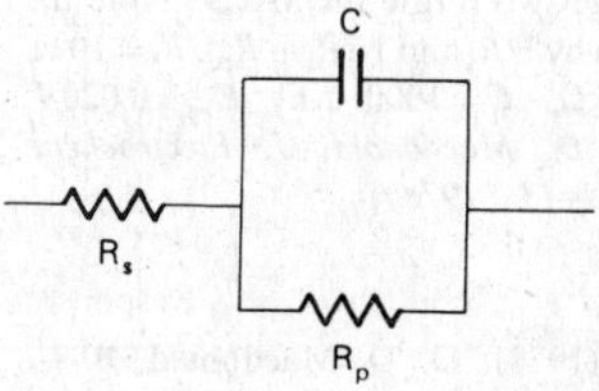

Figure 4-30 Simple electrical equivalent circuit for a corroding interface.

*D. D. Macdonald, *J. Electrochem. Soc.*, **125**:1443 (1978).

SACV has been applied extensively to the measurement of the polarization resistance of copper-nickel alloys in flowing seawater.* Typical voltammograms for 90:10 Cu:Ni are shown in Fig. 4-32 and plots of $1/R_{app}$ and $1/R_d$ as a function of voltage sweep rate are illustrated in Fig. 4-33. Both quantities extrapolate to a common value at $v=0$, which is identified with $1/(R_p+R_s) \approx 1/R_p$ since $R_p \gg R_s$. In this particular case it is evident that the lowest voltage sweep rate employed (0.1 mV/sec) is sufficiently small that reliable polarization resistances could be obtained without the necessity of extrapolating the reciprocal resistances to zero sweep rate. However, had the minimum voltage sweep rate employed been 10 mV/sec, the error in R_p would clearly have been significant (see Fig. 4-33).

4-29 Electronic Instrumentation

The revolution brought about by the introduction of electronic instrumentation into electrochemistry has also occurred in corrosion science. The most important instrument in this regard is the potentiostat, since it permits precise control over the driving force for charge transfer processes. However, other instruments such as galvanostats, zero-resistance ammeters, phase-sensitive detectors, and high-impedance electrometers have also contributed to modern corrosion science. Additional instruments include operational amplifiers, buffer amplifiers, inverting and noninverting

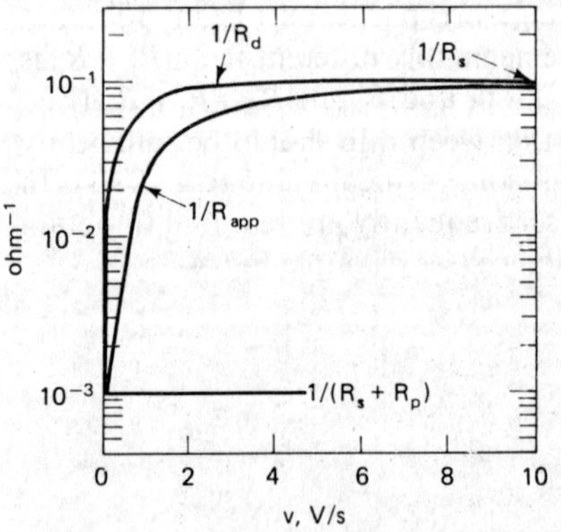

Figure 4-31 Variation of $1/R_{app}$ and $1/R_d$ with voltage sweep rate for SACV. Note the limits given by $1/R_s$ and $1/(R_s+R_p)$. $R_s = 10\,\Omega$, $R_p = 1000\ \Omega$, $C = 1000\ \mu F$, $E_m = 0.020V$. [*After D. D. Macdonald, J. Electrochem. Soc., 125:1443 (1978)*]

*D. D. Macdonald, *J. Electrochem. Soc.*, **125**:1443 (1978); D. D. Macdonald, B. C. Syrett, and S. S. Wing, *Corrosion* **34**:289 (1978); B. C. Syrett and D. D. Macdonald, *Corrosion*, **35**:505 (1979); D. D. Macdonald, B. C. Syrett, and S. S. Wing, *Corrosion* **35**:367 (1979); D. D. Macdonald, A. Moccari, and T. Y. Chen, "Development of Controlled Hydrodynamic Techniques for Corrosion Testing," Materials Technology Institute, 1985.

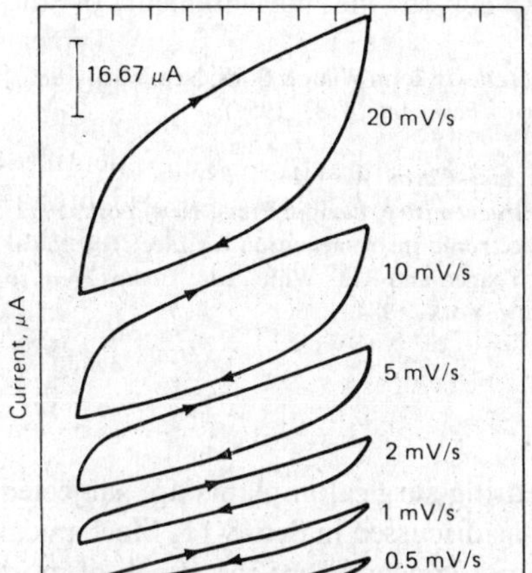

Figure 4-32 Small-amplitude cyclic voltammograms for 90.10 Cu:Ni alloy in flowing seawater. Flow velocity = 1.62 m/sec; $[O_2] = 0.045$ mg/liter; specimen area = 11.05 cm^2; $T = 26°C$; exposure time = 50 hr. [*After D. D. Macdonald, J. Electrochem Soc., 125: 1443(1978)*]

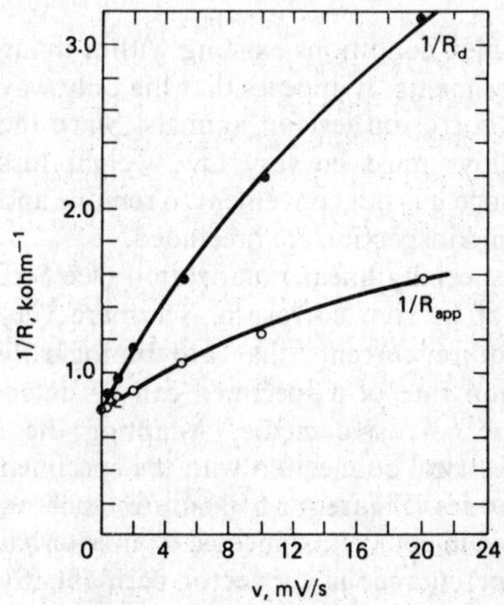

Figure 4-33 Plots of $1/R_d$ and $1/R_{app}$ for 90:10 Cu:Ni alloy in flowing seawater. Experimental conditions as listed in Fig. 4-31. [*After D. D. Macdonald, J. Electrochem. Soc., 125:1443 (1978)*]

amplifiers, voltage adders, integrators, current follower (voltage clamp), digital techniques, computer interfacers. Detailed descriptions and operating procedures are beyond the scope of this book. Interested readers are referred to the references listed below. Dr. Digby D. Macdonald of SRI International, Menlo Park, California, is perhaps the best expert in this field.

Suggested Readings

Bard, A. J., and L. R. Faulkner: *Electrochemical Methods*, John Wiley & Sons, New York (1980).
Bewick, A., M. Fleischmann, and M. Liler: *Electrochem. Acta*, **1**:83 (1959).
Britz, D.: *J. Electroanal. Chem.*, **88**:309 (1978).
Brown, E. R., D. E. Smith, and G. L. Booman: *Anal. Chem.*: **40**:1411 (1968).
Macdonald, D. D.: *Transient Techniques in Electrochemistry*, Plenum Press, New York, 1977.
McKubre, M. C. H., and D. D. Macdonald: Electronic Instrumentation for Electrochemical Studies, J. O. M. Bockis, B. E. Conway, E. Yeager, and R. E. White, eds., *Comp. Treatise of Electrochem*, Vol. 8, Plenum Press, New York, 1984.
Smith, D. E.: *Electroanal. Chem.*, **1**:1 (1966).

4-30 In Vivo Corrosion

Orthopedic devices and other load-bearing surgical implants are subjected to the corrosive action of body fluids, as discussed in Sec. 8-11. Mechanical damage due to corrosion is not the major concern since the attack of most alloys is relatively slow. Minute amounts of corrosion products released into the body may pose serious biological hazards. For example, nickel compounds are very carcinogenic, and there is evidence that gold and other noble metals may reduce local resistance to infection and retard wound healing. These effects are related to both the nature of the metallic compounds produced and their release rate (viz. the corrosion rate). Thus, there is a need for accurate measurement of the in vivo corrosion rates of various metals and alloys.

It is difficult to simulate the complex conditions existing within living organisms via laboratory in vitro experiments. It appears that the only way to ensure accurate data is by in vivo corrosion tests in animals. Since the corrosion rates of useful implant alloys must be very low, weight loss measurements cannot be used. Also, since it is not convenient to remove and replace specimens, periodic weighing and inspection are precluded.

Electrochemical measurements, especially linear polarization (see Sec. 10-9), are ideally suited for studies of in vivo corrosion. They are very rapid and sensitive and do not require currents that are biologically damaging. The instantaneous corrosion rate of a specimen can be determined without removing it from the corrosive media. Adapting these methods to in vivo studies requires electrical connection with the specimen and with counter and reference electrodes. Figure 4-34 illustrates such an arrangement. Here, a specimen mounted in epoxy is connected to an external power supply (a potentiostat) via a percutaneous connector permanently sutured at the base of a rat's head. A small platinum wire in the underside of the connector serves as a counter electrode, while the reference electrode is connected via an anal catheter. During measurements, the rat is anesthetized, the circuit is connected, and appropriate electrochemical data is obtained. This method permits long-term (two months or more) corrosion tests to be performed.

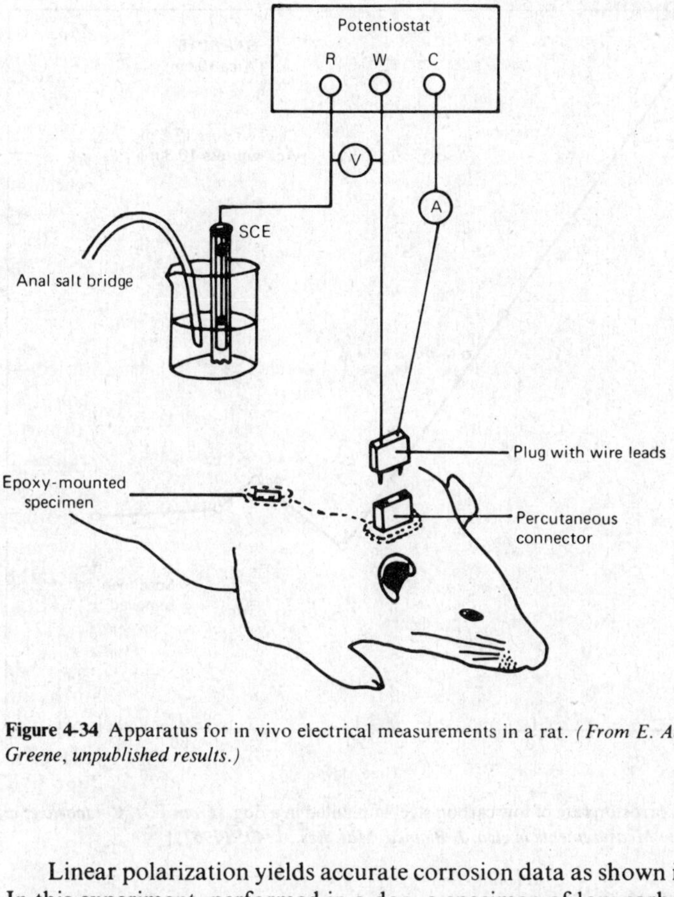

Figure 4-34 Apparatus for in vivo electrical measurements in a rat. *(From E. Abreu and N. D. Greene, unpublished results.)*

Linear polarization yields accurate corrosion data as shown in Fig. 4-35. In this experiment, performed in a dog, a specimen of low-carbon steel was implanted, and periodic linear polarization measurements were performed over a period of 35 days. At the end of the experiment, the specimen was surgically removed and weight loss was determined. Comparison of the total weight loss with the integrated rate-time data (area under the curve) shows good agreement.

Comparative corrosion tests show that:

1. Most metals and alloys corrode less rapidly in vivo than under simulated conditions in 0.9% (isotonic) saline solution.
2. The corrosion rates of most metals decrease with time (Fig. 4-35).

The lower corrosivity observed with implanted metals is probably due to the inhibiting effects of amino compounds present in body fluids.

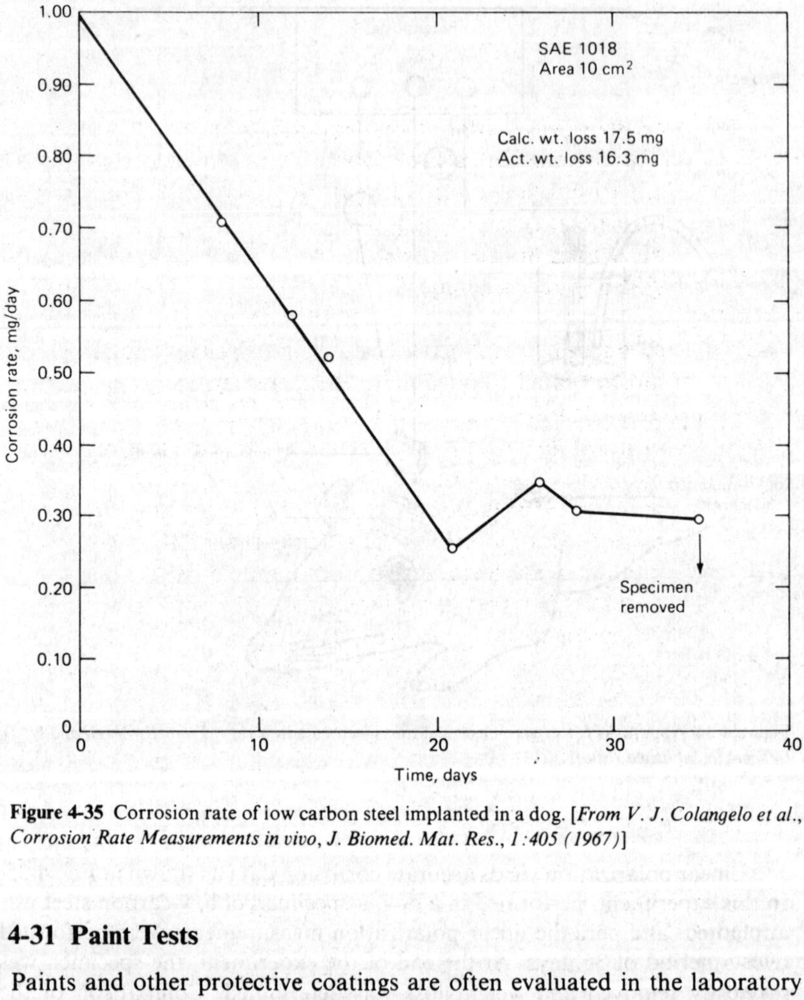

Figure 4-35 Corrosion rate of low carbon steel implanted in a dog. [*From V. J. Colangelo et al., Corrosion Rate Measurements in vivo, J. Biomed. Mat. Res., 1:405 (1967)*]

4-31 Paint Tests

Paints and other protective coatings are often evaluated in the laboratory and in the field by exposing sheet-metal panels which have been coated. These specimens are placed on racks and exposed to marine, industrial, or urban atmospheres and any particular environment of interest. These tests often last for several years and are frequently inspected for evaluation. Appearance of the coating, presence and extent of rusting, underfilm corrosion, and other factors are considered.

One disadvantage of the plain flat panel is that it does not contain crevices that are often present in equipment to be protected by coatings. Figure 4-36 shows a test panel developed by Kenneth Tator Associates for evaluating coatings. Experience has shown that results obtained on this panel correlate closely with results from coatings on actual equipment. Unsuitable coatings

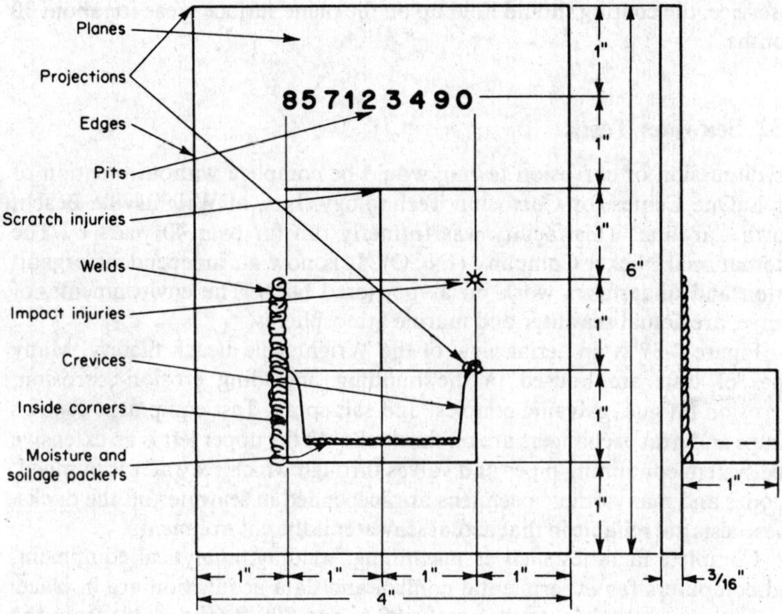

Figure 4-36 KTA coating test panel.

can be usually so judged after two to three months' exposure. Satisfactory panel life of six months usually indicates that the coating will give good performance in actual service.

To bridge the gap between results attained from test panels and coating life expectancy, a simple ratio between time for deterioration at some poorly protected spot and that for plane surfaces is established from observations on the test panels. The same ratio will hold true for the coating on any application where exposure is to comparable conditions. To judge the life of the coating in actual installations, therefore, it is necessary only to observe its behavior carefully at some point similar in shape to that part of the test panel where it first showed signs of failure. Such common surface features as sharp edges can be found on practically all types of industrial applications. Once breakdown is detected on a critical area, use of the above ratio will indicate how soon failure can be expected over all general plane areas as well. For example, if a ratio determined from test panel observations is

$$\frac{\text{Edge failure}}{\text{Plane failure}} = 0.2$$

and coating failure on sharp edges of equipment is detected after six months of service, the coating should hold up on the plane surface areas for about 30 months.

4-32 Seawater Tests

No discussion of corrosion testing would be complete without mention of the LaQue Center for Corrosion Technology, Inc., at Wrightsville Beach, North Carolina. This facility was formerly run for over 40 years by The International Nickel Company (INCO). It is now an independent organization and undertakes work on a sponsored basis. The environments, of course, are actual seawater and marine atmospheres.

Figure 4-37 is an aerial view of the Wrightsville Beach facility. Many types of tests are housed in the building, including erosion-corrosion, corrosion fatigue, galvanic couples, and salt spray. Test equipment such as pumps and heat exchangers are on the docks. At the upper left is an extensive pipe system containing pipes and valves through which seawater is pumped. Anodes and many other specimens are suspended in seawater off the docks. These tests are reliable in that actual seawater is the environment.

Complete facilities such as machining, welding, analytical equipment, and computers for experimental control and data acquisition are in place.

The atmospheric exposure racks 80 ft and 800 ft (Fig. 4-38) from the ocean are located nearby at Kure Beach. Here thousands of panels of metals

Figure 4-37 LaQue Center for Corrosion Technology test station at Harbor Island, N.C. *(Courtesy LCCT)*

Figure 4-38 LCCT corrosion test racks at Kure Beach, N.C. *(Courtesy LCCT)*

and alloys, paint- and metallic-coated specimens, and so on are exposed for various times, including many years.

4-33 Miscellaneous Tests for Metals

In addition to the tests described above, those listed in Table 4-6 and others mentioned elsewhere in this book, a variety of other laboratory and plant tests are utilized. Electrical resistance methods depend on increase in resistance of a thin-strip specimen or probe as they are thinned because of corrosion. Instruments for this type of test are available commercially for plant application and are useful for monitoring a process with continuous recording. An obvious advantage is that the specimens need not be removed

Table 4-7 Example of MTI recommended procedure
(Corrosion rates in mils per year (micrometers per year)

Test solution	Allcorr	Hastelloy+ C-276	Hastelloy+ B-2	Inconel++ 625	Incoloy + + 825	Carpenter 20Cb3+++	316 Stainless
90°C = 194°F*	0–6	0–6	0–6	12–24	12–24	24–60	24–60
1% HCl	(0–150)	(0–150)	(0–150)	(300–600)	(300–600)	(600–1,500)	(600–1,500)
100°C = 212°F*	0–6	12–24	0–6	24–60	12–24	12–24	> 200
10% H_2SO_4	(0–150)	(300–600)	(0–150)	(600–1,500)	(300–600)	(300–600)	(> 5,000)
100°C = 212°F*	0–6	0–6	0–6	6–12	0–6	12–24	12–24
50% H_3PO_4	(0–150)	(0–150)	(0–150)	(150–300)	(0–150)	(300–600)	(300–600)
122°C = 251°F*							
Boiling	6–12	60–200	> > > 200	24–60	6–12	24–60	12–24
65% HNO_3	(150–300)	(1,500–5,000)	(> > > 5,000)	(600–1,500)	(150–300)	(600–1,500)	(300–600)
118°C = 243°F*							
Boiling							
100% Acetic	0–6	0–6	0–6	0–6	60–200	6–12	0–6
Acid	(0–150)	(0–150)	(0–150)	(0–150)	(1,500–5,000)	(150–300)	0–150)
100°C = 212°F*							
50% Formic	0–6	0–6	0–6	6–12	12–24	6–12	6–12
Acid	(0–150)	(0–150)	(0–150)	(150–300)	(300–600)	(150–300)	(150–300)
100°C = 212°F*							

10% Oxalic Acid 50°C = 122°F†	0–6 (0–150)	6–12 (150–300)	0–6 (0–150)	0–6 (0–150)	12–24 (300–600)	12–24 (300–600)	24–60 (600–1,500)
9–10% $FeCl_3$ Boiling	0–6 (0–150)	0–6 (0–150)	>200 (>5,000)	12–24 (300–600)	>>200 (>>5,000)	>200 (>5,000)	>>>200 (>>>5,000)
5% $FeCl_3$ + 10% NaCl Boiling	0–6 (0–150)	0–6 (0–150)	>>200 (>>5,000)	>>200 (>>5,000)	>200 (>5,000)	>>>200 (>>>5,000)	>>>200 (>>>5,000)
50% H_2SO_4 + 3% $Fe_2(SO_4)_3$ ASTM A262 Solution Boiling	6–12 (150–300)	>200 (>5,000)	>200 (>5,000)	24–60 (600–1,500)	12–24 (300–600)	12–24 (300–600)	24–60 (600–1,500)
50% NaOH*	0–6 (0–150)	12–24 (300–600)	0–6 (0–150)	0–6 (0–150)	12–24 (300–600)	24–60 (600–1,500)	>200 (>5,000)

*Materials Technology Institute of the Chemical Process Industries, Inc.

†Crevice corrosion, but reported as average corrosion rate.

Trademark Acknowledgements

+Hastelloy—Registered Trademark of Cabot Wrought Products Division

++Incoloy, Inconel—Registered Trademarks of The Inco Family of Companies

+++20Cb3—Registered Trademark of Carpenter Technology

NOTE: Data presented here are average values and are not a guarantee of maximum or minimum values. They are based on laboratory and field test data from a number of sources and are representative only of the results obtained in such tests and should not be considered as maximums or minimums. Information presented here s for reference only.

Figure 4-39 Corrosion testing system with six-head tensile testing machine. *(Courtesy Korros-Data)*

for weighing. This method is not feasible for localized corrosion such as pitting.

Hydrogen-effusion methods consist of measuring the hydrogen, formed by corrosion, that diffuses through the specimen by collecting the hydrogen in a vacuum. The rate of hydrogen evolved is a direct measure of the corrosion rate.

Anodes for impressed currents can be tested in barrels or drums to determine anode consumption under various current densities.

For EMF measurements at elevated temperatures see Dobson.*

Figure 4-39 shows a computerized six-head tensile machine for testing metals under stress in corrosive environments.

To assist in the evaluation of *new*, wrought iron- and nickel-base corrosion-resistant alloys, the Materials Technology Institute devised a format based on five corrosion tests for producers of alloys to use in supplying data on their new alloys; see MTI Manual No. 3. A similar manual for titanium, zirconium, tantalum, niobium, and their alloys is also available. The data presented help the corrosion engineer determine services in which these alloys may be applicable. He or she may not have the time (or inclination) to conduct exploratory tests to get some idea of areas of applicability. Table 4-7 is an example of information supplied by a

*See J. V. Dobson, EMF Measurements at Elevated Temperatures and Pressures, in *Advances in Corrosion Science and Technology*, Vol. 7, Plenum Press, New York, 1980.

producer—in this case Teledyne Allvac (from an advertisement in *Chemical Engineering*, March 21, 1983). Allcorr is a single-phase nickel-base alloy (composition not given). This new alloy is compared to well-known older alloys. This gives the corrosion engineer clues on where to conduct tests in a given plant environment. This procedure should reduce the time lag between introduction of a new alloy and utilization.

Figure 4-40 shows a practical test used in the Fontana Corrosion Center for containers and other vessels. The specimens are small tanks made of various materials, and the tests are run for long times. One advantage here is simulation of actual construction; for example, welds of different types are represented. The containers—affectionately called "guinea pigs"—are sectioned for final examination. Pressure buildup in the containers is also observed.

Radioactive tracers may be used in corrosion tests. For example, automobile piston rings are made with a radioactive material and metal removed is determined by "counting" the lubricating oil.

A variety of acceptance or evaluation tests are utilized. Salt-spray tests are conducted on paints, electroplates, and other protective coatings. The specimen must resist exposure for a specified number of hours (e.g., 500) without rusting of the substrate steel in a salt-spray cabinet. Acidified chloride solutions are used in a similar manner for chromium plate and other electroplates, particularly by the automotive industry. The main value of these tests is to preclude poor workmanship and defective material.

Figure 4-40 Container tests for fuming nitric acid.

Much controversy exists concerning the value of these "accelerated" tests. One of the main difficulties is due to the fact that detailed procedures must be meticulously followed, and correlation with actual corrosion behavior is not always good.

The *Buehler Metal Digest*, Vol. 22, No. 2 (1983) is titled "Metallographic Preparation for Corrosion Studies." This is an excellent discussion of techniques for preparing corroded specimens for microscopic examination. It is useful for failure analysis and determining damage of material during corrosion tests—for example, encapsulation to preserve edges and fragile corrosion products. Color photography, is used to help identify corrosion products and provide clues for the mechanism of attack. Examples of the forms of corrosion are illustrated. A reprint of this volume of the digest is available *free* from Buehler Ltd., Lake Bluff, Illinois 60044.

4-34 Corrosion of Plastics and Elastomers

Polymeric materials corrode by processes quite different from those associated with metallic corrosion. Physicochemical processes, rather than electrochemical reactions, are responsible for the degradation of plastics and elastomers. Polymeric materials are attacked by:

1. Swelling
2. Dissolution
3. Bond rupture due to
   (a) Chemical reaction (e.g., oxidation)
   (b) Heat
   (c) Radiation (e.g., sunlight)

These reactions can occur singly or in combination. Swelling and dissolution, with and without chemical-bond breaking, are the chief causes of attack during exposure to liquids. This process is quite complex, as schematically shown in Fig. 4-41. A multilayer interface forms between the liquid and solid phases. Small quantities of the solvent penetrate the polymer, forming an infiltration layer with altered physical properties. Above this, swollen solid and gel layers containing larger amounts of solvent are usually present. A concentration gradient liquid layer extends out into the solution. All or some of these layers can be observed during liquid attack of polymers. The surface mechanical properties are altered; in some cases cracking of the swollen solid layer occurs due to internal stresses.

Considering the mechanisms of polymer corrosion, it is apparent that weight-loss measurements cannot be used to evaluate attack. In fact, it is difficult to express quantitatively the magnitude and nature of attack by any method. Table 4-8 lists some of the tests used to evaluate the chemical resistance of plastics. Similar measurements are used for hard and soft

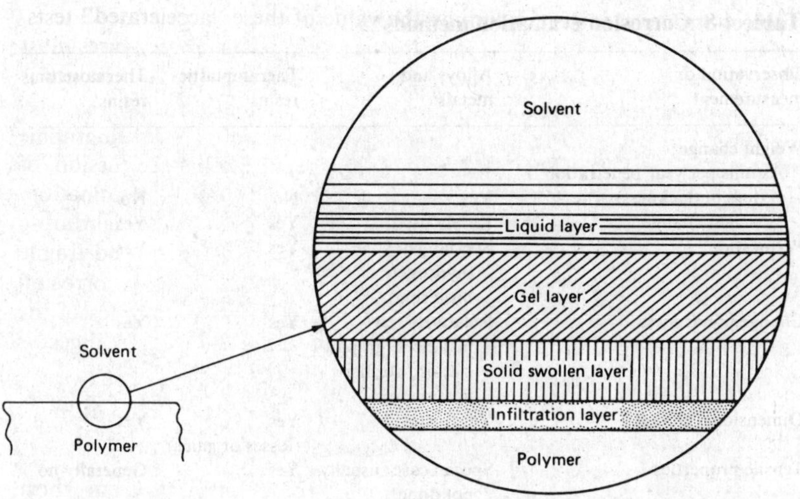

Figure 4-41 The polymer-solvent interface. [*From K. Uberreiter, The Solution Process, in J. Crank and G. S. Park, eds., Diffusion in Polymers, pp. 219–257, Academic Press, New York (1968)*]

rubber. The quantitative nature of polymeric corrosion data should be remembered when examining tabulated recommendations (as in this text). More specific information is available from many manufacturers and fabricators. Also, the *Modern Plastics Encyclopedia*, published annually by McGraw-Hill, provides an excellent source for the latest corrosion data and other engineering information.

ASTM Method C581-74(78) covers testing for chemical resistance of glass-fiber-reinforced thermosetting plastics (RTP) using immersed specimens. One big disadvantage of immersion tests concerns edge effects that are difficult to control. In addition, the immersion test does not simulate actual exposure conditions (edges usually not wetted).

Fenner and Fenner* eliminate the edge effects and provide simulation of actual conditions through use of open-end horizontal cylinders. The test panel closes the ends of the cylinder, and only the surface of the specimen is exposed to the environment. Partial filling of the cylinder permits liquid and vapor exposure. The usual tests for swelling, dissolution, loss of strength, and so forth are made after exposure. They also determine rate of permeation by inserting electrical-conductivity probes at different points in the laminate

*R. D. Fenner and O. H. Fenner, "Simulation of Plant Exposures of Rigid Plastics Under Laboratory One-Side Panel Studies," NACE Paper No. 20, CORROSION/84, New Orleans, Apr. 1984.

Table 4-8 Corrosion evaluation methods*

Observation or measurement	Alloys and metals	Thermoplastic resins	Thermosetting resins
Weight change:			
As mils per year penetration (loss in thickness)	Yes	No	No
% weight change	Rarely used	Yes	Yes
Permeation	Special gases (hydrogen blistering)	Yes	Yes
Change in hardness	Relatively few instances (graphitization, dezincification, etc.)	Yes	Yes
Dimensional changes	Yes (losses)	Yes (losses or gains)	Yes
Tensile properties	Special case; usually not done	Yes	Generally no
Compressive properties	Special case; usually not done	Generally no	Yes
Flexural properties	Special case; usually not done	Generally no	Yes
Elongation	No	Yes	No
Appearance change of test sample:	Yes	Yes	Yes
Surface attack (pitting, craters, crevices, blistering, etc.)	Yes	Yes	Yes
Laminar-wall attack	Yes	Yes	Yes
Color	Yes	Yes	Yes
Deleterious effects to environment	Yes	Yes	Yes

*Otto H. Fenner, Chemical and Environmental Properties of Plastics and Elastomers, in Charles A. Harper, ed., *Handbook of Plastics and Elastomers*, p. 4–4, McGraw-Hill Book Company, New York, 1975.

specimen wall. Results are presented, including comparison with immersion specimens, for a reinforced polyester and a vinylester in solvents, 28% hydrochloric acid, and 35% sulfuric acid. Environments aggressive to these materials are listed as (1) strongly oxidizing environments, (2) strong acids, (3) strong solvents, (4) water-miscible solvents, and (5) strong alkalies.

The Materials Technology Institute (MTI) and the Society of the Plastics Industry (SPI) have issued "Quality Assurance Report—RTP Corrosion-Resistant Equipment" (1981) in order to help ensure better quality equipment made of these materials. This should help both fabricators and users. This report should eventually result in standards promulgated by ASME (MTI does not issue standards).

4-35 Presenting and Summarizing Data

A variety of methods are used in reports, trade literature, books, technical literature, and other publications to present or summarize corrosion data. Some of these methods are quantitative and others qualitative in nature. In any case, a person seeking corrosion information is confronted with a variety of schemes and methods and is sometimes confused. Standardization, though desirable, is difficult because of the many different individual approaches to the problem. *Isocorrosion* charts are used here, in subsequent chapters, when sufficient data are available for their preparation.

Two very common methods for presenting corrosion data are tables and curves. These are usually used when reporting on a specific detailed investigation and when compilation or condensation of a large amount of data is not involved. For example, the results of tests on a number of materials in a particular plant liquor or environment are usually shown in a table. Tests on one or a few materials, where either the concentration or temperature of a given corrosive is varied, are often reported by means of curves. If duplicate tests do not provide check results, wide bands or shaded areas are sometimes used.

We believe that the best and most complete compilation of corrosion data on a variety of materials by a large number of corrosive media, is the report titled "Corrosion Data Survey" by George Nelson of the Shell Development Co.* A tremendous amount of information is compiled and condensed. Specific information is readily located in this report. Figure 4-42 is a sample of the method Nelson has used for many years to present corrosion information. This example shows the corrosion of lead by sulfuric acid as a function of temperature and concentration. The equivalent of Fig. 4-42 appears in the Shell report in a square a little over $\frac{1}{2}$ in. on the side.

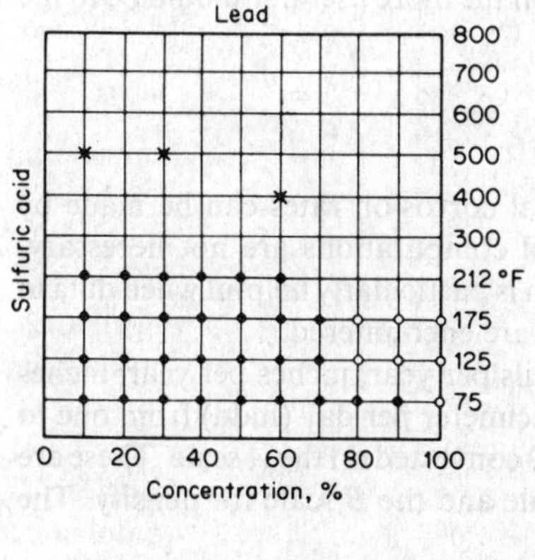

Figure 4-42 Nelson's method for summarizing corrosion data.

*Available from the National Association of Corrosion Engineers, Houston, Tex. 77219.

The symbols used represent corrosion rates as follows:

● = Corrosion rate less than 2 mpy
○ = Corrosion rate less than 20 mpy
□ = Corrosion rate from 20 to 50 mpy
× = Corrosion rate greater than 50 mpy

The square symbol does not happen to appear in this particular case. The information is clearly presented in the small area used for each material.

The *Corrosion Handbook** presents information in tabular form on chemical-resistant materials using ratings of class A, class B, and class C. In general, and with qualifications, class A means less than 5 mpy, class B from 5 to 50 mpy, and class C over 50 mpy, or generally unsuitable. One disadvantage of this classification is that the uninitiated may pick only class A materials, which may not be the economic ones.

Rabald, in his *Corrosion Guide*,† used four classifications:

+ = Practically resistant
(+) = Fairly resistant
(−) = Not particularly resistant but sometimes used
− = Unusable

He uses this same system in the German "Dechema-Werkstoffe-Tabelle," which presents a tremendous amount of corrosion information on materials and environments.

Producers and vendors of materials use terms such as excellent, fully resistant, good, satisfactorily resistant, fair, slightly resistant, poor, and not resistant.

The more quantitative the information the more useful it should be to the user.

4-36 Nomograph for Corrosion Rates

A rapid and ready conversion for several corrosion rates can be made by means of the nomograph. Mathematical computations are not necessary, and the accuracy is good. This nomograph is particularly helpful when data in milligrams per square decimeter per day are encountered.

Figure 4-43 permits conversion of mils per year, inches per year, inches per month, and milligrams per square decimeter per day (mdd) from one to another. The first three named are directly converted on the *A* scale. These are converted to mdd by means of the *C* scale and the *B* scale for density. The

*H. H. Uhlig, ed., *Corrosion Handbook*, John Wiley & Sons, Inc., New York, 1948.
†E. Rabald, *Corrosion Guide*, Elsevier Publishing Company, New York, 1951.

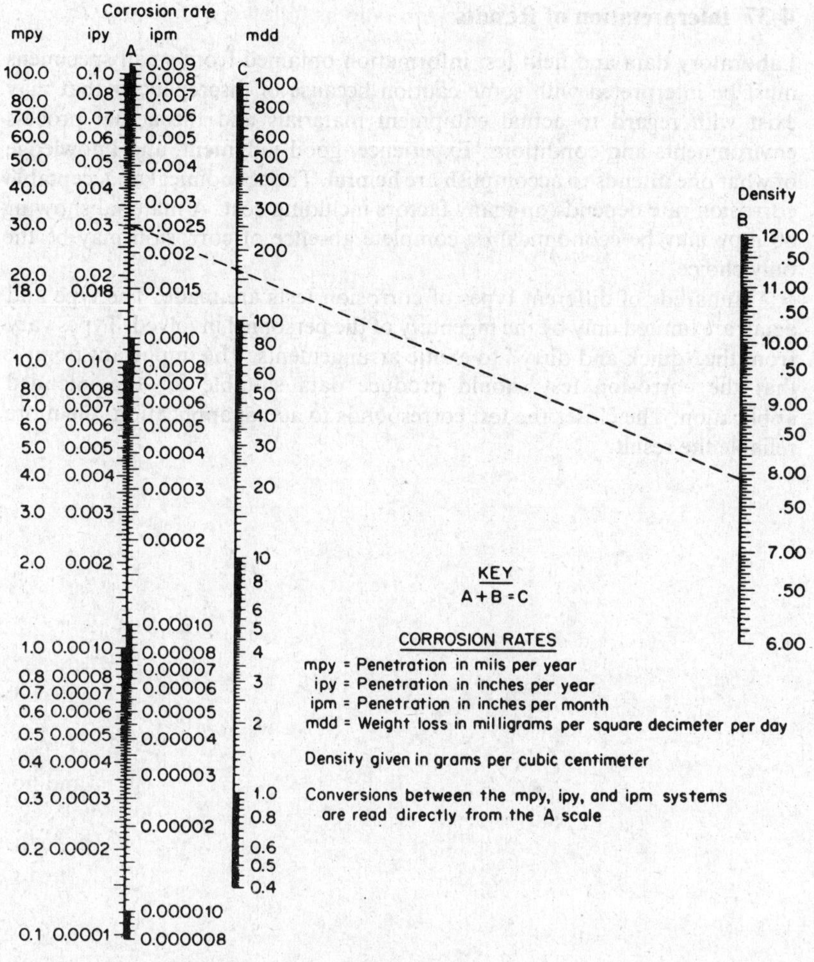

Figure 4-43 Nomograph for mpy, ipy, ipm, and mdd.

mdd unit does not consider or include the density or type of material involved. Density is given as grams per cubic centimeter. A straightedge is all that is needed to use the nomograph. The dotted-line example is for 18-8S stainless steel and a corrosion rate of 30 mpy. The heavy figures corresponding to 30 and 18 mpy on the *A* scale are the so-called maximum acceptable rates for cast and wrought 18-8S, respectively, in the Huey test—boiling 65% nitric acid.

4-37 Interpretation of Results

Laboratory data and field test information obtained from small specimens must be interpreted with some caution because of discrepancies that may exist with regard to actual equipment materials and the actual process environments and conditions. Experience, good judgment, and knowledge of what one intends to accomplish are helpful. The economical or acceptable corrosion rate depends on many factors including cost. A material showing 50 mpy may be economical or complete absence of corrosion may be the only choice.

Hundreds of different types of corrosion tests are made. The type and detail are limited only by the ingenuity of the personnel involved. Types vary from the "quick and dirty" to exotic arrangements. The important point is that the corrosion test should produce data suitable for the intended application. The closer the test corresponds to actual application, the more reliable the result.

CHAPTER

FIVE

MATERIALS

A large variety of materials, ranging from platinum to rocks, are used by the engineer to construct bridges, automobiles, process plant equipment, pipelines, power plants, etc. The corrosion engineer is primarily interested in the chemical properties (corrosion resistance) of materials, but must also have knowledge of mechanical, physical, and other properties to assure desired performance. The properties of engineering materials depend upon their physical structure and basic chemical composition.

5-1 Mechanical Properties

These properties are related to behavior under load or stress in tension, compression, or shear. Properties are determined by engineering tests under appropriate conditions. Commonly determined mechanical properties are tensile strength, yield point, elastic limit, creep strength, stress-rupture, fatigue, elongation (ductility), impact strength (toughness and brittleness), hardness, and modulus of elasticity (ratio of stress to elastic strain-rigidity). Strain may be elastic (present only during stressing) or plastic (permanent) deformation. These properties are helpful in determining whether or not a part can be produced in the desired shape and also resist the mechanical forces anticipated.

The words mechanical and physical are often erroneously used interchangeably. The above are mechanical properties. Sometimes modulus of elasticity is considered to be a physical property of a material because it is an *inherent* property that cannot be changed substantially by practical means such as heat treatment or cold-working.

5-2 Other Properties

The corrosion engineer is often required to consider one or more properties in addition to corrosion resistance and strength when selecting a material. These include density or specific gravity (needed to calculate corrosion rates); fluidity or castability; formability; thermal, electrical, optical, acoustical, magnetic properties; and resistance to atomic radiation. For example, a particular part must be castable into an intricate shape, possess good heat-transfer characteristics, and not be degraded by atomic radiation. In another case, the equipment must be a good insulator, reflect heat, and have low unit weight. Incidentally, radiation sometimes enhances properties of a material; e.g., the strength of polyethylene can be increased by controlled radiation.

Cost is not a property of a material, but it may be the overriding factor in selection of a material for engineering use, based on economic considerations.

METALS AND ALLOYS

5-3 Cast Irons

Cast iron is a generic term that applies to high carbon-iron alloys containing silicon. The common ones are designated as gray cast iron, white cast iron, malleable cast iron, and ductile or nodular cast iron. Ordinary *gray irons* contain about 2 to 4% carbon and 1 to 3% silicon. These are the least expensive of the engineering metals. The dull or grayish fracture is due to the free graphite flakes in the microstructure (Fig. 5-1). Gray cast irons can be readily cast into intricate shapes because of their excellent fluidity and relatively low melting points. They can be alloyed for improvement of corrosion resistance and strength.

ASTM (American Society for Testing Materials) specification A48 classifies gray irons as listed in Table 5-1. Their high compressive strength is indicated in the table. These materials are brittle and exhibit practically no ductility. They do not show a clearly defined yield point, but the yield strength is about 85% of the tensile strength. The modulus of elasticity in tension varies from 9.6 to 23.5 million lb/in.^{2} for these classes, with class 20 exhibiting the lowest value and class 60 the highest. (Theoretically, gray irons do not have a modulus of elasticity because the stress-strain curve is not a straight line.) Impact strength is generally low but is best for material with the highest ratio of tensile strength to hardness. These irons show high damping capacity (vibration damping). Specific gravity varies with carbon content and is in the range of 7 to 7.35. Thermal and electrical conductivities are lower than pure iron.

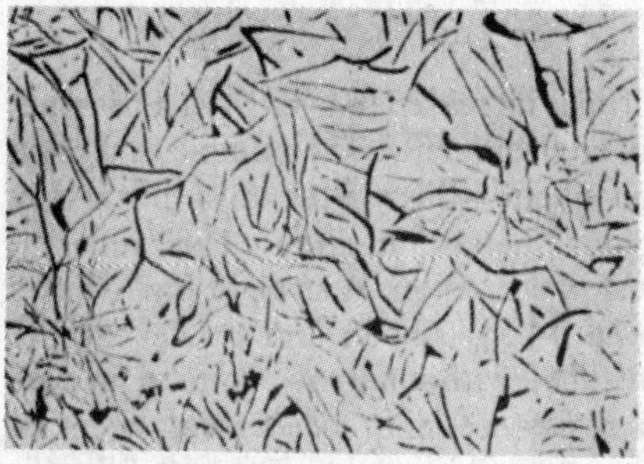

Figure 5-1 Microstructure of gray cast iron (as polished, 100 ×).

Table 5-1 Typical mechanical properties of gray iron test bars

ASTM class	Tensile strength,* lb/in.^{2}	Compressive strength, lb/in.^{2}	Fatigue limit, lb/in.^{2}	Brinell hardness
20	22,000	83,000	10,000	156
25	26,000	97,000	11,500	174
30	31,000	109,000	14,000	201
35	36,500	124,000	16,000	212
40	42,500	140,000	18,500	235
50	52,500	164,000	21,500	262
60	62,500	187,500	24,500	302

*Can be improved substantially by heat treatment.
Source: ASM Metals Handbook 1:354 (1961).

White cast irons White cast irons have practically all of the carbon in the from of iron carbide. These irons are extremely hard and brittle. Silicon content is low because this element promotes graphitization. Graphite formation is related to rate of cooling from the melt, so chilling can produce a white iron from one that would be normally gray.

Malleable irons These are produced by high-temperature heat treatment of white irons of proper composition. The graphite forms as rosettes or clusters instead of flakes and the material shows good ductility (hence the name malleable).

Ductile irons These materials exhibit ductility in the as-cast form. The graphite is present as nodules or spheroids as a result of a special treatment of the molten metal. The mechanical properties of ductile irons can be altered by heat treatment similar to ordinary steel.

5-4 High-Silicon Cast Irons

When the silicon content of gray cast iron is increased to over 14%, it becomes extremely corrosion resistant to many environments. The notable exception is hydrofluoric acid. In fact, these high-silicon irons are the most universally resistant of the commercial (nonprecious) metals and alloys. Their inherent hardness makes them resistant to erosion corrosion. A straight high-silicon iron, such as Duriron, contains about 14.5% silicon and 0.95% carbon. This composition must be closely controlled within narrow limits to provide the best combination of corrosion resistance and mechanical strength. The alloy is sometimes modified by the addition of 3% molybdenum (Durichlor) for increased resistance to hydrochloric acid, chlorides, and pitting. A newer alloy, Durichlor 51,* shows even better corrosion performance and has replaced the regular Durichlor composition. Durichlor 51 contains chromium, in addition to some molybdenum, which imparts improved resistance to oxidizing conditions. For instance, the presence of ferric or cupric chlorides in hydrochloric acid inhibits corrosion instead of causing severe selective attack as it does with most metals and alloys.

Duriron and Durichlor 51 exhibit tensile strengths around 20,000 $lb/in.^2$ and hardness of 520 Brinell. The specific gravity is 7.0. Both alloys can be machined only by grinding. Welding of the alloy is very difficult and, although simple shapes like pipe can be welded when proper precautions are observed, it is not practicable to weld complex shapes.

The excellent corrosion resistance of high-silicon irons is due to the formation of a passive SiO_2 surface layer that forms during exposure to the environment.

These alloys are available only in cast form for drain lines, pumps, valves, and other process equipment. They have found extensive use as anodes for impressed current cathodic protection. Figures 5-2 and 5-3 show

Figure 5-2 High-silicon iron pump. *(The Duriron Co., Inc.)*

*Durichlor 51, Patent No. 3,129,095.

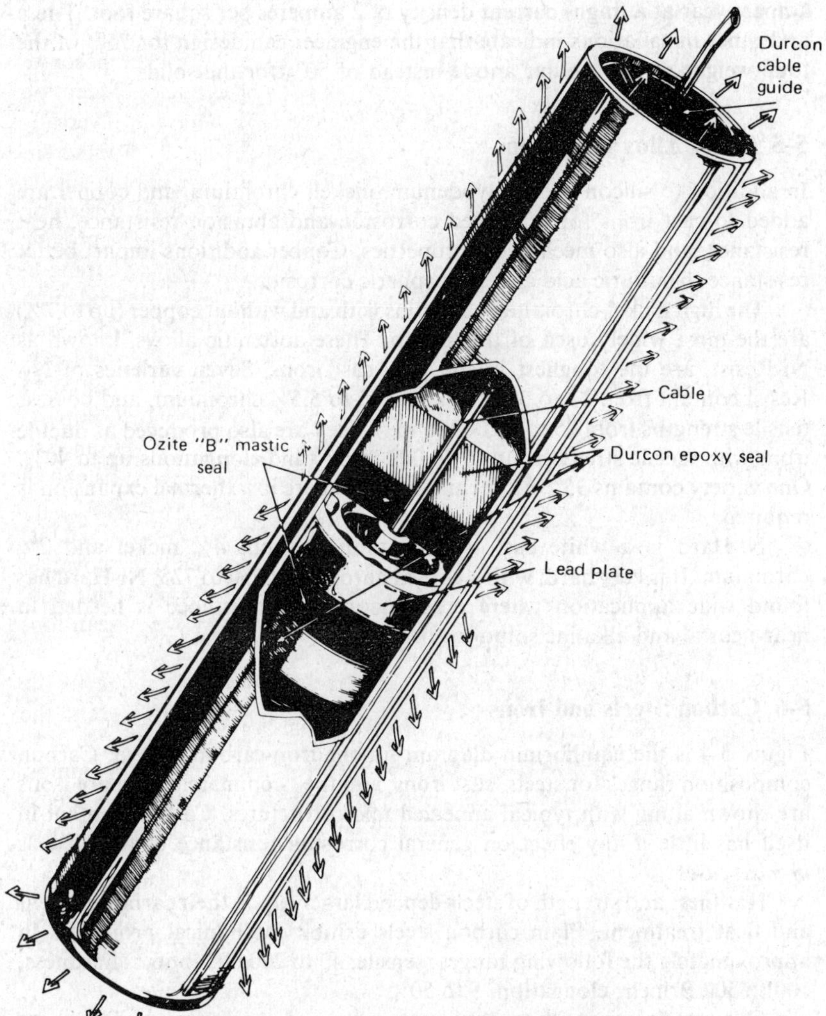

Figure 5-3 Durichlor 51 alloy tubular anode. *(The Duriron Co., Inc)*

a pump and a hollow or tubular anode made of Durichlor 51. The anodes are available in various sizes. Field experience during the past five years shows approximately one-half the consumption rate for tubular anodes (0.7 pound per ampere year) as compared to the solid type in a harbor dock installation. In another case, the consumption rate was 0.35 pound per ampere year at a (high) current density of 2 amperes per square foot. These and other installations indicate that the engineer can design for 75% of the total weight of the tubular anode instead of 50% for the solid.

5-5 Other Alloy Cast Irons

In addition to silicon and molybdenum, nickel, chromium, and copper are added to cast irons for improved corrosion and abrasion resistance, heat resistance, and also mechanical properties. Copper additions impart better resistance to sulfuric acid and atmospheric corrosion.

The high nickel-chromium cast irons with and without copper (up to 7%) are the most widely used of this group. These austenitic alloys, known as Ni-Resist, are the toughest of the gray cast irons. Seven varieties of Ni-Resist contain from 14 to 32% nickel, 1.75 to 5.5% chromium, and possess tensile strengths from 25 to 45,000 lb/in.^{2}. They are also produced as ductile irons with tensile strengths up to 70,000 lb/in.^{2} and elongations up to 40%. One variety contains 35% nickel and is used where low thermal expansion is required.

Ni-Hard is a white cast iron containing about 4% nickel and 2% chromium. It is very hard, with a Brinell hardness of 550 to 725. Ni-Hard has found wide application where erosion-corrosion resistance is needed in near-neutral and alkaline solutions or slurries.

5-6 Carbon Steels and Irons

Figure 5-4 is the equilibrium diagram for the iron-carbon system. Carbon composition ranges for steels, cast irons, and the "commercially" pure irons are shown along with typical annealed microstructures. Carbon content in itself has little if any effect on general corrosion resistance of these steels *in most cases*.

Hardness and strength of steels depend largely upon their carbon content and heat treatment. Plain carbon steels exhibit mechanical properties in approximately the following ranges: tensile, 40 to 200,000 lb/in.^{2}; hardness, 100 to 500 Brinell; elongation, 5 to 50%.

Commercial pure irons are ingot irons and Armco iron. These are relatively weak and not used where strength is a major requirement.

Wrought iron is a "mechanical" mixture of slag and low-carbon steel. Many claims of better corrosion resistance are made for this material, but each proposed application should be carefully studied to be sure the extra cost over ordinary steel is justified.

*R*ust is a term reserved for iron-base materials. Rust consists of iron oxides and usually is hydrated ferric oxide. When a TV commercial states that a nonferrous metal does not rust, it is correct, but that does not mean that the metal does not corrode.

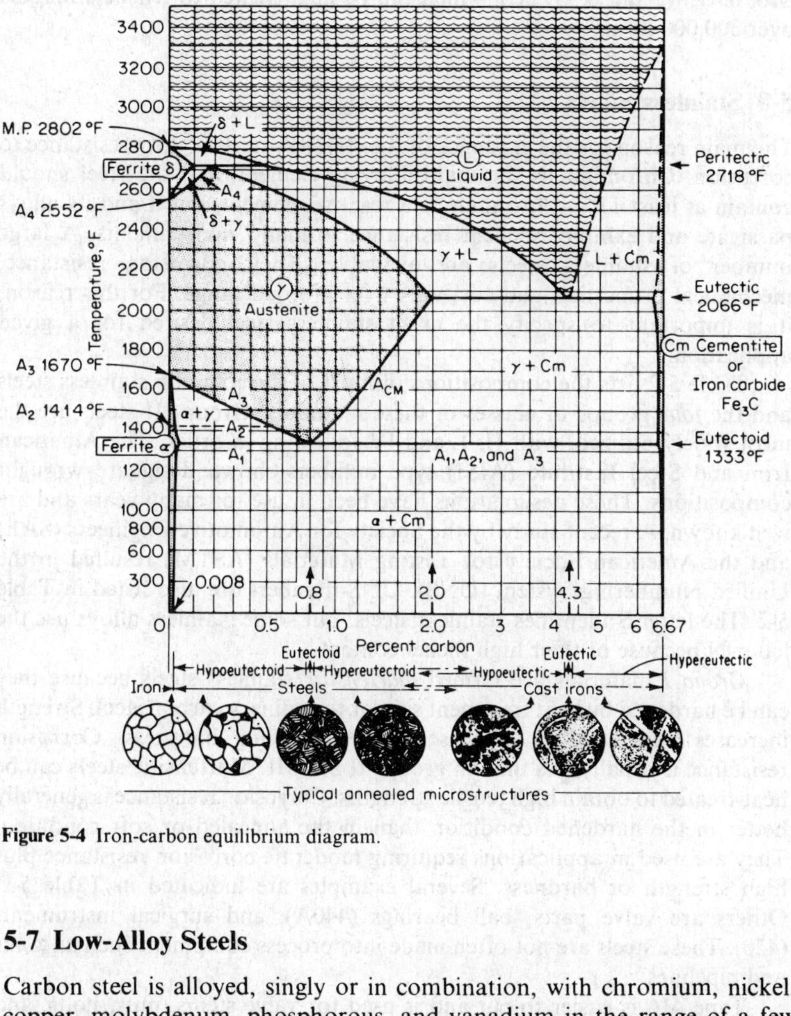

Figure 5-4 Iron-carbon equilibrium diagram.

5-7 Low-Alloy Steels

Carbon steel is alloyed, singly or in combination, with chromium, nickel, copper, molybdenum, phosphorous, and vanadium in the range of a few percent or less to produce low-alloy steels. The higher alloy additions are usually for better mechanical properties and hardenability. The lower range of about 2% total maximum is of greater interest from the corrosion standpoint. Strengths are appreciably higher than plain carbon steel, but the most important attribute is much better resistance to atmospheric corrosion. Appreciable advantage sometimes obtains in aqueous solutions.

Steels with very high strengths are of particular interest for aerospace applications. Here high strength-weight ratio is the overriding factor, with corrosion of lesser importance. A good example is H-11 steel (5% Cr, 1.5%

Mo, 0.4% V, and 0.35% C), which can be heat-treated to tensile strengths over 300,000 lb/in.^{2}.

5-8 Stainless Steels

The main reason for the existence of the stainless steels is their resistance to corrosion. Chromium is the main alloying element, and the steel should contain at least 11%. Chromium is a reactive element, but it and its alloys passivate and exhibit excellent resistance to many environments. A large number of stainless steels are available. Their corrosion resistance, mechanical properties, and cost vary over a broad range. For this reason, it is important to specify the exact stainless steel desired for a given application.

Table 5-2 lists the compositions of most of the common stainless steels and the *four* groups or classes of these materials. Group III steels are the most widely utilized, with II, I, and IV following in order. The American Iron and Steel Institute (AISI) type numbers shown designate wrought compositions. These designations have been in use for many years and are well known. A recent study by the Society for Automotive Engineers (SAE) and the American Society for Testing Materials (ASTM) resulted in the Unified Numbering System (UNS). UNS numbers are also listed in Table 5-2. The letter S identifies stainless steels, but some stainless alloys use the letter N because of their high nickel content.

Group I materials are termed *martensitic* stainless steels because they can be hardened by heat treatment similar to ordinary carbon steel. Strength increases and ductility decreases with increasing hardness. Corrosion resistance is usually less than in groups II and III. Martensitic steels can be heat-treated to obtain high tensile strengths. Corrosion resistance is generally better in the hardened condition than in the annealed or soft condition. They are used in applications requiring moderate corrosion resistance plus high strength or hardness. Several examples are indicated in Table 5-2. Others are valve parts, ball bearings (440A), and surgical instruments (420). These steels are not often made into process equipment such as tanks and pipelines.

Type 416 is easier to cut and is used for valve stems, nuts, bolts, and other parts to reduce machining costs.

An alloy* containing 9% Cr, 1% Mo and small amounts of columbium and vanadium has been developed for high temperature applications such as steam power plant tubes, breeder reactor tubes and pressure vessels. At temperatures below 600°C the alloy exhibits higher strength then Type 304 stainless steel.

Group II ferritic nonhardenable steels are so designated because they

* *Metal Progr.*, p. 12, Feb. 1982.

cannot be hardened by heat treatment. Figure 5-5 shows why this is the case. These steels fall outside the gamma loop and consist of alpha iron at temperatures up to the melting point. Ordinary carbon steels harden because of phase changes on cooling. As shown in Fig. 5-4, these steels are austenitic at temperatures above the A_3 line. Austenite is gamma iron, *nonmagnetic*, and has a face-centered cubic lattice. Upon cooling, it transforms to alpha iron or ferrite, which is magnetic and body-centered cubic. When cooled rapidly it ends up as hard, brittle martensite, which is magnetic but has a body-centered tetragonal lattice. Group I steels are magnetic and are hardenable because their compositions lie within the gamma loop. Type 405 in Table 5-2 appears to be an exception, but it does not harden because of the aluminum content.

Type 430 can be readily formed and has good corrosion resistance to the atmosphere. This is one reason why its most extensive use is for automobile trim. It is also used in ammonia oxidation plants for making nitric acid, and for tank cars and tanks for storage of nitric acid. The first chemical plant application of stainless steel was a type 430 tank car for shipping nitric acid. However, in these chemical applications, it has been largely displaced by 18-8 because of ease of welding and better ductility—plus better corrosion resistance if properly heat-treated. Other uses for type 430 are annealing baskets for brass and other relatively low-temperature heat treatments, oil-burner rings, window anchor bolts, and other decorative trim.

Types 442 and 446 find application where heat resistance is required, such as in furnace parts and heat-treating equipment. They possess good resistance to high-temperature oxidation and sulfur gas attack because of their high chromium content. These materials do not possess very good structural stability or high-temperature strength and should be selected with care.

One of the most interesting aspects of the group II steels is their resistance to stress corrosion. They do quite well in many cases where the 18-8 types fail, particularly in chloride-containing waters.

High-purity ferritic stainless steels Recently developed ferritic nonhardenable stainless steels containing very low carbon and nitrogen are now commercially available. Total of carbon plus nitrogen content is below 250 parts per million, although one variety can "tolerate" higher amounts. High purity is attainable through advanced steel-making technologies.

We have now gone full cycle. Straight chromium steels were the first stainless alloys but they exhibited brittleness and welding problems. The new breed of high-purity alloys possess good ductility and weldability if welds are properly made.

The first high-purity grade was Ebrite 26-1 (Airco) containing 26% Cr and 1% Mo. Another grade is 26-1S which contains titanium as a stabilizer.

Table 5-2 Chemical compositions of stainless steels

AISI type	UNS number	%C	%Cr	%Ni	% other elements	Remarks
			Group I Martensitic chromium steels			
410	S 41000	0.15 max	11.5–13.5	—	—	Turbine blades, valve trim
416	S 41600	0.15 max	12–14	—	Se, Mo, or Zr	"Free" machining
420	S 42000	0.35–0.45	12–14	—	—	Cutlery
431	S 43100	0.2 max	15–17	1.25–2.5	—	Improved ductility
440A	S 44002	0.60–0.75	16–18	—	—	Very hard; cutters
			Group II Ferritic nonhardenable steels			
405	S 40500	0.08 max	11.5–14.5	0.5 max	0.1–0.3 Al	Al prevents hardening
430	S 43000	0.12 max	14–18	0.5 max	—	Auto trim, tableware
442	S 44200	0.25 max	18–23	0.5 max	—	Resists O and S at high temperatures
446	S 44600	0.20 max	23–27	0.5 max	0.25*N* max	
			Group III Austenitic chromium-nickel steels			
201	S 20100	0.15 max	16–18	3.5–5.5	5.0–7.5 Mn 0.25*N* max	Mn substitute for Ni
202	S 20200	0.15 max	17–19	4–6	7.5–10 Mn 0.25*N* max	Mn substitute for Ni
301	S 30100	0.15 max	16–18	6–8	2 Mn max	Strain hardens
302	S 30200	0.15 max	17–19	8–10	2 Mn max	Architectural uses
302B	S 30215	0.15 max	17–19	8–10	2–3 Si	Si for high-temp. oxidation

304	S 30400	0.08 max	18–20	8–12	1 Si max	Continuous 18-8S
304L	S 30403	0.03 max	18–20	8–12	1 Si max	Very low carbon
308	S 30800	0.08 max	19–21	10–12	1 Si max	"High" 18-8
309	S 30900	0.2 max	22–24	12–15	1 Si max	25-12, heat resistance
309S	S 30908	0.08 max	22–24	12–15	1 Si max	Lower carbon
310	S 31000	0.25 max	24–26	19–22	1.5 Si max	25-20, heat resistance
310S	S 31008	0.08 max	24–26	19–22	1.5 Si max	Lower carbon
314	S 31400	0.25 max	23–26	19–22	1.5–3.0	Si for high-temp. oxidation
316	S 31600	0.10 max	16–18	10–14	2–3 Mo	18-8S Mo
316L	S 31603	0.03 max	16–18	10–14	2–3 Mo	Very low carbon
317	S 31700	0.08 max	18–20	11–14	3–4 Mo	Higher Mo
321	S 32100	0.08 max	17–19	8–11	Ti 4 × C(min)	Ti stabilized
347	S 34700	0.08 max	17–19	9–13	Cb + Ta 10 × C(min)	Cb stabilized
Alloy 20*	J 95150	0.07 max	29	20	3.25 Cu, 2.25 Mo	Best corrosion resistance
			Group IV Age-hardenable steels*			
322	—	0.07	17	7	0.07 Ti, 0.2 Al	
17-7PH†	S 17700	0.07	17	7	1.0 Al	
17-4PH†	S 17400	0.05	16.5	4.25	4.0 Cu	
14-8MoPH†	S 13800	0.05 max	14	8.5	2.5 Mo, 1% Al	
AM350†	S 35000	0.10	16.5	4.3	2.75 Mo	
CD4MCu‡	—	0.03	25	5	3.0 Cu, 2.0 Mo	

*Typical compositions
†Commercial designations
‡Cast form only

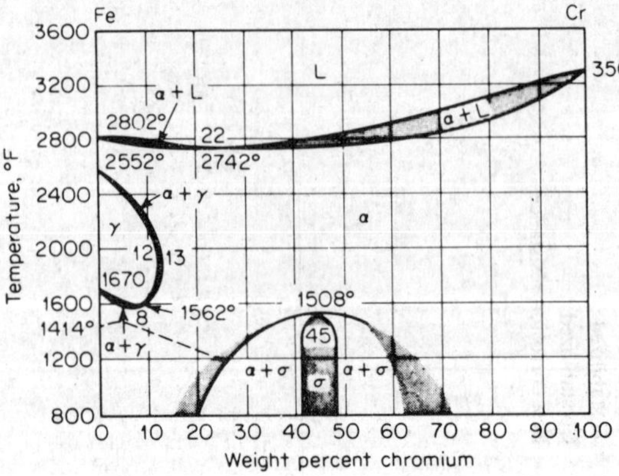

Figure 5-5 Iron-chromium equilibrium diagram.

Others are 29 Cr–4 Mo and 29 Cr–4 Mo–2 Ni (Du Pont). Welded specimens show no cracking during bend tests. Typical tensile strength, yield strength, and percent elongation for weld metal are 65–70,000 psi, 55–60,000 psi and 10–25% respectively, for 26-1; 70–80,000, 60–65,000 and 15–25% for 26-1S; and 90–95,000, 70–75,000 and 20–27% for 29-4.* Mechanical properties for wrought sheet are approximately similar.

The obvious advantage of these ferritic steels is resistance to stress-corrosion cracking in chloride environments where they are far superior to the austenitic stainless steels such as types 304 and 316. Pitting resistance is also better for the former. All of these alloys show excellent resistance to 65% nitric acid boiling. They also exhibit superior resistance to 20% acetic acid boiling and 45% formic acid boiling (organic acids) as compared to type 304.

The 29-4 alloy shows very good resistance to pitting and crevice corrosion.† For optimum ductility and resistance to pitting, intergranular corrosion, and stress corrosion this alloy should contain carbon not exceeding 0.01% and nitrogen below 0.020% (C + N less than 0.025%). The 29-4-2Ni alloy shows good resistance to 10% sulfuric acid. Caution should be exercised in the selection of this alloy for sulfuric and hydrochloric acid. None of the ferritic alloys should be considered to be a sulfuric acid alloy such as alloy 20. The 29-4 alloy exhibits better resistance to pitting and crevice corrosion than 26-1.

*R. A. Lula, Ferritic Stainless Steels: Corrosion Resistance Plus Economy, *Metal Progress*, **110**:24–29 (1976).

†M. A. Streicher, Development of Pitting Resistant Fe-Cr-Mo Alloys, *Corrosion*, **30**: 77–91 (1974).

Intergranular corrosion in the sensitized zone of welds was observed on early heats of 26-1. Attack is in the welds and adjacent metal as compared to austenitic steels wherein the attack is a distance away from the weld. Carbon plus nitrogen should be about 0.01% max except for 29-4, for which the maximum is 0.025%.

Care should be exercised in the selection of these alloys for use at low temperatures. Except for thin material, the ductile-to-brittle transition temperature (DBTT) could be at room temperature or above. Strength drops at high temperatures (i.e., above 1000°F, 540°C) and microstructural changes can occur (i.e., sigma phase) so these alloys cannot be considered as good high-temperature alloys.

Welding of these materials must be done properly and carefully in inert atmospheres to avoid contamination and embrittlement. It is not sufficient to supply the welder with instructions; he or she must be carefully supervised and the welds checked to be sure they are satisfactory. For example, momentarily pulling the weld rod out of the inert zone during the welding operation can result in degradation. These alloys are somewhat similar to titanium as far as guarding against contamination of the weld is concerned. Obviously, metal surfaces must be very clean before welding.

The nature of the mold environment is important for the production of castings. Inert mold materials must be used to avoid carbon pickup and degradation of the alloys (see Sec. 3-20).

The 29-4 alloy is not attacked by seawater, and its resistance in this environment is claimed to be comparable to Hastelloy C and titanium. This analogy should not be extended further for acids such as sulfuric acid. When titanium first appeared on the scene, it too was compared to Hastelloy C in seawater. In several cases they forgot about the seawater and assumed titanium was comparable to Hastelloy C in all environments! This is definitely not the case, and the same mistakes should not be made with 29-4. The high-purity ferritics are suitable for many applications, but knowledge of their properties is required for successful operation.

Group III austenitic stainless steels are essentially nonmagnetic and cannot be hardened by heat treatment. Like the ferritic steels, they are hardenable only by cold-working. Most of these steels contain nickel as the principal austenite former, but the relatively new ones like types 201 and 202 contain less nickel and substantial amounts of manganese.

The austenitic steels possess better corrosion resistance than the straight chromium (groups I and II) steels and generally the best resistance of any of the four groups except for CD-4MCu. For this reason austenitic steels are widely specified for the more severe corrosion conditions such as those encountered in the process industries. They are rust-resistant in the atmosphere and find wide use for architectural purposes, in the kitchen, in food manufacture and dispensing, and for applications where contamination (rust) is undesirable.

Table 5-3 Alloy Casting Institute standards designations and chemical composition ranges for heat and corrosion-resistant castings

Cast alloy designation	Wrought alloy type	Composition, % (balance Fe)					
		C	Mn max	Si max	Cr	Ni	Other elements
CA-15	410	0.15 max	1.00	1.50	11.5–14	1 max	Mo 0.5 max†
CA-40	420	0.20–0.40	1.00	1.50	11.5–14	1 max	Mo 0.5 max†
CB-30	431	0.30 max	1.00	1.00	18–22	2 max	—
CC-50	446	0.50 max	1.00	1.00	26–30	4 max	—
CD-4MCu	—	0.040 max	1.00	1.00	25–27	4.75–6.00	Mo 1.75–2.25, Cu 2.75–3.25
CE-30	—	0.30 max	1.50	2.00	26–30	8–11	—
CF-3	304L	0.03 max	1.50	2.00	17–21	8–12	—
CF-8	304	0.08 max	1.50	2.00	18–21	8–11	—
CF-20	302	0.20 max	1.50	2.00	18–21	8–11	—
CF-3M	316L	0.03 max	1.50	1.50	17–21	9–13	Mo 2.0–3.0
CF-8M	316	0.08 max	1.50	1.50	18–21	9–12	Mo 2.0–3.0
CF-12M	316	0.12 max	1.50	1.50	18–21	9–12	Mo 2.0–3.0
CF-8C	347	0.08 max	1.50	2.00	18–21	9–12	Cb 8 × C min, 1.0 max, or Cb-Ta 10 × C min, 1.35 max

CF-16F	303	0.16 max	1.50	2.00	18–21	9–12	Mo 1.5 max, Se 0.20–0.35
CG-8M	317	0.08 max	1.50	1.50	18–21	9–13	Mo 3.0–4.0
CH-20	309	0.20 max	1.50	2.00	22–26	12–15	—
CK-20	310	0.20 max	1.50	2.00	23–27	19–22	—
CN-7M	—	0.07 max	1.50	*	18–22	21–31	Mo-Cu*
HA	—	0.20 max	0.35–0.65	1.00	8–10	—	Mo 0.90–1.20
HC	446	0.50 max	1.00	2.00	26–30	4 max	Mo 0.5 max†
HD	327	0.50 max	1.50	2.00	26–30	4–7	Mo 0.5 max†
HE	—	0.20–0.50	2.00	2.00	26–30	8–11	Mo 0.5 max†
HF	302B	0.20–040	2.00	2.00	19–23	9–12	Mo 0.5 max†
HH	309	0.20–0.50	2.00	2.00	24–28	11–14	Mo 0.5 max† N 0.2 max
HI	—	0.20–0.50	2.00	2.00	26–30	14–18	Mo 0.5 max†
HK	310	0.20–0.60	2.00	2.00	24–28	18–22	Mo 0.5 max†
HL	—	0.20–0.60	2.00	2.00	28–32	18–22	Mo 0.5 max†
HN	—	0.20–0.50	2.00	2.00	19–23	23–27	Mo 0.5 max†
HT	330	0.35–0.75	2.00	2.50	13–17	33–37	Mo 0.5 max†
HU	—	0.35–0.75	2.00	2.50	17–21	37–41	Mo 0.5 max†
HW	—	0.35–0.75	2.00	2.50	10–14	58–62	Mo 0.5 max†
HX	—	0.35–0.75	2.00	2.50	15–19	64–68	Mo 0.5 max†

*There are several proprietary alloy compositions falling within the stated chromium and nickel ranges and containing varying amounts of silicon, molybdenum, and copper.

†Molybdenum not intentionally added.

Types 201 and 202 steels show about the same corrosion resistance as the type 302 grade. The "workhorses" for the process industries are types 304, 304L, 316, and 347. The effects of carbon and heat treatment were discussed in Chap. 3. The molybdenum-bearing steel, type 316, is considerably better in many applications than type 304. Type 316 exhibits much better resistance to pitting, sulfuric acid, and hot organic acids. Corrosion resistance and heat resistance generally increase with nickel and chromium contents. For instance, type 310, also called 25-20, is one of the better heat-resistant alloys.

Alloy 20 does not have an AISI type number, but it is listed in group III because of its extensive use in corrosion applications. It was developed by the author in 1935. This alloy is best known in wrought form as Carpenter 20 and in the cast form as Durimet 20. It possesses the best overall corrosion resistance of any of the alloys listed in Table 5-2. It is made with and without columbium additions.

Group IV consists of the age-hardened or precipitation-hardened steels. They are hardened and strengthened by solution-quenching, followed by heating for substantial times at temperatures in the approximate range of 800 to 1000°F. Some of these steels are listed in Table 5-2. Tensile strengths up to around 200,000 $lb/in.^2$ can be obtained. The first five listed find their widest application in the aircraft and missile industry under relatively mild corrosion conditions. Corrosion resistance to severe environments is generally less than 18-8 except for CD-4MCu, which is very much better than the other five listed. It is also superior to the 18-8 steels. This alloy was developed at The Ohio State University during a research program sponsored by the Alloy Casting Institute.

The higher hardness of group IV steels reduces the tendency for seizing and galling of rubbing parts such as valve seats and disks.

Cast stainless steels and alloys usually have somewhat different chemical compositions than their wrought counterparts. One difference is higher silicon content, which increases castability. Silicon content is usually 1% maximum in wrought stainless steels. This is necessary particularly for casting thin sections or small parts. A cast part is not rolled or deformed after casting into shape, so the microstructure can be varied considerably with regard to ferrite content for "duplex" alloys. High ferrite content in an austenitic matrix substantially increases strength. The amount and extent of ferrite phase present can be controlled by raising the percentage of ferrite formers (Cr and Mo) and by decreasing the percentage of austenite formers and stabilizers (Ni, N, and C). (Ordinary 18-8 has enough nickel added to "balance" the alloy—that is, make it completely austenitic.) For those and other reasons the Alloy Casting Institute uses ACI designations for cast grades as shown in Table 5-3. Castings should always be specified by ACI designations. AISI type number counterparts are given in the second column of Table 5-3. C indicates grades used primarily for aqueous environments and

H for heat-resistant applications. Alloy 20 falls into the CN-7M grade. For most environments the corrosion resistance of cast and wrought structures are considered equivalent, particularly in the absence of specific corrosion tests in the given environment. Cast and wrought forms are usually used together—for example, a cast valve, or pump, and a wrought pipe.

Aside from the *duplex* alloys described in the four preceding paragraphs, several such alloys have appeared on the market in recent years. CD4MCu (Table 5-2) is old and its structure is approximately 60% ferrite and 40% austenite. The newer wrought Ferralium (see Table 5-9) is about 50-50. These

Table 5-4 Nominal mechanical properties of stainless steels

Material	Condition	Tensile strength, lb/in.²	Yield point, lb/in.² 0.2% offset	Elongation, % in 2 in.	Hardness	
					Rockwell	Brinell
Type 410	Annealed	75,000	40,000	30	B82	155
Type 410	Hardened, tempered at 600°F	180,000	140,000	15	C39	375
Type 410	Hardened, tempered at 1000°F	145,000	115,000	20	C31	300
Type 420	Annealed	95,000	50,000	25	B92	195
Type 420	Hardened, tempered at 600°F	230,000	195,000	25	C50	500
Type 440A	Annealed	105,000	60,000	20	B95	215
Type 440A	Hardened, tempered at 600°F	260,000	240,000	5	C51	510
Type 430	Annealed	75,000	45,000	30	B82	155
Type 446	Annealed	80,000	50,000	23	B86	170
Type 301	Annealed	110,000	40,000	60	B85	165
Type 301	Cold-worked, ½ hard	150,000	110,000	15	C32	320
Type 304	Annealed	85,000	35,000	55	B80	150
CF-8	Annealed (15% ferrite)	87,000	47,000	52	—	150
Type 304L	Annealed	80,000	30,000	55	B76	140
Type 310	Annealed	95,000	40,000	45	B87	170
Type 347	Annealed	92,000	35,000	50	B84	160
Alloy 20	Annealed	85,000	35,000	50	B84	160
17-7PH	Annealed	130,000	40,000	35	B85	165
17-7PH	Aged 950°F	235,000	220,000	6	C48	480
17-4PH	aged at 900°F	200,000	178,000	12	C44	420
14-8MoPH	Annealed	130,000	50,000	30	B85	162
14-8MoPH	Cold-rolled, aged at 900°F	280,000	270,000	2	C52	520
AM350	Annealed	160,000	55,000	40	B95	215
AM350	Aged at 850°F	220,000	190,000	13	C45	450
CD4MCu	Annealed	105,000	85,000	20	C25	240
CD4MCu	Aged at 950°F	140,000	120,000	15	C31	310

two alloys are quite similar in chemical composition. Both alloys exhibit good resistance to stress corrosion.

See Fig. 3-71 for the effect of ferrite on resistance to stress corrosion by CF-3, CF-8, and CF-8M, compared to types 304 and 316.

The Materials Technology Institute published in 1981 two manuals titled *Performance and Reliability of Corrosion Resistant Castings*. Manual No. 5 covers Phase 1, "Causes of Unsatisfactory Performance", and No. 6 covers Phase 2, "Casting Discontinuities."

Table 5-4 lists *mechanical properties* of several stainless steels in each of the four groups. Note the wide range of properties available from these materials. The high-strength materials exhibit good strength-weight ratios for aircraft and missile applications. High hardness is desirable for wear and some applications where resistance to erosion is required (e.g., trim for high-pressure steam valves).

The 200 and 300 series stainless steels all exhibit roughly the same mechanical properties in the annealed condition. The major exception involves cast alloys with duplex microstructures. Note the higher yield strength (and accordingly higher design strength) for CF-8 containing 15% ferrite as compared with type 304. The austenitic steels retain good ductility and impact resistance at very low temperatures and are used for handling liquid oxygen and nitrogen. FCC metals and alloys (austenitic stainless, copper, and aluminum) all possess good properties at cryogenic temperatures. They only method available for hardening the austenitic stainless steels is cold-working. This usually decreases corrosion resistance only slightly, but in certain critical environments a galvanic cell could form between cold-worked and annealed material. Type 301 is utilized mostly in the cold-worked condition for such applications as train and truck bodies. The austenitic steels can be cold-rolled to strengths in the neighborhood of 300,000 $lb/in.^2$ in wire form. Types 301 and 302 are not used for severe corrosion applications. The former because of its lower Cr and Ni, and the latter because of higher carbon.

High-temperature properties are discussed in Chap. 11.

5-9 Aluminum and Its Alloys

Aluminum is a reactive metal, but it develops an aluminum oxide coating or film that protects it from corrosion in many environments. This film is quite stable in neutral and many acid solutions but is attacked by alkalies. This oxide film forms in many environments, but it can be artificially produced by passage of electric current. This process is called anodizing. The high-copper alloys are utilized mainly for structural purposes. The copper-free or low-copper alloys are used in the process industries or where better corrosion resistance is required.

In addition to corrosion resistance, other properties contributing to its widespread application are colorless and nontoxic corrosion products, appearance, electrical and thermal conductivity, reflectivity, and lightness or good strength-weight ratio.

Pure aluminum is soft and weak, but it can be alloyed and heat-treated to a broad range of mechanical properties. An example is the original high-strength aluminum alloy containing about 4% copper. Figure 5-6 shows the aluminum-rich portion of the Al-Cu phase diagram. The *K* phase is a solid solution of copper in aluminum and the θ phase is the compound $CuAl_2$. The heat-treating sequence consists of heating in the *K* range and cooling rapidly by quenching. The alloy is then aged to precipitate the θ phase. The solubility of copper is very low at lower temperatures. This dispersed phase hardens and strengthens the material, but it decreases corrosion resistance, particularly resistance to stress corrosion. This is the reason the structural alloys are Alclad or covered with a thin skin of pure aluminum.

Table 5-5 lists compositions of several wrought and cast alloys. Alloy 3003 is the workhorse of these alloys in the process industries. Alloys are sand-cast, die-cast, or cast into permanent molds.

Table 5-6 lists mechanical properties of some aluminum alloys. This table illustrates the wide range of properties available. For example, the tensile strength of annealed commercially pure aluminum is 13,000 lb/in.^{2} and 88,000 lb/in.^{2} for heat-treated 7178 alloy. Alloy 5052 exhibits the highest strength of non-heat-treatable alloys. Number 7178 is one of the highest-strength heat-treatable alloys utilized in aircraft and aerospace applications.

Aluminum alloys lose strength rapidly when exposed to temperatures of 350°F and higher. Aluminum shows excellent mechanical properties at cryogenic temperatures because it is a face-centered cubic material.

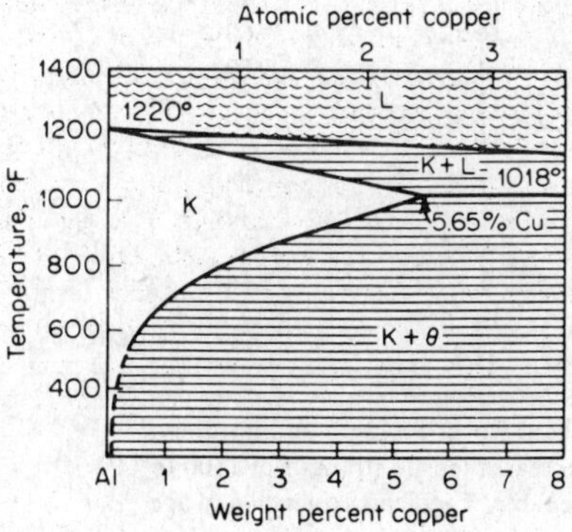

Figure 5-6 Aluminum end of the aluminum-copper diagram.

Table 5-5 Nominal composition of some wrought and cast aluminum alloys

Number	%Cu	%Si	%Mn	%Mg	%Cr	%Zn	%Fe	%Ti
				Wrought				
1100	Commercially pure—99.2% Al minimum							
2014	4.4	0.8	0.8	0.4	0.1	0.25	1.0	0.15
3003	0.2	0.6	1.2	—	—	0.1	0.7	—
5052	0.10	low	0.1	2.5	0.25	0.10	low	—
6061	0.25	0.6	0.15	1.0	0.25	0.25	0.7	0.15
7075	1.6	0.5	0.30	2.5	0.3	5.6	0.7	0.20
7178	2.0	0.50	0.3	2.7	0.3	6.8	0.7	0.20
				Cast				
43	—	5.0	—	—	—	—	—	—
195	4.5	0.8	—	—	—	—	—	—
220	—	—	—	10	—	—	—	—
356	—	7.0	—	0.3	—	—	—	—

Table 5-6 Nominal mechanical properties of aluminum alloys

Number	Temper*	Tensile strength, lb/in.^{2}	Yield strength, lb/in.^{2}	Elongation, %	Brinell hardness
			Wrought		
1100	O	13,000	5000	40	23
1100	H14	18,000	17,000	15	32
2014	O	27,000	14,000	18	45
2014	T6	70,000	60,000	13	135
3003	O	16,000	6000	35	28
3003	H14	22,000	21,000	12	40
5052	O	28,000	13,000	27	47
5052	H34	38,000	31,000	12	68
6061	O	18,000	8000	27	30
6061	T6	45,000	40,000	15	95
7075	O	33,000	15,000	16	60
7075	T6	83,000	73,000	11	150
7178	O	33,000	15,000	15	—
7178	T6	88,000	78,000	10	—
			Cast		
43	Sand cast	19,000	8000	8	40
43	Die cast	33,000	16,000	9	—
195	Sand, T6	36,000	24,000	5	75
220	—	46,000	25,000	14	—
356	Sand, T6	33,000	24,000	3.5	70

*O indicates annealed and H hardened by cold work. Strength and hardness increase with cold work; full cold work (H18) increases tensile strength of 1100 to 24,000 lb/in.^{2} 1100, 3003, and 5052 not heat-hardenable. T indicates quench and age heat treatment.

5-10 Magnesium and Its Alloys

Magnesium is one of the lightest commercial metals, specific gravity 1.74. It is utilized in trucks, automobile engines, ladders, portable saws, luggage, aircraft, and missiles because of its light weight and also good strength when alloyed. However, it is one of the least corrosion resistant and is accordingly used as sacrificial anodes for cathodic protection (Chap. 3) and dry-cell batteries. It is generally anodic to most other metals and alloys and must be insulated from them.

Magnesium exhibits good resistance to ordinary inland atmospheres due to the formation of a protective oxide film. This protection tends to break down (pits) in air contaminated with salt, and protective measures are required. These include coatings and "chrome" pickling, which also provides a good base for the coating. Corrosion resistance generally decreases with impurities and alloying. Alloys are quite susceptible to stress corrosion and must be protected. Presence of dissolved oxygen in water has no significant effect on corrosion. The metal is susceptible to erosion corrosion. Magnesium is much more resistant than aluminum to alkalies. It is attacked by most acids except chromic and hydrofluoric. The corrosion product in HF acts as a protective film.

Magnesium and its alloys are available in a variety of wrought forms and die castings. Tensile strengths in the approximate range 15,000 to 50,000 $lb/in.^2$ are obtainable.

5-11 Lead and Its Alloys

Lead is one of our oldest metals. It was used for water piping during the time of the Roman Empire, and some of it is still in operation. Lead ornaments and coins were utilized several thousand years ago. Lead forms protective films consisting of corrosion products such as sulfates, oxides, and phosphates. Most of the lead produced goes into corrosion applications; a large portion involves sulfuric acid. (See isocorrosion chart in Chap. 7.) Lead and its alloys are used as piping, sheet linings, solders (Pb-Sn), type metals, storage batteries, radiation shields, cable sheath, terneplate (steel coated with Pb-Sn alloy), bearings, roofing, and ammunition. Lead is soft, easily formed, and has a low melting point. Lead-lined steel is often made by "burning on" the lead. It is subject to erosion corrosion because of its softness.

When corrosion resistance is required for process equipment, chemical lead containing about 0.06% copper is specified, particularly for sulfuric acid. This lead is resistant to sulfuric, chromic, hydrofluoric, and phosphoric acids; neutral solutions; seawater; and soils. It is rapidly attacked by acetic acid and generally not used in nitric, hydrochloric, and organic acids.

Chemical lead exhibits a tensile strength of about 2300 $lb/in.^2$ at room temperature. Hard leads, containing 3 to 18% antimony, double this strength. However, the strength of both materials drops rapidly as temperature is increased, and they show about the same strength around 230°F. Design stress at appreciably higher temperatures drops to zero.

5-12 Copper and Its Alloys

Copper is different from most other metals in that it combines corrosion resistance with high electrical and heat conductivity, formability, machinability, and strength when alloyed, except for high temperatures. Copper exhibits good resistance to urban, marine, and industrial atmospheres and waters. Copper is a noble metal, and hydrogen evolution is not usually part of the corrosion process. For this reason it is not corroded by acids unless *oxygen* or other *oxidizing agents* (e.g., HNO_3) are present. For example, reaction between copper and sulfuric acid is not thermodynamically possible; but corrosion proceeds in the presence of oxygen, and the products are copper sulfate and water. Reduction of oxygen to form hydroxide ions is the predominant cathodic reaction for copper and its alloys. Copper-base alloys are resistant to neutral and slightly alkaline solutions with the exception of those containing ammonia, which cause stress corrosion and sometimes rapid general attack. In strongly reducing conditions at high temperatures (300 to 400°C), copper alloys are often superior to stainless steels and stainless alloys. Dezincification and stress corrosion are discussed in Chap. 3.

Table 5-7 lists chemical compositions and mechanical properties of some typical and common copper-base materials. Hundreds of compositions with a wide variety of mechanical properties are commercially available. For example a range of tensile strength from about 30,000 to 200,000 lb/in.^{2} is exhibited by pure copper and copper alloyed with 2% beryllium. The most common alloys are brasses (Cu-Zn), bronzes (Sn, Al, or Si additions to Cu), and cupronickels (Cu-Ni). Nickel silver does not contain silver but resembles it in appearance. This alloy is often the base for silver-plated tableware. Everdur possesses relatively high strength, and this property is utilized in hardware such as nuts, bolts, and valve stems for copper-base equipment, thus minimizing two-metal corrosion.

Copper and brasses are subject to erosion corrosion or impingement attack. The bronzes and aluminum brass are much better in this respect. The bronzes are stronger and harder. The cupronickels with small iron additions are also superior in erosion-corrosion resistance.

Copper and copper alloys are available as duplex tubing (inside one metal, outside another) in combination with steel, aluminum, and stainless steels. This construction solves many heat-exchanger materials problems. For example, tubing with ammonia on one side (steel) and brackish water on the outside (Admiralty Metal).

Copper and its alloys find extensive application as water piping, valves,

Table 5-7 Typical chemical composition and mechanical properties of some copper alloys

		Tensile strength, lb/in.^{2}		Yield strength, lb/in.^{2}		Elongation, %		Rockwell hardness	
Material	Composition, %	Hard*	Soft	Hard	Soft	Hard	Soft	Hard	Soft
High-purity copper	99.9$^+$ Cu	46,000	33,000	40,000	10,000	5	40	B-50	F-35
Beryllium copper	98 Cu, 1.9 Be, 0.2 Ni or Co	200,000†	70,000	150,000	30,000	2	35	C-38	B-65
Red brass	85 Cu, 15 Zn	70,000	40,000	55,000	15,000	7	45	B-76	B-5
Casting brass	85 Cu, 5 Zn, 5 Pb, 5 Sn	—	33,000	—	15,000	—	25	—	B-7
Cartridge brass	70 Cu, 30 Zn	76,000	48,000	60,000	17,000	10	65	B-83	B-20
Muntz metal	60 Cu, 40 Zn	80,000	54,000	60,000	20,000	15	45	B-87	B-45
Phosphor bronze A	95 Cu, 5 Sn, 0.25 P	80,000	48,000	65,000	20,000	8	50	B-86	B-28
Phosphor bronze D	90 Cu, 10 Sn, 0.25 P	102,000	66,000	70,000	28,000	12	65	B-98	B-55
Aluminum bronze	92 Cu, 8 Al	105,000	65,000	65,000	25,000	7	60	B-96	B-50
Everdur 1010	96 Cu, 3 Si, 1 Mn	95,000	58,000	60,000	22,000	7	60	B-92	B-35
Aluminum brass (As)	77 Cu, 21 Zn, 2 Al, 0.04 As	85,000	52,000	60,000	20,000	10	65	B-85	B-30
Admiralty arsenical	71 Cu, 28 Zn, 1 Sn, 0.04 As	—	48,000	—	18,000	—	65	—	B-25
Cupronickel 10	88 Cu, 10 Ni, 1 Fe, 0.4 Mn	60,000	40,000	57,000	22,000	15	46	B-68	B-25
Cupronickel 30	69 Cu, 30 Ni, 0.5 Fe, 0.6 Mn	70,000	55,000	60,000	22,000	10	45	B-80	B-35
Nickel silver	65 Cu, 25 Zn, 10 Ni	88,000	55,000	70,000	20,000	7	42	B-87	B-30

*Cold-rolled.

†Cold-rolled and age-hardened.

Table 5-8 Nominal composition and mechanical properties of some nickels and nickel alloys

Material	%Ni	%C	%Cr	%Mo	%Cu	%Fe	Other	Tensile strength, lb/in.^{2}	Yield point, lb/in.^{2}	Elongation, %	Rockwell hardness
Nickel 200	99.5	0.06	—	—	0.05	0.15		60,000	20,000	40	B-50
Cast nickel 210	95.6	0.80	—	—	0.05	0.50	1.6 Si, 0.9 Mn	50,000	25,000	20	B-55
Duranickel 301	94	0.15	—	—	0.05	0.15	4.5 Al, 0.5 Ti	170,000*	130,000	15	C-34
Monel 400	66	0.15	—	—	31	1.4		75,000	30,000	40	B-65
"K" Monel	66	0.15	—	—	29	0.9	3 Al	150,000*	110,000	25	C-28
"S" Monel (cast)	63	0.10	—	—	30	2	4 Si	120,000*	90,000	2	C-31
Inconel 600	76	0.08	16	—	0.2	8		85,000	35,000	40	B-70
Chlorimet 2 (cast)	62	0.03	—	32	—	3	1 Si	80,000	55,000	10	B-90
Hastelloy B (wrought)	62	0.10	—	28	—	5	1 Si	130,000	56,000	50	B-92
Chlorimet 3 (cast)	60	0.03	18	18	—	3	0.6 Si	75,000	50,000	15	B-90
Hastelloy C (wrought)	56	0.08	15	17	—	5	1 Si, 4 W	120,000	52,000	49	B-94
Hastelloy D (cast)	85	0.12	—	—	3	—	10 Si	115,000	115,000	0	C-34
Hastelloy F	47	0.05	22	7	0.1	17	1 W, 3 Co	102,000	46,000	45	B-86
Ni-o-nel 825	42		22	3	1.8	30	1 Ti	90,000	35,000	50	B-91
Ilium G (cast)	56	0.07	22.5	6.5	6.5	6		68,000	38,000	7	B-70
Nichrome	80	—	20	—	—	—		95,000	35,000	30	B-85

*Age-hardened.

Note: Most of the materials are listed by their registered trademark names and all are commonly so designated. Durimet 20 (Alloy 20) with 29% Ni is listed in Table 5-2.

heat-exchanger tubes and tube sheets, hardware, wire, screens, shafts, roofing, bearings, stills, tanks, and other vessels.

5-13 Nickel and Its Alloys

An important group of materials for corrosion applications is based on nickel. Nickel is resistant to many corrosives and is a natural for alkaline solutions. Most tough corrosion problems involving caustic and caustic solutions are handled with nickel. In fact, the corrosion resistance of alloys to sodium hydroxide is roughly proportioned to their nickel content. For example, 2% Ni cast iron is much superior to unalloyed cast iron. Another important attribute is the large and rapid increase in stress-corrosion resistance as the nickel content of stainless alloys increases above 10%. For example, Inconel shows excellent stress-corrosion resistance, and many tons of it are used for this reason. Nickel generally shows good resistance to neutral and slightly acid solutions. It is widely used in the food industry. It is not resistant to strongly oxidizing solutions, e.g., nitric acid, and ammonia solutions. Nickel forms a good base for alloys requiring strength at high temperatures. However, nickel and its alloys are attacked and embrittled by sulfur-bearing gases at elevated temperatures.

Table 5-8 lists some common nickels and nickel-base alloys. A wide range of compositions and mechanical properties are available. Aged Duranickel possesses very good mechanical properties and also good

Table 5-9 Approximate composition of newer alloys

Alloy	UNS number	%Ni	%Cr	%Mo	%C	% other
Carpenter 20Cb-3	N08020	35	20	2.5	0.03	3 Cu
Incoly 800	N08800	32	21	—	0.03	46 Fe
Incoloy 625	N06625	60	22	9	0.04	3.5 Cb
Incoloy 825	N08825	43	22	3	0.04	2 Cu
Hastelloy B-2	N10665	69	—	28	0.02	1 Fe
Hastelloy C-276	N10276	57	15	16	0.02	1 Fe
Hastelloy C-4	N06455	65	16	15	0.015	2 Fe
Hastelloy G-3	—	44	22	7	0.015	20 Fe, 2 Cu
Ferralium 255	S32550	5.5	26	3	0.04	1.7 Cu, 62 Fe
Chlorimet 3	—	60	18	18	0.015	3 Fe
Nitronic 50	S20910	13	22	2	0.05	5 Mn
Al-6X	N08366	24	21	6.5	0.03	2 Mn
JS-700	N08700	25	21	4	0.04	0.5 Cu, 0.4 Cb
JS-777	—	25	21	4.5	0.04	2.2 Cu

Trade marks: Carpenter, Carpenter Technology Corp.; Incoloy, International Nickel Co.; Hastelloy, Cabot Corp; Chlorimet 3, Duriron Co.; Nitronic, Armco, Inc.; Al-6X, Allegheny Ludlum Steel Corp.; JS-700 and JS-777, Jessop Steel Co.

resistance to many environments. Monel is a natural for hydrofluoric acid. Chlorimet 3 and Hastelloy C are two of the most generally corrosion-resistant alloys commercially available. Chlorimet 2 and Hastelloy B are very good in many cases where oxidizing conditions do not exist. Hastelloy D is brittle like the high-silicon irons. Nichrome is used for electrical resistors—heating elements. Nickel, high-nickel alloys, and alloys containing substantial amounts of nickel (8%) are the workhorses for most of the more severe corrosion problems.

Table 5-9 shows compositions of newer alloys that generally show better corrosion resistance than the 300 series stainless steels and are used for more severe corrosion services (for example, to combat pitting and stress corrosion). Most are high-nickel alloys. Apparently there is duplication in Tables 5-8 and 5-9 but those in Table 5-9 have much lower carbon contents made possible by steel making practices developed in recent years. Alloys in Table 5-9 involve extensive application for several years and are well established.

5-14 Zinc and Its Alloys

Zinc is not a corrosion-resistant metal, but it is utilized as a sacrificial metal for cathodic protection of steel. Its chief use is in galvanized (zinc-coated) steel for piping, fencing, nails, etc. It is also utilized in the form of bars or slabs as anodes to protect ship hulls, pipelines, and other structures.

Zinc alloy parts are made by die-casting because of their low melting points. Many automobile components such as grilles and door handles are die-cast but are usually plated with corrosion-resistant metals.

5-15 Tin and Tin Plate

Over half of the tin produced goes into coating other metals, primarily steel to make the "tin" can. In addition to offering corrosion resistance, tin-coated steel is easily formed and soldered, provides a good base for organic coatings, is nontoxic, and has a pleasant appearance. Tin coating is applied by dipping, but most of it is electroplated. Two advantages here are controlled "thinness" and the fact that steel can be plated with different thicknesses on each side of the sheet—usually thicker on the inside of the can than on the outside. Tin cans are used for food products, beverages, petroleum products, and paints. Alloys of tin with zinc, nickel, cadmium, or copper can be deposited as electroplates.

Tin should be cathodic to iron, but the potential reverses in most sealed cans containing food products and the tin acts as a sacrificial coating, thus protecting the steel. Complex-ion formation apparently causes this reversal. Tin is relatively inert, but in the presence of oxygen or other oxidizing agents it is attacked.

Tin shows excellent resistance to relatively pure water. Solid tin pipe and sheet and also tinned copper are utilized for producing and handling distilled water. Collapsible tubes are used for dentifrices and medicants. Tin foil is well known as a wrapping material but has lost most of this market to cheaper plastic films. Solders are another important outlet for tin. Tin and lead contents vary from 30 to 70%. Tin babbits are used for bearings.

Tin shows good resistance to atmospheric corrosion, dilute mineral acids in the absence of air, and many organic acids, but is corroded by strong mineral acids. It is generally not used for handling alkalies.

Pewter, an important tin alloy, is used for pitchers, flower vases, and other containers. It contains 6 to 7% antimony and 1 to 2% copper. Tin bronzes are discussed under copper alloys.

Tin is a weak, soft, and ductile metal. Tensile strength at room temperature is around 2500 lb/in.^{2} and this drops rapidly with increasing temperature.

5-16 Cadmium

Except for low-melting alloys and some bearings, cadmium is used almost exclusively as an electroplated coating. Its bright appearance and ease of soldering accounts for its use in electronic equipment and hardware. Cadmium is less electronegative than zinc and not as effective as zinc as a sacrificial anode in many cases. It is somewhat less protective than zinc for steel in industrial atmospheres. Cadmium is more expensive than zinc and its salts are toxic. Unlike zinc, cadmium has some measure of resistance to alkalies.

Cadmium plating is utilized on high-strength steels in aircraft because of improved resistance to corrosion fatigue. However, hydrogen embrittlement is a problem. Proper plating techniques and baking out the hydrogen after plating have solved this problem. At temperatures in the neighborhood of the melting point (610°F) steel can be attacked by cadmium.

Cadmium is a relatively weak metal with a tensile strength around 10,000 lb/in.^{2} and eleongation of 50%.

5-17 Titanium and Its Alloys

Titanium is a relative newcomer in that it was first used as a structural metal in 1952. It is strong and has a specific gravity of 4.5, about halfway between aluminum and steel. These materials have a high strength-to-weight ratio and accordingly they were first adapted for aircraft and ordnance. In addition, they are now utilized in missiles and in the chemical process industries. The author was involved as consultant in two of the largest tonnage applications of titanium in chemical plants. Titanium is a reactive metal and depends on a protective film (TiO_2) for corrosion resistance. Melting and welding must be done in inert environments or the metal becomes brittle due to absorbed gases. For this reason, it is not a high-temperature material.

Titanium possesses three outstanding characteristics that account for much of its application in corrosive services. These are resistance to (1) seawater and other chloride salt solutions, (2) hypochlorites and wet chlorine, and (3) nitric acid including fuming acids. Salts such as $FeCl_3$ and $CuCl_2$ that tend to pit most other metals and alloys actually inhibit corrosion of titanium. It is not resistant to relatively pure sulfuric and hydrochloric acids but does a good job in many of these acids when they are heavily contaminated with heavy metal ions such as ferric and cupric. Titanium shows surprisingly low two-metal effects because it readily passivates.

Titanium shows a pyrophoric tendency in red fuming nitric acid with high NO_2 and low water content and also in dry halogen gases.

Alloying with about 30% Mo greatly increases resistance to hydrochloric acid. Small amounts of tin reduces scaling losses during hot-rolling. Small additions of palladium, platinum, and other noble metals increase resistance to moderately reducing mediums. One such commercial titanium alloy contains about 0.15% palladium. Many elements alloy with titanium; commercial alloys include aluminum, chromium, iron, manganese, molybdenum, tin, vanadium, and zirconium.

Table 5-10 lists mechanical properties of titanium and a few commercial alloys. The alloys fall into three metallurgical types, as indicated in the table. Titanium has a close-packed hexagonal crystal structure (alpha) at room temperature. It transforms to body-centered cubic (beta) at 1625°F. Various elements tend to stabilize or favor one or the other structure. The alpha alloys are generally more ductile and weldable. Two-phase structures can be strengthened by heat treatment. Titanium has a low modulus of elasticity (16,800,000 lb/in.^{2}) as compared with iron (28,500,000 lb/in.^{2}).

Table 5-10 Nominal composition and mechanical properties of titanium and some alloys

Material	Condition	Tensile strength, lb/in.^{2}	Yield strength, lb/in.^{2}	Elongation, %
Commercial Ti	Annealed	85,000	70,000	26
		Alpha alloy		
5% Al, 2.5% Sn	Annealed	125,000	120,000	18
		Alpha-beta alloys		
8% Mn	Annealed	140,000	125,000	15
4 Al, 3 Mo, 1 V	Heat-treated	195,000	165,000	6
6 Al, 4 V	Annealed	135,000	120,000	11
6 Al, 4 V	Heat-treated	170,000	150,000	7
		Beta alloy		
3 Al, 13 V, 11 Cr	Heat-treated	180,000	170,000	6

With reference to Table 5-10, several grades of commerical titanium are available. The most widely used is ASTM Grade 2, which contains some oxygen. Grade 4 (listed as commercial Ti in the table) contains some oxygen and iron. Grade 7 is an alloy with 0.15% Pd that exhibits improved resistance to crevice corrosion but is infrequently used because of high cost. Grade 12 (also known as TiCode 12) contains 0.3% Mo and 0.8% Ni, in many cases has equivalent crevice corrosion resistance to Grade 7, and is 25% *stronger* than the other grades mentioned. This increased strength permits thinner sections and higher temperatures from the design standpoint. Some difficulty has been encountered in welding thick pieces of TiCode 12, but they were solved by pickling the surface before welding.*

Titanium and its alloys have found wide application as cast pumps and valves. Popular alloys are commercially pure Grade 3 and the stronger alloy, Grade 5 (Ti-6Al-4V).

Some words of caution are necessary to avoid unexpected failures (which have occurred). Titanium in contact with a corroding metal, such as steel, absorbs hydrogen and becomes brittle. Iron particles imbedded in the metal surface (improper cleaning or pickling) during forming results in severe pitting. Cleaning titanium equipment with inadequately inhibited hydrochloric acid showed rapid corrosion. Improper supports in heat exchangers resulted in fatigue, or corrosion fatigue, failures of titanium tubing. Titanium should not be used in straight (relatively pure) hydrochloric, sulfuric, and hydrofluoric acids or dry chlorine. Catastrophic failures of heat exchangers were reported due to contact with Teflon and a plastic containing lead.

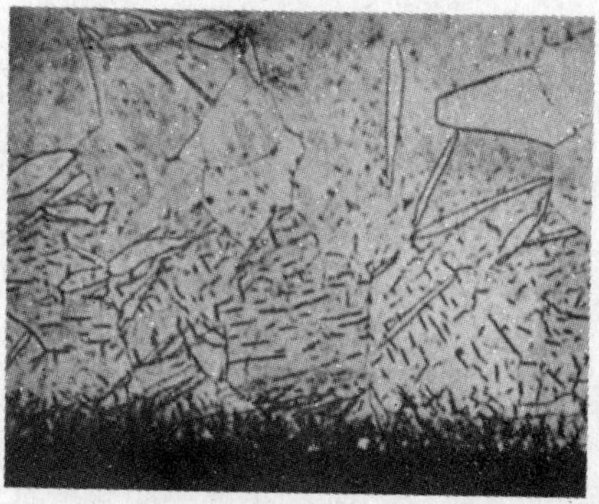

Figure 5-7 Hydride needles formed in metal because of acid corrosion of titanium.

*T. P. Oettinger, M. C. Hoffman, and M. G. Fontana. "Use of Titanium for Treatment of Toxic and Hazardous Wastes," ASTM Symposium on Titanium and Zirconium, Philadelphia, Oct. 10–11, 1984.

Figure 5-7 shows typical hydride needles formed because of acid corrosion of the titanium.*

5-18 Refractory Metals

These metals are characterized by very high melting points as compared with iron and steel. The jet engine and outer-space programs have provided the impetus in making these metals commercially available. Unfortunately, they show poor resistance to high-temperature oxidation, and protective measures are needed. They are discussed further in Chap. 11.

Aside from tantalum, columbium (niobium), molybdenum, tungsten, and zirconium are relative newcomers in the field of corrosion by aqueous solutions. Many alloys of these metals are commercially available.

Melting points of these metals are shown in Table 5-11.

Columbium This metal exhibits good corrosion resistance to organic and inorganic acids except hydrofluoric acid and hot concentrated sulfuric and hydrochloric acids. Apparently the formation of a Cb_2O_5 film results in protection. Columbium is poor in alkaline solutions.

Molybdenum Molybdenum shows good resistance to hydrofluoric, hydrochloric, and sulfuric acids, but oxidizing agents such as nitric acid cause rapid attack. It is good in aqueous alkaline solutions. This metal forms a volatile oxide (MoO_3) in air at temperatures above about 1300°F. One great advantage of molybdenum from the mechanical standpoint is its high modulus of elasticity—50,000,000 lb/in.^{2}. This means it is stiffer and will deflect much less than steel under a given load and cross section.

Table 5-11 Melting point of refractory metals

Metal	Melting point °F	Melting point °C
Columbium	4474	2468
Molybdenum	4730	2610
Tantalum	5425	2996
Tungsten	6170	3410
Zirconium	3366	1852
Iron (for comparison)	2798	1536

*Additional information on titanium can be found in S. R. Seagle and B. P. Bannon, Titanium: Its Properties and Uses, *Chem. Eng.*, pp. 111–113 (March 8, 1982); G. Bewer, H. Debrodt, and H. Herbst, Titanium for Electrochemical Processes, *J. Metals*, pp. 37–41 (Jan. 1982); and D. M. McCue, Titanium for Shell and Tube Exchangers, *Shell and Tube Heat Exchangers*, American Society for Metals, Metals Park, Ohio, pp. 45–51 (1982).

Tantalum Because of its superior resistance to most environments tantalum has been used for many years. A few exceptions include alkalies, hydrofluoric acid, and hot concentrated sulfuric acid. It is used in handling chemically pure solutions such as hydrochloric acid. Because of tantalum's wide spectrum of corrosion resistance it is utilized in repair of glass-lined equipment. Any evolution of hydrogen (corrosion reaction or otherwise) near tantalum will result in absorption and embrittlement of this metal. Desorption of hydrogen to restore ductility is not practical because of the high temperature and high vacuum required. Tantalum sheet is strong, so

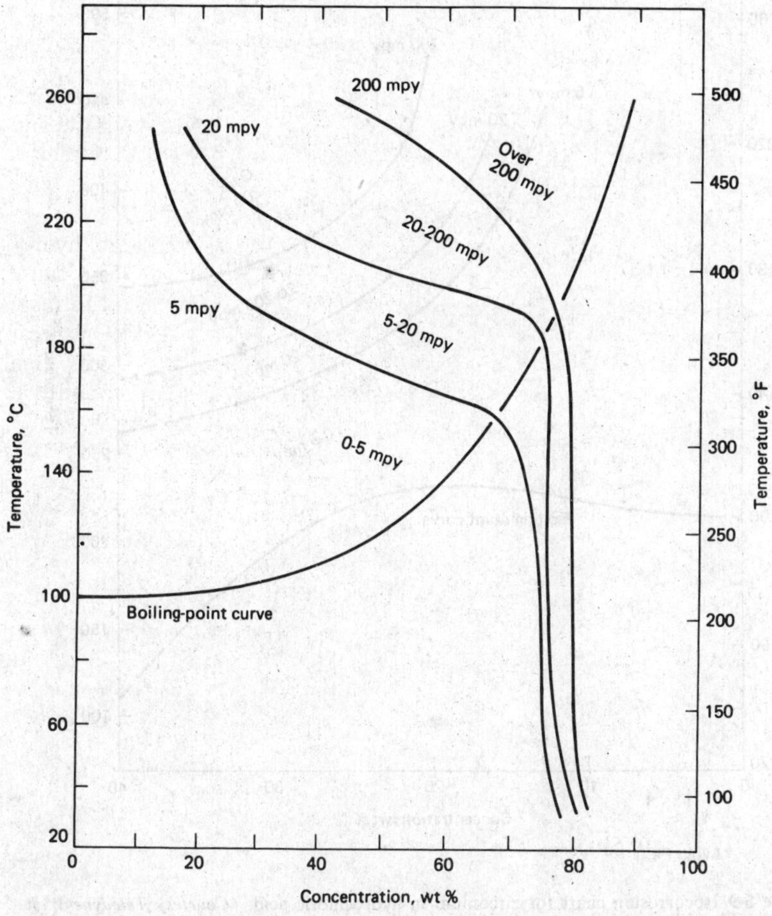

Figure 5-8 Isocorrosion chart for zirconium in sulfuric acid. *(Courtesy Teledyne Wah Chang, Albany, Ore.)*

cost is reduced through the use of thin sections. Tantalum is used for surgical implants.

A new Cabot Corporation Alloy, KBI-40, contains 60% tantalum and 40% columbium. It is *less expensive* than tantalum, its density is about 25% less, its tensile strength is one-third greater, it is ductile, and it is equal in corrosion resistance to tantalum in most of the severe environments where tantalum is used. KBI-40 is available in all wrought forms and suggested

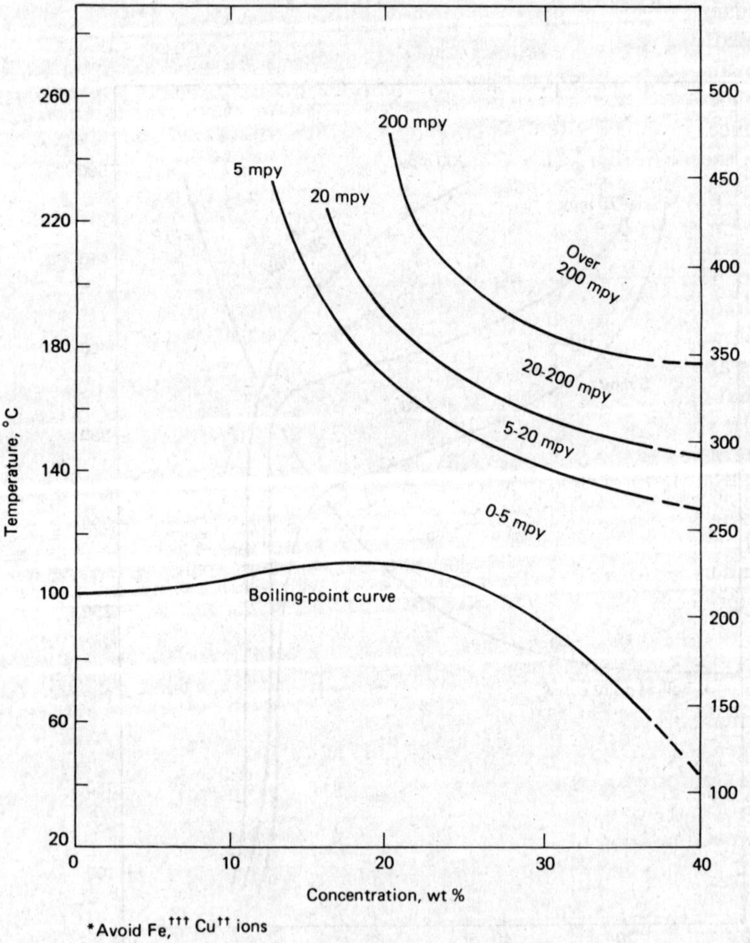

Figure 5-9 Isocorrosion chart for zirconium in hydrochloric acid. *(Courtesy Teledyne Wah Chang, Albany, Ore.)*

uses include heat exchanger tubes, coils, U-tubes, and linings for vessels and piping. It is generally much more corrosion-resistant than columbium.

Tungsten As indicated in Table 5-11, tungsten has the highest melting point of any metal. Its chief use involves strength at high temperatures. A common example is filaments in light bulbs. Tungsten shows good resistance to acids and alkalies, but it is not commonly used for aqueous solutions. Tungsten exhibits a tensile strength of 20,000 lb/in.^{2} at 3000°F.

Zirconium With the advent of the Atomic Age, zirconium has found increased use primarily because it has a very low thermal neutron cross section and resists high-temperature water and steam. Its excellent corrosion resistance is due to a protective oxide film. This metal exhibits good corrosion resistance to alkalies and acids (including hydriodic and hydrobromic acids), except for hydrofluoric acid and hot concentrated hydrochloric and sulfuric acids. Cupric and ferric chlorides cause pitting. Zirconium has found some application in hydrochloric acid service. Its corrosion resistance is affected by impurities in the metal such as nitrogen, aluminum, iron, and carbon.

Zirconium alloyed with amounts of tin, iron, chromium, and nickel (zircalloys) shows improved resistance to high-temperature water. Zirconium and its alloys in high-temperature water shows first a decreasing corrosion rate that may be followed by a rapid linear rate of attack (termed "break-away"). They also tend to pick up hydrogen (hydriding) from the corrosion reaction and becomes brittle. Zirconium possesses a tensile strength of about 16,000 lb/in.^{2} at 800°F and 80,000 lb/in.^{2} at room temperature. Its modulus of elasticity is 13,700,000 lb/in.^{2}.

Figures 5-8 and 5-9 are isocorrosion charts for zirconium in sulfuric acid and hydrochloric acid. Note the emphasis in Fig. 5-9 on avoiding ferric and cupric ions as mentioned two paragraphs above.*

5-19 Noble Metals

These materials are characterized by highly positive potentials relative to the hydrogen electrode, excellent corrosion resistance, unstable oxides (do not oxidize), and high cost. The latter accounts for the term *precious* metals. In most cases, formation of protective films (passivity) is not required. The noble metals are gold, silver, platinum, and other five "platinum" metals: iridium, osmium, palladium, rhodium, and ruthenium. Gold, silver,

*For much additional information on other corrosives and mechanical properties of zirconium and its alloys, see D. R. Knittel, "Zirconium," *Chem. Eng.*, pp. 95–98 (June 2, 1980), and R. T. Webster, "Applications of Zirconium and Niobium," *Shell and Tube Heat Exchangers*, American Society for Metals, Metals Park, Ohio, pp. 51–67 (1982).

platinum, and palladium are available in most commercial forms; the first three are used extensively in industry. The other noble metals are utilized chiefly as alloying elements, for example, Pt-Rh thermocouple wire. The use of noble metals for jewelry is well known. In spite of their high cost, the noble metals are the most economical for numerous corrosion applications. Their high scrap value is an advantageous characteristic. Cladding (veneer), linings, and coatings with inexpensive substrates providing strength are common combinations.

Gold Gold is one of the oldest metals utilized by man because it is found in nature in the pure state. Jewelry and coinage represent its first applications, with jewelry accounting for most of its consumption today. In these cases, gold is alloyed with copper because of the extreme softness of pure gold. Gold coins are rare. A karat is $\frac{1}{24}$ part, so fine or pure gold is 24 karats. Twelve karat would be 50% gold. In addition, gold and its alloys are used for dental inlays, electrical contacts, plating, tableware, process equipment (e.g., condensers and stills), printed circuits, gold leaf in surgery and body implants, and decorative purposes. A thin gold underplate (topped with cadmium) prevents hydrogen embrittlement of high-strength steels. Gold leaf and foil are readily produced because of the softness and excellent malleability of pure gold.

Gold is very stable in dilute nitric acid and hot strong sulfuric acid. It is attacked by aqua regia, concentrated nitric acid, chlorine and bromine, mercury, and alkaline cyanides.

Platinum Because it is resistant to many oxidizing environments and particularly air at high temperatures, platinum is utilized for thermocouples (Pt-PtRh), tanks and spinnerets for molten glass, crucibles for chemical analytical work, windings for electrical resistance furnaces (alloys up to 3200°F), and combustion or reaction chambers for extremely corrosive environments at temperatures over 1800°F. Sheet linings on a "base" alloy substrate are used for the latter. An applied oxide layer (e.g., Al_2O_3) prevents alloying or reaction with the substrate material. It has replaced fused quartz in several chemical applications of this type. Incidentally, contact with silica at very high temperatures embrittles the platinum. The inertness of platinum is attested by its extensive service as a catalyst.

Other uses for platinum and its alloys are spinnerets (70 Au, 30 Pt) for rayon, safety or frangible disks up to 900°F (silver is good only up to 300°F and gold to 160°F), sulfuric acid absorbers, electroplating anodes, impressed-current anodes for cathodic protection, other chemical equipment, and high-quality jewelry. It is not attacked by mercury.

Platinum exhibits good mechanical properties at high temperatures as shown in Table 5-12. Gold and silver are poor in this respect.

Platinum is attacked by aqua regia, hydriodic and hydrobromic acids, ferric chloride, and chlorine and bromine.

Table 5-12 Some mechanical and physical properties of noble metals

Material	Temperature, °F	Tensile strength, lb/in.^{2}	Yield strength, lb/in.^{2}	Elongation, %	Brinell hardness	Modulus of elasticity	Melting point, °F
Silver	Room	18,000	8000	55	26	11,000,000	1761
Gold	Room	19,000	nil	70	25	11,600,000	1945
70 Au-30 Pt	Room	29,000	3500	—	130	16,500,000	2640
Platinum	Room	21,000	<2000	40	40	21,000,000	3217
Platinum	750	13,000					
Platinum	1830	4000					
Platinum	2190	2400					

Silver Silver is best known for its use as coinage and tableware, in solid and plated form. Sterling silver contains not more than 7.5% copper for hardness. Silver loses its "nobility" in contact with sulfur compounds—tarnishing is well known. It serves as electrical contacts, electrical bus bars (even at red heat), brazes, solders, and dental alloys with mercury. Silver is widely used in the chemical industry as solid silver and also as loose, clad, or brazed linings. Applications include stills, heating coils, and condensers for pure hydrofluoric acid, evaporating pans for production of chemically pure anhydrous sodium hydroxide, autoclaves for production of urea, and all kinds of equipment for production of foods and drugs where purity of product is paramount. It is highly resistant to organic acids.

Silver is attacked by nitric acid, hot hydrochloric acid, hydriodic and hydrobromic acids, mercury, and alkaline cyanides and may be corroded by reducing acids if oxidizing agents are present.

5-20 Metallic Glasses

This new class of materials, sometimes called amorphous alloys, became commercially available in the mid 1970s. These alloys can be quenched from the liquid state without crystallization and possess undercooled liquid structures similar to those of inorganic glasses. Typical alloys include at least one transition or noble metal and one or more metalloids (boron, carbon, phosphorus, silicon, etc.). The compositions are adjusted to be close to low-melting, stable eutectics that promote noncrystalline solidification during rapid quenching at rates near 10^6 degrees per second. This is accomplished by introducing the liquid alloy between two rotating metal drums yielding thin foils or wires.

Metallic glasses are usually very strong, ductile, and stiff (high modulus of elasticity). In addition, some of the iron-based glasses have corrosion resistances approaching tantalum or the noble metals. These materials passivate very easily at low chromium contents as shown in Fig. 5-10. Type 304 stainless steel is included here for comparison. Note that the 5%

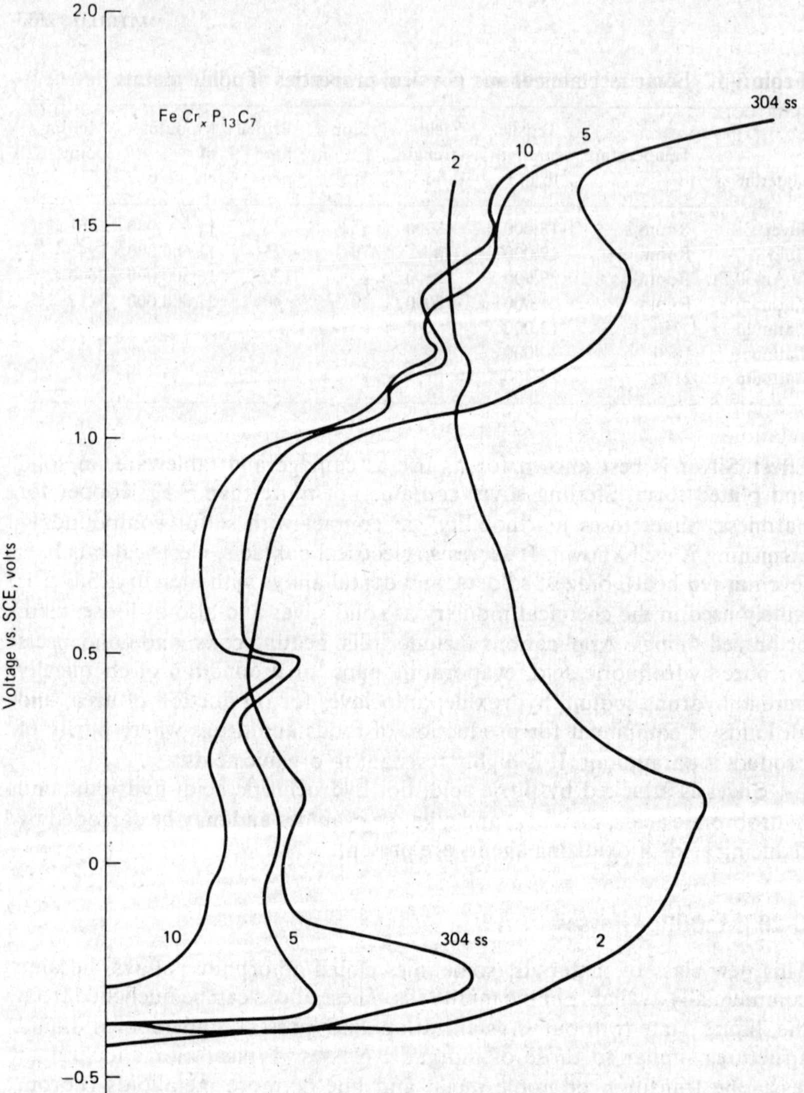

Figure 5-10 Potentiostatic anodic polarization of metallic glasses in $1\bar{N}\ H_2SO_4$ at room temperature [*After M. Naka et al., J. Japan Inst. Metals*, **38**:*835 (1974) and Corrosion*, **32**:*146 (1976)*.]

chromium glass passivates easier (i.e. lower critical anodic current density) than conventional 18-8 stainless steel. Further increases in chromium produce additional improvement.

The iron-based metallic glasses also exhibit outstanding resistance to hot, acid chloride solutions as shown in Table 5-13. At 8% chromium, the glasses are superior to conventional stainless steel alloys—their pitting resistance is equal to or greater than that of the high-nickel alloys (Hastelloy alloy C) and titanium.

The outstanding corrosion resistance of the metallic glasses is appreciably reduced by heat treatments which cause crystallization. This suggests that the unique corrosion properties of these materials are due to their amorphous structures. However, their high concentration of inert metalloids is probably a contributing factor also.

There are some limitations to applications of metallic glasses. A recent report indicates they are susceptible to hydrogen embrittlement under certain conditions. More important, their preparation dictates that one dimension must always be very small, usually less than 1 to 2 mils. Thus only thin foils and wires can be produced. Since there is no limit to the width of foils that can be manufactured, it is likely that large sheets of metallic glasses bonded to substrate metals will be available in the future.

Table 5-13 Corrosion resistance of stainless steels and metallic glasses in 6% ferric chloride at 60°C (20-hour exposure)

Alloy	Composition[a]	Weight loss	Remarks
Type 430[b]	Fe 18Cr	87%	Extreme pitting
Type 304[b]	Fe 20Cr 8Ni	6%	Pitting
Type 316[b]	Fe 18Cr 10Ni 2Mo	1.5%	Pitting
Metglas 2826 A[b]	Fe 14Cr 35Ni 12P 6B	nil	No attack
—[c]	Fe 8Cr 13P 7C	nil	No attack
—[c]	Fe 10Cr 13P 7C	nil	No attack
—[c]	Fe 15Cr 13P 7C	nil	No attack
—[c]	Fe 20Cr 13P 7C	nil	No attack
—[c]	Fe 10Cr 5Ni 13P 7C	nil	No attack
—[c]	Fe 10Cr 10Ni 13P 7C	nil	No attack
—[c]	Fe 10Cr 20Ni 13P 7C	nil	No attack

[a]Composition in atomic %. Note that atomic and weight % are nearly identical for the stainless steels.

[b]Data from P. Sexton and N. D. Greene, "Corrosion Resistance of Metallic Glasses," AIME meeting, Niagara Falls, N.Y. (September 20–23, 1976).

[c]Data from M. Naka, K. Hashimoto, and T. Masumoto, *J. Japan Inst. Metals*, **38**:835 (1974) and *Corrosion*, **32**:146 (1976).

Suggested Reading

Diegle, R. B., Sorensen, N. R., Tsuru, T., and Latanision, R. M.: The Corrosion Resistance of Glassy Alloys. *Treatise on Materials Science and Technology*, J. C. Scully, ed., Academic Press, New York, 1983.

Gilman, J. J.: *Phys. Today*, **28**, (May 1975); general description.

——: *J. Appl. Phys.*, **46**: 1625 (1975); mechanical properties.

Naka, M., K. Hashimoto, and T. Masumoto: *J. Japan Inst. Metals*, **38**: 835 (1974) and *Corrosion*, **32**: 146 (1976); corrosion resistance.

Tsuru, T., and Latanision, R. M.: Corrosion Resistance of Microcrystalline Stainless Steels, *J. Electrochem. Soc.* **129**: 1402–1408 (1982).

Viswanadham, R. K., J. A. S. Green, and W. G. Montague: *Scripta Met.*, **10**:229 (1976); hydrogen embrittlement.

5-21 Metallic Composites*

There has been intense research and development in the area of high-strength metallic composites. These consist of a metallic matrix strengthened by metallic or nonmetallic fibers, filaments, or whiskers. Composites are stronger and stiffer (greater modulus of elasticity) than the matrix metal and are often more resistant to crack growth. Boron, graphite, glass, and metal fibers are employed to improve the properties of various metals. The principle applications have been lightweight composites (e.g., aluminum-boron) for aerospace uses. However, composite metals are finding increased uses in automobiles, high-speed machinery, and sporting equipment.†

Reinforcing filaments within a composite can be exposed to the surrounding environments either by edge exposure or by accidental surface damage. Selective corrosion of these filaments is dangerous because it removes the reinforcing components, and the damage is hidden from direct view. Selective corrosion of the matrix material is also undesirable, but it is more easily detected and produces less structural weakening.

Corrosion of a composite material is controlled by two factors: (1) the specificity of a given corrosive toward the individual components and (2) galvanic interactions between them. The corrosivity of an environment varies from metal to metal. For example, tungsten is rapidly attacked by hydroxides and other basic solutions, whereas nickel is highly resistant to these media. As discussed in Chap. 3, if dissimilar metals are placed in electrical contact, galvanic corrosion is possible. After coupling, the corrosion rate of the more active metal is accelerated, while that of the noble metal is reduced. This effect becomes increasingly pronounced as the separation between the two metals in the galvanic series (Table 3-2) is

*Portions of this section are excerpts from N. D. Greene and N. Ahmed, *Materials Protection*, **9**:16 (1970).

†For further information about composite metallics and their applications see K. K. Chawla, Composite Materials: Recent Developments, *J. Metals*, 82–86 (Mar. 1983).

increased. Frequently, galvanic corrosion is most intense at the junction between dissimilar metals, especially if very active and very noble metals are connected in high-resistance electrolytes.

Corrosive specificity and galvanic corrosion are separate and independent effects. Thus, if the matrix and filaments of a composite are close together in the galvanic series, galvanic effects are negligible.

Some examples of composite corrosion are shown in Figures 5-11 to 5-13 and summarized in Table 5-14. Either the matrix or the filament is attacked, depending on the corrosive. The copper matrix of a copper/tantalum filament composite is quickly attacked by brief exposure to concentrated nitric acid (Figure 5-11). A similar effect is produced when it is exposed to an oxidizing ammonium hydroxide solution (Table 5-14). However, when this composite is exposed to concentrated hydrofluoric acid, the tantalum filaments are rapidly corroded (Figure 5-12). Note the complete absence of corrosion of the copper matrix indicated by the existence of polishing scratches. Due to the closeness of copper and tantalum in the galvanic series, there is probably little galvanic corrosion in this system. Similar effects can be observed with a composite consisting of tungsten fibers in a nickel matrix. The nickel matrix is attacked during exposure to a

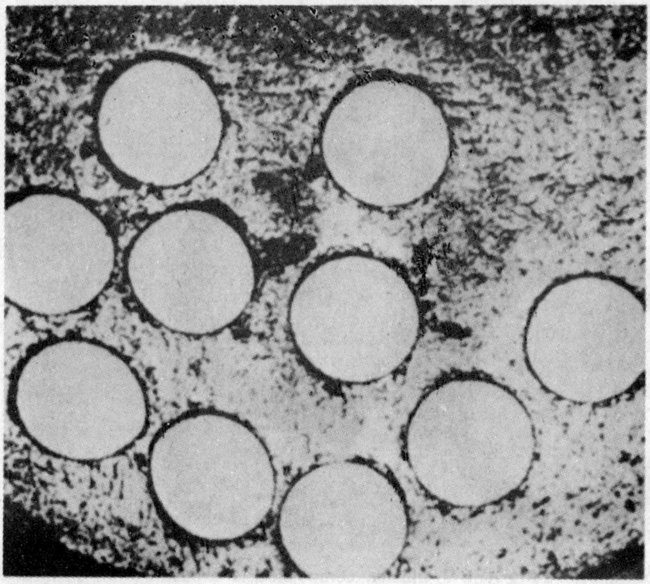

Figure 5-11 Selective attack of copper matrix in a Cu/Ta filament composite after brief exposure to 70% nitric acid (70 ×).

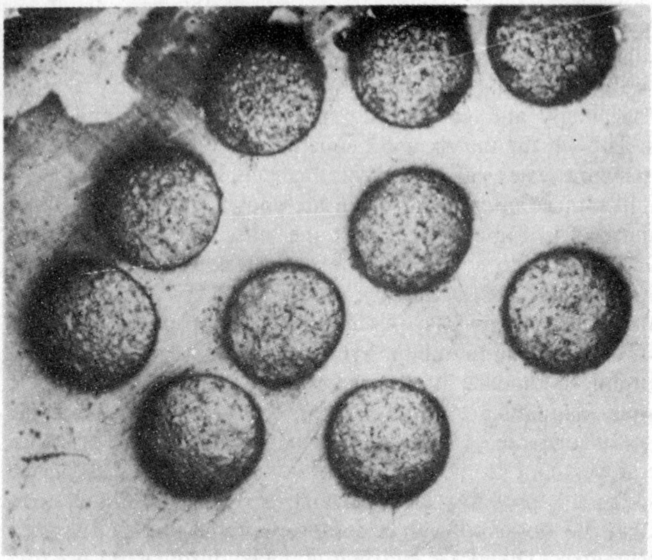

Figure 5-12 Selective corrosion of tantalum in a Cu/Ta filament composite after brief exposure to 45% hydrofluoric acid (70 ×).

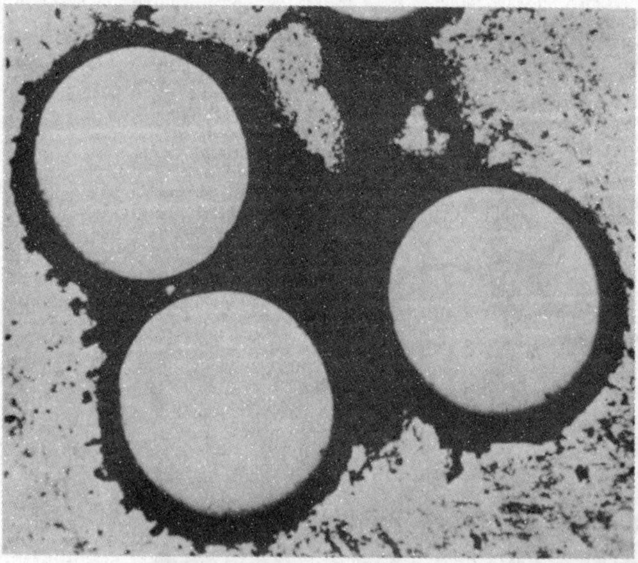

Figure 5-13 Appearance of a Mg/AFC-77 filament composite after exposure to 2% oxalic acid solution. Localized galvanic corrosion of the magnesium matrix occurs at the filament interface (150 ×).

Table 5-14 Corrosion characteristics of composites

Composite	Corrosive	Component selectively attacked
Cu/Ta filaments	75% HNO_3	Copper
Cu/Ta filaments	45% HF	Tantalum
Cu/Ta filaments	1:1:1 volume mixture 30% NH_4OH, 3% H_2O_2, H_2O	Copper
Ni/W filaments	1:1 volume mixture concentrated nitric and acetic acids	Nickel
Ni/W filaments	1:1 volume mixture 10% NaOH and 30% $K_3Fe(CN)_6$	Tungsten
Mg/AFC-77* filament	2% oxalic acid	Magnesium

*Composition: 14.5% Cr; 13.5% Co: 5.0% Mo: balance Fe.

nitric acid mixture, while tungsten filaments are selectively dissolved by an oxidizing alkaline solution (Table 5-14).

Nonmetallic filament reinforced metal matrix composites should behave similarly to composites containing metal filaments. However, since these inorganic solids are very resistant to most aqueous media, their selective dissolution is less likely. Also, galvanic corrosion is not likely except for those filaments possessing electrical conductivity such as graphite and boron.

Composites should be carefully evaluated in a number of environments which might be encountered in practice. Also, in designing composite structures, consideration of the specific resistances of the components and their relative position in the galvanic series should be considered as important as the mechanical and physical properties of the individual elements.

The American Society for Metals handbooks contain a wealth of information on properties (including corrosion) of metals and alloys: Vol. 1, *Iron and Steel* (1978), Vol. 2, *Non-ferrous Alloys* and *Pure Metals* (1979), and Vol. 3, *Stainless Steels and Special Purpose Metals* (1980).

In addition, the Mid-June 1983 issue of *Metal Progress*, "1983 Materials and Processing Handbook," has four sections: (1) Materials Selection, (2) Process Engineering, (3) Fabrication Technology, and (4) Testing and Inspection. Detailed information is presented on most commercial metals and alloys, including "Superalloys."

NONMETALLICS

5-22 Natural and Synthetic Rubbers

The outstanding characteristic of rubber and elastomers is resilience, or low modulus of elasticity. Flexibility accounts for most applications such as

Table 5-15 Property comparisons of elastomers*

		Physical properties				Resistance to:															
Material	Chemical name	Adhesion to metals	Tear resistance	Abrasion resistance	Resistance to gas diffusion	Dilute acids	Concentrated acids	Aliphatic hydrocarbons	Aromatic hydrocarbons	Ketones, etc.	Lacquer solvent	Oil and gasoline	Animal, vegetable oils	Water absorption	Oxidation	Ozone	Sunlight aging	Heat aging	Low temperature	Flame
Natural rubber	Polyisoprene	E	VG	E	F	F/G	F/G	P	P	G	P	P	P/G	VG	G	P	P	F	VG	P
Buna S (SBR, GR-S)	Butadiene styrene	E	F	G/E	F	F/G	F/G	P	P	G	P	P	P/G	G/VG	F	P	P	F/G	VG	P

Butyl	Isobu tylene-isoprene	G	G	G	E	E	G	P	P	G	F/G	P	VG	VG	E	E	VG	VG	G	P
Nitrile (Buna N)	Butadiene-acrylonitrile	E	F	G	G	G	G	E	G	P	F	E	VG	G	G	F	P	G	F/G	P
Neoprene	Chloroprene	E	G	E	G	E	G	F/G	F	P/F	P	G	G	G	VG	VG	VG	G	G	G
Silicone	Siloxane polymer	E	P	P	F	E	F	P	P	F	P	F	G/E	E	E	E	E	O	O	F/G
Hypalon	Chlorosulfonated polyethylene	E	F	E	G/E	E	G	F/G	F	P/F	P	G	G	G	E	O	O	VG	G	G
Nordel (EDPM)	Ethylene propylene	F	F	G/E	F	E	G	P	P	E	F/G	P	G/E	E	E	O	O	E	E	P/G
Hydrin	Epichlorhydrin	F/G	F/G	F/G	F/G	F/G	F	E	G	P	F	E	E	G	G	E	G	VG	G/VG	P/F
	Chlorinated polyethylene	F/G	F	G	E	E	G	G	F	F	F	G	G	VG	E	E	O	VG	G	G
Viton	Fluoroelastomer	F/G	F/G	G	E	G/E	E	E	E	P	P/F	E	E	VG	O	O	O	O	F/G	E
Adiprene, Estane	Urethane rubber	E	E	O	F	F	P	F/G	P	P	P	G	G	VG	E	E	VG	F/G	E	F/G
Hytrel	Polyester elastomer	G	O	E	E	G	P	E	G	G	G	VG	G	VG	E	E	VG	VG	O	F/G

RATINGS: O = outstanding; E = excellent; VG = very good; G = good; F = fair; P = poor.

**Dupont Elastomers Notebook*, p. 189, (April 1976), E. I. duPont de Nemours, Wilmington, Delaware.

tubing, belting, and automobile tires. However, chemical and abrasion resistance, and good insulating qualities, result in many corrosion applications. Rubber and hydrochloric acid form a natural combination in that rubber-lined steel pipes and tanks have been "standard" for this service for many years.

Generally speaking, the natural rubbers have better mechanical properties (resiliency and resistance to cuts and their propagation) than the synthetic or artificial rubbers, but the synthetics have better corrosion resistance.

Natural rubber Organically speaking, rubber is a long-chain molecule of isoprene (polyisoprene). It comes from trees as a liquid latex. The coillike structure of these molecules accounts for the high elasticity (100 to 1000% elongation). Soft rubber has a temperature limitation of around 160°F. This temperature limit can be raised to about 180°F by hardening through "alloying" (compounding). Adding sulfur and heating makes the rubber harder and more brittle. Charles Goodyear first discovered this process, called *vulcanization*, in 1839. About 50% sulfur results in a hard product known as ebonite, which is used to make bowling balls. Semihard and hard rubbers are used for tires, tank linings, and many other items. Corrosion resistance usually increases with hardness. In the case of lining pipes and tanks the rubber is usually applied soft and then cured in place (for large items) or in autoclaves. Modulus of elasticity varies roughly from 500 to 500,000 lb/in.^{2} for soft and hard rubber, respectively.

Synthetic rubbers At the start of World War II when the main sources of natural rubber were taken over by the enemy, there was great interest in developing substitutes. Neoprene was developed in the early thirties by Du Pont, but it was in small supply in 1941. Neoprene was one of the five strategic materials in World War II—the other four were metals.

A wide variety of synthetic rubbers is available, including combinations with plastics. Neoprene and nitrile rubber possess resistance to oils and gasoline. One of the first extensive applications of neoprene was, and is, gasoline hoses. An outstanding characteristic of butyl rubber is impermeability to gases. This accounts for its use as inner tubes and process equipment such as a seal for floating-top storage tanks. Butyl rubber exhibits better resistance to oxidizing environments such as air and dilute nitric acid. Neoprene-lined vessels handle strong sodium hydroxide.

Rubbers can be compounded to produce a wide range of mechanical properties. Hardness can vary from "bone hard" (bowling ball) to soft (rubber band). Tensile strengths can be varied from 900 to 4,500 psi and elongation from zero to 1000%. Table 5-15 shows property comparisons for rubbers and other elastomers.

Soft rubbers are best for abrasion resistance. A common mistake made, because of metals-oriented thinking, is to use hard rubber for erosion-corrosion conditions. Linings may also consist of hard and soft layers.

The chemistry and compounding of all types of rubbers are quite complex, and information is not readily available to consumers. The best procedure is to discuss the situation at hand with the suppliers in order to obtain the proper material for the given problem. If corrosion testing is involved it is imperative that the specimens be representative of the actual installation. Further, if lining is involved, the test specimens should be covered—sometimes bond failure is the primary cause of unsatisfactory performance. All of the statements in this paragraph apply also to plastics.

5-23 Other Elastomers

Table 5-15 lists some elastomers (with their trade names) and their properties. These are summarized as follows:

Chlorosulfonated polyethylene (*Hypalon*) closely resembles neoprene with greater resistance to heat, ozone, and chemicals.

Ethylene propylene (*Nordel*) is generally similar to butyl rubber, with the additional advantages of excellent resistance to ozone, heat, water, and sunlight together with low cost.

Epichlorhydrin (*Hydrin*) has the good properties of neoprene and nitrile rubbers. It has excellent resistance to swelling in fuels and oils, good heat-aging properties, and it resists ozone.

Fluorelastomers (*Viton*) have excellent resistance to a variety of corrosives and hydrocarbons and outstanding resistance to sunlight, ozone, and heat-aging. They are useful at high temperatures.

Urethane (*Adiprene*) possesses good tensile strength, wear resistance, and good resistance to oils, ozone, and oxidation.

MTI Manual No. 7,*Practical Guide to the Use of Elastomeric Linings* (May 1983) is a useful reference.

5-24 Plastics

Production and utilization of plastics have increased tremendously during the past 30 years. One of the early incentives for development of these materials was an inexpensive substitute for ivory billiard balls. Plastics are now available in a wide variety of forms including billiard balls, pumps, valves, pipe, fans, nose cones, airplane canopies, telephones, hoisery, radio cabinets, insulation, bushings, drawer rollers, waxes, pot handles, and heart valves and other body implants. Plastics are produced by casting, molding, extrusion, and calendering. They are available as solid parts, linings, coatings, foams, fibers, and films.

The American Society for Testing Materials states: "A plastic is a

Table 5-16 Mechanical and physical properties of some plastics

Material	Tensile strength, lb/in.^{2}	Elongation, %	Hardness, Rockwell R	Impact Izod, ft-lb	Modulus of elasticity, lb/in.^{2} × 10^3	Specific gravity	Heat distortion temperature, °F/264 lb/in.^{2}
Thermoplastics							
Fluorocarbons	2500	100–350	70	4	60	2.13	270
Methyl methacrylate	8000	5	220	0.5	420	1.19	200
Nylon	10,000	45	110	1.5	400	1.14	325
Polyether-chlorinate	6000	130	100	0.4	150	1.4	210
Polyethylene (low density)	2000	90–800	10	16	25	0.92	—
Polyethylene (high density)	4000	15–100	40	1–12	120	0.95	120
Polypropylene	5000	10–700	90	1–11	200	0.91	150
Polystyrene	7000	1–2	75	0.3	450	1.05	180
Rigid polyvinyl chloride	6000	2–30	110	1	400	1.4	150
Vinyls (chloride)	2500	100–450	80	good	low	1.18	145
Thermosetters							
Epoxy (cast)	10,000	nil	90	0.8	1000	1.1	350
Phenolics	7500	nil	125	0.3	1000	1.4	300
Polyesters	4000	nil	100	0.4	1000	1.1	350
Silicones	3500	nil	89	0.3	1200	1.75	350–900
Ureas	7000	nil	115	0.3	1500	1.48	265

material that contains as an essential ingredient an organic substance of large molecular weight, is solid in its finished state, and at some stage in its manufacture or in its processing into finished articles, can be shaped by flow." In other words, they are high-molecular-weight organic materials that can be formed into useful articles. Some occur in nature, but most are produced synthetically. In general, plastics, as compared with metals and alloys, are much weaker, softer, more resistant to chloride ions and hydrochloric acid, less resistant to concentrated sulfuric and oxidizing acids such as nitric, less resistant to solvents, and have definitely lower temperature limitations. Cold flow, or creep, at ambient temperatures is a problem particularly with the thermoplastics.

Plastics are readily divided into two classes, *thermoplastics* and *thermosetters*. The former soften with increasing temperature and return to their original hardness when cooled. Most are meltable; for example, nylon is extruded into fibers or filaments from the molten state. The thermosetters harden when heated and retain hardness when cooled. They "set" into permanent shape when heated under pressure. Generally they cannot be reworked as scrap.

Tables 5-16 and 5-17 list some properties and corrosion resistance of several well-known plastics. A wide range of properties are available. These properties can be changed considerably through plasticizers, fillers, and hardeners.

Plastics do not generally dissolve like metals. They are degraded or corroded because of swelling, loss in mechanical properties, softening, hardening, spalling, and discoloration.

Table 5-18 compares trade names and chemical names of several plastics.

THERMOPLASTICS

5-25 Fluorocarbons

Teflon,* Kel F,* and other fluorocarbons are the "noble metals" of the plastics in that are corrosion resistant to practically all environments up to 550°F. They consist of carbon and fluorine. The first polytetrafluorethylene was produced by Du Pont and designated Teflon (TFE). The author made some of the first industrial applications of Teflon during World War II. Similiar materials, with the composition modified, or copolymers, are Kel F, Viton (elastomer), and PFA.

In addition to corrosion resistance, its low coefficient of friction is the basis for a very successful application of Teflon, illustrated in Fig. 5-14. A confined Teflon sleeve separates the plug and body of this plug valve and

*These and others are trade names.

Table 5-17 Effect of various environments on plastics

Material	Acids		Alkalis		Organic solvents	Water absorption, %/24 hr	Continuous heat resistance, °F
	Weak	Strong	Weak	Strong			
ABS (acrylonitrile-butadiene-styrene)	N	AO	N	N	A	0.4	190–230
Acetals (copolymers)	V	A	N	N	R	0.2	185–220
Acrylics							
Methyl methacrylate	R	AO	R	A	A	0.3	140–200
MMA alpha-methylstyrene copolymer	R	AO	R	R	A	0.2	200–230
Acrylic-PVC alloy	N	N	N	N	A	0.1	
Allyl resins							
Allyl diglycol carbonate	N	AO	N	SA	R	0.2	212
Diallyl phthalate (mineral-/glass-filled)	N	SA	SA	SA	N	0.3	300–400
Cellulosic compounds							
Ethyl cellulose	SA	A	N	SA	A	1	115–185
Cellulose acetrate, nitrate, propionate, and acetate butyrate	SA	A	SA	A	A	1–7	140–220
Chlorinated polyether	N	AO	N	N	R	0.01	290
Epoxy resins							
Cast	N	V	N	SA	R	0.1	250–550
Glass-fiber-filled molding compounds	N	R	N	N	N	0.1	300–500
Fluoroplastics							
Polychlorotrifluoroethylene	N	N	N	N	R	0.0	350–390
Polytetrafluoroethylene	N	N	N	N	N	0.0	550
FEP fluoroplastic	N	N	N	N	N	0.01	400
Polyvinylidene fluoride	N	R	N	N	R	0.04	300
Furan (asbestos-filled)	R	AO	N	R	R	0.1	265–330
Melamine-formaldehyde							
Asbestos filler	R	A	R	SA	N	0.1	250–400
Glass-fiber filler	N	A	N	R	N	0.1	300–400

Methyl-pentene polymer	N	AO	N	R	A	0.01	250–320
Nylons							
Type 6, cast	R	A	N	R	R	1.0	180–250
Type 12, unfilled	N	A	N	N	R	0.3	175–260
Phenol-formaldehyde, -furfural compounds	R	AO	SA	A	R	0.2	250
Phenylene oxides	N	N	N	N	A	0.07	175–220
Polyallomer	N	AO	N	R	R	0.01	
Polyaryl ether	N	V	N	N	A	0.3	250–270
Polyarylsulfone	N	N	N	N	R	2.0	500
Polybutylene	R	AO	R	R	—	0.01	225
Polycarbonates							
Unfilled	N	SA	V	A	A	0.1	250
ABS-polycarbonate alloy	N	AO	N	SA	A	0.3	220–250
Polyesters							
Linear aromatic	SA	SA	SA	A	R	0.02	600
alkyd, asbestos filled	N	SA	N	SA	N	0.1	450
Polyethylenes (low-, medium-, high-density)	R	AO	R	R	R	0.01	180–250
Polyimides	R	R	SA	A	R	0.3	500
Polyphenylene sulfides	N	AO	N	N	R	—	400–500
Polypropylenes	N	AO	N	R	R	0.01	190–300
Polystyrenes	N	AO	N	N	A	0.1	150–200
Polysulfone	N	N	N	N	R	0.2	300–345
Silicones	R	V	R	V	V	0.1	500–600
Urea-Formaldehyde	A	A	V	A	R	0.5	170
Urethanes (cast)	SA	A	SA	V	V	1.0	190–250
Vinyl polymers and copolymers							
Vinyl chloride & vinyl chloride-acetate	N	R	N	N	V	0.5	150–175
Vinylidene chloride	N	R	R	R	R	0.1	160–200
Chlorinated polyvinyl chloride	N	N	N	N	N	0.1	230

Code: N = no effect; R = resistant, generally; SA = slightly attacked; A = attacked; AO = attacked by oxidizing acids; V = variable behavior, depending on specific media.

Source: *Modern Plastics Encyclopedia*, McGraw-Hill, New York, 1970, 1974, 1975.

Table 5-18 Trade and chemical names of some polymers

Generic or trade name	Chemical name or composition
Epoxy, Epon	Epoxide, epoxy
Phenolic	Phenol-formaldehyde
ABS	Acrylonitrile-butadiene styrene
Silicone	Siloxane
Kel-F	Monochlorotrifluoroethylene
PFA	Polyfluoroalkoxy
Teflon (TFE)	Polytetrafluoroethylene
Teflon (FEP)	Fluorinated ethylene propylene
Tenite	Cellulose acetate butyrate
Marlex, Hi-Fax	Polyethylene (linear or high density)
Geon, PVC, Tygon	Polyvinyl chloride
Teslar, Tedlar, PVF	Polyvinyl fluoride
Kynar	Polyvinylidene fluoride
Acrylic, Plexiglas, Lucite	Polymethyl methacrylate
Lexan	Polycarbonate
Irrathene	Irradiated polyethylene
Saran	Polyvinylidene chloride, polyvinyl chloride (copolymers)
Mylar	Polyethylene terephthalate
Dacron fiber	Polyethylene terephthalate
Orlon fiber, Acrilan	Polyacrylonitrile
Nylon (66)	Hexamethylene adipamide
H-Film, HT-1 Fiber	Polyimide
Delrin	Polyacetal
TAC	Triallylcyanurate
DAP	Diallylphthalate

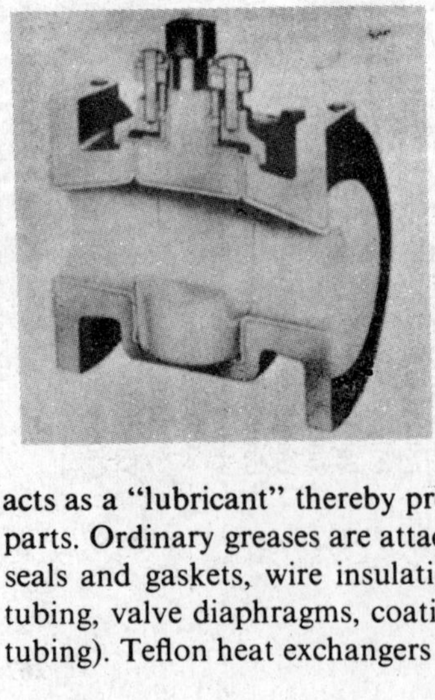

Figure 5-14 PTFE lined valve (T-line) with PFTE or PFA plug. *(Courtesy The Duriron Co., Inc.)*

acts as a "lubricant" thereby preventing sticking or "freezing" of the metal parts. Ordinary greases are attacked by many environments. Other uses are seals and gaskets, wire insulation, expansion joints, lining for pipe, tape, tubing, valve diaphragms, coatings, and even heat exchangers (thin-walled tubing). Teflon heat exchangers have found extensive application in spite of the fact that Teflon is an "insulator." Since corrosion does not occur, corrosion products do not form and stick to the tubing wall, and scaling (deposits) is not a problem because of Teflon's "nonstick" properties (domestic example: frying pans). Therefore, overall heat transfer is good.

5-26 Acrylics

Methyl methacrylate (Lucite and Plexiglas) is the best known of this group. It is used for brush and purse handles, transparent displays, working models, airplane canopies, and taillights. It is pretty, and transmits and conducts light. The acrylics are soft, easily scratched, and not very temperature-resistant.

5-27 Nylon

This material is best known for hosiery, but also finds much use for strictly corrosion-resistant applications. Its main applications depend on strength, low coefficient of friction, and wear resistance. These are gears, drawer and shelf rollers, fishline, sutures, tennis-racket strings, dentures, automobile door catches, and brush bristles. It is also utilized for electrical insulation since it permits higher-temperature operation than rubber. The author made the first industrial application of nylon as a bushing for spinning buckets (7700 rpm) in the manufacture of rayon.

5-28 Chlorinated Polyether

Penton, a chlorinated polyether, is one of the newer plastics; it is widely used for handling corrosives, including aggressive environments. Solid parts such as valve and pipe and also linings and coatings are available. Coatings can be applied by the fluidized-bed process up to 40 mils thick.

5-29 Polyethylenes

Alathon, Aeroflex, Polythene, and other polyethylene materials represent the largest production of the plastics and are marketed by many producers. They are used for packaging film, sheet, squeeze bottles, pans, tumblers, ice trays, and many other household items. Piping for water and other chemicals is common. Certain solvents produce stress-corrosion cracking of polyethylene. Considerable latitude in properties is possible, as indicated by Tables 5-16 and 5-17.

Ultrahigh-molecular-weight polyethylene (UHMWPE) exhibits very good resistance to erosion corrosion. Figure 5-15 shows a butterfly valve lined with UHMWPE. This valve is also made with a lined gate, but a photo of this valve is not presented because it would not reproduce well (too much white). Pumps lined with UHMWPE are also available, as are pump impellers encapsulated in UHMWPE, Teflon, and other plastics.

Figure 5-15 Butterfly valve lined with ultrahigh-molecular-weight polyethylene—UHMWPE. *(Courtesy The Duriron Co., Inc.)*

5-30 Polypropylene

Polypropylene (Moplen, Pro-Fax, Escon) originated in Italy. It exhibits better heat and corrosion resistance and is stiffer than polyethylene. It is made into valves, bottles sterilized by heat, pipe, fittings, and scooter parts.

5-31 Polystyrene

Polystyrene (Styron, Lustrex) is used for wall tile, battery cases, flowmeters, radio cabinets, bottle closures, and refrigeration equipment. It possesses good chemical resistance but is too brittle for many structural applications. Polystyrene shows good resistance to hydrofluoric acid.

5-32 Rigid Polyvinyl Chloride (PVC)

This material is basically rigid but can be softened by additions of polyvinyl acetate and plasticizers to vary mechanical properties. PVC is used for pipe and fittings, ducts, fans, sheet, containers, and linings.

5-33 Vinyls

Tygon, Vinylite, Plioflex, Saran, and other vinyls are a well known and versatile group of materials. Most of them are copolymers of vinyl chloride and vinyl acetate. Some uses are pipe and tubing, packaging film, floor tile, fibers, phonograph records, raincoats, garden hose, and insulation. Laboratory tubing is a common example. They can be joined by heat-sealing, fusion-welding, and adhesives. The introduction of Saran into the industrial picture around 1943 was timely because copper tubing and insect screen were scarce. Saran is different from most vinyls in that it consists primarily of vinylidene chloride.

5-34 Other Thermoplastics

These are ABS, acetals (Delrin), polycarbonate (Lexan), polyimide, polyimide, polysulfone, and polyurethane (also a thermosetter). See Tables 5-17 and 5-18.

THERMOSETTERS

5-35 Epoxies

Epon, Durcon, Araldite, and other epoxies represent perhaps the best combination of corrosion resistance and mechanical properties. Epoxies are available as castings, extrusions, sheet, adhesives, and coatings. They are used as sinks, bench tops, pipe, valves, pumps, small tanks, potting compounds, adhesives, linings, protective coatings, dies for forming metal, printed circuits, insulation, and containers. Liquid resins can be cast into molds (as in a foundry) and then set by heating.

5-36 Phenolics

Phenolic materials (Bakelite, Durez, Resinox) are among the earliest and best-known plastics. They are mostly based on phenol formaldehyde. Bakelite was developed by Leo Baekeland shortly after the turn of the century. Applications include radio cabinets, telephones, electrical sockets and plugs, pumps, valves, trays, auto distributors, rollers, and coatings.

5-37 Polyesters

Polyester plastics (Mylar, Dacron, Dypol, Vibrin) are best known to the general public for their use in the "space mirror" satellite (Echo). The satellite was made of thin, reflective Mylar film. Corrosion resistance of the polyesters compares unfavorably with many other plastics. One of the main uses of polyesters involves reinforced material, as discussed in Sec. 5-40. A popular application is luggage. The body of the Corvette automobile is made of fiber-glass-reinforced polyester.

5-38 Silicones

Silicones (Dow Corning) offer outstanding heat resistance. Mechanical properties change little with variations in temperature. These plastics differ from most plastics in that an important ingredient is inorganic silicon. Silicones are used for molding compounds, laminating resins, and insulation for electric motors and electronic equipment. Resistance to attack by chemicals is not outstanding.

5-39 Ureas

Ureas (Lauxite, Beetle, Avisco) were the second important class, after the phenolics, of the thermosetting resins to be developed. They are based on urea and formaldehyde. Corrosion resistance is not good. Uses include kitchen dishware and utensils, electrical fixtures, radio cabinets, closures, and adhesives (e.g., bonding plywood).

5-40 Laminates and Reinforced Plastics

These materials usually consist of thermosetting resins "filled" or laminated with cloth, mats, paper, chopped fabrics, or fibers such as fiber glass, which is commonly used. The main advantage is that tensile strengths can be increased to as high as around 50,000 lb/in.^{2}. This results in a high strength-weight ratio for these light materials. Availability and uses include tanks, pipe, ducts, sheets, rod, car bodies (Corvette), boats, and missile and satellite parts.

Another important product is the widely used fiber-reinforced plastic (FRP) pump, as shown in Fig. 5-16. Mallinson* discusses abrasion of FRP in corrosive environments. A case history involving a large elbow handling "crystal-laden acidic vapors" is described. Also discussion of particle size and hardness, and importance of design. An erosion-corrosion allowance (like with metals) is suggested in some cases.

In a paper titled "Growth of Composites is Bolstered by New Fibers,"† carbon fibers, alumina fibers, an aramid fiber and other high strength plastics are discussed by Chowdhury. Carbon-fiber-reinforced plastics are under consideration for structural parts of automobiles and other equipment (when the cost is economical). In the late 1970s the price was $300/lb; at this writing the price is about $20.

The use of graphite fiber composite cases would reduce the boosters' weight by 38 tons in the Space Shuttle.

During the late 1970s FRP structures for severe corrosion applications

*J. H. Mallinson, *Chem. Eng.*, 143–146 (May 31, 1982).
†J. Chowdhury, *Chem. Eng.*, 37–41 (Aug. 20, 1984).

Figure 5-16 Durco Mark II self-priming FRP pump. *(Courtesy The Duriron Co. Inc.)*

fell into some disrepute because of poor quality in many cases that resulted in costly failures (some tanks were replaced with stainless steel). The Materials Technology Institute and the Society of the Plastics Industry joined in a venture, at MTI's urging and financial support, to solve these problems. Fabricators, consumers, and resin producers collaborated in this project. This resulted in "Report on Development of Quality Assurance Practices and Procedures for Reinforced Thermoset Plastic (RTP) Corrosion Resistant Equipment," April 1981. This involves corrosion-resistant RTP for relatively severe corrosives. MTI does not set standards, but the information is the proposed basis for an ASME standard—not available at this writing. This all should result in more reliable equipment for chemical plant service.

One of the first appearances in the literature on corrosion of a FRP (glass-phenolic) is in a paper by M. G. Fontana, titled "Synthetic Rubber and Plastics as Materials of Construction in the Chemical Industry," *Trans. AIChE*, 42:359–377 (1946). This was also the first paper on this subject. Teflon was included in the manuscript but it was deleted by company "management."

Suggested Reading

Harper, C. A., ed.: *Handbook of Plastics and Elastomers*, McGraw-Hill Book Company, New York, 1975.

Managing Corrosion with Plastics, Vol. 5, NACE, 1983.

Modern Plastics Encyclopedia, McGraw-Hill Book Company New York (latest available edition).

OTHER NONMETALLICS

5-41 Ceramics

Ceramic materials consist of compounds of metallic and nonmetallic elements. A simple example is MgO or magnesia. Other ceramics include brick, stone, fused silica, stoneware, glass, clay tile, porcelain, concrete, abrasives, mortars, and high-temperature refractories. In general, compared with metals, ceramics resist higher temperatures, have better corrosion and abrasion resistance, including erosion-corrosion resistance, and are better insulators; but the ceramics are brittle, weaker in tension, and subject to thermal shock. Most ceramic materials exhibit good resistance to chemicals, with the main exceptions of hydrofluoric acid and caustic. Parts are formed by pressing, extrusion, or slip-casting.

Much research* is ongoing to produce "ultratough" ceramics such as silicon carbide, silicon nitride, alumina, and zirconia. The main thrust is to reduce brittleness so these materials could be used in diesel engines, gas turbines, etc.

A large comprehensive report is MTI Manual No. 9, *Refractories for the Chemical Process Industries* (1984) and its supplement, *Applicable ASTM Standards*.

Acidbrick This material is made from fireclay with a silica content about 10% greater than ordinary firebrick. A common application is lining of tanks and other vessels to resist corrosion by hot acids or erosion corrosion. A brick-lined steel tank usually contains an intermediate lining of lead, rubber, or a plastic. Acid-resistant cements and mortars join the brick. Floors subject to acid spillage are made of acidbrick. In addition to the MTI manual (above), Pierce and Semler† and Sheppard‡ describe use of acidbrick linings.

Stoneware and porcelain Both of these find many applications because of their good corrosion resistance. Porcelain parts are usually smaller in size than stoneware, and the porcelain is less porous. Both can be glazed for easy cleaning. Stoneware and porcelain show tensile strengths of about 2000 and 5000 lb/in.^{2}, respectively. Stoneware sinks, crocks, and other vessels, absorption towers, pipes, valves, and pumps are available. Porcelain can be made into similar equipment (e.g., acid nozzles) and is widely used as insulators and spark plugs.

*G. W. Graff, Ceramics: Primed for a Broader Industrial Role, *Chem. Eng.*, 14–17 (July 25, 1983).

†R. R. Pierce and C. E. Semler, Ceramic and Refractory Linings for Acid Condensation, *Chem. Eng.*, (Part I: Dec. 12, 1983; Part II: Jan. 23, 1984).

‡W. L. Sheppard, Spotting and Avoiding Problems with Acid-Resistant Brick, *Chem. Eng.*, 107–110 (May 3, 1982).

Structural clay These clay products include building, fire, sewer, and paving brick, terra-cottas, pipe, and roofing and wall tile. Hot acids sometimes attack these materials.

Glass Glass is an amorphous inorganic oxide, mostly silica, cooled to a rigid condition without crystallization. Glass laboratory ware, such as Pyrex, and containers are well known. Piping and pumps are available. Transparency is utilized for equipment such as flowmeters. Glass fibers are widely used for air filters, insulation, and reinforced plastics. Hydrofluoric acid and caustic attack glass, and it shows slight attack in hot water.

There are a wide variety of glass compositions. Modifiers (e.g., CaO), fluxes (e.g., B_2O_3), and stabilizers (e.g., Al_2O_3) are added to obtain various properties including corrosion resistance. Properties of glass are also affected by thermal history (e.g., heat treatment).

Corrosion tests are made on powder wherein the amount of alkali extracted is measured. Weight loss tests (similar to metals) are also conducted. Uniform or selective and localized attack can occur (because of phase separation). When hot water attacks glass, it is not dissolved in the usual sense but is hydrolytically decomposed. Resistance to water varies from excellent to poor depending on the composition of the glass.

For an in-depth understanding of glass the reader is referred to A. M. Filbert and M. L. Hair, Surface Chemistry and Corrosion of Glass, in *Advances in Corrosion Science and Technology*, vol. 5, pp. 1–54, Plenum Press, New York, 1976. This paper presents much detailed information, including corrosion data, compositions, structures, and 107 references.

An interesting application of glass concerns advanced flexible reusable surface insulation (AFRSI) on the space shuttle where temperatures of 1500°F are encountered.* The 36 × 36-inch "blankets" consist of silica glass on the outside, a microquartz felt in the middle, and glass cloth on the outside. The layers are held together with a silicone adhesive.

Vitreous silica This material, also called fused quartz (almost pure silica), has better thermal properties than most ceramics and excellent corrosion resistance at high temperatures. It is used for furnace muffles, burners, reaction chambers, absorbers, piping, etc., particularly where contamination of product is undesirable.

Concrete Tanks and pipes made of concrete are well known for handling mild corrosives. If aggresive environments are involved, the concrete is protected by coatings or linings.

Table 5-19 lists some properties of a few very-high-temperature refractories. These are also called *super refractories*. They are used to resist

*R. M. Hoppe, *Chem. Eng.*, 29–32 (Aug. 9, 1982).

Table 5-19 Some properties of high-temperature refractories

	Magnesia*	Mullite	Silicon carbide	Stabilized zirconia	Bonded 99% Al_2O_3
Fusion point, °F	4800	3300	—	4700	3650
Use limit, °F	4170 oxid. 3100 red.	3000	3000	4400	3300
Modulus of rupture, lb/in.^{2}	2500	1500	2000‡	1900	2000
Moh's hardness†	6	6.5	9.6	7	9
Thermal shock resistance	Poor	Good	Good	Fair	Fair
Relative cost	2.8	1	2.1	10	3.1

*Basic refractories have poor resistance to hot acids.
†Scale 1 to 10. Talc = 1, low carbon steel = 4, diamond = 10.
‡At 2500°F.

molten metals, slags, and hot gases. Alumina has also found application for pump and valve seats because of its high hardness and good wear and corrosion resistance. Silicon carbide performs well as spray nozzles for hot sulfuric acid. Additional discussion of high-temperature refractories appears in Chap. 11.

5-42 Carbon and Graphite

These are unique nonmetallics in that they are good conductors of heat and electricity. High thermal conductivity results in excellent thermal shock resistance. They are used for heat exchangers, columns, pumps, and impressed-current anodes. Carbon and graphite are inert to many corrosive environments. They are weak and brittle as compared with metals. Tensile strength varies between about 500 and 3000 lb/in.^{2}, and impact resistance is nil. Abrasion resistance is poor. High-temperature stability is good and they can be used at temperatures up to 4000 or 5000°F if protected from oxidation (burning). Silicon-base coatings (silicides or silicon carbide) and iridium coatings are claimed to give protection up to around 2900°F.

Carbon exhibits good resistance to alkalies and most acids. Oxidizing acids such as nitric, concentrated sulfuric, and chromic acid attack it. Fluorine, iodine, bromine, chlorine, and chlorine dioxide are likely to attack carbon. Karbate (trade name), a resin-bonded graphite, has found wide application in the chemical process industries. Graphite in nuclear reactors is quite well known.

Pyrolytic graphite, which is a dense, anisotropic material, has better strength and oxidation resistance than the more common types of carbon.

5-43 Wood

Cypress, pine, oak, and redwood are the main woods used for corrosion applications. Filter-press frames, structural members of buildings, barrels, and tanks are often made of wood. Containers must be kept wet or the staves will shrink, warp, and leak. Generally speaking, wood is limited to water and dilute chemicals. Strong acids, oxidizing acids, and dilute alkalies attack woods. They are also subject to biological attack. Impregnation with waxes and plastics helps reduce chemical and biological attack. Wood is a favorite for water cooling towers.

CHAPTER

SIX

CORROSION PREVENTION

MATERIALS SELECTION

6-1 Metals and Alloys

The most common method of preventing corrosion is the selection of the proper metal or alloy for a particular corrosive service. Since this is the most important method of preventing or reducing corrosion damage, Chaps. 5, 7, 8, and 11 are also devoted to this topic, and only brief mention of some general rules will be presented here. One of the most popular misconceptions to those not familiar with metallurgy or corrosion engineering concerns the uses and characteristics of stainless steel. Stainless steel is not stainless, it is not the most corrosion-resistant material, and it is not a specific alloy. Stainless steel is the generic name for a series of more than 30 different alloys containing from 11.5 to 30% chromium and 0 to 22% nickel, together with other alloy additions. Stainless steels have widespread application in resisting corrosion, but it should be remembered that they do not resist all corrosives. In fact, under certain conditions, such as chloride-containing mediums and stressed structures, stainless steels are less resistant than ordinary structural steel. Stainless alloys are more susceptible to localized corrosion such as intergranular corrosion, stress-corrosion cracking, and pitting attack than ordinary structural steels. Frequently, the quality of stainless steels is checked with a magnet. This is based on the belief that nonmagnetic stainless steels represent "good" alloys, and stainless steels that are magnetic are inferior. This test has no basis and, in fact, is misleading. Many stainless steel alloys are magnetic, and many of the cast austenitic stainless steels show some ferromagnetic properties. There is no correlation between magnetic susceptibility and corrosion resistance. Under certain conditions many of the magnetic stainless steels are superior to the nonmagnetic varieties. In summary, a large number of corrosion failures can be directly attributed to the indiscriminate selection of stainless steels for construction on the basis that they are the "best." Stainless steels represent a class of highly corrosion-resistant materials of relatively low cost that should be carefully used.

In alloy selection, there are several "natural" metal-corrosive combinations. These combinations of metal and corrosive usually represent the maximum amount of corrosion resistance for the least amount of money. Some of these natural combinations are as follows:

1. Stainless steels–nitric acid
2. Nickel and nickel alloys–caustic
3. Monel–hydrofluoric acid
4. Hastelloys (Chlorimets)–hot hydrochloric acid
5. Lead–dilute sulfuric acid
6. Aluminum–nonstaining atmospheric exposure
7. Tin–distilled water
8. Titanium–hot strong oxidizing solutions
9. Tantalum–ultimate resistance
10. Steel–concentrated sulfuric acid

The above list does not represent the only material-corrosive combinations. In many instances, cheaper materials or more resistant materials are available. For nitric acid service, the stainless steels are usually considered first, as these have excellent resistance to this medium under a wide range of exposure conditions. Tin or tin coatings are almost always chosen as a container or piping material for very pure distilled water. For many years, tantalum has been considered and used as an "ultimate" corrosion-resistant material. Tantalum is resistant to most acids at all concentrations and temperatures and is generally used under conditions where minimal corrosion is required, such as implants in the human body. An interesting feature about tantalum is that it almost exactly parallels the corrosion resistance of glass. Both glass and tantalum are resistant to virtually all mediums except hydrofluoric acid and caustic solutions. For this reason, manufacturers of glass-lined equipment use tantalum plugs to seal defects since this material matches the resistance of glass.

There are some general, but usually accurate, rules that may be applied to the resistance of metals and alloys. For reducing or nonoxidizing environments, such as air-free acids and aqueous solutions, nickel, copper, and their alloys are frequently employed. For oxidizing conditions, chromium-containing alloys are used. For extremely powerful oxidizing conditions, titanium and its alloys have shown superior resistance. This generalized rule is demonstrated in Chaps. 7 and 8.

6-2 Metal Purification

The corrosion resistance of a pure metal is usually better than that of one containing impurities or small amounts of other elements. However, pure metals are usually expensive and are relatively soft and weak. In general, this category is used in relatively few cases which are more or less special.

Aluminum is a good example because it is not expensive in a fairly pure state—99.5% plus. The commercially pure metal is used for handling hydrogen peroxide, where the presence of other elements may cause decomposition because of catalytic effects. In another case, localized attack of aluminum equipment occurred because of segregation of impurity iron in the alloy. Reduction of the maximum iron content, agreeable to both producer and user, eliminated the localized attack and satisfactory performance of the equipment was obtained without added cost of material.

Another example is arc-melted zirconium, which is more corrosion resistant than induction-melted zirconium because of more impurities in the latter. This is a special case in an atomic-energy application where a little corrosion is too much.

6-3 Nonmetallics

This category involves integral or solid nonmetallic construction (mainly self-supporting) and also sheet linings or coverings of substantial thickness (to differentiate from paint coatings). The five general classes of nonmetallics are (1) rubbers, natural and synthetic; (2) plastics; (3) ceramics; (4) carbon and graphite; and (5) wood. Their mechanical properties and corrosion resistance are described in Chap. 5.

In general, rubbers and plastics, as compared with metals and alloys, are much weaker, softer, more resistant to chloride ions and hydrochloric acid, less resistant to strong sulfuric acid and oxidizing acids such as nitric, less resistant to solvents, and have relatively low temperature limitations (170 to 200°F for most). Ceramics possess excellent corrosion and high-temperature resistance, with the main disadvantages being brittleness and lower tensile strength. Carbons show good corrosion resistance, electric and heat conductivity, but they are fragile. Wood is attacked by aggressive environments.

Materials Technology Institute Manual No. 7, *Practical Guide to the Use of Elastomeric Linings* (May 1983), covers rubber linings for tanks, lines, fans, filters, scrubbers, etc. This manual contains sections on rubbers used, selection, testing, design, fabrication, preparation of equipment to be lined, processing, inspection, acceptance, maintenance, chemical resistance, and cost factors. Good and bad practices are described. Hard and soft natural rubbers, polychloroprene and butyl rubbers are included.

ALTERATION OF ENVIRONMENT

6-4 Changing Mediums

Altering the environment provides a versatile means for reducing corrosion. Typical changes in the medium that are often employed are (1) lowering temperature, (2) decreasing velocity, (3) removing oxygen or oxidizers, and (4) changing concentration. In many cases, these changes can significantly reduce corrosion, but they must be done with care. The effects produced by these changes vary depending on the particular system, as discussed in Chap. 2.

Lowering temperature This usually causes a pronounced decrease in corrosion rate. However, under some conditions, temperature changes have little effect on corrosion rate (see Sec. 2-8). In other cases, increasing temperature decreases attack. This phenomenon occurs as hot, fresh or salt water is raised to the boiling point and is the result of the decrease in oxygen solubility with temperature. Boiling seawater is therefore less corrosive than hot seawater (e.g., 150°F).

Decreasing velocity This is often used as a practical method of corrosion control. As discussed in Sec. 2-7, velocity generally increases corrosive attack, although there are some important exceptions. Metals and alloys that passivate, such as stainless steels, generally have better resistance to flowing mediums than stagnant solutions. Very high velocities should be always avoided where possible, because of erosion-corrosion effects (Chap. 3).

Removing oxygen or oxidizers This is a very old corrosion control technique. Boiler feedwater was deaerated by passing it through a large mass of scrap steel. In modern practice this is accomplished by vacuum treatment, inert gas sparging, or through the use of oxygen scavengers (see Sec. 6-5). Hydrochloric acid that has contacted steel during its manufacture or storage contains ferric chloride as an oxidizer impurity. This impure acid, termed *muriatic acid* in commerce, rapidly corrodes nickel-molybdenum alloys (Hastelloy B, Chlorimet 2), whereas these materials possess excellent resistance in pure hydrochloric acid (Chap. 7). Although deaeration finds widespread application, it is not recommended for active-passive metals or alloys. These materials require oxidizers to form and maintain their protective films and usually possess poor resistance to reducing or nonoxidizing environments. For additional discussion of the effects of oxidizers, see Sec. 2-6.

Changing concentration This and its effect on corrosion have been described (Sec. 2-9). Decreasing corrosive concentration is usually effective. In many processes, the presence of a corrosive is accidental. For example, corrosion by the water coolant in nuclear reactors is reduced by eliminating chloride ions. Many acids such as sulfuric and phosphoric are virtually inert at high concentrations at moderate temperatures. In these cases, corrosion can be reduced by *increasing* acid concentration.

No discussion of corrosion control would be complete without mentioning the *magic devices* or *water-conditioning gadgets* that have been and continue to be widely sold for purposes of controlling water corrosion. These gadgets are usually promoted on the basis that they will "stop corrosion," "prevent scaling," "destroy bacteria," "improve taste and odor," or "reduce water hardness." Some manufacturers make all of the above claims for their product! In every case, the device is based on some pseudoscientific principle, is simply constructed, quite expensive, and totally worthless. Several of them consist merely of a pipe coupling that looks identical to those available in any hardware store. Surprisingly, large numbers of these gadgets are installed each year by trained engineers.

Magic devices should not be confused with the water-softening, water-treating and cathodic protection apparatus and systems sold by reputable manufacturers. The worthless device is easily spotted by a number of clues: (1) It is based on a questionable or a "secret" new principle. (2) The advertising contains an excessive number of testimonials (usually from untrained persons). (3) The promotion makes no mention of any limitations—the device will work in any kind of water and protect any size system. (4) The device is always sold with a complete guarantee.*

6-5 Inhibitors

An inhibitor is a substance that, when added in small concentrations to an environment, decreases the corrosion rate. In a sense, an inhibitor can be considered as a retarding catalyst. There are numerous inhibitor types and compositions. Most inhibitors have been developed by empirical experimentation, and many inhibitors are proprietary in nature and thus their composition is not disclosed. Inhibition is not completely understood because of these reasons, but it is possible to classify inhibitors according to their mechanism and composition.

Adsorption-type inhibitors These represent the largest class of inhibiting substances. In general, these are organic compounds which adsorb on the metal surface and suppress metal dissolution and reduction reactions. In most cases, it appears that adsorption inhibitors affect both the anodic and cathodic processes, although in many cases the effect is unequal. Typical of this class of inhibitors are the organic amines.

*For an excellent survey of magic devices including a historical tabulation of those sold since 1865, see B. Q. Welder and E. P. Partridge, *Ind. Eng. Chem.*, **46**:954 (1954).

Hydrogen-evolution poisons These substances, such as arsenic and antimony ions, specifically retard the hydrogen-evolution reaction. As a consequence, these substances are very effective in acid solutions but are ineffective in environments where other reduction processes such as oxygen reduction are the controlling cathodic reactions.

Scavengers These substances act by removing corrosive reagents from solution. Examples of this type of inhibitor are sodium sulfite and hydrazine, which remove dissolved oxygen from aqueous solutions as indicated in Eqs. (6.1) and (6.2) below:

$$2Na_2SO_3 + O_2 \rightarrow 2Na_2SO_4 \tag{6.1}$$

$$N_2H_4 + O_2 \rightarrow N_2 + 2H_2O \tag{6.2}$$

It is apparent that such inhibitors will work very effectively in solutions where oxygen reduction is the controlling corrosion cathodic reaction but will not be effective in strong acid solutions.

Oxidizers Such substances as chromate, nitrate, and ferric salts also act as inhibitors in many systems. In general, they are primarily used to inhibit the corrosion of metals and alloys that demonstrate active-passive transitions, such as iron and its alloys and stainless steels.

Vapor-phase inhibitors These are very similar to the organic adsorption-type inhibitors and possess a very high vapor pressure. As a consequence, these materials can be used to inhibit atmospheric corrosion of metals without being placed in direct contact with the metal surface. In use, such inhibitors are placed in the vicinity of the metal to be protected, and they are transferred by sublimation and condensation to the metal surface. The vapor-phase inhibitors are usually only effective if used in closed spaces such as inside packages or on the interior of machinery during shipment.

Table 6-1 lists some important inhibitors, their applications, and their sources. Examples of all the above-mentioned types of inhibitors appear in this table. It is important to remember that inhibitors are specific in terms of metal, environment, temperature, and concentration range. As mentioned above, the concentration and type of inhibitor to be used in a specific corrosive is usually determined by empirical testing, and this information is usually available from manufacturers. It is important to use enough inhibitor, since many inhibiting agents accelerate corrosion, particularly localized attack such as pitting, when present in small concentrations. Hence, too little inhibitor is less desirable than none at all. To avoid this possibility, inhibitors should be added in excess and their concentration checked periodically. When two or more inhibiting substances are added to a corrosive system, the inhibiting effect is sometimes greater than that which

Table 6-1 Corrosion inhibitor reference list

Metal	Environment	Inhibitor	Reference
Admiralty	Ammonia, 5%	0.5% hydrofluoric acid	54
Admiralty	Sodium hydroxide, 4° Be	0.6 moles H_2S per mole NaOH	71
Aluminum	Acid hydrochloric, 1*N*	0.003*M* α phenylacridine, β naphthoquinone, acridine, thiourea or 2-phenylquinoline	39
Aluminum	Acid nitric, 2–5%	0.05% hexamethylene tetramine	22
Aluminum	Acid nitric, 10%	0.1% hexamethylene tetramine	22
Aluminum	Acid nitric, 10%	0.1% alkali chromate	16
Aluminum	Acid nitric, 20%	0.5 hexamethylene tetramine	22
Aluminum	Acid phosphoric	Alkali chromates	52
Aluminum	Acid phosphoric, 20%	0.5% sodium chromate	16, 60
Aluminum	Acid phosphoric, 20–80%	1.0% sodium chromate	16, 60
Aluminum	Acid sulphuric, conc.	5.0% sodium chromate	45
Aluminum	Alcohol anti-freeze	Sodium nitrite and sodium molybdate	6
Aluminum	Bromine water	Sodium silicate	10
Aluminum	Bromoform	Amines	44
Aluminum	Carbon tetrachloride	0.05% formamide	55
Aluminum	Chlorinated aromatics	0.1–2.0% nitrochlorobenzene	21
Aluminum	Chlorine water	Sodium silicate	10
Aluminum	Calcium chloride, sat	Alkali silicates	59
Aluminum	Ethanol, hot	Potassium dichromate	52
Aluminum	Ethanol, commercial	0.03% alkali carbonates, lactates, acetates or borates	50
Aluminum	Ethylene glycol	Sodium tungstate or sodium molybdate	41
Aluminum	Ethylene glycol	Alkali borates and phosphates	52
Aluminum	Ethylene glycol	0.01–1.0% sodium nitrate	7
Aluminum	Hydrogen peroxide, alkaline	Sodium silicate	75
Aluminum	Hydrogen peroxide	Alkali metal nitrates	20
Aluminum	Hydrogen peroxide	Sodium metasilicate	59
Aluminum	Methyl alcohol	Sodium chlorate plus sodium nitrite	42
Aluminum	Methyl chloride	Water	72
Aluminum	Polyoxyalkene glycol fluids	2% Emery's dimer acid (dilinoleic acid), 1.25% $N(CHMe_2)_3$, 0.05–0.2% mercaptobenzothiazole	43
Aluminum	Seawater	0.75% sec. amyl stearate	5
Aluminum	Sodium carbonate, dilute	Sodium fluosilicate	67
Aluminum	Sodium hyroxide, 1%	Alkali silicates	59
Aluminum	Sodium hydroxide, 1%	3–4% potassium permanganate	17
Aluminum	Sodium hydroxide, 4%	18% glucose	57
Aluminum	Sodium hypochlorite contained in bleaches	Sodium silicate	58
Aluminum	Sodium acetate	Alkali silicates	59
Aluminum	Sodium chloride, 3.5%	1% sodium chromate	22
Aluminum	Sodium carbonate, 1%	0.2% sodium silicate	28

(continued)

Table 6-1 Corrosion inhibitor reference list (continued)

Metal	Environment	Inhibitor	Reference
Aluminum	Sodium carbonate, 10%	0.05% sodium silicate	28
Aluminum	Sodium sulfide	Sulfur	46
Aluminum	Sodium sulfide	1% sodium metasilicate	59
Aluminum	50% sodium trichloracetate soln.	0.5% sodium dichromate	1
Aluminum	Tetrahydrofurfuryl alcohol	1% sodium nitrate or 0.3% sodium chromate	15
Aluminum	Triethanolaimine	1% sodium metasilicate	22
Brass	Carbon tetrachloride, wet	0.001–0.1 aniline	53
Brass	Furfural	0.1% mercaptobenzothiazole	36
Brass	Polyoxyalkene glycol fluids	2.0% Emery's acid (dilinoleic acid), 1.25% $N(CHMe_2)_3$, 0.05–0.2% mercaptobenzothiazole	43
Brass	50% sodium trichloracetate soln.	0.5% sodium dichromate	1
Cadium plated steel	55/45 ethylene glycol—water	1% sodium fluorophosphate	12
Copper	Fatty acids as acetic	H_2SO_4, $(COOH)_2$ or H_2SiF_6	63
Copper	Hydrocarbons containing sulfur	P-hydroxybenzophenone	61
Copper	Polyoxyalkene glycol fluids	2% Emery's acid (dilinoleic acid), 1.25% $N(CHMe_2)_3$, 0.05–0.2% mercaptobenzothiazole	43
Copper & brass	Acid sulfuric, dil	Benzyl thiocyanate	68
Copper & brass	Ethylene glycol	Alkali borates & phosphates	22
Copper & brass	Polyhydric alcohol anti-freeze	0.4–1.6% Na_3PO_4 plus 0.3–0.6 sodium silicate plus 0.2–0.6% sodium mercaptobenzothiazole	62
Copper & brass	Rapeseed soil	Succinic acid	9
Copper & brass	Sulfur in benzene solution	0.2% 9, 10 anthraquinone	29
Copper & brass	Terahydrofurfuryl alcohol	1% sodium nitrate or 0.3% sodium chromate	15
Copper & brass	Water-alcohol	0.25% benzoic acid, or 0.25% sodium benzoate at a pH of 7.5–10	23
Galvanized iron	Distilled water	15 ppm. mixture calcium and zinc metaphosphate	77
Galvanized iron	55/45 ethylene glycol—water	0.025% trisodium phosphate	12
Iron	Nitroarylamines	Dibenzylaniline	19
Lead	Carbon tetrachloride, wet	0.001–0.1% aniline	53
Magnesium	Alcohol	Alkaline metal sulfides	16
Magnesium	Alcohol, methyl	1% oleic or stearic acid neutralized with ammonia	13
Magnesium	Alcohols, polyhydric	Soluble fluorides at pH 8–10	26
Magnesium	Glycerine	Alkaline metal sulfides	16

(continued)

Table 6-1 Corrosion inhibitor reference list (continued)

Metal	Environment	Inhibitor	Reference
Magnesium	Glycol	Alkaline metal sulfides	16
Magnesium	Trichlorethylene	0.05% formamide	55
Magnesium	Water	1% potassium dichromate	8
Monel	Carbon tetrachloride, wet	0.001–0.1% aniline	53
Monel	Sodium chloride, 0.1%	0.1% sodium nitrite	72
Monel	Tap water	0.1% sodium nitrite	72
Nickel & silver	Sodium hypochlorite contained in bleaches	Sodium silicate	58
Stainless steel	Acid sulfuric, 2.5%	5–20 ppm. $CaSO_4.5H_2O$	35
Stainless steel	Cyanamide	50–500 ppm. ammonium phosphate	65
Stainless steel, 18–8	Potassium permanganate contained in bleaches	Sodium silicate	58
Stainless steel, 18–8	Sodium chloride, 4%	0.8% sodium hydroxide	49
Steel	Acid citric	Cadmium salts	37
Steel	Acid sulfuric, dil	Aromatic amines	51
Steel	Acid sulfuric, 60–70%	Arsenic	74
Steel	Acid sulfuric, 80%	2% boron trifluoride	4
Steel	Aluminum chloride—hydrocarbon complexes formed during isomerization	0.2–2.0% iodine, hydriodic acid or hydrocarbon iodide	30
Steel	Ammoniacal ammonium nitrate	0.2% thiourea	40
Steel	Ammonium nitrate—urea solns.	0.05–0.10% ammonia 0.1% ammonium thiocyanate	18
Steel	Brine containing oxygen	0.001–3.0 methyl, ethyl or propyl substituted dithiocarbamates	48
Steel	Carbon tetrachloride, wet	0.001–0.1% aniline	53
Steel	Caustic—cresylate solution as in regeneration of refinery caustic wash solutions, 240–260F	0.1–1.0% trisodium phosphate	3
Steel	Ethyl alcohol, aqueous or pure	0.03% ethylamine or diethylamine	25
Steel	55/45 ethylene glycol—water	0.025% trisodium phosphate	12
Steel	Ethylene glycol	Alkali borates & phosphates	22
Steel	Ethylene glycol	Guanidine or guanidine carbonate	
Steel	Ethyl alcohol, 70%	0.15% ammonium carbonate plus 1% ammonium hydroxide	56
Steel	Furfural	0.1% mercaptobenzothiazole	36
Steel	Halogenated dielectric fluids	0.05–4% $(\gamma C_4H_3S)_4$ Sn $\gamma(C_4H_3)_2$Sn or γC_4H_3S $SnPh_3$	24

(continued)

Metal	Environment	Inhibitor	Reference
Steel	Halogenated organic insulating materials as chlorinated dipheryl	0.1% 2, 4$(NH_2)_2C_6H_3NHPh$, $o-MeH_4NH_2$ or $p-NO_2C_6H_4NH_2$	31
Steel	Herbicides as 2, 4 dinitro—6—alkyl phenols in aromatic oils	1.0–1.5% furfural	32
Steel	Isopropanol, 30%	0.03% sodium nitrite plus 0.015% oleic acid	72
Steel	1:4 methanol—water	To 4 l. water and 1 l. methanol add 1 g. pyridine and 0.05 g. pyragallol	66
Steel	Nitrogen fertilizer solutions	0.1% ammonium thiocyanate	2
Steel	Phosphoric acid, conc.	0.01–0.5% dodecylamine or 2 amino bicyclohexyl and 0.001% potassium iodide, potassium iodate or iodacetic acid	47
Steel	Polyoxyalklene glycol fluids	2% Emery's acid (dilinoleic acid) 1.25% $N(CHMe_2)_3$ 0.05–0.2% mercaptobenzothiazole	43
Steel	Sodium chloride, 0.05%	0.2% sodium nitrite	72
Steel	50% sodium trichloracetate soln.	0.5% sodium dichromate	1
Steel	Sulfide containing brine	Formaldehyde	14
Steel	Tetrahydrofurfuryl alcohol	1% sodium nitrate or 0.3% sodium chromate	15
Steel	Water	Benzoic acid	70
Steel	Water for flooding operations	Rosin amine	38
Steel	Water saturated hydrocarbons	Sodium nitrite	73
Steel	Water, distilled	Aerosol (an ionic wetting agent)	27
Tin	Carbon tetrachloride, wet	0.001–0.1% aniline	53
Tin	Chlorinated aromatics	0.1–2.0% nitrochlorobenzene	21
Tinned copper	Sodium hypochlorite contained bleaches	Sodium silicate	58
Tin plate	Alkali cleaning agents as trisodium phosphate, sodium carbonate, etc.	Diethylene diaminocobaltic nitrate	34
Tin plate	Alkaline soap	0.1% sodium nitrite	64
Tin plate	Carbon tetrachloride	2% mesityl oxide, 0.001% diphenylamine	76
Tin plate	Sodium chloride, 0.05%	0.2% sodium nitrite	72
Titanium	Hydrochloric acid	Oxidizing agents as chromic acid or copper sulfate	33
Titanium	Sulfuric acid	Oxidizing agents or inorganic sulfates	33
Zinc	Distilled water	15 ppm. mixture calcium and metaphosphates	77

Source: Maxey Brooke, Corrosion Inhibitor Checklist, *Chem. Eng.*, 230–234 (Dec. 1954).

REFERENCES

1. Alquist, F. N. and Wasco, J. L., Corrosion, 8, 410–12 (1952).
2. J. T. Baker Chemical Co. Product Bull. 101, p. 6.
3. Baker, R. A. U.S. Pat. 2,572,301 (Oct. 23, 1951).
4. Baer, M. S., U.S. Pat. 2,513,131 (June 27, 1950).
5. Baldeschwigler, E. L. and Zimmer, J. C., U.S. Pat. 2,396, 236 (March 12, 1946).
6. Bayes, A. L., U.S. Pat. 2,147, 395 (Feb. 14, 1939).
7. Bayes, A. L., U.S. Pat. 2,153, 952 (Apr. 11, 1939).
8. Beck, A. and Dibelka, H., U.S. Pat. 1,876,131 (Sept. 6, 1933).
9. Bhatnagar, S. S. and Krishnamurth, K. G., J. Sci. Ind. Res. (India) 4,238–40 (1945), C.A. 40, 2968.
10. Bohner, Hauszeit v., A. W. v. Erftwerk, A. G., Aluminum, 3, 347–8 (1931).
11. Brooke, M., Chem. Eng., 59, 286–7 (Sept. 1952).
12. Brophy, J. E. et al., Ind. Eng. Chem., 43, 884–96 (1951).
13. Bushrod, V. J., U.S. Pat. 2,325,304 (July 27, 1944).
14. Clay, J. A., Jr., Pet. Engr., 18, No. 2, 111–14 (1946).
15. Clendening, K. A., Can. J. Research, 26 F, 209–20 (1948).
16. Colegate, G. T., Metallurgia, 39, 316–18 (1949), C.A. 43, 4622 c
17. Collari, N. and Fongi, N., Alluminio, 16, 13–21 (1947), C.A. 41, 4759 f.
18. Crittenden, E. D., U.S. Pat 2,549,430 (Apr. 17, 1951).
19. E. I. du Pont de Nemours and Co., Brit. Pat. 439,055 (Nov. 28, 1935).
20. E. I. du Pont de Nemours and Co., Fr. Pat. 792,106 (Dec. 23, 1935).
21. Egerton, L., U.S. Pat. 2,391,685 (Dec. 25, 1945).
22. Eldredge, G. G. and Mears, R. B., Ind. Eng. Chem., 37, 736–41 (1945).
23. Elder, J. A. Jr., U.S. Pats. 2,487,755–6 (Aug. 9, 1949).
24. Ellenburg, A. M., U.S. Pat. 2,573, 894 (Nov. 6, 1951).
25. L'Etat Francais Represente par le Ministre de la Marine, Fr. Pat. 785,177 (Aug. 2, 1935).
26. I. G. Farbenind, Ger. Pat. 636,912 (Oct. 17, 1936).
27. Fontana, M. G., Ind. Eng. Chem., 42, 7, 65A–66A (July, 1950).
28. Geller, Vereningte Aluminum-Werke, A. G. Lautawerk, Ger. Office of Technical Services, PB-70016, Frames 6609–6610 (July 1937).
29. Gindin & Sil's Doklady Akad. Nauk USSR, 63, No. 6, 685–8 (1948). C.A. 43, 4207 c.
30. Glassmire, W. F. and Smith, W. R., U.S. Pat. 2,586,323 (Feb. 19, 1952).
31. Hardy, E. E. and Jackson, E. F., U.S. Pat. 2,617,770 (Nov. 11, 1952).
32. Hanson, W. J. and Nex, R. W., U.S. Pat. 2,623,818 (Dec. 30, 1952).
33. Harple, W. W. and Kiefer, G., Chem. Eng. Prog., 49,566 (1953).
34. Henkle et Cie, G.M.B.H., Brit. Pat. 415,672 (Aug. 30, 1934).
35. Hetherington, H. C., U.S. Pat. 2,462,638 (Feb. 22, 1949).
36. Hillyer, J. C. and Nicewander, D. C., U.S. Pat. 2,473,750 (June 21, 1949).
37. Hoar, T. P. and Havenhand, D. J., Iron Steel Inst. (London) 133, 252 (1936).
38. Howell, W. E., Barton, J. K. and Heck, E. T., Producers Monthly, 13, No. 7, 27–34 (1949).

39. Jenckel & Woltmann, Z. anorg. allgem. Chem., 217, 298 (1934).
40. Keener, F. G., U.S. Pat. 2,238,651 (April 15, 1941).
41. Lamprey, H., U.S. Pat. 2,147,409 (Feb. 14, 1939).
42. Lamprey, H., U.S. Pat. 2,153,961 (Apr. 11, 1939).
43. Langer, T. W. and Mago, B. F., U.S. Pat. 2,624,708 (Jan. 6, 1953).
44. Lichtenberg, H., Aluminum, 19, 504–9 (1937).
45. Lichtenberg, H., Aluminum, 20, 264–5 (1938).
46. Lichtenberg, H. and Geier, Aluminum, 20, 784 (1938).
47. Malowan, J. E., U.S. Pat. 2,567,156 (Sept. 4, 1951).
48. Marsh, G. A., U.S. Pat. 2,574,576 (March 13, 1951).
49. Mathews, J. W. and Uhlig, H. H., Corrosion, 7, 419–22 (1951).
50. McDermatt, F. A., U.S. Pat. 1,927,842 (Sept. 26, 1933).
51. Mann, C. A., Lauer, B. E. and Hultin, C. T., Ind. Eng. Chem., 28, 1048–51 (1936).
52. Mears, R. B. and Eldredge, G. G., Trans. Electrochem. Soc., 83, 403 (1943).
53. Ohlmann, E. O., U.S. Pat. 2,387,284 (Oct. 23, 1945).
54. Phillips, C. Jr., U.S. Pat. 2,550,425 (Apr. 24, 1951).
55. Plueddemann, E. P. and Rathmann, R., U.S. Pat. 2,423,343 (July 1, 1947).
56. Rae, Mfg. Chem., 16, 394 (1945).
57. Rhodes, F. H. and Berner, F. W., Ind. Eng. Chem., 25, 1336–7 (1933).
58. Robinson, E. A., Soap and Sanitary Chem., 28, 34–6 (Jan. 1952).
59. Rohrig, H., Aluminum, 17, 559–62 (1935).
60. Rohrig, H. and Gaier, K., Aluminum, 19, 448–50 (1973).
61. Schultz, W. A., U.S. Pat. 2,089,579 (Aug. 10, 1937).
62. Smith, W. R., U.S. Pat. 2,524,484 (Oct. 3, 1950).
63. Soc. Continentale Parker, Fr. Pat. 732,230 (Apr. 27, 1931).
64. Stone, I., U.S. Pat. 1,168,722 (July 31, 1934).
65. Swain, R. C. and Paden, J. H., U.S. Pat. 2,512,590 (June 20, 1950).
66. Swiss Pat. 247,835 (Dec. 16, 1947).
67. Thomas, J. F. J., Can. J. Research, 21B, 43–53 (1943).
68. Uhlig, H. H., Corrosion Handbook, 1st. Ed., John Wiley & Sons.
69. Unknown, Light Metals, 12, No. 134, 165–8 (1949), see Corrosion 6, 42 (1950).
70. Vernon, W. H. J., Corrosion, 4, 141–8 (1948).
71. Viles, P. S. and Walsh, D. C., U.S. Pat. 2,550,434 (Apr. 24, 1951).
72. Wachter, A., Ind. Eng. Chem., 37, 749–51 (1945).
73. Wachter, A. and Smith, S. S., Ind. Eng. Chem., 35, 358–67 (1943).
74. Wachter, Traseder & Weber, Corrosion 3, 406 (1947).
75. Weiler, W., Dent. Farber-Ztg., 74, 27 (1938).
76. Williams, U. S. Signal Laboratory, Fort Monmouth, N.J., Technical Memorandum No. M-1156 (1948).
77. Worspop, F. E. and Kingsburf, A., Chem. Eng. Mining Rev. (Australia), 173–176, (1950).

would be achieved by either of the two (or more) substances alone. This is called a *synergistic* effect. Many of the inhibitors listed in Table 6-1 are synergistic combinations of two or more inhibiting agents. At present, the mechanism of the synergistic effect is not completely understood.

Although inhibitors can be used to great advantage to suppress the corrosion of metals in many environments, there are certain limitations of this type of corrosion prevention which should be recognized. First, it may not be possible to add inhibitors to all corrosive systems because they may contaminate the environment. Further, many inhibitors are toxic, and their application is limited to those mediums that will not be used directly or indirectly in the preparation of food or other products that will come in contact with humans. Arsenic salts, which exert a powerful inhibiting effect in strong acids, have limited application for this reason. Inhibitors are primarily used in closed systems where the corrosive environment is either contained for long periods or recirculated. Inhibitors are usually not practical in "once-through" systems. Finally, inhibitors generally rapidly lose their effectiveness as the concentration and temperature of the environment increase.

Additional *general* information including *theoretical* aspects and also other examples of *applications* of various types of inhibitors are discussed here.

Rosenfeld* in his NACE Whitney Award Lecture 1981 presents a theoretical discussion and also the use of quantum chemistry calculations, x-ray photoelectron spectroscopy, radioactive tracers (i.e., ^{36}Cl), and chemical polarization as "tools" for studying inhibition. Inhibition can be caused by both adsorption and phase layers on the metal surface. He shows that adsorption inhibition prevents pitting of stainless steels. Protection is related to the adsorption displacement of activator (Cl^-) from the surface by passivating ions. It is related to this displacement power and not the oxidizing power of the anion. NO_3^- is more effective than CrO_4^{2-}, although the latter is a stronger oxidizer. He indicates that protection of steel from corrosion by hydrogen sulfide is due to the emergence of phase films containing iron, sulfur, and inhibitor components on the steel surface. He also shows that only a small part (in one case 4%) of the surface needs to be covered by the adsorbed inhibitor to stop corrosion. Adsorption in some cases has a chemical nature. Rosenfeld's main point "is that the inhibitor mechanism, irrespective of the amount adsorbed by the surface, consists in changing the electrophysical properties of surface atoms by the donor-acceptor inhibitor/metal reaction, rather than in screening the metal surface from the corrosive environment."

*I. L. Rosenfeld, New Data on the Mechanism of Metals Protection with Inhibitors, *Corrosion*, **37**:371–377 (July 1981).

An International Conference on Corrosion Inhibition was held during May 1983. Evidence presented during recent years indicates that many organic and inorganic inhibitors become effective through interaction with one of several corrosion products to form a new protective phase rather than by absorption on the metal surface. The conference focuses on distinction between *interface* and *interphase* inhibitors. The Proceedings of this Conference (about 30 papers) will be published by NACE in 1984.

One commercial inhibitor and rust remover known as KO-99 is claimed to be of the interphase type.

Greene* discusses mechanisms, applications, and limitations of oxidizing inhibitors. They must be used with extreme caution in the presence of halides and other ions associated with localized corrosion.

A review paper that gives 161 literature references is by G. Trabanelli and V. Carassiti, "Mechanism and Phenomenology of Organic Inhibitors," in *Advances in Corrosion Science and Technology*, vol. 1, pp. 147–228, Plenum Press, New York, 1970. Structural formulas of organic inhibitors are included and are helpful.

An important cause of corrosion in overhead streams in oil refineries is attack by hydrogen sulfide and hydrochloric acid. Film-forming inhibitors are suggested by Nathan and Perugini.† Case histories are described. Inhibitors are also helpful in reducing hydrogen blistering.

Environmental considerations have focused attention on nonheavy metal inhibitors (i.e., chromates are banned). A paper by Breske‡ contains information on testing and field experience with these inhibitors for corrosion by waters. Fouling (deposition) is also discussed. Brunn¶ replaced chromate with a triazole inhibitor. See also Chapter 8 under cooling waters.

Vukasovich and Sullivan§ studied combined and separate effects of molybdate, phosphate, and borate used with nitrate and tolyltriazole for cooling water in auto engines and describe test procedures.

Addition of 0.2% water to anhydrous ammonia has stopped attack on steel. Inhibitors are also discussed in Chapter 8 under the appropriate environments and in Chapter 3 as well.

*N. D. Greene, Mechanism and Application of Oxidizing Inhibitors, *Materials Performance*, **21**:20–22 (Mar. 1982).

†C. C. Nathan and J. J. Perugini, Prevention of Corrosion in Refinery Overhead Streams by Use of Neutralizing and Film-Forming Inhibitors, *Materials Performance*, **13**:29–33 (Sept. 1974).

‡T. C. Breske, Testing and Field Experience with Nonheavy Corrosion Inhibitors, *Materials Performance*, **16**:17–24 (Feb. 1977).

¶A. F. Brunn, Jr., Chromate Inhibitor Successfully Replaced in Copper-Nickel Plant Cooling Tower, *Materials Performance* **22**:13–16 (June 1983).

§M. S. Vukasovich and F. J. Sullivan, Evaluation of Molybdate as an Inhibitor in Automotive Engine Coolants, *Materials Performance* **22**: 25–33 (Aug. 1983).

DESIGN

The design of a structure is frequently as important as the choice of materials of construction. Design should consider mechanical and strength requirements together with an allowance for corrosion. In all cases, the mechanical design of a component should be based on the material of construction. This is important to recognize, since materials of construction used for corrosion resistance vary widely in their mechanical characteristics.

6-6 Wall Thickness

Since corrosion is a penetrating action, it is necessary to make allowances for this reduction in thickness in designing pipes, tanks, and other components. In general, wall thickness is usually made twice the thickness that would give the desired life. If a 10-year life is required for a given tank, and the best estimate of corrosion rate is $\frac{1}{8}$ in. in 10 years (corrosion rate about 12 mpy), the tank would be designed with a wall thickness of $\frac{1}{4}$ in. Such a design factor allows for some variation in the depth of penetration during uniform corrosion, which in most cases is not completely uniform. Of course, the wall thickness must meet mechanical requirements such as pressure, weight, and stress considerations.

This general rule may not apply where *reliable* corrosion data and *effective* monitoring are involved. Wallace and Webb* present an interesting discussion on realistic corrosion allowances. They ask the question, "Are you lining your vessels with cold, hard cash?" In other words, excessive corrosion allowances add weight and costs. They also suggest, for example, different corrosion allowances in the upper and lower zones of a tall vertical vessel. The upper portion could be subject to less corrosion than the lower half. I used this idea over 40 years ago when specifying heavier top portions of steel tank cars for shipping sulfuric acid—the upper area corroded many times faster than the lower.

6-7 Design Rules

There are many design rules which should be followed for best corrosion resistance. When I started working for the Du Pont Co. in 1934 I had to write a report every month and state the dollars saved because of my efforts. After one year or so we convinced the design people that "preventive medicine" was cost effective. In other words, "design corrosion out" of the system instead of waiting until the equipment *fails* in service. (It costs much less to change some lines on a drawing.) This resulted in close communication between designers and corrosion engineers and all major

*A. E. Wallace and W. P. Webb, Cut Vessel Costs with Realistic Corrosion Allowances, *Chem. Eng.*, 123–126 (Aug. 24, 1981).

projects included funds for utilization of corrosion engineers. I firmly believe the corrosion engineer should "sign off" on equipment drawings and not only the design engineer.

Some of the design rules that should be followed are listed below. It would be helpful if the designer had a good background in corrosion but unfortunately this is usually not the case.

1. Weld rather than rivet tanks and other containers. Riveted joints provide sites for crevice corrosion (see Chap. 3).
2. Design tanks and other containers for easy draining and easy cleaning. Tank bottoms should be sloped toward drain holes so that liquids cannot collect after the tank is emptied. Concentrated sulfuric acid is only negligibly corrosive toward steel. However, if a steel sulfuric acid tank is incompletely drained and the remaining liquid is exposed to the air, the acid tends to absorb moisture, resulting in dilution, and rapid attack occurs.
3. Design systems for the easy replacement of components that are expected to fail rapidly in service. Frequently, pumps in chemical plants are designed so that they can be readily removed from a piping system.
4. Avoid excessive mechanical stresses and stress concentrations in components exposed to corrosive mediums. Mechanical or residual stresses are one of the requirements for stress-corrosion cracking (see Chap. 3). This rule should be followed especially when using materials susceptible to stress-corrosion cracking.
5. Avoid electrical contact between dissimilar metals to prevent galvanic corrosion (see Chap. 3). If possible, use similar materials throughout the entire structure, or insulate different materials from one another.
6. Avoid sharp bends in piping systems when high velocities and/or solids in suspension are involved (erosion corrosion).
7. Provide thicker structures to take care of impingement effects.
8. Make sure materials are properly selected.
9. List complete specifications for all materials of construction and provide instructions to be sure the specs are followed all the way through to final inspection. Specify quality control procedures if relevant.
10. Be sure all relevant codes and standards are met.
11. Set realistic and scheduled dates for delivery of equipment.
12. Specify procedures for testing and storage of parts and equipment. For example, after hydraulic testing do not let the equipment sit full or partially full of water for any extended period of time. This could result in microbial corrosion, pitting, and stress corrosion. With regard to storage, spare stainless steel tubing showed stress-corrosion cracking when stored near the seacoast.
13. Specify operating and maintenance procedures (i.e., scheduled shutdowns).

14. Properly design against excessive vibration, not only for rotating parts but also, for example, for heat exchanger tubes.
15. Provide for "blanketing" with dry air or inert gas if vessels "inhale" moist marine atmosphere while being emptied.
16. Select plant site upwind from other "polluting" plants or atmosphere if relevant and/or feasible.
17. Avoid hot spots during heat-transfer operations. Heat exchangers and other heat-transfer devices should be designed to ensure uniform temperature gradients. Uneven temperature distribution leads to local heating and high corrosion rates. Further, hot spots tend to produce stresses that may produce stress-corrosion cracking failures.
18. Design to exclude air. Oxygen reduction is one of the most common cathodic reactions during corrosion, and if oxygen is eliminated, corrosion can often be reduced or prevented. In designing chemical plant equipment, particular attention should be paid to agitators, liquid inlets, and other points where air entrainment is a possibility. Exceptions to this rule are active-passive metals and alloys. Titanium and stainless steels are more resistant to acids containing dissolved air or other oxidizers.
19. The most general rule for design is: *avoid heterogeneity*. Dissimilar metals, vapor spaces, uneven heat and stress distributions, and other differences between points in the system lead to corrosion damage. Hence, in design, attempt to make all conditions as uniform as possible throughout the entire system.

CATHODIC AND ANODIC PROTECTION

6-8 Cathodic Protection

Cathodic protection was employed before the science of electrochemistry had been developed. Humphrey Davy used cathodic protection on British naval ships in 1824. The principles of cathodic protection may be explained by considering the corrosion of a typical metal M in an acid environment. Electrochemical reactions occurring are the dissolution of the metal and the evolution of hydrogen gas; for example,

$$M \rightarrow M^{n+} + ne \tag{6.3}$$

$$2H^{+} + 2e \rightarrow H_2 \tag{6.4}$$

Cathodic protection is achieved by supplying electrons to the metal structure to be protected. Examination of Eqs. (6.3) and (6.4) indicates that the addition of electrons to the structure will tend to suppress metal dissolution and increase the rate of hydrogen evolution. If current is considered to flow from (+) to (−), as in conventional electrical theory, then a structure is protected if current enters it from the electrolyte. Conversely, accelerated corrosion occurs if current passes from the metal to the electrolyte. This current convention has been adopted in cathodic protection technology and is used here for consistency.

There are two ways to cathodically protect a structure: (1) by an external power supply or, (2) by appropriate galvanic coupling. Figure 6-1 illustrates cathodic protection by impressed current. Here, an external dc power supply is connected to an underground tank. The negative terminal of the power supply is connected to the tank, and the positive to an inert anode such as graphite or Duriron. The electric leads to the tank and the inert electrode are carefully insulated to prevent current leakage. The anode is usually surrounded by backfill consisting of coke breeze, gypsum, or bentonite, which improves electric contact between the anode and the surrounding soil. As shown in Fig. 6-1, current passes to the metallic structure, and corrosion is suppressed.

Cathodic protection by galvanic coupling to magnesium is shown in Fig. 6-2. As discussed in Chap. 3, magnesium is anodic with respect to steel

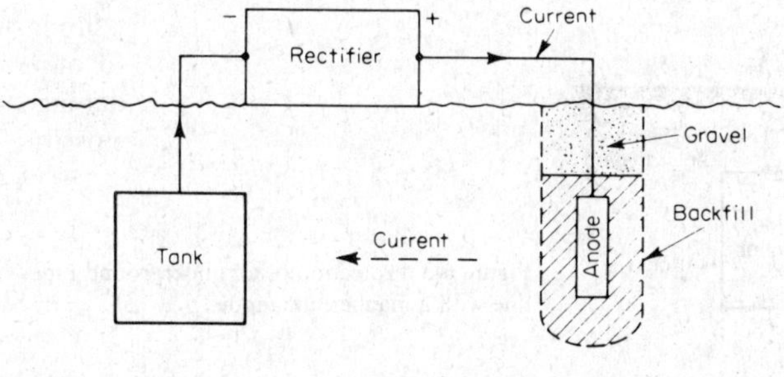

Figure 6-1 Cathodic protection of an underground tank using impressed currents.

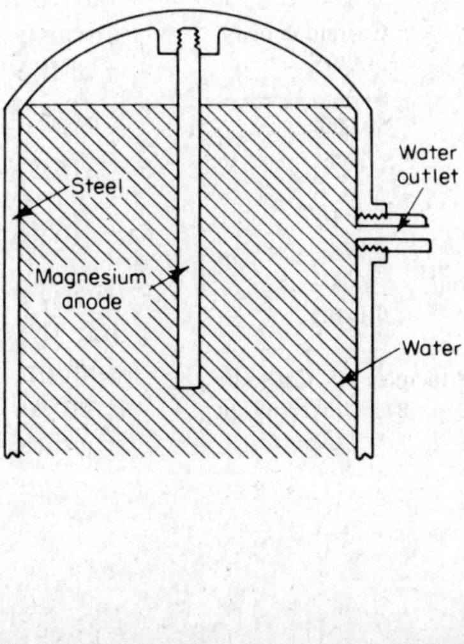

Figure 6-2 Cathodic protection of a domestic hot-water tank using a sacrificial anode.

and corrodes preferentially when galvanically coupled. The anode in this case is called a sacrificial anode since it is consumed during the protection of the steel structure. Cathodic protection using sacrificial anodes can also be used to protect buried pipelines, as shown in Fig. 6-3. The anodes are spaced along the pipe to ensure uniform current distribution.

Protective currents are usually determined empirically, and some typical values are listed in Table 6-2. Aggressive corrosives such as hot acids require prohibitively high currents, whereas much lower currents are needed to protect steel in less severe environments (concrete). Table 6-2 indicates typical average values of protective currents. Specific applications can deviate from these values. For example, in certain very acidic soils, 10 to 15 mA is often needed to reduce the corrosion of steel structures to tolerable levels. Also, pipes with organic coatings require much lower currents since the only areas requiring protection are defects or "holidays" in the protective layer. In such cases, trial-and-error adjustments of anode size or applied current can be made until satisfactory protection is achieved. A more accurate and less time-consuming approach is to measure the potential of the protected structure with a suitable reference electrode.

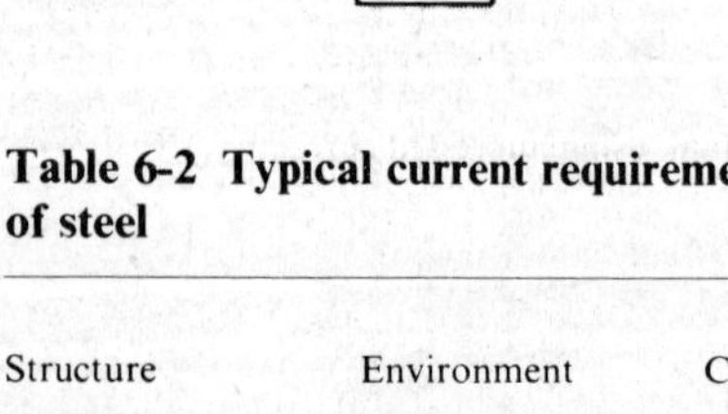

Figure 6-3 Protection of an underground pipeline with a magnesium anode.

Table 6-2 Typical current requirements for cathodic protection of steel

Structure	Environment	Conditions	Current density, mA/ft^2
Tank	Hot H_2SO_4	Static	50,000
Pipelines and storage tanks	Underground (soil)	Static	1–3
Pipelines	Fresh water	Flowing	5–10
Water heaters	Hot, fresh water	Slow flow	1–3
Pilings	Seawater	Tidal motion	6–8
Reinforcing rods	Concrete	Static	0.1–0.5

Source: Some data taken from M. Stern, Principles of Cathodic Protection, *Symposium on Corrosion Fundamentals*, p. 84, University of Tennessee Press, 1956.

Steel structures exposed to soils, fresh and brackish water, and seawater are protected if they are polarized to a potential of −0.85 volt versus a copper/copper sulfate reference electrode. Figure 6-4 shows such an electrode designed for cathodic-protection surveys. This electrode has the advantages of low cost, good accuracy, and ruggedness. The potential of a structure is determined with a high-resestance voltmeter as shown in Fig. 6-5. During this measurement, the reference electrode is placed in the ground or on a sponge soaked in brine to make electrical contact. The cathodic current density necessary to polarize the pipe to −0.85 volt can be readily determined. In cases where sacrificial anodes (e.g., magnesium) are used, this same measurement is used to indicate the number and size of anodes needed for full protection. On long pipes or large, complex structures, reference-electrode surveying is utilized to determine uniformity of applied currents.

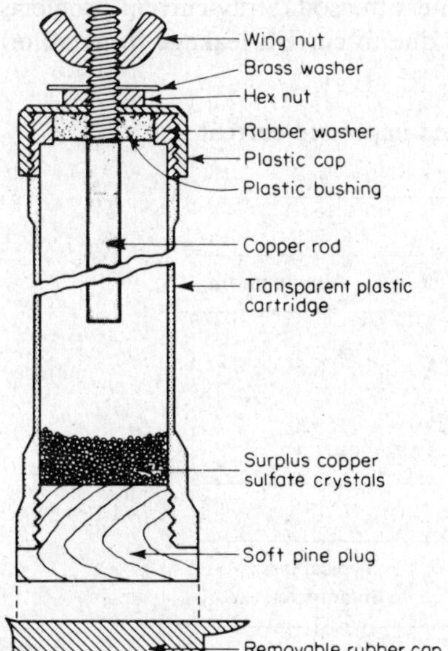

Figure 6-4 Copper–copper sulfate reference electrode.

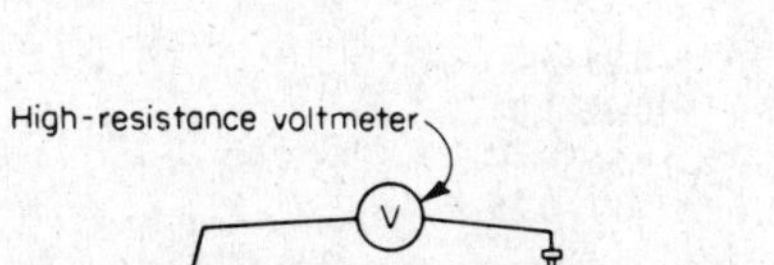

Figure 6-5 Pipe potential measurements with a reference electrode.

Anode selection for cathodic protection is based on engineering and economic considerations. Table 6-3 compares several types of sacrificial and impressed-current anodes. Of the sacrificial anodes, magnesium is the most widely used. Although its efficiency is low (about 50%), this is more than offset by its very negative potential, which provides high current output.

There is a considerable variety of impressed-current anodes ranging from low-cost scrap steel, which suffers relatively large losses, to the inert platinized titanium which is both efficient and expensive. Steel, graphite, and silicon-iron are the most widely used anode materials, with lead and platinized titanium finding increased applications in marine environments.

Stray-current effects are often encountered in cathodic-protection systems. The term *stray current* refers to extraneous direct currents in the earth. If a metallic object is placed in a strong current field, a potential difference develops across it and accelerated corrosion occurs at points where current leaves the object and enters the soil. Stray-current problems were quite common in previous years due to current leakage from trolley

Table 6-3 Comparisons of sacrificial and impressed-current anodes for cathodic protection

	Sacrificial anodes		
	Magnesium	Zinc	Aluminum-tin
Theoretical consumption, lb/ampere-year	9	23	6.5
Actual consumption, lb/ampere-year	18	25	16–20
Potential vs. $Cu/CuSO_4$	−1.7	−1.15	−1.3

Impressed-current anodes		
Material	Typical applications	Typical loss, lb/ampere-year
Scrap steel	Soil, fresh- and sea-water	20
Aluminum	Soil, fresh- and sea-water	10–12
Graphite	Soil and fresh water	0.25–5.0
High-silicon iron and Si-Cr iron	Soil, fresh water, and seawater	0.25–1.0
Lead	Seawater	0.1–0.25
Plantinized titanium	Seawater	nil

Source: Modified from J. H. Morgan, *Cathodic Protection*, The Macmillan Company, New York, 1960.

tracks. Pipelines and tanks under tracks were rapidly corroded. However, since this type of transportation is now obsolete, stray currents from this source are no longer a problem. A more common source of stray currents is from cathodic-protection systems. This is especially pronounced in densely populated oil production fields and within industrial complexes containing numerous buried pipelines.

Figure 6-6 illustrates stray currents resulting from a cathodic-protection system. The owner of the buried tank installed cathodic protection. He did not know of the nearby pipeline that failed rapidly due to the stray-current field. If the owner of the pipeline applies cathodic protection, it is possible to prevent stray-current attack of his pipe, but it will produce stray-current attack of the buried tank. It is easy to see how stray-current corrosion tends to escalate. As each owner adds or increases protective currents to his structures, he increases the current requirements on other adjacent structures. In one industrial area containing a high density of protected underground structures, protective current requirements rapidly rose to 20 mA/ft^2 in several areas! The solution to this problem is cooperation between operators. For example, the stray-current problem shown in Fig. 6-6 could be prevented by electrically connecting the tank and pipe by a bus connector and rearranging anodes as shown in Fig. 6-7. Here, both pipe and tank are protected without stray-current effects, with the owners sharing the cost of installation and operation.

Cathodic protection involving impressed current anodes to reduce highway and other bridge maintenance costs is being done and substantial savings realized. Corrosion of the reinforcing steel is reduced. Solar panels to supply current for bridges in rural areas has been suggested.

Cathodic protection is still more of an art than a science, and past experience is the best basis for judgment. This remark was made by a corrosion engineer who retired after 30 years of experience, mostly on cathodic

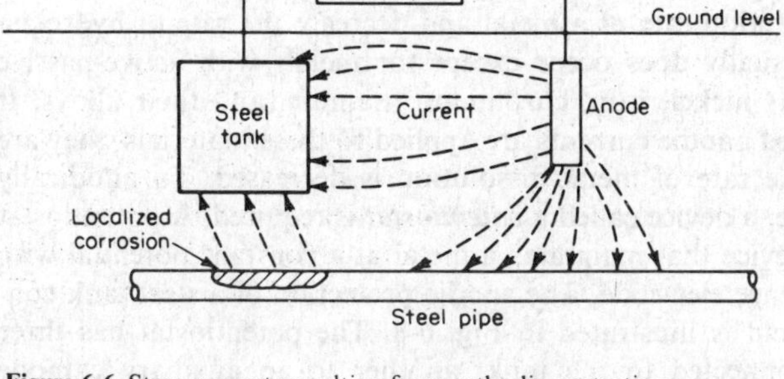

Figure 6-6 Stray currents resulting from cathodic protection.

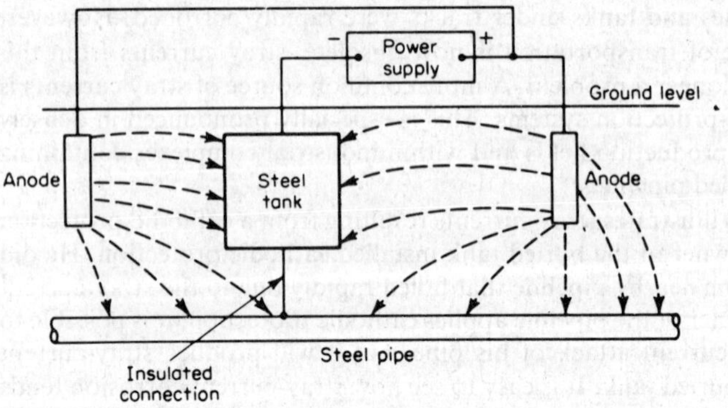

Figure 6-7 Prevention of stray-current corrosion by proper design.

protection of pipelines. There are many firms in the cathodic-protection business, and they should be consulted if you have a major project.

One of the most extensive bibliographies on any subject is on underground corrosion and Part 1 appeared in *Materials Performance*, pp. 41–43 (Jan. 1982). A total of 10 parts was published in consecutive issues. This bibliography contains 1172 English-language journal articles. Subjects include underground and soil corrosion, bacterial corrosion, cathodic protection and protective coatings.

6-9 Anodic Protection

In contrast to cathodic protection, anodic protection is relatively new; it was first suggested by Edeleanu in 1954. This technique was developed using electrode kinetics principles and is somewhat difficult to describe without introducing advanced concepts of electrochemical theory. Simply, anodic protection is based on the formation of a protective film on metals by externally applied anodic currents. Considering Eqs. (6.3) and (6.4), it appears that the application of anodic current to a structure should tend to increase the dissolution rate of a metal and decrease the rate of hydrogen evolution. This usually does occur except for metals with active-passive transitions such as nickel, iron, chromium, titanium, and their alloys. If carefully controlled anodic currents are applied to these materials, they are passivated and the rate of metal dissolution is decreased. To anodically protect a structure, a device called a *potentiostat* is required. A potentiostat is an electronic device that maintains a metal at a constant potential with respect to a reference electrode. The anodic protection of a steel tank containing sulfuric acid is illustrated in Fig. 6-8. The potentiostat has three terminals, one connected to the tank, another to an auxiliary cathode

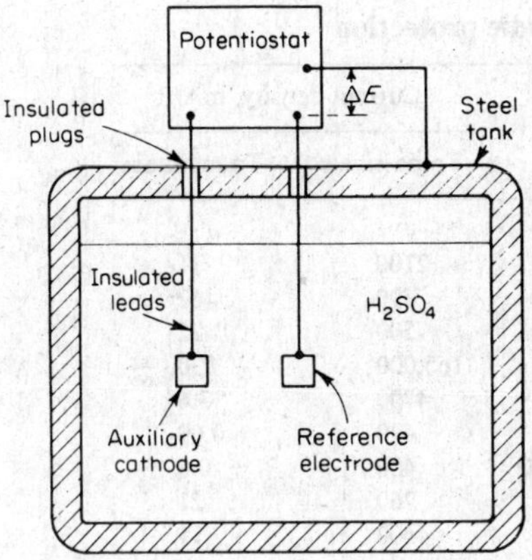

Figure 6-8 Anodic protection of a steel storage tank containing sulfuric acid.

(a platinum or platinum-clad electrode), and the third to a reference electrode (e.g., calomel cell). In operation, the potentiostat maintains a constant potential between the tank and the reference electrode. The optimum potential for protection is determined by electrochemical measurements.

Anodic protection can decrease corrosion rate substantially. Table 6-4 lists the corrosion rates of austenitic stainless steel in sulfuric acid solutions containing chloride ions with and without anodic protection. Examination of the table shows that anodic protection causes a 100,000-fold decrease in corrosive attack in some systems. Although anodic protection is limited to passive metals and alloys, most structural materials of modern technology contain these elements. Thus, this restriction is not as important as it first might seem. Table 6-5 lists several systems where anodic protection has been

Table 6-4 Anodic protection of austenitic stainless steel at 30°C

Protected at 0.500 volt vs. saturated calomel electrode

		Corrosion rate, mpy	
Alloy type	Environment (air exposed)	Unprotected	Anodically protected
	$N\ H_2SO_4 + 10^{-5} M$ NaCl	14	0.025
304	$N\ H_2SO_4 + 10^{-3} M$ NaCl	2.9	0.045
(19Cr-9Ni)	$N\ H_2SO_4 + 10^{-1} M$ NaCl	3.2	0.20
	$10N\ H_2SO_4 + 10^{-5} M$ NaCl	1930	0.016
	$10N\ H_2SO_4 + 10^{-3} M$ NaCl	1125	0.04
	$10N\ H_2SO_4 + 10^{-1} M$ NaCl	77	0.21

Source: S. J. Acello and N. D. Greene, *Corrosion*, **18**:286*t* (1962).

Table 6-5 Current requirements for anodic protection

Fluid and concentration	Temperature, °F	Metal	Current density, mA/ft² To passivate	To maintain
H_2SO_4				
1 molar	75	316SS	2100	11
15%	75	304	390	67
30%	75	304	500	22
45%	150	304	165,000	830
67%	75	304	4700	3.6
67%	75	316	470	0.09
67%	75	Carpenter 20	400	0.8
93%	75	Mild steel	260	21
Oleum	75	Mild steel	4400	11
H_3PO_4				
75%	75	Mild steel	38,000	19,000
115%	180	304SS	0.03	0.00014
NaOH				
20%	75	304SS	4400	9.4

Source: C. E. Locke et al., *Chem. Eng. Progr.*, **56**:50 (1960).

applied successfully. The primary advantages of anodic protection are its applicability in extremely corrosive environments and its low current requirements. (For additional information concerning the principles and mechanism of anodic protection, see Chap. 10.)

An interesting and economical application of anodic protection is the use of type 316 stainless pipes for cooling acid in sulfuric acid plants. The pipes have anodic protection. This arrangement replaces the old thick-wall cast-iron heat exchangers.

The reader is referred to an excellent book, *Anodic Protection—Theory and Practice in the Prevention of Corrosion*, by O. L. Riggs and C. E. Locke, Plenum Press, New York (1981). This extensively covers *all phases* of this subject.

6-10 Comparison of Anodic and Cathodic Protection

Some of the important differences between anodic and cathodic protection are listed in Table 6-6. Each method has advantages and disadvantages, and anodic and cathodic protection tend to complement one another. Anodic protection can be utilized in corrosives ranging from weak to very aggresive, while cathodic protection is restricted to moderately corrosive conditions because of its high current requirement, which increases as the corrosivity of the environment increases. (Compare steel in H_2SO_4 in Tables 6-2 and 6-5.) Hence, it is not practical to cathodically protect metals in

Table 6-6 Comparison of anodic and cathodic protection

	Anodic protection	Cathodic protection
Applicability		
Metals	Active-passive metals only	All metals
Corrosives	Weak to aggressive	Weak to moderate
Relative cost		
Installation	High	Low
Operation	Very low	Mediums to high
Throwing power	Very high	Low
Significance of applied current	Often a direct measure of protected corrosion rate	Complex—does not indicate corrosion rate
Operating conditions	Can be accurately and rapidly determined by electrochemical measurements	Must usually be determined by empirical testing

very aggressive mediums. Anodic protection, on the other hand, uses very small applied currents, and it can be utilized in strong corrosive mediums.

The installation of a cathodic-protection system is relatively inexpensive since the components are simple and easily installed. Anodic protection requires complex instrumentation including a potentiostat and reference electrode, and its installation cost is high. The operating costs of the two systems differ because of the difference in current requirements noted above. Throwing power, or the uniformity of current-density distribution, varies between the two types of protection. The throwing power of cathodic protection is generally low, which requires numerous closely spaced electrodes to achieve uniform protection. Anodic-protection systems have throwing power, and consequently, a single auxiliary cathode can be utilized to protect long lengths of pipe.

Anodic protection possesses two unique advantages. First, the applied current is usually equal to the corrosion rate of the protected system. Thus, anodic protection not only protects but offers a means for monitoring instantaneous corrosion rate. Secondly, operating conditions for anodic protection can be precisely established by laboratory polarization measurements. In contrast, the operating limits for cathodic protection are usually established by empirical trial-and-error tests. Although various rapid evaluation methods for estimating the current requirements for cathodic protection have been suggested, all of these have proved to be unreliable to a greater or lesser degree, and the final choice is usually based on past experience.

The concept of anodic protection is based on sound scientific principles, as discussed in Chap. 10, and has been successfully applied to industrial corrosion problems. However, the incorporation of anodic protection into corrosion engineering practice has occurred very slowly since its introduction. The reluctance of corrosion engineers to utilize this new method of preventing corrosion is probably due in large measure to their personal corrosion experiences and to classic corrosion literature. The disastrous effects that are produced if a cathodic-protection system is connected with reverse polarity, the rapid deterioration of an anode in a galvanic couple, and the generalized rule of classic corrosion literature, which states that impressed anodic currents accelerate corrosion, all tend to suppress the introduction of this protection technique. In essence, anodic protection represents an exception to the general rule that impressed anodic current or the removal of electrons from a metal accelerates corrosion. In the future, anodic protection will probably revolutionize many current practices of corrosion engineering. Utilizing this technique, it is possible to reduce the alloy requirements for a particular corrosion service. Anodic protection can be classed as one of the most significant advances in the entire history of corrosion science.

COATINGS

6-11 Metallic and Other Inorganic Coatings

Relatively thin coatings of metallic and inorganic materials can provide a satisfactory barrier between metal and its invironment. The chief function of such coatings is (aside from sacrificial coatings such as zinc) to provide an effective barrier. Metal coatings are applied by electrodeposition, flame spraying, cladding, hot dipping, and vapor deposition. Inorganics are applied or formed by spraying, diffusion, or chemical conversion. Spraying is usually followed by baking or firing at elevated temperatures. Metal coatings usually exhibit some formability, whereas the inorganics are brittle. In both cases a complete barrier must be provided. Porosity or other defects can result in accelerated localized attack on the basic metal because of two-metal effects.

Examples of metal-coated articles are automobile bumpers and trim, household appliances and fixtures, silverware, galvanized steel, and tin cans. Bathtubs and "glassed" steel tanks are representative of ceramic coatings. Examples of conversion and diffusion coatings are anodized aluminum and chromized steel, respectively.

Electrodeposition This process, also called electroplating, consists of immersing a part to be coated in a solution of the metal to be plated and passing direct current between the part and another electrode. The character of the deposit depends on many factors including temperature, current density, time, and composition of the bath. These variables can be adjusted to produce coatings that are thick (say 20 mils) or thin (thousandths of a mil in the case of some tin plate), dull or bright, soft (lead) or hard (chromium), and ductile or brittle. Hard platings are utilized to combat erosion corrosion. The electroplate can be a single metal, layers of several metals, or even an alloy composition (e.g., brass). For example, an automobile bumper has an inner flash plate of copper (for good adhesion), an intermediate layer of nickel (for corrosion protection), and a thin top layer of chromium (primarily for appearance). Zinc, nickel, tin, and cadmium, in that order, are plated on the largest tonnage basis. Gold, silver, and platinum plates are common. The majority of the metals can be applied by electrodeposition.

Flame spraying This process, also called metallizing, consists of feeding a metal wire or powder through a melting flame so that the metal, in finely divided liquid particles, is blown onto the surface to be protected. Oxygen and acetylene or propane are commonly used for the melting flame. The coatings are usually porous and are not protective under severe wet corrosive conditions. Generally the porosity decreases with the melting point of the metal—zinc, tin, and lead are better from this standpoint than steel or stainless steel. The surface to be sprayed must be roughened (sandblasted) to obtain a mechanical bond. Sometimes a paint coating is applied over the sprayed metal to fill the voids and provide a better barrier. The porous metal makes a good base for the paint and a good bond is obtained. Flame spraying is an economical way of building up worn surfaces on parts such as shafting. High-melting metals may be deposited by plasma-jet spraying.

Flame-sprayed applications include tank cars and vessels of all kinds, bridges, ship hulls and superstructures, refrigeration equipment, and many fabricated steel products. Exhaust stacks sprayed with aluminum and sealed with a silicone-aluminum-organic coating are protected up to 900°F. Sprayed stainless (18-8) steels covered with sprayed aluminum gives protection in air up to 1500°F. Average costs for a 5-mil coating are 6¢/ft^2 for aluminum and 12¢/ft^2 for zinc.

Cladding This involves a surface layer of sheet metal usually put on by rolling two sheets of metal together. For example, a nickel and a steel sheet are hot-rolled together to produce a composite sheet with, say, $\frac{1}{8}$ in. of nickel and 1 in. of steel. The cladding is usually thin in relation to the other material. High-strength aluminum alloys are often clad with a commercially pure aluminum skin to provide the corrosion barrier because the alloy is susceptible to stress corrosion.

Sometimes a thin liner is spot-welded to the walls of a steel tank. Nickel, aluminum, copper, titanium, stainless steels, and other materials are often used as cladding for steel.

Development of the very-low-carbon stainless steels (type 304L) has increased the utilization of clad vessels. A stainless clad steel tank cannot be quench-annealed. A higher alloy rod is necessary for welding clad parts to avoid dilution of the weld deposit and loss of corrosion resistance. For example, type 310 stainless steel weld rod is used to join type 304L clad steel plate.

Cladding presents a great economic advantage in that the corrosion barrier or expensive material is relatively thin and is backed up by inexpensive steel. A good example is a high-pressure vessel that could have a $\frac{1}{16}$- or $\frac{1}{8}$-in. clad on 3 in. of steel. Costs might be astronomical if the entire wall were made of highly corrosion-resistant material.

Hot dipping Hot dip coatings are applied to metals by immersing them in a molten metal bath of low-melting-point metals, chiefly zinc, tin, lead, and aluminum. Hot dipping is one of the oldest methods for coating with metal. Galvanized steel is a popular example. Thickness of the coating is much greater than electroplates because very thin dip coatings are difficult to produce. Coated parts can be heat-treated to form an alloy bond between the coating and the substrate.

Vapor deposition This is accomplished in a high-vacuum chamber. The coating metal is vaporized by heating electrically, and the vapor deposits on the parts to be coated. This method is more expensive than others and generally limited to "critical" parts, for example, high-strength parts for missiles and rockets. The exception to this statement is a new process (U.S. Steel Corporation) for depositing aluminum on steel on a production basis.

Diffusion Diffusion coatings involve heat treatment to cause alloy formation by diffusion of one metal into the other. For this reason the process is also termed "surface alloying." Parts to be coated are packed in solid materials or exposed to gaseous environments containing the metal that forms the coating. Sherardizing (zinc), chromizing (chromium) and calorizing or alonizing (aluminum) are examples. In the latter, a protective layer of Al_2O_3 forms over the iron-aluminum alloyed surface. Carbon steel, low-alloy steels, and stainless steel are alonized. This provides the strength of the treated metal and good corrosion resistance to air and sulfur gases at high temperatures ($\sim 1650°F$). These aluminum-treated steels are used in the petroleum and chemical industries (e.g., production of sulfuric acid) as heat exchangers and other hot surfaces. This process is not generally recommended for aqueous environments.

Alonized steels are used on a competitive basis, but they would be utilized extensively as substitutes in the event of a shortage of chromium.

Chemical conversion Coatings from chemical conversion are produced by "corroding" the metal surface to form an adherent and protective corrosion product. *Anodizing* consists of anodic oxidation in an acid bath to build up an oxide layer. The best-known product is anodized aluminum, wherein the protective film is Al_2O_3. Great improvement in corrosion resistance is not obtained, so anodized aluminum should not be used where untreated aluminum would show rapid attack. The surface layer is porous and provides good adherence for paints. The anodized surface can be "sealed" by exposing to boiling water. Anodized aluminum is used for many architectural purposes (e.g., building wall panels) and others where a pleasing appearance is of prime importance. In other words, anodized aluminum could be considered as "controlled weathering" to produce a uniform surface.

Additional examples are Bonderizing and Parkerizing (*phosphatizing* in a phosphoric acid bath), *chromatizing* (exposure to chromic acid and dichromates), and *oxide* or heat coatings for steel. Automobile bodies are the best known example of phosphatizing. Here the treatment provides a good base for the paint and also provides some time before rusting occurs if the finish is damaged. Chromate treatments are applied to magnesium and zinc parts and offer some measure of corrosion resistance, although the parts are normally painted. Oxide coatings are produced on steel by heating in air or by exposing to a hot liquid. These coatings must be treated with a petroleum product to avoid rusting—the colored oxide is present primarily for appearance. The author "blues" his guns by immersing them in a hot caustic solution and then rubbing the oxidized metal with raw linseed oil while it is still hot.

Surface modification Surface treatments involving directed energy beams are receiving increasing attention. Eventually these treatments may be desirable when alloying elements such as chromium may not be readily available.

Laser surface alloying is described in an excellent review paper by C. W. Draper.* Techniques, basic considerations, and results are described. CO_2 lasers are most popular. An AIME book† covers the effects of directed beam processing and resultant behavior of various alloys of steel, aluminum and copper (Fe-Al bronze). McCafferty et al.‡ show that laser treatment improves resistance to uniform corrosion in hydrochloric acid and citrate solutions, and has no effect on pitting potential in 0.1M NaCl for this 3003

*C. W. Draper, Laser Surface Alloying: The State of the Art, *J. Metals*, **34**(6):24–32 (June 1982).

†C. R. Clayton and C. M. Preece, eds, *Corrosion of Metals Processed by Directed Energy Beams*, Metals Society of AIME, New York (1982).

‡E. McCafferty, P. G. Moore, and G. T. Pease, Effect of Laser Surface Melting on the Electrochemical Behavior of an Al–1% Mn Alloy, *J. Electrochem. Soc.* **129**:9–17 (Jan. 1982).

aluminum alloy. Moore and McCafferty* show passivation for steel alloyed with 5, 20, and 80% chromium. Lasers are also used to impregnate metals with hard particles (i.e., WC and TiC) for improved wear resistance.

Ion implantation Application of ion beams to modify surfaces is another method that may eventually have practical application. Potter et al.† in a review paper describe beneficial results on corrosion resistance.

Glassed steel or glass-lined steel is an important material of construction for the process industries and also in the home,—in hot-water tanks, for example. The smooth surface is an advantage when ease of cleaning is a requirement or sticky materials, such as latex, are being handled. Glassed

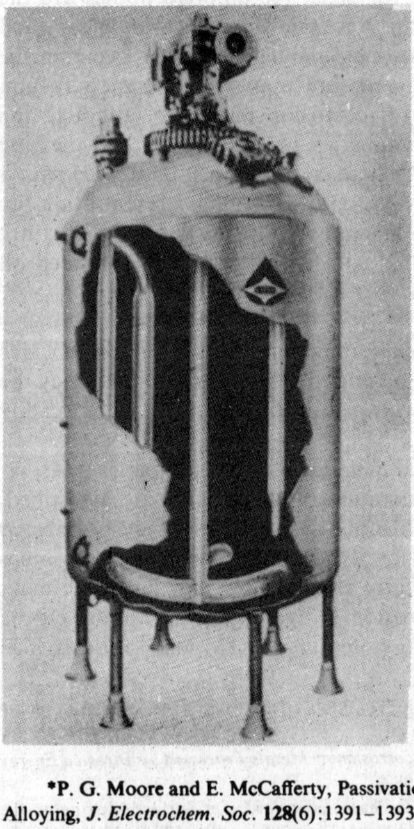

Figure 6-9 Glassed steel vessel and internals. *(Pfaudler Co.)*

*P. G. Moore and E. McCafferty, Passivation of Fe-Cr Alloys Prepared by Laser Surface Alloying, *J. Electrochem. Soc.* **128**(6):1391–1393 (June 1981).

†D. I. Potter, M. Ahmed, and S. Lamond, Metallurgical Surfaces Produced by Ion Implantation, *J. Metals*, 17–22 (Aug. 1983).

steel is widely used in the drug industry, in pilot plants, in wine, brewery, and food plants, and in many others where severe corrosives or contamination of product are involved. Figure 6-9 shows glassed-steel equipment.

Concrete is utilized for many corrosion applications. Examples are encasing structural steel (also for fireproofing), concrete-lined pipe, and concrete vessels.

6-12 Organic Coatings

These involve a relatively thin barrier between substrate material and the environment. Paints, varnishes, lacquers, and similar coatings doubtless protect more metal on a tonnage basis than any other method for combating corrosion. Exterior surfaces are most familiar, but inner coatings or linings are also widely utilized. Approximately 2 billion dollars per year are expended in the United States on organic coatings. A myriad of types and products are involved, and some are accompanied by outlandish claims. Substantial knowledge of this complex field is required for successful performance. The best procedure for the uninitiated is to consult with a reputable producer of organic coatings. As a general rule, these coatings should not be used where the environment would rapidly attack the substrate material. For example, a paint would not be used to line the inside of a tank car for shipping hydrochloric acid. One defect or a small area of exposed metal would result in rapid perforation. An evaluation test program is also recommended.

Aside from proper application, the three main factors to consider for organic coatings, listed in order of importance, are (1) surface preparation, (2) selection of primer or priming coat, and (3) selection of top coat or coats. If the metal surface is not properly prepared, the paint may peel off because of poor bonding. If the primer does not have good adherence or is not compatible with the top coat, early failure occurs. If the first two factors are wrong, the system will fail regardless of the top coat used. Poor paint performance is, in most cases, due to poor application and surface preparation.

Surface preparation involves surface roughening to obtain mechanical bonding ("teeth") as well as removal of dirt, rust, mill scale, oil, grease, welding flux, crayon marks, wax, and other impurities. In other words a clean, rough surface is needed. The best method is to grit-blast or sandblast the steel surface. Other methods are pickling and other types of chemical treatments, scraping, wire brushing, flame cleaning (heat with torch and scrape off dirt and scale), chiseling, and chipping. A study of surface preparation on paint life showed 10.3 years for sandblasting, 9.6 years for pickling, and 2.3 years for weathering and then hand cleaning. Pinholes in welds and sharp edges should be ground out to ensure contact between the paint and the metal. Other chemical methods are solvent degreasing, hot or cold alkali treatments, phosphatizing, chromate treatment, and electrochemical treatments such as anodizing and cathodic cleaning.

In addition to economic considerations, the selection of surface preparation method depends upon the metal to be painted; the shape, size and accessibility of the structure; the coating system; and the service conditions.

Primers can contain rust-inhibitive pigments such as zinc chromate and zinc dust and thereby provide another function in addition to acting as barriers. Wettability is needed so that crevices and other surface defects will be filled rather than bridged. Short drying times are advantageous to preclude contamination before the top coats are applied, particularly in field applications.

Top-coat selection is important. Use of cheap paints is false economy because the majority of the cost of a paint system is in application. Many times paint is applied primarily for appearance—it might be cheaper to provide corrosion allowance by making the steel thicker in the first place. However, who wants a rusty tank on their property? An important point here is that good appearance and good corrosion protection, even in severe atmospheres, can be obtained at very little *extra* cost (a fraction of a cent per square foot per year) by selecting a good top-coat material.

The coating thickness must be such that no bare metal is exposed. It is almost impossible to apply one coat of a paint and have it completely free of pinholes or other defects ("holidays" in the trade). Multiple coats are needed so that a pinhole in one coat will be covered by a complete film of another. Thickness is important also because paint deteriorates or weathers with time.

Various methods are available to reduce maintenance painting costs. One is to institute a touch-up program (you should do this to your car) to cover bad spots early instead of waiting until the coating is so bad that a repaint job is required. It would be a rare case where a coating system fails all over at the same time. Another method is to apply the paint by a hot-spray method. In this case the higher temperature permits higher solids content (with less thinner) and good sprayability, which means a thicker film per coat. Another method consists of applying tape on edges so that the edge is protected; edges are of course the hardest to protect. Another method is to design the structure such that minimum surface area and edges are presented (use a channel or pipe instead of an I beam).

Many companies let all of their paint jobs to outside contractors on a bid basis. This may result in low first cost, but it is usually more costly in the long run. If a plant has sufficient work, it should hire and train its own painters. One company made this change and as a result saved many thousands of dollars per year.

To sum up, a good paint job consists of proper surface preparation, proper coating selection, and proper application. A tremendous variety of paints are available, but it is beyond the scope of this book to go into detail.

Asphalts and bituminous paints are often used on pipelines. Sometimes a cloth wrapping is used with the coating for reinforcement. Alkyds, glyptols, concrete, red lead, iron oxide, phenolics, lithopones, titanium dioxide paints, and chlorinated rubber are just a few examples. Vinyl and epoxy paints have been widely adopted for corrosion applications.

Figure 6-10 is helpful for initial selection of coating types for a given application.*

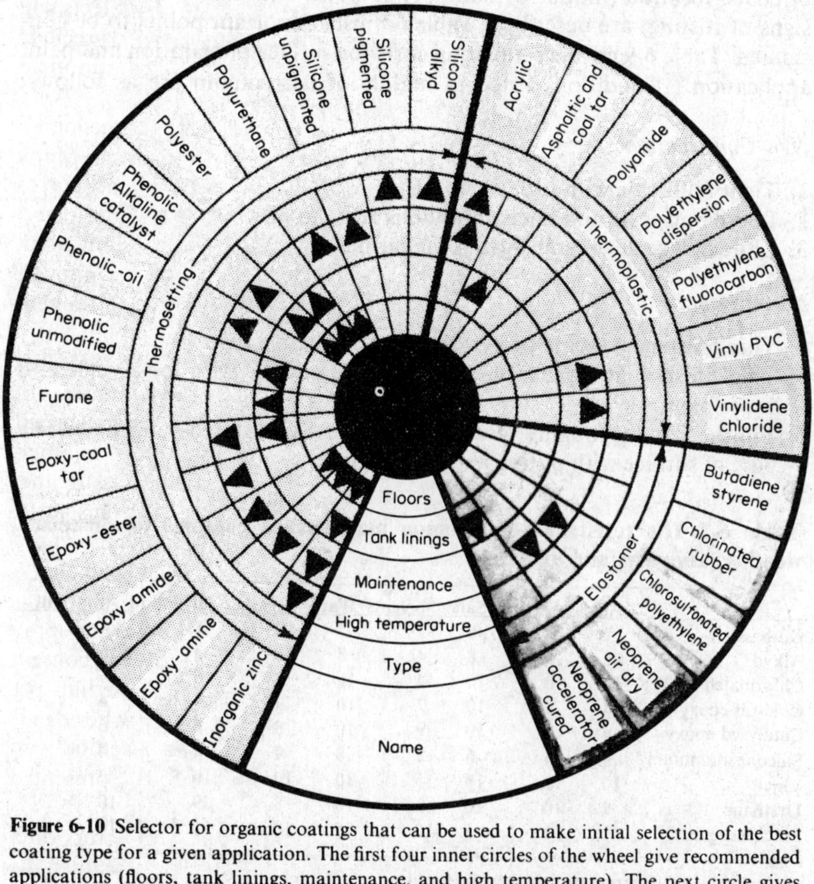

Figure 6-10 Selector for organic coatings that can be used to make initial selection of the best coating type for a given application. The first four inner circles of the wheel give recommended applications (floors, tank linings, maintenance, and high temperature). The next circle gives three basic classifications of coatings: thermosetting, thermoplastic, and elastomer. The outside circle gives the generic types of coatings within the three main types. For example, the following coatings are recommended for high-temperature applications (starting at the 7 o'clock position and reading clockwise): inorganic zinc, silicone pigmented, silicone alkyd, and acrylic.

*R. M. Garrett, How to Choose the Right Protective Coating, *Material Protection*, **3**:8–13 (Mar. 1964).

An excellent reference manual on paints and protective coatings prepared for the Departments of the Army, Navy, and the Air Force, available at low cost from the United States Government, is highly recommended.*

Kirby† discusses maintenance painting in a two-part series. He includes many of the subjects described above. Table 6-7 shows the resistance of coatings exposed to several environments. Required film thickness, number of coats, location (indoor vs outdoor) of work, and time to paint (at first signs of rusting) are described. Table 6-8 lists important points to be considered. Table 6-9 covers estimating costs for surface preparation and paint application.‡ Conditions listed in middle of left column are as follows:

New Construction:

1. Tight mill scale with little or no rust.
2. Loose mill scale and fine, amorphous rust, no pits.
3. Little mill scale, overall rust, some pitting.

Maintenance:

4. Thin finish coat, some primer showing, negligible rust.
5. Thin finish coat, primer showing, 10% of surface with loose scale and loose paint.
6. Finish coat thoroughly weathered and badly blistered; approximately 30% of surface with pitting and hard scale.

Table 6-7 The resistance of common maintenance coatings to chemicals, weather, abrasion, and oxidation

	Acids	Alkalies	Salts	Solvents	Water	Weather	Oxidation	Abrasion
Oil-base	1	1	6	2	7	10	1	4
Alkyd	6	6	8	4	8	10	3	6
Chlorinated rubber	10	10	10	4	10	8	6	6
Coal-tar epoxy	8	8	10	7	10	4	5	4
Catalyzed epoxy	9	10	10	9	10	8	6	6
Silicone aluminum	4	3	6	2	8	9	4	4
Vinyl	10	10	10	5	10	10	10	7
Urethane	9	10	10	9	10	8	9	10
Zinc (inorganic)	1	1	5	10	5	10	10	10

A value of 10 represents the best protection.

* *Paints and Protective Coatings* (Army TM5-618, Navfac MO-110, Air Force AFM 85-3), Superintendent of Documents, USGPO, Washington, D.C. 20402.

†Gary N. Kirby, What You Need to Know About Maintenance Painting—Part I, *Chem. Eng.*, 75–78 (July 26, 1982); Part II, 113–116 (Aug. 23, 1982).

‡See also S. P. Thompson, Managing a Maintenance Painting Program to Reduce Costs, *Materials Performance*, 48–51 (Oct. 1982). (Could save up to 50% of annual painting costs.)

Table 6-8 Importance points to know about the various coatings you can select

Coating	Advantages	Remember
Oil-base	Good rust penetration. Good substrate wetting.	Poor acid, alkali, solvent and oxidation resistance. Slow-drying, embrittles and yellows with aging.
Alkyd		
Short-oil	Inexpensive, fast-drying, good adhesion, easy recoating.	Do not use over zinc-rich primers. Fair acid, alkali, solvent and oxidation resistance. Fair impact resistance.
Long-oil	Good durability, good weathering, flexible, easy recoating	Poor acid, alkali, solvent and oxidation resistance.
Chlorinated rubber	Good water, acid, alkali resistance. Easy recoating.	Poor aromatic solvent resistance; "strings" when brushed. Moderate temperature resistance.
Epoxy	Versatile.	May surface-chalk in sunlight. Use light colors outdoors.
Coal-tar	Excellent resistance to water, seawater and soil immersion.	Check touch-up and recoating limits. Check flexibility and abrasion resistance. Dark colors only.
Amine	Good water, acid, alkali, solvent resistance. Hard, good temperature resistance.	Skin sensitizer. Check recoating limits.
Polyamide	Good water, alkali resistance. Tough, good temperature resistance.	Check recoating limits.
Ketimine	Indefinite recoating time. Wide range of chemical and solvent resistance.	Ketone fumes while drying.
Urethane (2-component, aliphatic)	Good water, acid, alkali, solvent resistance. Good abrasion resistance. Good gloss and color retention. Low-temperature application.	Check temperature resistance. Check solvent fumes. Expensive. Check recoatability.
Vinyl	Good water, acid, alkali resistance. Tough, good recoatability, rapid drying. Low-temperature (5°F) application.	Low temperature resistance (150°F). Poor aromatic ketone solvent-resistance. Low percent solids. Needs good surface. Mostly sprayed. Ketone fumes.
Zinc-rich primers (untop-coated)		
Inorganic	The best organic-solvent resistance. Good abrasion resistance. Temperature resistance to 1,000°F.	Requires good surface preparation. Limited (untop-coated), to pH of 5 to 10. Generally sprayed (conventional). Film thickness limited.

Table 6-9 Estimating costs for surface preparation and paint application for various conditions

Surface preparation

Sandblasting	Blasting rate, 100 psi 5/16-in. nozzle		Cost, $/ft^2
SSPC	Ideal	Condition 6	
SP-5 White metal	100 ft^2/hr	83 ft^2/hr	0.75–1.50
SP-6 Commercial blast	150	125	0.50–1.00

Hand cleaning

	Rate of cleaning				
	Structural steel		Tanks		
Condition of steel	ft^2/hr	ft^2/8 hr	ft^2/hr	ft^2/ 8 hr	Cost, $/ft^2
1	150	1,200	187.5	1,500	
2	75	600	125	1,000	0.25–0.50
3	50	400	62.5	500	0.75
4	125	1,000	500	4,000	Do not paint
5	87.5	700	137.5	1,100	0.25
6	25	200	62.5	500	0.50–0.75
7	12.5	100	31.25	250	1.00

Application

	Application rate, ft^2/coat/man-hour				
	Flat surfaces	Medium structural steel (250 ft^2/ton)			
		Oil-base	Chlorinated rubber	Epoxy	Vinyl
Spray	300–600	210–240	180–210	160–195	145–180
Roller	200–400	150–180	125–150	100–125	100–125
Brush	100–200	70–80	65–70	60–65	60–65

7. Badly pitted with rust nodules.

Table 6-10 shows a sample specification.

Suggested Reading

Berger, Dean M.: Preparing Concrete Surfaces for Painting. *Chem. Eng.*, 141–148 (Oct. 25, 1976).

———: Specifying Zinc Rich Primers, *Chem. Eng.*, 101–104 (Dec. 14, 1981).

Leidheiser, H. ed.: *Corrosion Control by Organic Coatings*, NACE, 1981.

Tator, Kenneth B.: Coating Developments, Problems, and Trends in the Chemical Industry, *Materials Performance*, 33–36 (Sept. 1974).

Metals Handbook, Vol. 5., 9th Ed., *Surface Cleaning, Finishing and Coating*. American Society for Metals, Metals Park, Ohio, 1982.

Paints and Protective Coatings (Army TM5-618, Navfac MO-110, Air Force AFM 85-3), Superintendent of Documents, USGPO, Washington, D.C. 20402.

Table 6-10 Sample specification

A. Surface preparation
Sandblast to a commercial blast finish, according to the Steel Structures Painting Council Specification SSPC-SP-6, Commercial Blast Cleaning (add recommended blast profile, if given).

B. Primer
Within 4 hr after blasting, brush or spray (airless preferred) on 3 to 5 mils dry-film thickness of:
Brand A "Product Name" aluminum polyamide epoxy primer
The dry-film thickness to be measured by the:
Brand B dry-film thickness gage
Allow to dry at least 16 hr, but not more than 5 days.

C. Intermediate coat
Airless spray (only) 2 to 3 mils dry-film thickness of:
Brand A "Product Name" vinyl coating, gray color
The dry-film thickness to be measured by the:
Brand B dry-film thickness gage. Allow to dry 2 h.

D. Finish coat
Airless spray (only) 2 to 3 mils dry-film thickness of:
Brand A "Product Name" vinyl coating, white
The dry-film thickness to be measured by the:
Brand B dry-film thickness gage

E. Application
1. Do not apply paint at temperatures under 50°F.
2. All surfaces to be painted shall be dry and free of dirt, rust, grease, and other contaminants.

F. Manufacturer's directions
All mixing, thinning and application shall be in accordance with the manufacturer's recommendations.

G. Safety precautions
1. Follow manufacturer's safety recommendations.
2. Keep all flammable materials away from heat, sparks, and open flames.
3. Keep containers tightly closed and upright.

H. Cleanup
After completion of the work, remove all painting materials, tools, scaffolds and ladders.

6-13 Corrosion Control Standards

The National Association of Corrosion Engineers has prepared standards (recommended practices) for a number of corrosion situations. These are listed in Table 6-11.

6-14 Failure Analysis

A substantial portion of this book is devoted to failure analysis (causes and cures) of corrosion problems. MTI sponsored a project on this subject and the results are in an MTI publication, *An Atlas of Corrosion and Related Failures* (1985). The reader is referred to this comprehenisve report. Step-by-step procedures are described starting with request to investigate and

Table 6-11 NACE Standards—Recommended Practices

RP-01-70 Protection of Austenitic Stainless Steel in Refineries Against Stress Corrosion Cracking by Use of Neutralizing Solution During Shut Down
RP-01-72 Surface Preparation of Steel and Other Hard Materials by Water Blasting Prior to Coating or Recoating
RP-02-72 Direct Calculation of Economic Appraisals of Corrosion Control Measure
RP-03-72 Method for Lining Lease Production Tanks with Coal Tar Epoxy
RP-04-72 Methods and Controls to Prevent In-Service Cracking of Carbon Steel (P-1) Welds in Corrosive Petroleum Refining Environments
RP-05-72 Design, Installation, Operation, and Maintenance of Impressed Current Deep Groundbeds
RP-01-73 Collection and Identification of Corrosion Products
RP-02-73 Handling and Proper Usage of Inhibited Oilfield Acids
RP-01-74 Corrosion Control of Electric Underground Residential Distribution Systems
RP-02-74 High Voltage Electrical Inspection of Pipeline Coatings Prior to Installation
RP-01-75 Control of Internal Corrosion in Steel Pipelines and Piping Systems
RP-02-75 Application of Organic Coatings to the External Surface of Steel Pipe for Underground Service
RP-03-75 Application and Handling of Wax-Type Protective Coatings and Wrapper Systems for Underground Pipelines
RP-04-75 Selection of Metallic Materials to be Used in All Phrases of Water Handling for Injection into Oil Bearing Formations
RP-05-75 Design, Installation, Operation, and Maintenance of Internal Cathodic Protection Systems in Oil Treating Vessels
RP-06-75 Corrosion Control of Offshore Steel Pipelines
RP-07-75 Preparation and Installation of Corrosion Coupons and Interpretation of Test Data in Oil production Practice
RP-01-76 Control of Corrosion for Steel Fixed Offshore Platforms Associated with Petroleum Production
RP-02-76 Extruded Asphalt Mastic Type Protective Coatings for Underground Pipelines
RP-03-76 Monolithic Organic Corrosion-resistant Floor Surfacings
RP-01-77 Mitigation of Alternating Current and Lightning Effects on Metallic Structures and Corrosion Control Systems
RP-01-78 Design, Fabrication, and Surface Finish of Metal Tanks and Vessels to Be Lined for Chemical Immersion Service
RP-02-78 Design and Operation of Stripping Columns for Removal of Oxygen from Water
RP-01-80 Cathodic Protection of Pulp and Paper Mill Effluent Clarifiers
RP-01-82 Initial Conditioning of Cooling-Water Equipment
RP-01-83* (Revision of RP-01-69) Control of External Corrosion on Underground or Submerged Metallic Piping Systems

*Published in *Materials Performance*, 34–56 (Apr. 1983). RP-01-83 contains a bibliography and good discussion of coatings.

Request current information on price from National Association of Corrosion Engineers, P.O. Box 218340, Houston, Tex. 77218.

ending with the final report. Step by step for *on-site* investigation and also for *laboratory* examination are included. A "fault tree" is shown in chart form. Numerous excellent photographs of typical failures are included. Importance of examining specimens before cleaning is emphasized, as are proper documentation and reporting.

CHAPTER

SEVEN

MINERAL ACIDS

Most of the severe corrosion problems encountered involve the mineral acids or their derivatives. In this chapter we shall describe the production, use, and effects of sulfuric, nitric, hydrochloric, hydrofluoric, and phosphoric acids. The widespread use of these acids places them in an important position with regard to costs and destruction by corrosion. In some cases corrosion increases with concentration of the acid and in others it decreases. For these reasons it is important to have a good picture of corrosion by various acids. Sulfuric, nitric, and hydrochloric acids are the three most important inorganic acids. (Sulfuric and sulfamic acids exhibit about the same corrosive behavior.)

SULFURIC ACID

More sulfuric acid is produced than any other chemical in the world. It is used directly or indirectly in nearly all industries and is a vital commodity in our national economy. The consumption rate of sulfuric acid, like steel production or electric power, could be used as a yardstick to judge economic conditions in our nation. The principal uses of sulfuric acid are for production of hydrochloric acid, other chemicals, and their derivatives; pickling of steel and other metals; manufacture of fertilizers, dyes, drugs, pigments, explosives, synthetic detergents, rayon, and other textiles; petroleum refining; storage batteries; metal refining; and production of rubbers. Corrosion problems occur in plants for making the acid and also in consumers' plants when it is utilized under a wide variety of conditions.

Sulfuric acid is made by the contact process or the lead-chamber process, with the former accounting for about 70% of total production. Figure 7-1

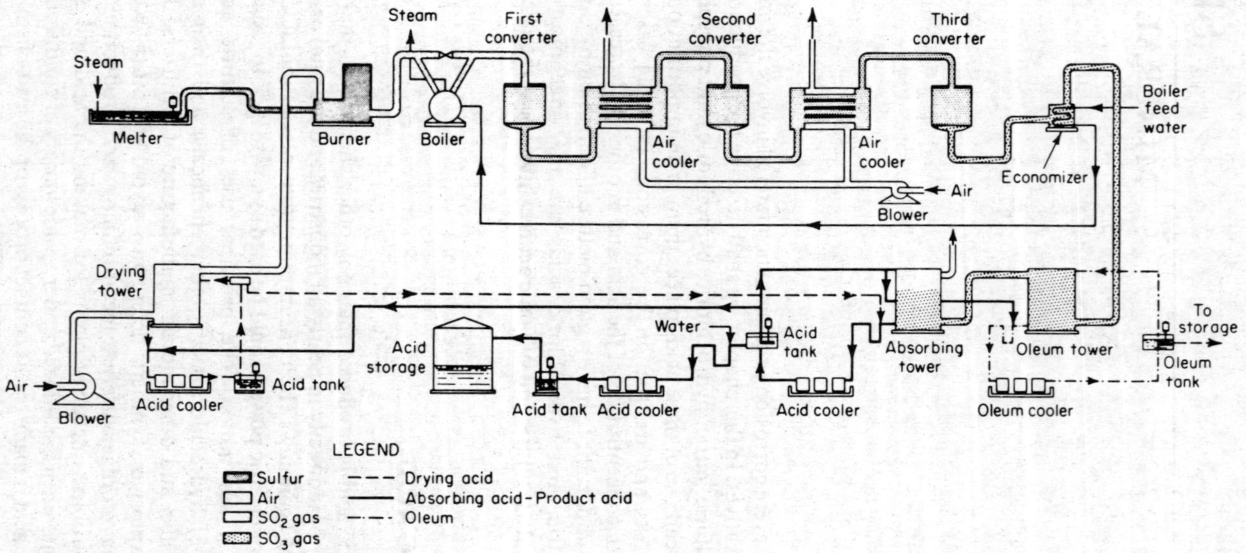

Figure 7-1 Typical contact process for producing sulfuric acid. *(General Chemical Co.)*

shows schematically a typical contact-process plant. Sulfur, or sulfur compounds such as copper sulfide ore, is burned to form SO_2, which is converted to SO_3 in the presence of oxygen and a catalyst. Absorption of this gas by water produces sulfuric acid. Fuming acid (oleum) is made by dissolving SO_3 in 100% acid, and these acids are designated according to the percent free sulfur trioxide or the equivalent percent H_2SO_4. For example, acid containing 20% free SO_3 is called either 20% oleum or 104.5% sulfuric acid. Most of the acid produced is shipped to the customer in three concentrations, namely, 78%, 93%, and oleum.

7-1 Steel

Ordinary carbon steel is widely used for sulfuric acid in concentrations over 70%. Storage tanks, pipelines, tank cars, and shipping drums made of steel commonly handle 78%, 93%, 98% acids and oleum. Most of the equipment shown in Fig. 7-1 is made of steel.

Figure 7-2 shows corrosion of steel by "strong" sulfuric acid as a function of temperature and concentration. More dilute acids attack steel very rapidly. Figure 7-2 is an *isocorrosion* chart. This is a streamlined method devised by the author for presenting corrosion data, but it can be utilized only when a large amount of data are available to delineate the curves.

The curves in Fig. 7-2 represent corrosion rates of 5, 20, 50, and 200 mpy. These are isocorrosion or constant-corrosion lines. In other words, the outlined areas represent regions where corrosion rates of 0 to 5, 5 to 20, 20 to 50, 50 to 200, and over 200 mpy would be expected. The corrosion of

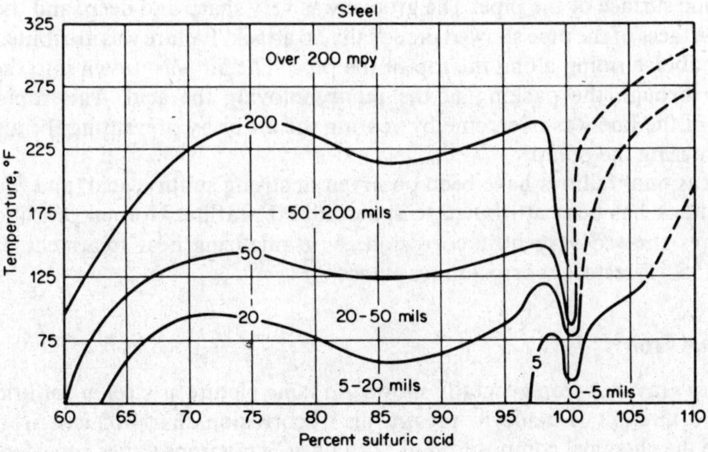

Figure 7-2 Corrosion of steel by sulfuric acid as a function of concentration and temperature.

steel by strong sulfuric acid is complicated because of the peculiar dips in the curves or the rapid increase in corrosion in the neighborhood of 101% acid. The narrowness of this range means that the acids must be carefully analyzed in order to obtain reliable corrosion data. The dips or increased attack around 85% are more gradual and less difficult to establish.

Figure 7-2 indicates that steel would not be suitable in concentrations below about 65% at any temperature. Above 70% strength, steel can be used, depending on the temperature involved. Steel is generally unsuitable above 175°F at concentrations up to about 100%. Applications involving corrosion rates in the range of 50 to 200 mpy would involve relatively short life for steel and should be considered carefully; steel should be used only if economics demands the use of steel in spite of fairly frequent replacement. Steel shows comparatively high rates of corrosion around 101% acid. Corrosion data in this region are not plentiful, particularly in acid strengths just above 101%, so the shape of the right-hand portion of the dips is not well established. The dotted lines are more or less conjecture, and additional data are required. The 200-mpy curve stops at 100% acid because one hesitates to guess at the shape of this curve in stronger acids.

High-*velocity* acid would increase corrosion over that shown in Fig. 7-2. For example, steel pumps would not be satisfactory. Few data are available on the effects of velocity, but it seems safe to say that a few feet per second should not alter the situation. Solids in suspension may also result in erosion corrosion.

Aeration in itself has little effect on corrosion of steel by concentrated acids because they are inherently oxidizing in nature. Entrained air may have a destructive effect on steel in strong-acid service. A long steel line handling 93% sulfuric acid failed prematurely because of grooving of the top inside surface of the pipe. The groove was very sharp and deep, and the other surfaces of the pipe showed practically no attack. Failure was attributed to air bubbles riding along the top of the pipe. The air was drawn into the system through the packing of the pumps moving the acid. This rapid failure of the line was overcome by venting the air or by preventing the air from entering the pump.

Occasional failures have been observed in strong sulfuric acid, and the rapid attack has been attributed to spheroidized pearlite. Normal pearlitic structures showed very little corrosion. A normalizing heat treatment of about 1550°F restores corrosion resistance.

7-2 Cast Iron

Ordinary gray cast iron generally shows the same picture as steel in sulfuric acid. No attempt is made to present an isocorrosion chart for cast iron because the chemical composition of "ordinary" gray irons varies considerably. Cast iron shows somewhat better corrosion resistance in hot strong acids, and in very hot and very strong acid it is better than most materials, including stainless high alloys. However, the corrosion rates are high. The better resistance in this case is probably due to the graphite network interfering with the reaction between the acid and the metallic matrix.

A peculiar phenomenon occurs in oleum, and ordinary gray iron is not generally recommended for this service. The weight loss or corrosion rate is low but the metal may split open in service. Apparently the acid penetrates the metal along the graphite flakes, and a little corrosion in these confined areas builds up enough pressure form the corrosion products to split the iron. This is similar to the wedging action described under stress corrosion in Chap. 3. A specimen of gray iron exposed to oleum at room temperature will gain weight. If this specimen is washed and dried and then let stand, acid will reappear on the surface. In other words, the metal acts somewhat like a sponge. It is believed that interconnecting graphite flakes are required for cracking to occur. Malleable and nodular iron or ductile iron should be satisfactory for oleum service.

There is some evidence of galvanic corrosion between cast iron and steel in about 100% sulfuric acid.

Steel is generally preferred over gray iron primarily for safety reasons in applications where line breakage would result in acid contact with personnel. Cast iron lines should be well supported because of the brittleness of this material.

7-3 Chemical Lead

Lead is used extensively for sulfuric acid in the lower concentration ranges. Lead and steel complement each other as illustrated by Fig. 7-3. Lead takes over below 70% acid where steel is attacked, and vice versa.

Figure 7-4 is an isocorrosion chart for lead. Corrosion is practically nil in the lower concentrations but increases as temperature and concentration increase. Rapid attack occurs in concentrated acids because the lead sulfate surface film is soluble. This chart is based on chemical lead (ASTM B29-55),

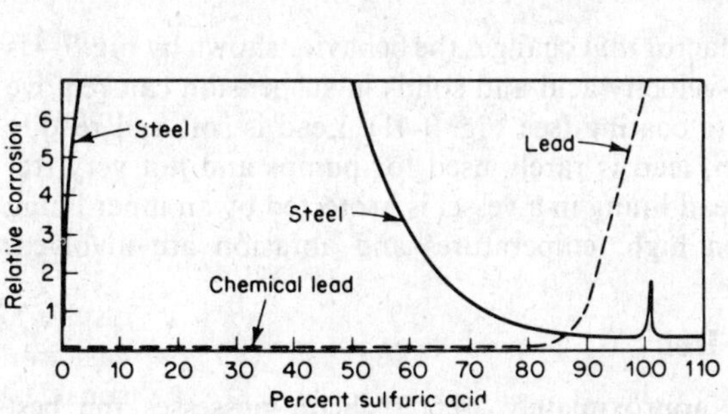

Figure 7-3 Corrosion of steel and chemical lead at atmospheric temperatures.

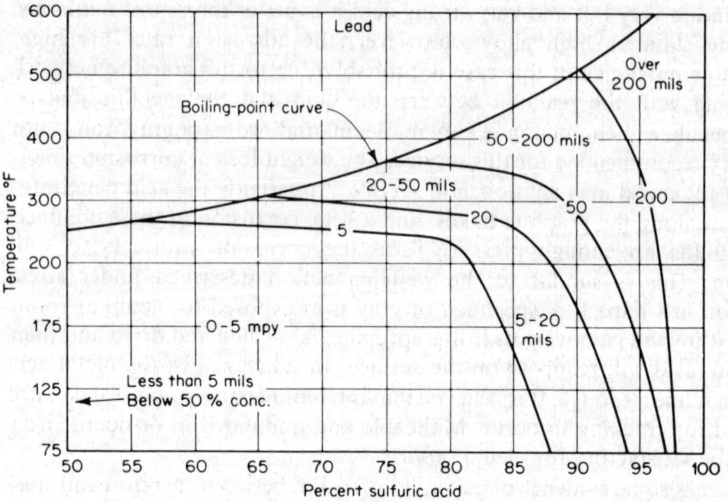

Figure 7-4 Corrosion of chemical lead by sulfuric acid as a function of concentration and temperature.

which contains about 0.06% copper. This is the lead used for corrosion applications. High-purity lead is less resistant particularly in the stronger and hotter acids and also exhibits poorer mechanical properties. Tellurium lead is more corrosion resistant to hot concentrated acids but is equal to or inferior to chemical lead for all other acid concentrations, based on extensive laboratory tests and actual plant installations. Hard lead (4 to 15% antimony) is often used where a stronger material than chemical lead is required such as in castings. However, the strength differential disappears at about 190°F. Corrosion resistance of hard lead is good in the more dilute acids, but chemical lead is better for more aggressive conditions involving sulfuric acid.

The most important factor that changes the behavior shown by Fig. 7-4 is erosion corrosion. High-velocity acid and solids in suspension can remove the protective lead sulfate coating (see Fig. 3-41). Lead is soft and readily abraded. For this reason, lead is rarely used for pumps and not very frequently for valves. The lead lining in a vessel is protected by an inner lining of acidproof brick when high temperatures and abrasion are involved.

7-4 High-Silicon Cast Iron

A cast iron containing approximately 14.5% silicon possesses the best all-round corrosion resistance, over the range 0 to 100% sulfuric acid, of all

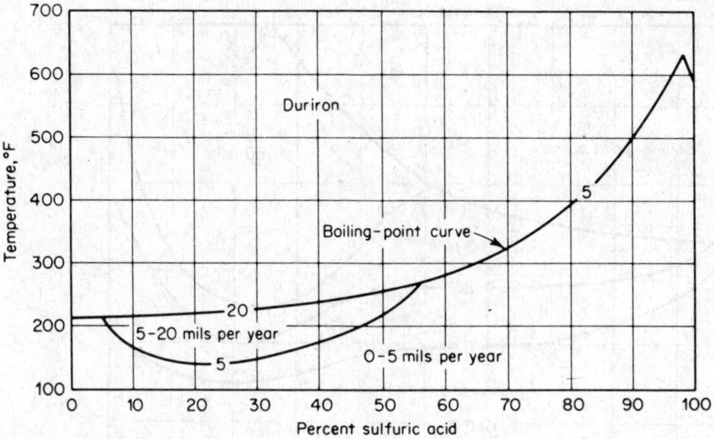

Figure 7-5 Corrosion of Duriron by sulfuric acid as a function of concentration and temperature.

"commercial" metals and alloys. It is widely used for many sulfuric acid applications. It is hard, brittle, susceptible to severe thermal shock, and available only in cast form, but it is relatively inexpensive, does not contain "strategic" alloying elements, is not affected by aeration, and is very resistant to erosion corrosion, particularly when solids are in suspension in the liquid handled, because of its inherent hardness. Equipment made of this alloy includes pumps, valves, heat exchangers, sulfuric acid concentrator heating tubes, spargers, fans, small vessels, tank outlets, pipe and fittings, impressed-current anodes, bubble caps, and tower sections. The alloy used most extensively goes under the trade name of Duriron.

Figure 7-5 is an isocorrosion chart for Duriron. This chart is different from the others in that the boiling-point curve serves as the 5 and 20 mpy lines. This indicates good resistance up to and including boiling temperatures. Corrosion rates in strong hot acids are less than 5 mpy and usually zero. Like other gray cast irons, Duriron is not recommended for fuming acids or for over 100% acid. Sulfur dioxide and fluorides, when present in sulfuric acid, may increase corrosion considerably. Duriron has been used at temperatures as high as 1000°F, but it must be heated and cooled slowly to avoid thermal shock.

7-5 Durimet 20

Figure 7-6 is an isocorrosion chart for an alloy widely used for applications involving sulfuric acid. (See Table 5-2 for composition.) It was originally

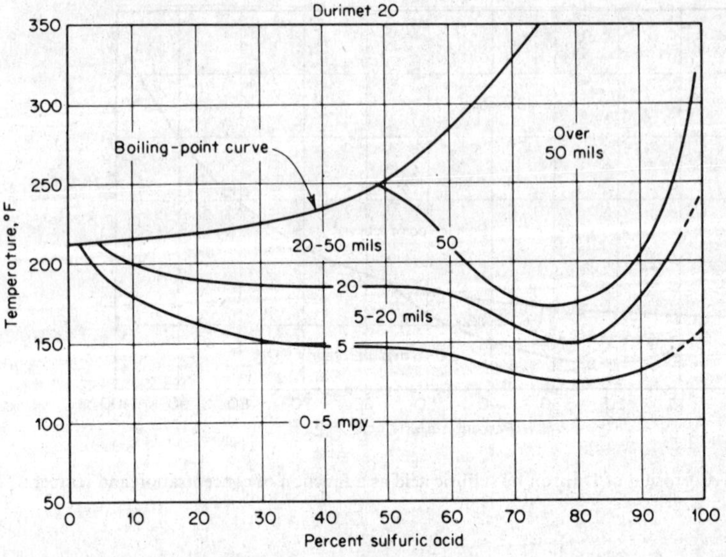

Figure 7-6 Corrosion of Durimet 20 by sulfuric acid as a function of concentration and temperature.

developed for sulfuric acid service but is used for many other environments. The greatest tonnage has been cast by The Duriron Co. as Durimet 20. Other foundries apply their own trade names, but the material is identified always by the number 20 (e.g., Esco 20). In wrought form it is known as Carpenter 20. FA-20 is an austenitic and ductile alloy with *mechanical* properties similar to those of the 18-8 stainless steels.

This alloy differs from lead and steel in that it is used over the entire concentration range. Corrosion resistance in oleum is very good. Highest corrosion rates occur in about 78% sulfuric acid. Ferric sulfate and copper sulfate in the acid act as inhibitors and decrease attack. Ferric chloride and cupric chloride in appreciable concentrations cause pitting.

Pumps and valves represent the major use of Durimet 20 for sulfuric acid services. Many industries including contact-process acid plants, have standardized on this alloy for these items. Steel and cast iron valves give trouble because of sulfating and sticking of moving parts. Lead is subject to erosion corrosion in pumps and valves.

7-6 Nickel-Molybdenum and Nickel-Molybdenum-Chromium Alloys

Figure 7-7 is an isocorrosion chart for cast Chlorimet 2, which consists of approximately two-thirds nickel and one-third molybdenum. Cast and

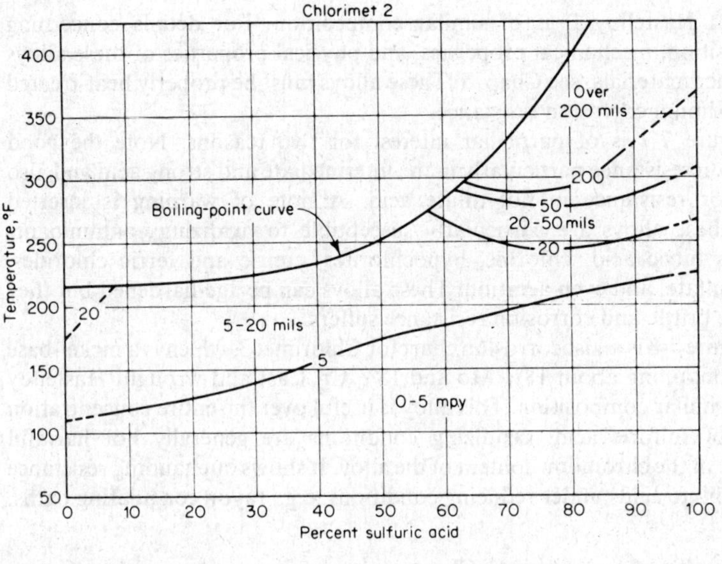

Figure 7-7 Corrosion of Chlorimet 2 and Hastelloy B by sulfuric acid as a function of concentration and temperature.

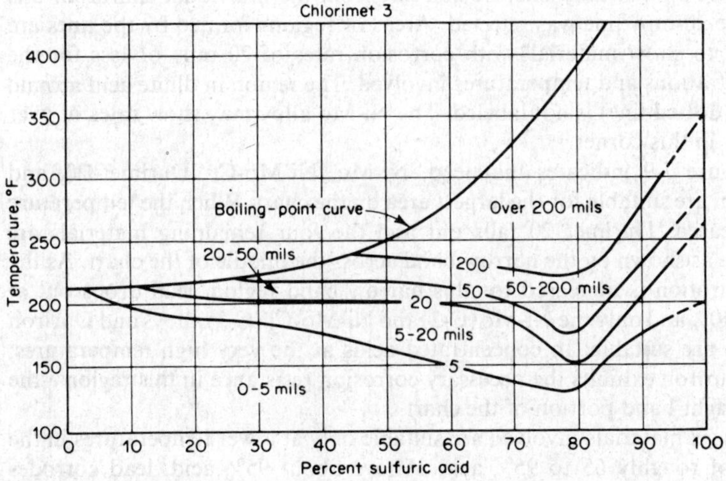

Figure 7-8 Corrosion of Chlorimet 3 by sulfuric acid as a function of concentration and temperature.

wrought Hastelloy B is of similar composition. For details concerning composition, mechanical properties, and physical properties of these alloys and other materials, see Chap. 5. These alloys must be properly heat-treated for maximum corrosion resistance.

Figure 7-7 is of particular interest for two reasons. Note the good corrosion resistance particularly in the intermediate and strong acid and also the poor resistance in hot dilute acid. A note of warning is inserted here—these alloys are particularly susceptible to oxidizing contaminants such as nitric acid, chlorine, hypochlorites, cupric and ferric chlorides, ferric sulfate, and even aeration. These alloys can be age-hardened but they become brittle and corrosion resistance suffers.

Figure 7-8 is an isocorrosion chart for Chlorimet 3, which is a nickel-base alloy containing about 18% Mo and 18% Cr. Cast and wrought Hastelloy C is of similar composition. This alloy is useful over the entire concentration range of sulfuric acid. Oxidizing conditions are generally not harmful because of the chromium content of the alloy. It shows outstanding resistance to hot dilute acids under reducing conditions, e.g., rayon coagulating baths.

7-7 Combined Isocorrosion Chart

The six materials (and gray cast iron) discussed in Secs. 7-1 to 7-6 cover by far most of the installations or equipment for handling sulfuric acid. Presentation of data can be further streamlined by combining individual isocorrosion charts as shown by Fig. 7-9. This is accomplished by placing on one chart a particular isocorrosion line from the individual charts. In this case the 20-mpy line was selected. Areas or regions formed by the lines are labeled to show materials with corrosion rates of 20 mpy or less for the concentrations and temperatures involved. The region in dilute acid around 200°F (dotted line) is not labeled. The Ni-Mo alloy may show rates of over 20 mpy in this corner.

Figure 7-9 indicates that lead, Ni-Mo, Ni-Mo-Cr, Durimet 20, and Duriron are suitable for the largest area on the chart. When the temperature is increased, Durimet 20 falls out and the four remaining materials are suitable as shown by the narrow band across the middle of the chart. As the concentration is increased for this narrow band region, lead drops out at about 90% and only the Ni-Mo (C-2) and Ni-Mo-Cr (C-3) alloys and Duriron (DUR) are suitable. In concentrated acids at the very high temperatures, only Duriron exhibits the necessary corrosion resistance in this region—the upper right-hand portion of the chart.

All six materials involved are suitable only at lower temperatures in the range of roughly 65 to 95% acid. Above about 95% acid, lead corrodes rapidly. Note the region in the lower right-hand corner of the chart designated "all except Pb."

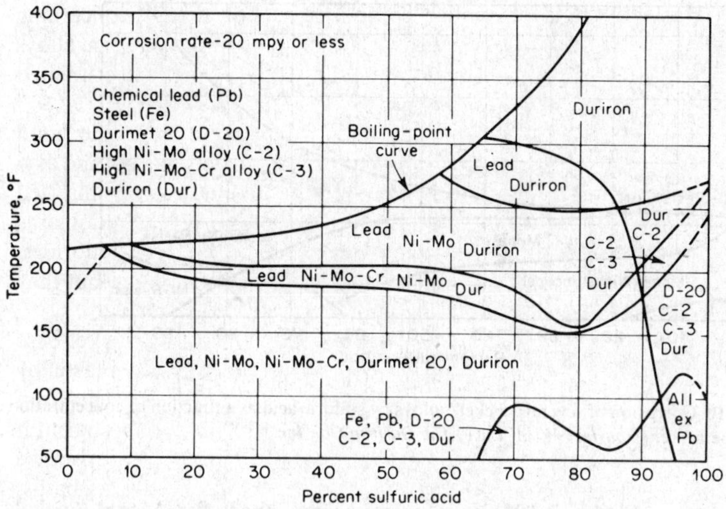

Figure 7-9 Combined chart for corrosion of six alloys by sulfuric acid.

7-8 Conventional Stainless Steels

The common stainless steels are generally not used for handling sulfuric acid. Type 316 is occasionally utilized for cold very dilute acid solutions containing other ingredients and under conditions that are not strongly reducing in nature. This alloy exhibits borderline passivity, and reliable and reproducible data are difficult to obtain. An isocorrosion chart for type 316 in sulfuric acid consisted entirely of dotted lines!

The stainless steels show good resistance in the strong acids, but they generally cannot be justified over ordinary steel unless appreciable iron contamination of the acid must be avoided. An exception to this statement concerns around 101% acid. The early failure of steel pipelines in this acid prompted investigation that resulted in the information presented in Figs. 7-2 and 7-3. This problem was solved through the use of type 430 stainless steel. Practically all other stainless steels would resist this condition.

7-9 Monel, Nickel, Inconel, and Ni-Resist

These nickel-base and nickel-containing materials find application in sulfuric acid under essentially reducing conditions and at moderate

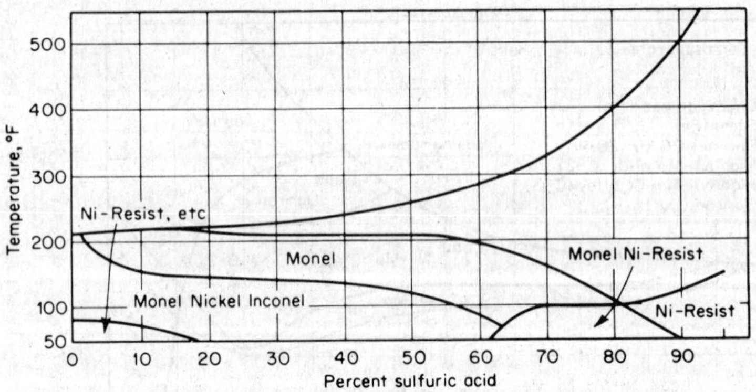

Figure 7-10 Corrosion of several "nickel" alloys by sulfuric acid as a function of concentration and temperature. *(Courtesy W. A. Luce, The Duriron Co., Inc.)*

temperatures. Figure 7-10* is an isocorrosion chart for Monel, nickel, Inconel, and Ni-Resist (20 mpy lines). Strongly oxidizing conditions drastically change the situation.

7-10 Copper and Its Alloys

Copper and copper-base alloys are not used extensively for handling sulfuric acid because of susceptibility to aeration and other oxidizing conditions. Corrosion at the water line can be serious. Brasses suffer dezincification. Tin bronzes exhibit acceptable corrosion resistance up to 60% concentration and 175°F. Silicon bronze (Everdur) may handle slightly more aggressive conditions. Aluminum bronzes are perhaps the only alloys of this category used for sulfuric acid. They are more resistant to erosion corrosion and not as susceptible to oxidizing conditions. Figure 7-11 is an isocorrosion chart for a 10% aluminum copper alloy. Aluminum bronze does a good job handling sulfuric acid containing carbonaceous sludges in oil refineries.

7-11 Other Metals and Alloys

The noble metals such as gold and platinum show excellent resistance to sulfuric acid. They are used where *no* corrosion can be tolerated (e.g., spinnerettes for rayon) or no contamination is allowable (e.g., absorbers for chemically pure H_2SO_4). Tantalum is excellent except in very concentrated and very hot acid. Zirconium is suitable up to 60% at boiling

*W. A. Luce, Corrosion by Sulfuric Acid, *Proceedings of Short Course on Process Industry Corrosion*, OSU and NACE, 81 (1960).

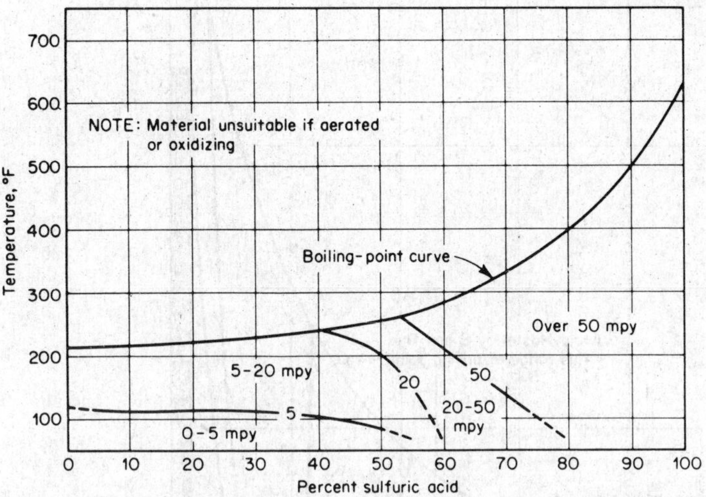

Figure 7-11 Corrosion of aluminum bronze (90-10) by unaerated sulfuric acid as function of concentration and temperature. *(Courtesy W. A. Luce, The Duriron Co., Inc.)*

temperatures and up to about 200° for stronger acids. Titanium is rapidly attacked by all concentrations except quite dilute acid. However, impurities may act as inhibitors. For example, perhaps the major application on a tonnage basis for titanium involves the attacking of a nickel ore by hot H_2SO_4. Table 10-1 lists corrosion data for titanium alloys. Aluminum is used for applications involving only dilute acid.

Molybdenum exhibits excellent resistance to sulfuric acid. It shows less than 1 mpy in 10 to 95% acid at 160°F, and less than 5 mpy in up to 50% boiling acid. Aeration has no appreciable effect. At 400°F molybdenum showed less than 1 mpy in 10% and less than 4 mpy in 20% acid.

7-12 Summary Chart

Perhaps the best summary chart for sulfuric acid materials including some nonmetallics is presented by George A. Nelson.* This chart is shown as Fig. 7-12, and the code appears in the table on the facing page. This chart is for corrosion rates of less than 20 mpy. It is the result of collection of innumerable data and much plant experience. Compositions of the materials are listed and discussed in Chap. 5. The reader should be cautioned that this, like other charts, is a streamlined summary and indicates materials that may

*George A. Nelson, *Corrosion Data Survey*, Shell Development Co., Emeryville, Calif., 1960.

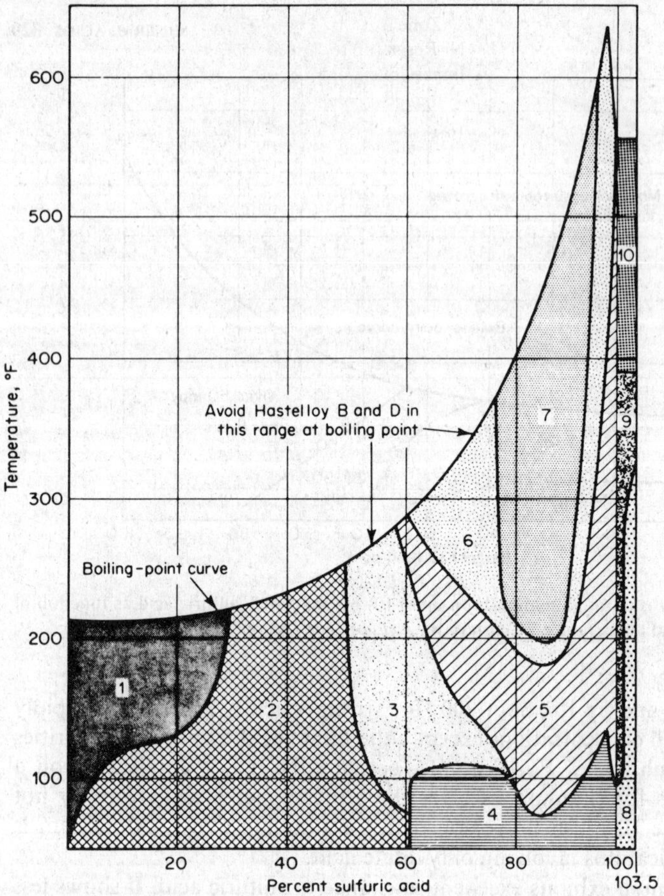

Figure 7-12 Corrosion resistance of materials to sulfuric acid—corrosion rate less than 20 mpy. *(Courtesy G. A. Nelson, Shell Development Co.)*

Code for sulfuric acid chart shown in Fig. 7-12

Materials in shaded zones having reported corrosion rate less than 20 mpy

Zone 1

10% aluminum bronze (air free)	Impervious graphite
Illium G	Tantalum
Glass	Gold
Hastelloy B and D	Platinum
Durimet 20	Silver
Worthite	Zirconium
Lead	Nionel
Copper (air free)	Tungsten
Monel (air free)	Molybdenum
Haveg 43	Type 316 stainless (up to 10% aerated)
Rubber (up to 170°F)	

Glass
Silicon iron
Hastelloy B and D
Durimet 20 (up to 150°F)
Worthite (up to 150°F)
Lead
Copper (air free)
Monel (air free)
Haveg 43
Rubber (up to 170°F)
10% aluminum bronze (air free)

Glass
Silicon iron
Hastelloy B and D
Durimet 20 (up to 150°F)
Worthite (up to 150°F)
Lead
Monel (air free)

Steel
Glass
Silicon iron
Hastelloy B and D
Lead (up to 96% H_2SO_4)
Durimet 20
Worthite

Glass
Silicon iron
Hastelloy B and D
Durimet 20 (up to 150°F)
Worthite (up to 150°F)

Glass
Silicon iron
Hastelloy B and D (20–50 mpy)

Glass
Silicon iron
Tantalum

Glass
Steel
18 Cr-8 Ni
Durimet 20

Glass
18 Cr-8 Ni
Durimet 20

Glass
Gold

Zone 2
Ni-Resist (up to 20% at 75°F)
Impervious graphite
Tantalum
Gold
Platinum
Silver
Zirconium
Nionel
Tungsten
Molybdenum
Type 316 stainless (up to 25% at 75°F) aerated

Zone 3
Impervious graphite
Tantalum
Gold
Platinum
Zirconium
Molybdenum

Zone 4
Ni-Resist
Type 316 stainless (above 80%)
Impervious graphite (up to 96% H_2SO_4)
Tantalum
Gold
Platinum
Zirconium

Zone 5
Lead (up to 175°F and 96% H_2SO_4)
Impervious graphite (up to 175°F and 96% H_2SO_4)
Tantalum
Gold
Platinum

Zone 6
Tantalum
Gold
Platinum

Zone 7
Gold
Platinum

Zone 8
Worthite
Hastelloy C
Gold
Platinum

Zone 9
Worthite
Gold
Platinum

Zone 10
Platinum

Table 7-1 Summary of materials of construction for equipment and piping for standard strengths and grades of sulfuric acid and oleum at ambient temperatures

Products	Storage tank materials	Pipe and fittings	Valves	Gaskets	Pumps	Pump packings
Sulfuric acid 54°–60°Be′ 66° Be′-100%	Heavy steel	Steel, schedule 80 pipe, welding fittings and 150-lb welding flanges Cast-iron pipe, with cast-iron flanged 250-lb fittings	FA-20 alloy (‡) with TFE plastic packing and 150-lb flanges Plug type, FA-20 alloy, TFE plastic sleeve and 150-lb flanges	TFE plastic (*) CFE plastic (†) Compressed asbestos Blue African asbestos	Centrifugal-type FA-20 alloy All iron with FA-20 alloy impeller High-silicon cast iron	Packless or mechanical seals TFE plastic CFE plastic Graphited-lubricated blue African asbestos
Sulfuric acid electrolyte and other grades up to 60°Be′	Steel or wood with chemical lead lining	Chemical lead pipe, $\frac{1}{4}$-in. minimum wall thickness, with 6% antimony lead 125-lb flanges or lap-joint flanges High-silicon cast-iron pipe and fittings Rigid, unplasticized, normal impact polyvinyl chloride (PVC) schedule 80 pipe and fittings	FA-20 alloy or high-silicon cast iron with TFE plastic packing and 150-lb flanges 6% antimony lead with TFE plastic packing and 125-lb flanges Plug type, FA-20 alloy, TFE plastic sleeve and 150-lb flanges	TFE plastic CFE plastic Compressed asbestos Blue African asbestos	Centrifugal type FA-20 alloy High-silicon cast iron	Packless or mechanical seals TFE plastic CFE plastic Graphited-lubricated blue African asbestos
Sulfuric acid electrolyte 66° Be′	Steel or wood with chemical lead lining	Chemical lead pipe, $\frac{1}{4}$-in. minimum wall thickness, with 6% antimony lead 125-lb	FA-20 alloy or high-silicon cast iron with TFE plastic packing and 150-lb flanges	TFE plastic CFE plastic Compressed asbestos Blue African asbestos	Centrifugal type FA-20 alloy High-silicon cast iron	Packless or mechanical seals TFE plastic CFE plastic

	Steel with baked phenolic lining such as "Heresite" P-403	flanges or lap-joint flanges High-silicon cast-iron pipe and fittings	6% antimony lead with TFE plastic packing and 125-lb flanges Plug type, FA-20 alloy, TFE plastic sleeve and 150-lb flanges			Graphited-lubricated blue African asbestos
Sulfuric acid chemically pure (reagent ACS) 95.5-96.5%	Glass-lined steel	"Pyrex" glass TFE plastic-lined hose	Porcelain "Y" valves with TFE plastic discs and 150-lb flanges	TFE plastic envelope type	Centrifugal type glass lined Diaphragm type with TFE or CFE plastic diaphragm	TFE plastic CFE plastic
Oleum all strengths	Heavy steel	Steel, schedule 80 pipe, welding fittings and 150-lb welding flanges (§)	FA-20 alloy with TFE plastic packing and 150-lb flanges	TFE plastic CFE plastic Compressed asbestos Blue African asbestos	Centrifugal type FA-20 alloy Meehanite CB-3	Packless or mechanical seals TFE plastic CFE plastic Graphited-lubricated blue African asbestos
Sulfan (stabilized sulfuric anhydride)	Steel	Steel, schedule 80 pipe, welding fittings and 150-lb welding flanges (§)	FA-20 alloy with TFE plastic packing and 150-lb flanges	TFE plastic CFE plastic	Centrifugal type FA-20 alloy Meehanite CB-3	Packless or mechanical seals TFE plastic CFE plastic Graphited-lubricated blue African asbestos

*TFE—polytetrafluoroethylene—representative trade name, Teflon.
†CFE—polychlorotrifluoroethylene—representative trade names, Genetron Plastic VK, Kel-F and Fluorothene.
‡FA-20 Alloy—representative trade names, Durimet 20, Aloyco 20, Carpenter 20.
§Cast iron and high-silicon cast iron must not be used for oleum.

be applicable to a given problem. Factors such as erosion corrosion and contaminants in the acid may change the picture drastically.

7-13 Equipment at Ambient Temperatures

Table 7-1 is a summary of materials of construction for equipment and piping for standard strengths and grades of sulfuric acid and oleum at ordinary temperatures. This table is taken verbatim from the work of R. W. King.*

7-14 Sulfuric Acid Plant Equipment

Figure 7-1 is representative of a contact-process plant, although many variations exist. The sulfur melters use steel, cast iron, or aluminum for the heating tubes in a brick-lined pit. Molten sulfur is transferred by stainless alloy (e.g., FA-20) pumps and valves. Solid sulfur is fed to burners in ordinary cast iron or Ni-Hard screw conveyors. Hot gases are transported from the burners to the waste-heat boilers in large cast iron ducts. These boilers contain steel tubes with hot gas inside the tubes. Gases are moved by cast iron (2% copper cast iron preferred) blowers. Converters contain steel or calorized steel tubes. Tube life was substantially increased by going from 13- to 10-gage tubing. Absorbers are brick-lined towers with ceramic rings or steel with steel cooling coils. Strong acid is cooled in Duriron heat exchangers. FA-20 valves and pumps for contact acid are practically standard. Piping is steel or cast iron. Storage tanks and tank cars are made of steel. Heresite lining (see Table 7-1) is used for battery acid to minimize iron pickup. Life of tank cars is prolonged by designing for thicker dome sheets where most of the corrosion occurs.

Corrosion of the tops of storage tanks (in the "vapor" area) is combated by using a thicker steel top or by coating. Water-line attack can be minimized by agitation to prevent formation of a top layer of more dilute acid (absorption of moisture from the air).

The chamber process for manufacture of sulfuric acid uses lead chambers and ducts and towers lined with steel, lead, and brick. Blowers are lead with lead-covered steel impellers. High-silicon iron pumps are widely used. The chamber process generally produces acid up to 78%. The contact plants produce strong acid and oleum. This accounts for the extensive use of lead in the former and steel in the latter.

Many types of concentrators are utilized. Heating tubes are usually Duriron and sometimes Hastelloy D. The latter alloy is subject to attack at around 57% sulfuric acid as indicated in Fig. 7-12. Vessels are usually brick-lined steel with an intermediate lining of lead.

*R. W. King, Chap. 23, in W. W. Duecker and J. R. West (eds.), *Manufacture of Sulfuric Acid*, Reinhold Publishing Corporation, New York, 1959.

Dean and Grab* developed a calculation method based on hydrodynamic principles for corrosion rates in strong sulfuric acid. This method assumes that the steel is covered with ferrous sulfate and the corrosion rate is controlled by diffusion of ferrous sulfate from the surface. Good correlation is obtained between calculated rates and actual corrosion tests.

The Materials Technology Institute is sponsoring a project on corrosion by flowing acid: concentration; 93–98.5%; temperature, 20°–70°C; and velocity; 0–6 meters per second (19.5 ft/sec). The report should be published in 1985.

The International Nickel Co. in 1983 published Corrosion Engineering Bulletin (CEB-1), *The Corrosion Resistance of Nickel-Containing Alloys in Sulfuric Acid and Related Compounds*. This contains a wealth of information.

NACE Committee T-5A-18 plans to publish in 1985 a Standard

Table 7-2 Concentration and temperature limits in sulfuric acid for some nonmetallics

Material	Temperature, °F	
	Ambient	Elevated
Polyvinyl chloride (PVC—Type 1)	93%	93% at 150°
Polyvinyl chloride (PVC—Type 2)	93%	93% at 130°
Polyvinylidene chloride (Saran)	60%	50% at 125°
Polyethylene (high density)	98%	98% at 125°
Polypropylene	98%	? at 200°
Polyether (Penton)	98%	60% at 220°
Acrylonitrile-butadiene-styrene (Kralastic)	60%	10% at 140°
Fluoropolymers (Teflon and Kel F)	100%	100% at 400°
Polyesters	80%	75% at 150°
Epoxies (Durcon)*	90%	75% at 200°
Phenolics (Bakelite)	90%	70% at 200°
Butyl rubber	75%	50% at 175°
Neoprene	75%	50% at 185°
Hypalon	93%	—
Pyrex	All conc.	All conc. at 600°
Acidproof brick	All conc.	All conc. at 600°
Haveg 43	60%	50% at 250°
Impervious graphite	100%	96% at 175°
Natural rubber (soft)	70%	60% at 170°
Natural rubber (hard)	80%	60% at 180°
Wood (cypress and redwood)	8%	5% at 125°

*See Fig. 7-14.

Source: W. A. Luce, *Proceedings of Short Course on Process Industry Corrosion*, OSU and NACE, 1960, with additions by M. G. Fontana.

*S. W. Dean and G. D. Grab, "Corrosion of Mild Steel by Concentrated Sulfuric Acid," Paper No. 147, NACE CORROSION/84 (1984).

Recommended Practice on handling and storing concentrated sulfuric acid. Information is essentially the same as presented in this chapter.

7-15 Nonmetallics

The corrosion resistance of several nonmetallic materials is included in Fig. 7-12. Table 7-2 indicates the corrosion resistance of a variety of nonmetallic materials to sulfuric acid. The relatively poor mechanical properties, particularly with regard to creep and flow at elevated temperatures, and also brittleness, should be emphasized here. For example Teflon *cold-flows* under stress even at room temperature.

Figure 7-13 shows corrosion resistance of Pfaudler glass-lined steel to sulfuric acid as a function of concentration and temperature.

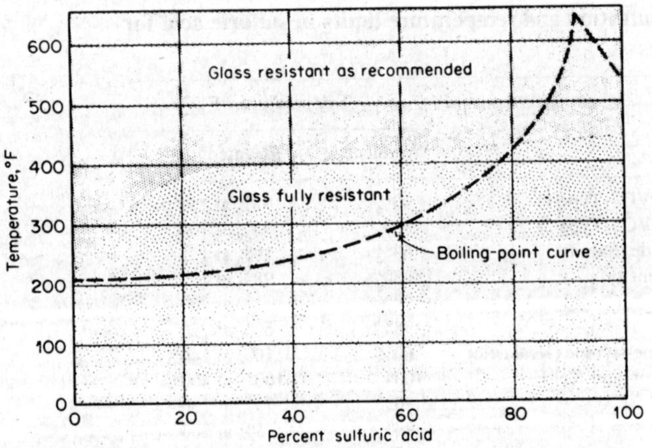

Figure 7-13 Corrosion resistance of Pfaudler 53 glass to sulfuric acid. *(Pfaudler Co.)*

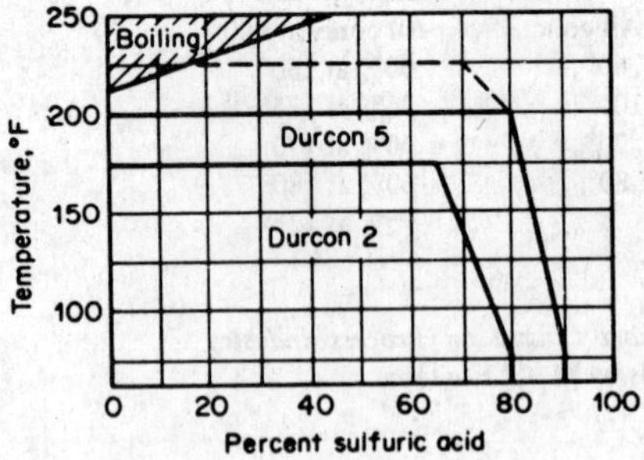

Figure 7-14 Corrosion resistance of Durcons 2 and 5 to sulfuric acid.

Another point to be emphasized is the fact that chemical resistance may vary with formulation (plasticized or unplasticized and filler used), particularly for the plastics, rubbers, and elastomers.

Figure 7-14 ullustrates this point. Note the difference in resistance between Durcons 2 and 5. These are both epoxy resins. Materials should be tested for a given application. Nonmetallics are discussed further in Chap. 6.

NITRIC ACID

The choice of metals and alloys for nitric acid services is quite limited as far as variety is concerned. For most plant applications the choice is usually between only two general classes of materials—the stainless steels and alloys and the high-silicon irons. The choice is further limited because a minimum chromium content is required in the stainless materials, and the high-silicon alloys are available only in cast form. A simple classification of materials for nitric acid service is difficult, as is usually the case for any corrosive environment. However, for purposes of this discussion the following classifications are used: class 1—generally used and suitable for a variety of conditions of temperature and concentration; class 2—used under certain conditions only because of high cost, limited corrosion resistance, resistance only to specific concentrations, or a combination of these; and class 3—generally not used or not suitable primarily because of insufficient corrosion resistance. These classifications are based primarily on practical considerations.

7-16 Stainless Steels

The stainless steels are the most widely used of the metals and alloys on a tonnage basis. There are a large number of stainless steels and alloys, and the choice for a given application will depend largely upon chromium content, fabrication considerations, and cost. Table 7-3 shows the corrosion resistance

Table 7-3 Influence of chromium on resistance of low-carbon steel to boiling 65% nitric acid

% Cr	Average corrosion rate, mpy
4.5	155,000
8.0	1700
12.0	120
18.0	30
25.0	8

of iron-chromium alloys to boiling 65% nitric acid. These data indicate the reason why most of the stainless steels and alloys (with and without nickel) used for nitric acid service contain a minimum of 15% chromium. The addition of nickel increases the corrosion resistance; for example, the addition of 8% nickel to an 18% chromium-iron alloy decreases the corrosion rate in boiling 65% acid to about 12 mpy.

7-17 Class 1 Materials

Metals and alloys in class 1 are high-silicon iron (about 14.5% silicon), 18-8S (type 304), 17% chromium (type 430) stainless steel, and titanium. The high-silicon irons are outstanding with respect to corrosion resistance, but they are available in cast form only and have relatively poor mechanical properties. Despite these limitations, these materials are used because they are relatively inexpensive and also because other materials are unsuitable under severe conditions involving high temperatures, high concentrations, or erosion corrosion.

18-8S, the common stainless steel for corrosion services, is the most widely used of all the materials for handling nitric acids. Figure 7-15 is an isocorrosion chart for 18-8S in nitric acid. This chart applies to properly heat-treated types 304, 304L, 321, 347, and CF-8, CF-3, CF-8T, and CF-8C, which are all 18-8 alloys. Type 316 and CF-8M or 18-8SMo are slightly less resistant to nitric acid and are more expensive, so they are not selected specifically for nitric acid service.

Figure 7-15 shows that 18-8S exhibits excellent corrosion resistance to nitric acid of all concentrations at room temperatures up to 80°F and also to

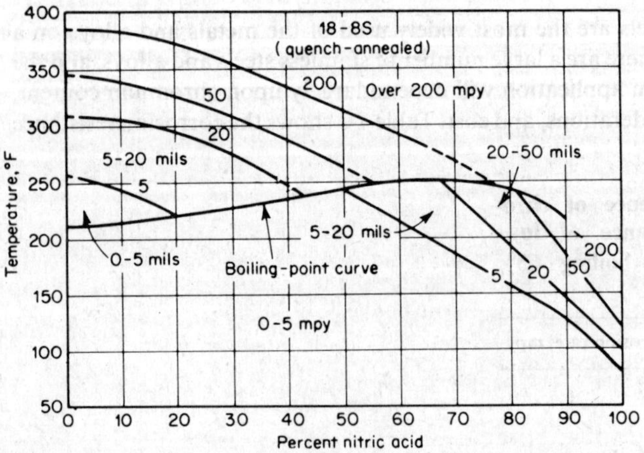

Figure 7-15 Corrosion of quench-annealed 18-8S by nitric acid including elevated temperatures and pressures.

boiling acids up to 50% strength. The corrosion resistance decreases as the concentration and temperature are increased. 18-8S shows poor corrosion resistance to hot very strong acids including fuming nitric acid but excellent corrosion resistance to red and white fuming nitric acids at normal room temperatures.

Figure 7-15 differs from previous charts presented in that isocorrosion lines are presented for temperatures above the atmospheric boiling points of the acid. These tests were run in closed autocalves, so the pressure developed is the equilibrium pressure for the temperature involved. 18-8S shows practically no corrosion in 5 and 20% boiling acids, but fairly rapid rates of attack occur at temperatures around 300°F and above. Corrosion by 65% acid increases rapidly as the temperature is increased above boiling; for example, a rate of 500 mpy is observed at 320°F. Note the tremendous shrinkage at 0 to 5 mpy as temperatures are increased above boiling. The high corrosion at these elevated temperatures is confirmed by actual plant experiences.

An iron-base alloy containing 15 to 16% chromium (type 430) was one of the first stainless steels to be used for chemical plant construction. A tank car for shipping nitric acid and a synthetic nitric acid plant were constructed in the 1920s utilizing this alloy. The corrosion resistance of type 430 and its cast counterpart is somewhat less than that of 18-8S. The former gives out earlier than 18-8S as the temperature or concentration is increased. Some of the main limitations for the straight chromium alloy are brittleness of castings and brittleness of wrought material near welds. For these reasons, the austenitic stainless steels have replaced type 430 for many nitric acid applications wherein the corrosion resistance of all these materials was satisfactory.

Figure 7-16 is an isocorrosion chart for a 14.5% silicon cast iron known commercially as Duriron. Duriron exhibits excellent corrosion resistance in concentrations over 45% at temperatures including boiling. In fact, corrosion rates in stronger acids are practically nil even at high temperatures. Corrosion resistance is very good in *all* concentrations below about 160°F.

The data represented by the curves in the dilute acids are new. Previously it was widely assumed that high-silicon iron possessed very good corrosion resistance for all concentrations (including dilute) and temperatures. Figure 7-16 shows that corrosion may be substantial in the dilute acids at the higher temperatures. Note that the corrosion resistance of Duriron *increases* as the concentration of acid is increased. See also zones 6 and 7 in Fig. 7-12.

Another high-silicon iron known as Durichlor contains about 3% molybdenum. The corrosion resistance of Durichlor to nitric acid is not better than that of Duriron and is probably less. In general, molybdenum as an alloying element does not increase corrosion resistance of metals to nitric acid and usually reduces it, as mentioned above in this section.

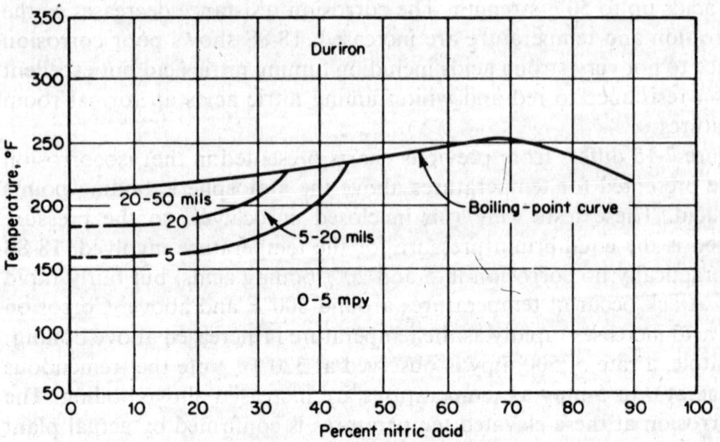

Figure 7-16 Corrosion of high-silicon iron by nitric acid as a function of concentration and temperature.

Duriron is available in a variety of cast forms for chemical plant work including pumps, valves, heat exchangers, fans, pipe, and small vessels. High-silicon-iron equipment has been used for many years in the manufacture and handling of nitric acid. In the older saltpeter process, retorts, condensers, and coolers were made of high-silicon iron. It is also used extensively in the ammonia-oxidation processes. Concentrators, dehydrating towers, condenser coils, tower packing, fans, pumps, lines, and valves are often made of Duriron. Hot fuming nitric acid is cooled in Duriron equipment during manufacture.

The high-silicon irons are also used for equipment handling nitric acid mixed with other chemicals such as sulfuric acid, sulfates, and nitrates.

In connection with a new process involving the use of about 70% nitric acid at temperatures considerably above the atmospheric boiling point, a plant tried a stainless alloy, which failed in a very short time. Successful operation was obtained with high-silicon iron.

Durimet 20, the cast FA-20 alloy, could be placed in class 1 not only because of its good corrosion resistance but also because in some cases it can be obtained as finished equipment for the same price as 18-8S because of production considerations. In other words, this alloy is a borderline case with respect to the above classification. If the cost of a given part or piece of equipment was essentially the same for Durimet 20 and 18-8S, the alloy would be in class 1. If the cost of the former was substantially higher, it would fall into class 2.

Carpenter 20, the wrought alloy, is in class 2 because in this form the alloy is considerably more expensive than 18-8S. The exception to these general statements concerns very concentrated acid, wherein the Durimet 20 and Carpenter 20 show considerably better corrosion resistance than 18-8S above ordinary room temperatures. These statements apply directly to straight nitric acid conditions. In some cases, where other constituents are present in the acid that may increase corrosion of 18-8S, the higher alloyed material may be more advantageous.

Figure 7-17 is an isocorrosion chart for Durimet 20 and Carpenter 20. These materials exhibit excellent corrosion resistance in all concentrations up to about 100°F and also to boiling acids up to about 50% strength. Corrosion increases as concentrations and temperatures are increased as shown by the right-hand portion of the chart.

Comparison of Figs. 7-15 and 7-17 shows much the same corrosion picture for the 0- to 5-mpy region. However, the 20-type alloy is much better in the stronger and hotter acids. For example, the 50-mpy line in Fig. 7-17 is located in approximately the same position as the *200-mpy* line for 18-8S. In other words, the 20-type alloy finds application under conditions represented by the upper right-hand regions of the chart, whereas 18-8S is unsuitable.

Questions concerning the comparative corrosion resistance of cast and wrought or rolled alloys are sometimes raised. In general, their corrosion resistance can be considered equal. In some cases one may show lower corrosion rates in a given environment and vice versa in other mediums.

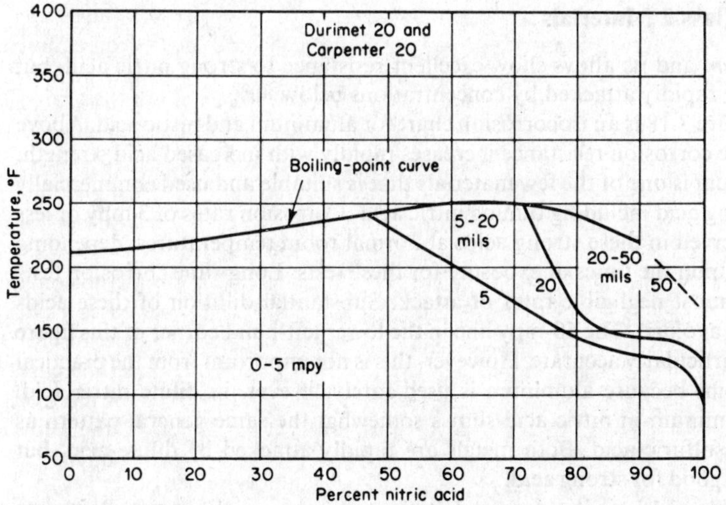

Figure 7-17 Corrosion of Durimet 20 and Carpenter 20 by nitric acid as a function of concentration and temperature.

The point to be emphasized is that these questions are generally of academic interest only because the selection of a casting or a rolled materials for a given application is usually based on other engineering considerations. For example, wrought pump casings and impellers would be extremely expensive to produce. On the other hand, a flexible connection or bellows that requires very then material would not be made from cast metal.

*Titanium** exhibits outstanding resistance to nitric acid, including all concentrations and temperatures well above the atmospheric boiling points. For example, it shows less than 5 mpy in 65% acid at 350°F. This relatively new commercial metal fills the gap in hot acids with a wrought and ductile metal. Although it is still somewhat expensive, in some cases it is the only material that will do the job. Titanium exhibits excellent resistance to fuming nitric acid, which is used as an oxidizer in fuel systems for rockets and other space vehicles. However, titanium has shown pyrophoric tendencies in red fuming nitric and is not recommended if the water content of the acid is below 1.5% and the nitrogen dioxide content is above 2.5%. Titanium and its alloys are generally very good with regard to resistance to stress corrosion.

Presence of oxidizing ions in acids tends to decrease corrosion of titanium. For example, presence of chromium ions in nitric acid decreases (already low) corrosion by nitric acid. This is opposite to the effect on stainless steels, where hexavalent chromium ions markedly increase corrosion in hot nitric acid.

7-18 Class 2 Materials

Aluminum and its alloys show excellent resistance to strong nitric acid, but they are rapidly attacked by concentrations below 80%.

Figure 7-18 is an isocorrosion chart for aluminum and nitric acid. Above 80%, the corrosion resistance increases rapidly with increased acid strength. Aluminum is one of the few materials that is suitable and used commercially for strong acid including fuming nitric acid. Corrosion rates of 3 mpy or less are observed in these strong acids at normal room temperatures, depending largely upon the times of exposure for these tests. Long-time corrosion tests show almost negligible rates of attack. Substantial dilution of these acids must be avoided. The 50-mpy line in the lower left-hand corner of this figure is not particularly accurate. However, this is not important from the practical standpoint because aluminum is used rarely, if ever, in dilute nitric acid.

Aluminum in nitric acid shows somewhat the same general pattern as steel in sulfuric acid. Both metals are rapidly attacked by dilute acid, but they are good for strong acid.

Figure 7-18 applies most directly to commercially pure aluminum

*Titanium was class 2, but is now class 1 because its cost has been reduced considerably.

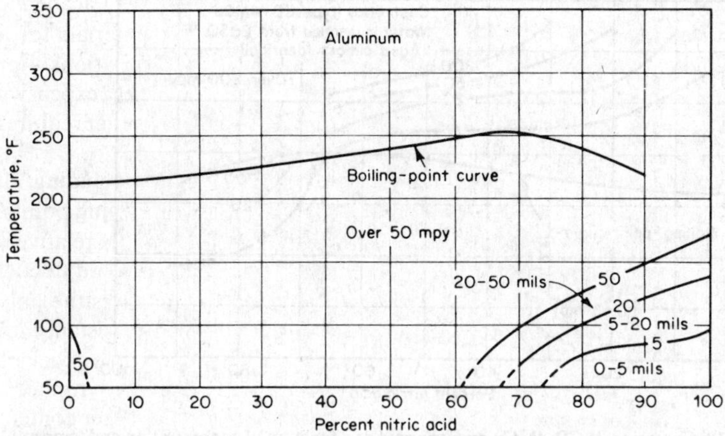

Figure 7-18 Corrosion of aluminum by nitric acid as a function of concentration and temperature.

1100; to high-purity commercial aluminum, 99.6% minimum; and the well-known and stronger 3003 alloy, which is the one commonly used for chemical process equipment. However, this chart should also apply to the 3004, 5052, and 6061 alloys. The high-strength alloys such as 2014, 2017, and 2024 or the "Dural" type materials fall into this pattern only for the strong acid at the lower temperatures.

Comparison of Figs. 7-15 and 7-18 shows that the stainless steels are superior in lower concentrations, but aluminum is better in the concentrated acid at slightly above room temperatures. Increased velocity of fuming nitric acid *decreases* corrosion of stainless steel and increases attack on aluminum (see Sec. 3-30). An ususual galvanic effect was observed between 18-8S and aluminum in fuming nitric acid. When coupled, the rate on aluminum was very high and very low on the stainless steel. This is contrary to what one would expect based on corrosion resistance of the uncoupled metals. This behavior indicates that aluminum could be used as a sacrificial anode to protect austenitic stainless steel in fuming nitric acid.

Aluminum equipment is used in the manufacture and handling of strong nitric acids including the fuming varieties. Equipment includes cooling coils, condensers, piping, hoods, ducts, storage tanks, tank cars, and drums. The presence of lower oxides of nitrogen does not appreciably affect corrosion.

Alloy CD-4MCu (Alloy Casting Institute designation) shows excellent resistance to nitric acid, as shown in Fig. 7-19. This is perhaps the only age-hardenable stainless-type alloy that shows good corrosion resistance to nitric acid in the hardened condition.

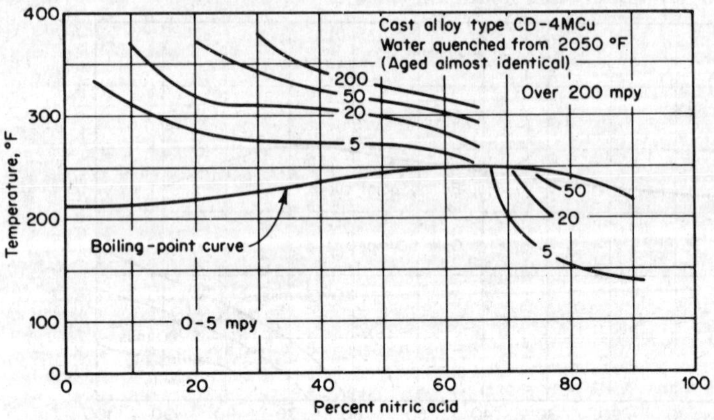

Figure 7-19 Corrosion of CD-4MCu by nitric acid as a function of concentration and temperature.

The high-alloy austenitic stainless, such as types 309 and 310, are slightly better than 18-8S, but they are more expensive. The lower chromium stainless steels, such as type 410, are rarely used in nitric acid and only in dilute acids at low temperatures.

Inconel, Illium, Durco D-10, Hastelloy C, and Chlorimet 3, which are nickel-base high alloys containing chromium, do not increase corrosion resistance enough (usually worse than 18-8S) to justify their cost. Haynes 25 shows good corrosion resistance to fuming nitric acid, but it is expensive.

Table 7-4 Concentration and temperature limits in nitric acid for some nonmetallics

	Temperature, °F	
Material	Ambient	Elevated
Teflon	100%	100% up to 500°
Polyethylene	60%	20% at 100°
PVC (unplasticized)	50%	40% at 140°
Butyl rubber	50%	30% at 150°
Saran	10%	5% at 100°
Karbate	30%	10% at 185°
Penton	70%	30% at 250°
Durcon	60%	40% at 150°

Source: L. R. Honnaker, *Proceedings of Short Course on Process Industry Corrosion*, OSU and NACE, 1960, with additions by M. G. Fontana.

Gold, tantalum, and platinum exhibit excellent resistance to nitric acid but are used only in special cases wherein the high cost is justified.

Table 7-4 indicates the resistance of several nonmetallics to nitric acid. Teflon shows outstanding corrosion resistance. Glass-filled Teflon is widely used for rotating rings in mechanical seals for nitric acid pumps. Aside from Teflon, the organic or organic-bonded nonmetallics are attacked by stronger acids. Butyl rubber is much better than other rubbers and elastomers, which are quite poor in nitric acid. The plastics are not generally used for piping and fabricated equipment, primarily because the stainless steels do such a fine job in the concentration ranges where plastics show good resistance.

7-19 Class 3 Materials

Ordinary cast iron, nickel cast irons, magnesium, steel, and low-alloy steels are rapidly attacked by most concentrations of nitric acid. Occasionally cast iron and steel are used in very strong acid at room temperatures when danger from dilution is not present. Copper, nickel, and copper- and nickel-base alloys such as brass, bronze, Monel, and cupronickels show high corrosion rates in nitric acid. Silver exhibits poor corrosion resistance. The nickel-molybdenum alloys such as Hastelloy B and Chlorimet 2 are readily corroded. Hastelloy D, a nickel-base silicon-copper alloy, is generally

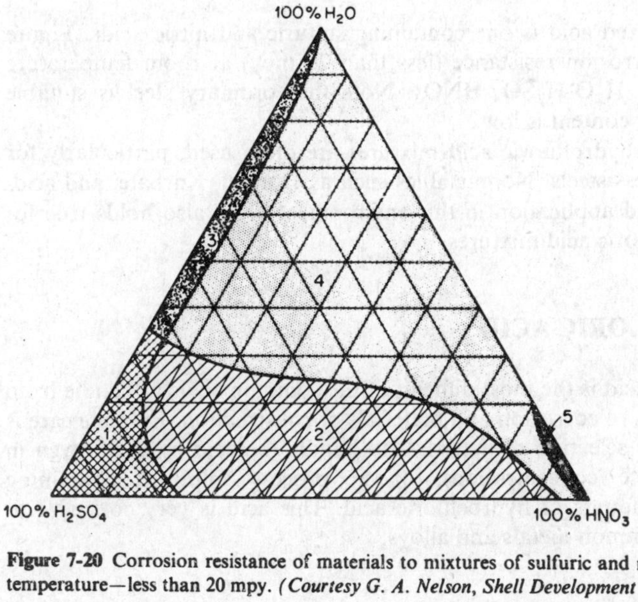

Figure 7-20 Corrosion resistance of materials to mixtures of sulfuric and nitric acid at room temperature—less than 20 mpy. *(Courtesy G. A. Nelson, Shell Development Co.)*

Code for mixed acids chart shown in Fig. 7-20
Materials in shaded zones reported corrosion rate less than 20 mpy

		Zone 1	
Steel	Glass	Tantalum	Gold
Durimet 20	Silicon iron	Platinum	
Worthite			
		Zone 2	
Cast iron	Durimet 20	Silicon iron	Gold
Steel	Worthite	Tantalum	Lead
18 Cr-8 Ni	Glass	Platinum	
		Zone 3	
Durimet 20	Glass	Tantalum	Gold
Worthite	Silicon iron	Platinum	
		Zone 4	
18 Cr-8 Ni	Worthite	Silicon iron	Platinum
Durimet 20	Glass	Tantalum	Gold
		Zone 5	
18 Cr-8 Ni	Glass	Tantalum	Gold
Durimet 20	Silicon iron	Platinum	Aluminum
Worthite			

not recommended for nitric acid services. Lead is not used for nitric acid of appreciable concentration.

7-20 Mixed Acids

A common mixed acid is one containing sulfuric and nitric acids. Figure 7-20 shows corrosion resistance (less than 20 mpy) at room temperature for the system H_2O-H_2SO_4-HNO_3. Note that ordinary steel is suitable when the water content is low.

Nitric and hydrofluoric acid mixtures are often used, particularly for pickling stainless steels. Nonmetallics such as Durcon, Karbate, and acid-proof brick find application in this environment. This also holds true for nitric-hydrochloric acid mixtures.

HYDROCHLORIC ACID

Hydrochloric acid is the most difficult of the common acids to handle from the standpoints of corrosion and materials of construction. Extreme care is required in the selection of materials to handle the acid by itself, even in relatively dilute concentrations, or in process solutions containing appreciable amounts of hydrochloric acid. This acid is very corrosive to most of the common metals and alloys.

When aeration or *oxidizing agents* are also present, corrosive conditions may be very rugged. Many unexpected failures in service are explained by the presence of ostensibly minor impurities. The commercial or muriatic acid often behaves very differently from the chemically pure or reagent grade hydrochloric acid as far as corrosion is concerned. More and more so-called by-product acid is appearing on the market and being used in the process industries. These by-product acids could contain any number of impurities, depending on the parent process. Almost invariably some iron is present, and ferric chloride is often a destructive pitting agent. In fact, *hot* strong acid containing substantial amounts of ferric chloride (or cupric chloride) presents a gap that cannot be filled with commercial metals and alloys and many nonmetallics.

The limits for so-called acceptable corrosion rates are usually raised when considering materials of construction for handling hydrochloric acid. Materials that show very low rates of corrosion are often not economically feasible. Good judgment is required to obtain a good balance between service life and cost of equipment. Where contamination is a problem, expensive materials such as tantalum are the only ones that can be utilized.

Metals and alloys are placed in three classifications: (1) generally used and suitable for most applications; (2) used with caution and under specific conditions; and (3) generally unsuitable under any conditions and recommended only for trace amounts of acid.

7-21 Class 1 Metals and Alloys

Chlorimet 2, Chlorimet 3, Hastelloy B, Hastelloy C, Durichlor, tantalum, zirconium, and molybdenum are in class 1. Molybdenum is an important constituent of the alloys. Tantalum is expensive, but it is often used where contamination is to be avoided—e.g., in steam tubes for heating chemically pure acid. This metal has high strength and thin-wa ed tubes (2 mils) withstand substantial stem pressures. Equipment that requires *no* corrosion, such as a measuring device, can be made of tantalum because of its excellent corrosion resistance.

Figure 7-21, and its code on the facing page, is an excellent summary of the corrosion resistance of a wide variety of materials. Based on additional work, molybdenum should be added to zone 1. Aeration showed no appreciable effects in these tests, which included up to 20% acid.

Durichlor is a high-silicon iron containing molybdenum and is much more corrosion resistant to hydrochloric acid than the alloy without molybdenum. Durichlor shows marked passivity effects in that initial (few hours) corrosion rates may be high, but attack decreases rapidly with time. This alloy is used in industry for all concentrations of hydrochloric and muriatic acids at moderate temperatures. Strong boiling acid shows rapid attack.

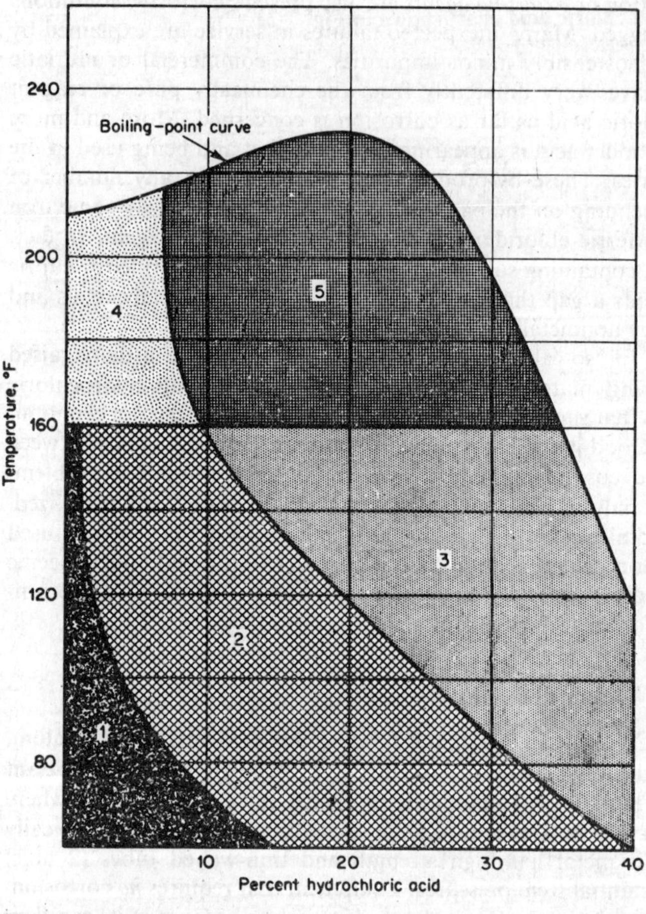

Figure 7-21 Corrosion resistance of materials to hydrochloric acid—less than 20mpy. *(Courtesy G. A. Nelson, Shell Development Co.)*

Aeration does not affect corrosion resistance. Durichlor pumps give satisfactory service in 30% hydrochloric acid and also sludges containing 10% acid at ambient temperatures. The new alloy Durichlor C is equivalent to standard Durichlor.

The *Chlorimets* and *Hastelloys* are nickel-base alloys with large molybdenum contents. Chlorimet 2 and Hastelloy B show good corrosion resistance to all concentrations of hydrochloric acid up to boiling temperatures. However, these alloys are attacked if aeration or oxidizing ions are present. Chlorimet 3 and Hastelloy C are good in dilute acids at moderate temperatures and exhibit better resistance to oxidizing environments because of their high chromium contents.

Code for hydrochloric acid chart shown in Fig. 7-21

Materials in shaded zones having reported corrosion rate less than 20 mpy

	Zone 1
Chlorimet 2	Rubber
Glass	Silicon bronze (air free)
Silver	Copper (air free)
Platinum	Nickel (air free)
Tantalum	Monel (air free)
Hastelloy B	Zirconium
Durichlor ($FeCl_3$ free)	Tungsten
Haveg	Titanium—up to 10% HCl at room temperature
Saran	
	Zone 2
Chlorimet 2	Haveg
Glass	Saran
Silver	Rubber
Platinum	Silicon bronze (air free)
Tantalum	Zirconium
Hastelloy B	Molybdenum
Durichlor ($FeCl_3$ free)	Impervious graphite
	Zone 3
Chlorimet 2	Haveg
Glass	Saran
Silver	Rubber
Platinum	Molybdenum
Tantalum	Zirconium
Hastelloy B (chlorine free)	Impervious graphite
Durichlor ($FeCl_3$ free)	
	Zone 4
Chlorimet 2	Durichlor ($FeCl_3$ free)
Glass	Monel (air free, up to 0.5% HCl)
Silver	Zirconium
Platinum	Impervious graphite
Tantalum	Tungsten
Hastelloy B (chlorine free)	
	Zone 5
Chlorimet 2	Tantalum
Glass	Hastelloy B (chlorine free)
Silver	Zirconium
Platinum	Impervious graphite

7-22 Class 2 Metals and Alloys

In this class are copper, the various bronzes, cupronickels, Monel, nickel, Inconel, Ni-Resist, Hastelloy D, Duriron, type 316 stainless steel, and stainless high alloys. The first five listed are susceptible to aeration effects (see code for Fig. 7-21). Monel is slightly better than nickel and Inconel. Ni-Resist is suitable only for low concentrations at room temperature. Type 316 could be used in very low concentrations, and the stainless high alloys, such as Durimet 20 and Worthite, are not much better. Hastelloy D is fairly good at room temperature in concentrations up to 15%. Duriron can be used in all concentrations at room temperature, but increasing temperatures lower the critical concentration. Durichlor is preferred over Duriron. Titanium is good for up to 10% at room temperature, but corrosion by hydrochloric acid is *decreased* when ferric and cupric chlorides are present.

The materials in class 2 are not generally considered to be hydrochloric acid alloys. Because they are susceptible to influences other than the acid itself, they must be used with caution and only when the specific conditions are definitely known. Materials in the first classification are preferred.

7-23 Class 3 Metals and Alloys

Ordinary carbon steels and cast irons are never used for hydrochloric acid service. Rapid corrosion occurs at pH 4 or 5 or below, particularly if appreciable solution velocities are involved. Aeration or oxidizing conditions result in destructive attack, even in very dilute solutions. Inhibitors are required when iron and steel equipment is cleaned by pickling with hydrochloric acid. Zinc and magnesium are rapidly attacked by this acid. Tin and tin plate are used for solutions containing small amounts of acid such as food products. Aeration is also destructive to tin.

Aluminum and its alloys are not recommended for hydrochloric acid. The protective oxide surface films on aluminum are readily destroyed by this acid, and active corrosion occurs. Aeration is not an important factor here. In addition, the presence of metallic ions such as iron and copper tends to accelerate attack. Aluminum of very high purity shows some corrosion resistance to hydrochloric acid, but its use is not practical.

Lead and its alloys are not recommended, but hard lead is somewhat better than chemical lead. Brasses dezincify, and the straight chromium stainless steels and type 304 either dissolve or pit.

7-24 Aeration and Oxidizing Agents

It is important to emphasize the effects of aeration and oxidizing agents on corrosion by hydrochloric acid because of their great influence on a large number of materials, particularly copper and high-copper alloys. These materials show good corrosion resistance to hydrochloric acid under reducing conditions but are rapidly attacked under oxidizing conditions. Copper completely immersed in hydrochloric acid will show a low corrosion rate, but a semi-immersed specimen will show rapid localized corrosion at the liquid line because of the presence of air in this area. These remarks also apply to the non-chromium-bearing materials such as nickel and the nickel-molybdenum alloys. Accordingly, these alloys should not be used in hydrochloric acid if appreciable aeration or oxidizing agents like nitric acid or ferric chloride are present. Ferric chloride in small amounts is more destructive than the oxygen usually found in process acids.

Two general conditions should be considered: acid with and without aeration or oxidizing agents. It is generally difficult to obtain air-free conditions in process plants because the acid is usually exposed to air in the process. A common cause of aeration is found in pumps, which often draw in air through the packing during operation.

An excellent example of the destructive effect of aeration was found in a plant handling a process liquor with dilute hydrochloric acid present. Equipment made of Chlorimet 2 and Hastelloy B failed rapidly, even though these nickel-molybdenum alloys show excellent resistance to hydrochloric acid at all concentrations and including boiling temperatures. It was found that the process liquor was agitated by sparging with air prior to coming in contact with the high-alloy equipment. Satisfactory service life was obtained by replacing the Chlorimet 2 with Chlorimet 3, which contains 18% chromium. Another plant experienced rapid failure of Chlorimet 2 pumps in hot dilute hydrochloric acid. Investigation showed that the acid contained small amounts of nitric acid, which was not even mentioned by the plant because it was thought that such small amounts could not be harmful to this high-alloy material.

7-25 Nonmetallic Materials

The nonmetallics have found widespread use in hydrochloric applications because of their good resistance and immunity to attack by oxidizing ions.

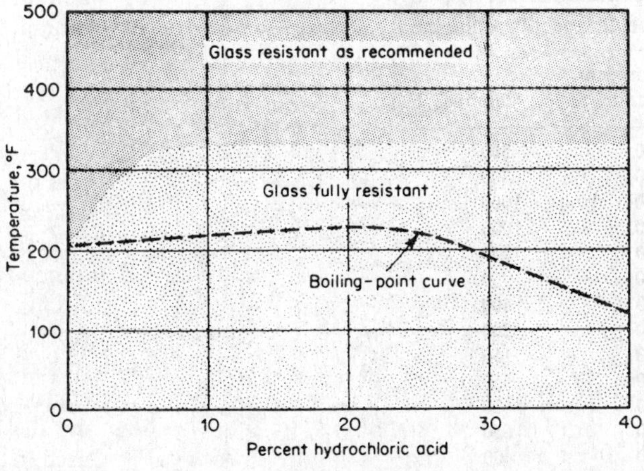

Figure 7-22 Corrosion resistance of Pfaudler 53 glass to hydrochloric acid. *(Pfaudler Co.)*

Cost also enters the picture because many of the suitable metals and alloys are expensive. Several nonmetallics are included in Fig. 7-21. Subject to their temperature limitations most of the plastics and rubbers are suitable for all concentrations of hydrochloric acid. Rubber-lined steel has been used for many years for vessels and piping for hydrochloric acid service. Wood finds application as an inexpensive material for dilute acid.

Figure 7-22 shows the corrosion resistance of glass-lined steel to hydrochloric acid.

7-26 Hydrogen Chloride and Chlorine

Almost any construction metal including steel is suitable for *dry* hydrogen chloride gases at room temperature. However, the presence of moisture changes the behavior to that of corrosion by hydrochloric acid. Table 7-5 indicates upper temperature limits for a variety of metals and alloys for continuous service in dry hydrogen chloride and chlorine.

Titanium is resistant to wet chlorine but not to dry chlorine. Zirconium is resistant to dry chlorine but not to the wet gas.

HYDROFLUORIC ACID

Hydrofluoric acid is unique in its corrosion behavior. High-silicon cast irons, stoneware, and glass are generally resistant to most acids, but all of these

Table 7-5 Maximum suggested temperature for continuous service in dry hydrogen chloride and dry chlorine

Material	Hydrogen chloride, °F	Chlorine, °F
Platinum	2200	500
Gold	1600	300
Nickel	950	1000
Inconel	900	1000
Hastelloy B	850	1000
Hastelloy C	850	950
Carbon steel	500	400
Monel	450	800
Silver	450	150
Cast iron	400	350
18-8	750	600
18-8Mo	750	650
Copper	200	400

Source: M. H. Brown, W. B. DeLong, and J. R. Auld, *Ind. Eng. Chem.*, **39**:839–844 (1947).

materials are readily attacked by hydrofluoric acid. Magnesium shows rather poor corrosion resistance to many acids, but this metal resists attack by hydrofluoric acid; in fact, some shipping containers for this acid are made of magnesium. Below 1% concentration some attack occurs, but in concentrations of 5% and over the magnesium is practically immune to corrosion because of the formation of a surface fluoride film, and corrosion of the metal is retarded. This same surface effect also applies to other metals and alloys that resist corrosion by hydrofluoric acid.

Hydrofluoric acid and fluorine are toxic, and extreme care is mandatory in the handling of these materials. Hydrofluoric acid burns are painful and heal slowly. A thorough washing of the hands and face with soap and water is recommended even if contact of the acid with the skin is not suspected.

7-27 Aqueous Hydrofluoric Acid

Steel is suitable for handling concentrations from about 60 to 100%. Corrosion of steel increases rapidly as the concentration decreases below 60%. Pure acid shows appreciable attack at 62% strength but commercial 60% acid contains sulfuric acid and fluosilicic acid as impurities that reduce corrosion to a satisfactory level. In some cases, particularly at higher temperatures, the silicon content of ordinary carbon steel is important. Increasing silicon from essentially zero to a few tenths of a percent increases attack. Killed steels are preferred to avoid the possibility of hydrogen blistering.

Wrought *Monel* is an outstanding material in that it resists all concentrations of hydrofluoric acid at all temperatures up to and including boiling. Here we have a natural combination, Monel and hydrofluoric acid. Aeration and the presence of oxidizing salts increase the corrosion of Monel. The indications are that Monel is not appreciably affected by the impurities usually present in hydrocarbon alkylation processes involving the use of hydrofluoric acid. Monel-clad equipment is sometimes used in pressure vessels to decrease the cost of the equipment. Monel castings are also suitable for hydrofluoric acid service, provided the castings are sound and of good quality. S-Monel (hardenable by heat treatment) is sometimes used where resistance to galling and abrasion is required.

Silver is used for more severe services, such as boiling strong acid.

Copper is generally suitable for hot and cold dilute solutions and for high strengths up to approximately 150°F. Copper is considerably more sensitive than Monel to aeration and oxidizing salts in the hydrofluoric acid and also to erosion corrosion. The attack on copper is substantially increased if high-velocity flow or solids in suspension are involved. These remarks on copper also apply in general to brass and bronze. Brass is susceptible to stress corrosion in hydrofluoric acid. *Cupronickels* are better than copper but not as resistant as Monel.

Lead shows fairly good resistance to hydrofluoric acid in concentrations below 60% at room temperature. Higher temperatures in the upper side of this range increase the corrosion of lead. For example, lead in 25% acid at 176°F showed a corrosion rate of 9 mpy, but the rate of attack in 50% acid at 176°F was 28 mpy. Stronger acids, particularly anhydrous, attack lead.

Aerated hydrofluoric acid or acid containing oxidizing salts, such as ferric or cupric ions, presents a problem, since the copper alloys are not suitable. In these cases the use of nonmetallics and high-alloy *stainless* materials should be considered. Some of these alloys are 25-20 (Cr-Ni) and Durimet 20. The ordinary straight chromium stainless steels and the

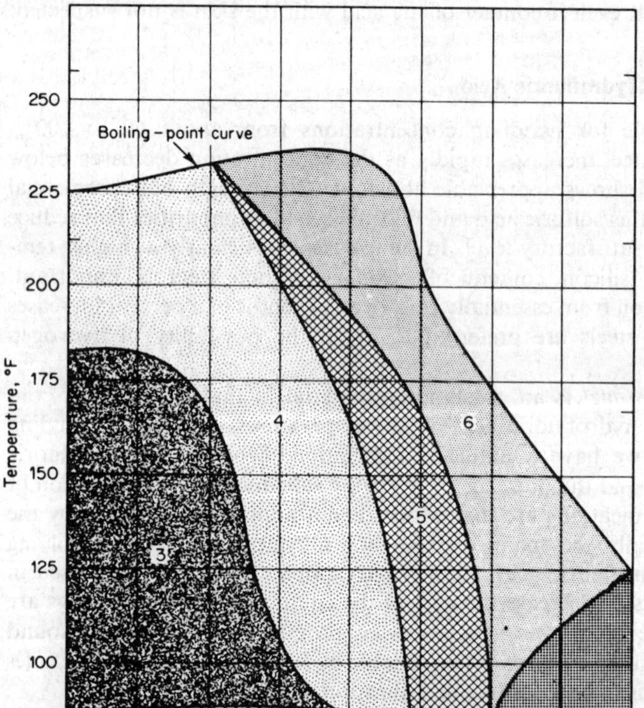

Figure 7-23 Corrosion resistance of materials to hydrofluoric acid—less than 20 mpy. *(Courtesy G. A. Nelson, Shell Development Co.)*

Code for hydrofluoric acid chart shown in Fig. 7-23

Materials in shaded zones having reported corrosion rate less than 20 mpy

Zone 1	
Monel (air free)	Platinum
Copper (air free)	Silver
70 Cu-30 Ni (air free)	Gold
Lead (air free)	Impervious graphite
Nickel (air free)	Haveg 43
Alloy 20	Rubber
Ni-Resist	25 Cr-20 Ni steel
Hastelloy C	
Zone 2	
Monel (air free)	Platinum
70 Cu-30 Ni (air free)	Silver
Copper (air free)	Gold
Lead (air free)	Impervious graphite
Nickel (air free)	Rubber
Alloy 20	Haveg 43
Hastelloy C	
Zone 3	
Monel (air free)	Platinum
70 Cu-30 Ni (air free)	Silver
Copper (air free)	Gold
Lead (air free)	Impervious graphite
Alloy 20	Haveg 43
Hastelloy C	Rubber
Zone 4	
Monel (air free)	Platinum
70 Cu-30 Ni (air free)	Silver
Copper (air free)	Gold
Lead (air free)	Impervious graphite
Hastelloy C	Haveg 43
Zone 5	
Monel (air free)	Silver
70 Cu-30 Ni (air free)	Gold
Lead (air free)	Impervious graphite
Hastelloy C	Haveg 43
Platinum	
Zone 6	
Monel (air free)	Silver
Hastelloy C	Gold
Platinum	Haveg 43
Zone 7	
Carbon steel	Silver
Monel (air free)	Gold
Hastelloy C	Haveg 43
Platinum	

Table 7-6 Concentration and temperature limits in hydrofluoric acid for some nonmetallics

	Temperature	
Material	Ambient	Elevated
Teflon	100%	100% up to 500°F
Polyethylene	60%	50% at 100
Karbate	60%	60% at 185, 48% boiling
PVC	60%	60% at 120
Haveg (carbon filled)	60%	60% at 150
Saran	60%	50% at 100
Soft natural rubber	10%	10% at 100
Flexible hard rubber	50%	50% at 150
Butyl rubber	60%	60% at 150
Neoprene	60%	60% at 150
Penton	100%	60% at 220, 30% at 250
Durcon 5	60%	50% at 200

Source: L. R. Honnaker, *Proceedings of Short Course on Process Industry Corrosion*, OSU and NACE, 1960, with additions by M. G. Fontana.

austenitic stainless steels such as 18-8 are probably not suitable because of their susceptibility to pitting. Type 410 stainless is sensitive to erosion corrosion.

Commercial hydrofluoric acid, particularly in the stronger concentrations, often contains a substantial amount of solids in suspension. Settling out some of the impurities in a tank or convenient vessel is desirable to avoid the deposition of salts on piping and other equipment. Depositions of this type on metals and alloys are undesirable because pits tend to form under the deposits. One method of minimizing these difficulties consists of positioning the outlet pipe in a storage tank such that this outlet is a substantial distance above the bottom of the tank. This arrangement permits the sludge to settle and remain more or less undisturbed. Occasional cleaning of the tank is required.

Figure 7-23 is a Shell chart showing resistance of various materials to hydrofluoric acid. Table 7-6 lists information on several *nonmetallics*.

7-28 Anhydrous Hydrofluoric Acid

This acid is not particularly corrosive, and ordinary steel is extensively used. Critical parts such as valve stems, valve trim, and pump shafts are made of Monel. The stainless steels and alloys offer no particular advantage over carbon steel. Graphitization can occur with cast iron. At temperatures above 150°F, Monel and cupronickels are preferred over steel, which shows substantial attack. At very high temperatures (1000°F) nickel, Monel, and cupronickels exhibit good resistance. Carbon steel and the stainless steels are rapidly attacked.

7-29 Fluorine

Dry fluorine gas, like chlorine, is practically noncorrosive to metals and alloys. Electrolytic cells used for the production of fluorine (from anhydrous hydrofluoric acid and fused potassium bifluoride) are made of welded carbon steel. Two steel cathodes and a steel gas barrier are also parts of these cells. Fluorine is stored and shipped in steel cylinders under pressure. On the other hand, moist fluorine or aqueous solutions of this halogen are extremely corrosive. They attack practically all metals except gold and platinum. Fluorine solutions cause pitting of stainless steels similar to the action of chlorine and bromine. Moist or dry fluorine attacks steel at high temperatures. At a red heat steel burns rapidly in an atmosphere of fluorine. Nickel, Monel, and aluminum show good resistance at temperatures up to 400°F, where the stainless steels are rapidly corroded.

PHOSPHORIC ACID

Corrosion by phosphoric acid depends somewhat on the methods of manufacture and the impurities present in the commercial finished product. Fluorides, chlorides, and sulfuric acids are the main impurities present in the manufacturing processes and in some marketed acids. For example, the high-silicon irons, the austenitic stainless steels without molybdenum (pitting), ceramics, and tantalum are appreciably affected by the presence of hydrofluoric acid.

7-30 Materials of Construction

Two of the most widely used alloys are type 316 stainless steel and Durimet 20 (and similar alloys). These alloys show very little attack in concentrations up to 85% and temperatures including boiling. Lead and its alloys are also used at temperatures up to 200°C at concentrations up to 80% for pure and 85% for impure acid. Lead forms an insoluble phosphate that provides protection. High-silicon irons, glass, and stoneware show good resistance to pure acids. Table 7-7 lists data for some nonmetallics.

Copper and high-copper alloys are not widely used in phosphoric acid. High nickel-molybdenum alloys exhibit good resistance to pure acids but are attacked when aeration and oxidizing impurities are present. Aluminum, cast iron, steel, brass, and the ferritic and martensitic stainless steels exhibit poor corrosion resistance.

Table 7-7 Concentration and temperature limits in phosphoric acid for some nonmetallics

Material	Concentration and temperature
Teflon	85% up to 500°F
PVC unplasticized	85% at 140
Polyethylene	85% at 120
Saran	85% at 150
Karbate	85% boiling
Soft natural rubber	85% at 150
Hard natural rubber	85% at 160
Butyl rubber	85% at 150
Neoprene	85% at 160
Haveg 41	85% at 250
Penton	85% at 250
Carbon and graphite	85% to 500 plus
Durcon	85% at 200

Source: L. R. Honnaker, *Proceedings of Short Course on Process Industry Corrosion*, OSU and NACE, 1960, with additions by M. G. Fontana.

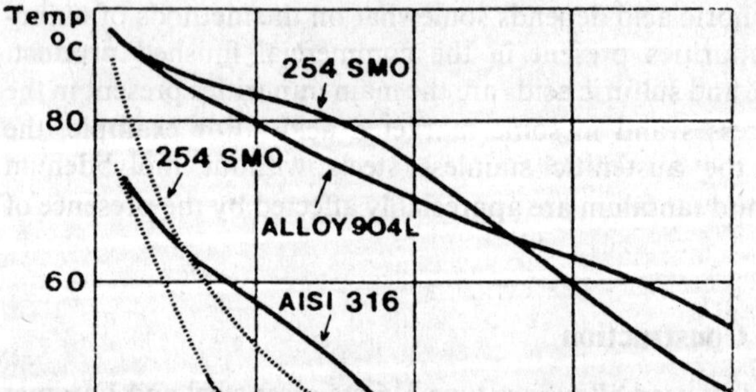

Figure 7-24 Isocorrosion curves, 0.1 mm/yr (4 mpy) in pure sulfuric acid (solid lines) and in sulfuric acid containing 2000 ppm Cl^- (dotted lines). 254 SMO = UNS S31254.

7-31 Miscellaneous

Corrosion by mineral acids is also discussed in other chapters. Some of the high alloys in Table 5-9 are used in sulfuric and phosphoric acid services. Titanium is discussed in connection with Table 5-10, including precautions to avoid failures. Figures 5-8 and 5-9 are isocorrosion charts for zirconium in sulfuric and hydrochloric acids. NACE Publication 5A171 (1983) describes materials for handling hydrofluoric acid. NACE Committee TP5A-23 produced this report. Corrosion by halogens appears in Chapter 8.

The effect of high chloride content in sulfuric acid on corrosion* of three alloys is shown in Fig. 7-24. Alloy 904L contains 20% Cr, 25% Ni,4.5% Mo, 1.5% Cu, and 0.02% C. Alloy 254 SMO contains 20% Cr, 18% Ni, 6% Mo, 0.7% Cu, and 0.02% c.

*B. Wallen, M. Liljas, and J. Olsson, Performance of a High Molybdenum Steel in Gas Cleaning Systems with Particular Reference to the Pulp and Paper Industry, *Materials Performance*, 40–48 (June 1982).

CHAPTER

EIGHT

OTHER ENVIRONMENTS

In addition to the mineral acids described in Chap. 7, most other environments are corrosive to a greater or lesser degree. Supposedly mild corrosives can cause severe problems under certain conditions.

8-1 Organic Acids

Acetic acid is the most important organic acid from the standpoint of quantity produced. Many other organic acids show similar corrosion behavior and, in the absence of data, one must assume that they all behave alike. Types 316 and 304 stainless steels, copper and bronzes, 1100 and 3003 aluminum, Durimet 20, Duriron, and Hastelloy C are widely utilized for handling acetic acid. Type 316 is preferred for the more severe conditions involving glacial (98%+) acid or elevated temperatures. Aluminum, copper, and type 304 are good for room-temperature glacial acid and for more dilute acid. Copper was the early work horse for acetic acid but has lost much ground to the stainless steels, partly because of the reduced cost differential between these materials. Durimet 20 is used for pumps and Duriron for pumps, lines, and columns. Hastelloy C and Chlorimet 3 cannot be justified except for the most severe conditions.

Acetic acid exhibits unusual corrosion behavior with regard to effects of temperature. Copper and stainless steel switch positions as temperature is increased. Both are good at room temperature, 316 is better near boiling temperatures, but copper shows very little attack at temperatures above boiling (no aeration) and here the stainless steels can exhibit very rapid attack. A most unusual corrosion problem involved a high-temperature high-pressure heat exchanger. Rapid corrosion of 316 occurred at the hot end and no corrosion at the cooler end. Replacement tubes of copper showed more corrosion at the colder end but less at the hot end. The *net* result was longer life for the copper tubes. Temperature drop along the tube length involved the critical temperature switch mentioned above.

Copper is sometimes subject to attack by acetic acid when cupric ions are in solution. The reaction is $Cu^{2+} + Cu \rightarrow 2Cu^{+}$.

Organic acids are "weaker" than the inorganic acids because they are only slightly ionized. Formic acid is one of the strongest organics and is the most corrosive. Aluminum is not suitable for formic acid. Maleic and lactic acids are more aggressive than acetic with regard to intergranular attack on stainless steels. The fatty acids, e.g., stearic, are less corrosive, but 316 is required at high temperatures. Naphthenic acid presents a corrosion problem in petroleum refining mainly because of the high temperatures involved. Citric and tartaric acid are found in food products. Table 8-1 lists some data for corrosion by organic acids.

In addition to decadence and corruption, corrosion was an important contributor to the decline and fall of the Roman Empire. The Romans made wine and drank it from lead containers. Acetic acid in wine attacks lead

Table 8-1 Corrosion by organic acids

Acid	Concentration	Temperature, °F	Aluminum*	Copper & bronze†	Type 304	Type 316	Durimet 20	Duriron
Acetic	50%	75	●	●	○	●	●	●
Acetic	50%	212	×	○	□	●	●	●
Acetic	Glacial	75	●	●	●	●	●	●
Acetic	Glacial	212	○	×	×	○	○	●
Citric	50%	75	○	□	○	○	●	●
Citric	50%	212	□	□	×	○	○	●
Formic	80%	75	○	○	○	●	●	●
Formic	80%	212	×	○	×	○	○	●
Lactic	50%	75	○	○	○	●	●	○
Lactic	50%	212	×	○	×	○	○	○
Maleic	50%	75	○	□	○	○	●	●
Maleic	50%	212	×	—	○	○	○	○
Naphthenic	100%	75	○	○	●	●	●	—
Naphthenic	100%	212	○	×	●	●	●	—
Tartaric	50%	75	○	□	●	●	●	●
Tartaric	50%	212	×	—	●	●	●	●
Fatty	100%	212	●	□	○	●	●	●

Legend: ● Less than 2 mpy, ○ less than 20 mpy, □ from 20 to 50 mpy, × over 50 mpy.
*More than 1% water for napthenic and fatty acids.
†Aeration greatly increases corrosion rate.
Source: Corrosion Data Survey, NACE.

Table 8-2 Typical corrosion failures and their prevention[a]

Solvent	Type[a]	Conditions,[b] °C	Components, impurities	Metal corroded	Type of attack[c]	Prevention	Ref.
				Alcohols			
Methanol	p; 1C	BP	$<0.05\%$ H_2O	Al	Uniform + pitting	Add 1% H_2O	58
Methanol	p; 2C	RT	Chlorides	Ti	SCC + HE	Eliminate Cl, add H_2O	59, 81
Methanol	p; 2C	RT	Methyl formiate	Carbon steel, Zn	Uniform	Eliminate methyl formiate	146
Ethanol	p; 2C	RT	20% HCl	Ti	5 mm/yr	9–30% H_2O	61
Ethanol	p; 1C	BP	$<0.05\%$ H_2O	Al	Pitting	Add 1% H_2O	58
Ethanol	p; 3C	RT–BP	Halogenides, oxygen	Steel, cast iron	Uniform, pitting	Deaeration, eliminate chlorides	147
Glycol monomethyl ether	p; 1C	40	—	Al	Pitting	Add 1% H_2O	58
Ethylene glycol, glycerol	p; 3C	140	NaCl, H_2O	AISI 304, 316	Pitting, crevice, corrosion	Use Monel, Inconel, nickel	148
Phenol	p; 2C	120–180	$<0.3\%$ H_2O	Al	Pitting	Add $>0.3\%$ H_2O	97, 149
	p; 2C	180	Na phenolate	Carbon steel	Uniform	Use AISI 304	147
Cresole	ap; 3C	125	$H_2O + HCl$	Monel, Si bronze	Uniform	Enamel, porcelain, glass	166

Acids and compounds							
Formic acid	p; 2C	RT–BP	Oxygen	Cu, Cu alloys	Uniform	Deaeration, temperature decrease, use Carpenter 20	147, 150
Formic acid	p; 2C	BP	3–10% H_2O	AISI 304, 316	Uniform 1 mm/yr	Hastelloy C <0.1% H_2O	147, 150 158
Formic acid + acetic acid	p; 3C	110	27% acetic acid 3% H_2O	Ti	Uniform + HE	Hastelloy C	146
Acetic acid	p; 2C	BP	Acetic anhydride	Al, Ti	Uniform	Hastelloy C, add H_2O	143, 147
Acetic acid	p; 2C	RT–BP	Oxygen	Cu, Cu alloys	Uniform 7 mm/yr	Deaeration, temperature decrease	147
Acetic acid	p; 2C	BP	H_2O	AISI 316, 317, Inconel	Uniform	Ti Hastelloy C	150, 151
Acetic acid	p; 2C	BP	Chlorides, inorganic, organic	AISI 316	SCC	Ta; eliminate Cl, temperature decrease	152
Acetic acid + CO	p; 2C	250	Co acetate: KI	AISI 316	Uniform	Ti, Zr, Ta, Hastelloy B	153
Acetic acid	p; 3C	RT–90	1% Chloride 1% H_2O	AISI 304	Pitting	Ti; Eliminate Cl	150, 147
Acetic acid	p; 2C	RT–BP	Fluorides	Ti	Uniform	Hastelloy C	154
Phthalic acid anhydride	p; 3C	210	Maleic acid, H_2O	AISI 304	Uniform	AISI 316	147
Formic acid + acetic acid + ethyl acetate	p; 5C	90	H_2O, HCOOH propionic acid	AISI 316 Monel	Uniform	Hastelloy C, Ti	150, 147 155

(continued)

Table 8-2 Typical corrosion failures and their prevention (continued)

Solvent	Type[a]	Conditions,[b] °C	Components, impurities	Metal corroded	Type of attack[c]	Prevention	Ref.
			Acids and compounds (continued)				
Acetic acid + toluene	p; 3C	90	Oxygen	C or 17% Cr steel	0.5 mm/yr 0.2 mm/yr	AISI 316	157
Mono- + tri-chloroacetic acid	p; 2C	RT		AISI 316 Ti, Zr	Uniform	Silicon iron	158, 159
Mono- + di- + trichloro-acetic acid	p; 5C	150	$Cl_2 + HCl$	All metals	Uniform	Enamel coating	160
42% Butyl-acetate + 30% acetic acid + 18% propionic acid + 9% propylacetate	p; 4C	BP		AISI 316	Pitting	Ti	156, 158
Cyanogen chloride + dichloro-methane	ap; 4C	20	$Cl_2 + HCl$	AISI 316	4 mm/yr	Nickel plating	161
Dimethyl-formamide	ap; 4C	100	H_2O + formic acid + di-methylamine	Carbon steel	Uniform	Eliminate H_2O $< 0.1\%$ H_2O	162
			Aldehydes, esters, amines, mercaptans				
Butyric aldehyde	ap; 3C	RT	Air + butyric acid	Cu	Discoloration	Eliminate air + acid	150

Benzaldehyde	ap; 2C	RT	Chlorides	Al, Al alloys	Uniform	Eliminate Cl	147
Benzaldehyde	ap; 2C	BP	Chlorides	AISI 316	Pitting	Eliminate Cl	147
Methylethyl ketone	ap; 2C	RT	H_2O	Carbon steel, steel, cast iron	Discoloration	Deaeration, Drying, use galvanized steel	147
Eethyl acetate	ap; 2C	RT	Acetic acid	Carbon steel, cast iron	Discoloration	Deaeration drying, use galvanized steel	147
Methylamines	p; 2C	RT	Oxygen	Al, Cu, Cu alloys	Uniform tSCC	Carbon steel, cast iron	57, 164
Triethanol-amine	p; 2C	RT	Oxygen	Cu, Cu alloys	Uniform tSCC	Carbon steel, cast iron	163, 165
Aniline	p; 2C	BP	Oxygen	Al, steel	Uniform	AISI 316, eliminate air	165
Pyridine sulfate	ap; XC	102	H_2SO_4 + carbolic oil	AISI 304, 316	Pitting, crevice corrosion	Nionel, Carpenter 20	147
Ethylene oxide	ap; 1C	RT		Cu, Cu alloys	Explosive	AISI 304	147
Ethyl mercaptan	p; 1C	100		Steel, cast iron, Cu, Cu alloys, Ni, Pb, Ag	Uniform, voluminous products	Al, Al alloys, Cr plating	147, 167
Mercaptan in naphtha	p; XC	RT	H_2O, oxygen	Cu, Cu alloys, Ni, Ni alloys, Ag	Uniform	Al, Al alloys, Cr plating	147
Hydrocarbons, halogenated hydrocarbons							
Chloroform	ap; 3C	BP	$H_2O + HCl$	Al, Al alloys, AISI 304, 316	Pitting tSCC	Stabilization (0.1% aniline, < 0.05% H_2O)	66, 168
Epichlorhydrin	ap; 3C	RT	$H_2O + HCl$	Cu alloys, AISI 304, 316	Pitting tSCC	Enamel coating	146

(continued)

Table 8-2 Typical corrosion failures and their prevention (continued)

Solvent	Type[a]	Conditions,[b] °C	Components, impurities	Metal corroded	Type of attack[c]	Prevention	Ref.
Ethyl ether	ap; 2C	RT–BP	Acids + H_2O	Steel, cast iron	Discoloration	Galvanized steel	147
Carbon tetrachloride trichloroethylene	ap; 3C	BP	$H_2O + HCl$	Al, Cu, Cu alloys, steel 304, 316, Ti	Explosive, pitting, SCC, HE	Stabilization, inhibition, neutralization, drying	147, 168
Ethylene dibromide	ap; 3C	BP	$H_2O + HBr$	Al, steel, Cu, 304, 316, Ni, Hastelloy B, C	Explosive, uniform pitting	$< 0.02\%$ H_2O stabilization, neutralization, 17% Cr steel	146, 147
Dichloroethane + ethylene	ap; 4C	150	$H_2O + HCl$	Ni, Ni alloys, AISI 316	Uniform tSCC	Ta plating	147
Hexachloroethane + perchloroethylene + Cl_2	ap; 4C	RT	$H_2O + FeCl_3$	Steel, alloys, cast iron, Ni, Ni alloys, Hastelloy C	9 mm/yr 3.5 mm/yr 1.3 mm/yr 2 mm/yr	Lead plating, enamel coating	147
Dichlorobenzene	ap; 3C	BP	$H_2O + HCl$	17% Cr steel, AISI 304	11 mm/yr Pitting tSCC	Stabilization, eliminate H_2O and HCl	146

Benzene	ap; 3C	RT	Butyric acid, oxygen	Carbon steel	0.7 mm/yr	Eliminate oxygen	169
Hydrocarbons	ap; 3C	RT	$H_2O + HCl$	Carbon steel	Uniform	Eliminate HCl, $H_2O <$ $0.3 \cdot H_2O$ sat.	67, 170
Hydrocarbons + aqueous solution	p; 4C 2P	RT	$H_2O + NaCl$ + oxygen	Carbon steel	Uniform, crevice-corrosion	Inhibition	174, 175 176
Crude oil	ap; XC	220	Naphthenic acids	Steel, 6% Cr steel, AISI 304	Uniform, pitting, erosion corrosion	Neutralization of acids + AISI 316, Cr 25 Ni 20 steel Ni alloys	171
Diphenyl	ap; 3C	230	H_2O + chlorides	AISI 304	Rusting, pitting, tSCC	Eliminate Cl and H_2O	143

[a]Key: p, protic; ap, aprotic; 1C, one component; 2C, two components; 3C, three components; XC, components unknown; 2P, two phases.

[b]RT, room temperature; BP, boiling point.

[c]HE, hydrogen embrittlement; iSCC, intergranular stress corrosion cracking; tSCC, transgranular stress corrosion cracking.

Monel: 63–70% Ni, max. 2.5% Fe, 0.3% C, remainder Cu.

Inconel 600: min. 72% Ni, 14–17% Cr, 6–10% Fe, max. 0.15% C.

Carpenter 20: 20% Cr, 29% Ni, 2% Mo, 3% Cu, 1% Si.

Hastelloy C: 14.5–15.5% Cr, 15–17% Mo, 4–7% Fe, max. 0.08 C, remainder Ni.

Hastelloy B: 26–30% Mo, 4–6% Fe, max. 0.05% C, max. 1% Cr, remainder Ni.

Source: Ewald Heitz, Corrosion of Metals in Organic Solvents, in *Advances in Corrosion Science and Technology*, vol. 4, pp. 226–229, Plenum Press, New York, 1974.

and the citizens' brains became addled from lead poisoning. Lead acetate is made by dissolving lead in acetic acid.

In addition to organic acids, organic *solvents* can cause corrosion when water and acids are present. In the pure state these solvents do not attack ordinary steel but plastics and rubber are vulnerable. There is an excellent review by Ewald Heitz, "Corrosion of Metals in Organic Solvents," in *Advances in Corrosion Science and Technology*, vol. 4, pp. 149–243, Plenum Press, New York (1974). Heitz defines organic solvents as all liquid organic compounds independent of practical use as a solvent, so many organic acids are included. This is an in-depth article and includes chemical and electrochemical corrosion, classification of compounds, effect of multiple phases, and a large number of case histories involving many environments and metals and alloys. Also correlation between weight-loss and polarization-resistance tests. Uniform attack, pitting, crevice corrosion, stress corrosion, erosion corrosion and hydrogen embrittlement are observed in the same way as aqueous corrosion.

Table 8-2 shows typical corrosion failures and their prevention. Lack of space precludes listing of the references because of their large number (178). Conclusions in the Heitz paper are quoted as follows:

1. Corrosion in organic solvents exhibits a number of unexpected effects, such as a higher corrosion rate in alcoholic acid solutions as compared to aqueous solutions and corrosion processes with autocatalytic characteristics.

2. For a good understanding of corrosion in organic solvents classification of the solvents according to their physicochemical properties and composition is advantageous. In classifying solvents, their protic or aprotic character, the number of components, and the number of phases are crucial.

3. The discussion of thermodynamic quantities gives evidence on the pH value in organic solvents and shows that there is a considerable lack of knowledge on electrode potentials and corrosion potentials. Stoichiometric equations generally can be formulated in analogy to aqueous solutions.

4. Parameters of the corrosion rate are the solvent composition and structure, especially the protolytic action, solubility of corrosion products, and oxygen solubility. Decisive for the active-passive behavior of metals are traces of water and acids in the organic solvent.

5. A great number of reaction mechanisms can be assumed to be electrochemical. Some mechanisms, for example, the reaction of base metals with halogenated hydrocarbons, are radical and have no analogy to aqueous solutions. Other processes resemble high-temperature corrosion in their mechanism.

6. A collection of case histories and reported corrosion failures from practice shows a great variety of corrosion problems and phenomena. In some cases it is difficult to find the cause; for others practical solutions have been found, but theoretical explanations are lacking; while for others corrosion can be prevented on the basis of purely theoretical considerations.

7. Corrosion protection and prevention can be achieved by application of principles known from aqueous systems. Literature for current materials selection is presented. Emphasis is laid on prevention measures based on consideration of the medium, such as drying, deaeration, and neutralization of the solvent. Application of electrochemical

protection methods is sometimes restricted because of the low electrical conductivity of the medium. Corrosion protection by organic coatings is limited by the reciprocal solubility of the solid organic material and the organic solvent.

8-2 Alkalies

The common alkalies such as caustic soda (NaOH) and caustic potash (KOH) are not particularly corrosive and can be handled in steel in most applications where contamination is not a problem. However, one must guard against stress corrosion in certain concentrations and temperatures, as described in Chap. 3. Rubber-base and other coatings and linings are applied to steel equipment to prevent iron contamination.

Nickel and nickel alloys are extensively used for combating corrosion by caustic. Nickel is suitable under practically all conditions of concentration and temperature. In fact, the corrosion resistance to caustic is almost directly proportional to the nickel content of an alloy. As little as 2% nickel in cast iron is beneficial. Monel (70% Ni), the austenitic stainless steels (8 to 20% Ni), and other nickel-bearing alloys are in many applications involving high temperatures or control of contamination.

Figure 3-65 is a chart for steel versus temperature and concentration of NaOH. The austenitic stainless steels are better than carbon steel, but for the tough jobs (high concentrations and temperatures) Inconel 600 (UNS NO6600) is one of the best. This alloy is superior when chlorides are present, which is often the case.

DeAngelo and Wilhemi* describe a system for pollution control of waste caustic solutions.

Aluminum is a very poor material for handling caustic. This metal and its alloy are rapidly attacked even by dilute solutions.

Many metals and alloys exhibit low corrosion rates in caustic, but steel, cast iron, nickel, and nickel-containing alloys are suitable for a large majority of applications.

Ammonia and ammoniacal solutions generally do not present difficult corrosion problems. In manufacture and handling, steel and cast iron are satisfactory except for high temperatures, where types 430 and 304 stainless steels are required. Aluminum is often adopted in refrigeration systems and storage tanks. The major warning is not to use copper and copper-base alloys because even traces of ammonia can cause stress corrosion. For example, decomposition of organic materials containing nitrogen has caused cracking. Ordinary dissolution also occurs in ammonium solutions with the formation of complex ions.

B. E. Wilde, in a private communication, describes stress-corrosion cracking of carbon steel by anhydrous ammonia (used as a fertilizer).

*D. J. DeAngelo and A. R. Wilhemi, Wet Air Oxidation of Spent Caustic Liquors, *Chem. Eng. Progr.*, 68–73 (Mar. 1983).

Table 8-3 Corrosiveness of atmospheres at 37 test sites relative to that at State College, Pa. (taken as unity); based on losses of two 4- by 6-in. specimens

			Steel				Zinc			
			Grams				Grams			
	Location	Type of atmosphere	Avg loss	Range*	1960–1961 rating	1948–1956† rating	Avg loss	Range*	1960–1961 rating	1948–1956† rating
1	Norman Wells, N.W.T.	Rural	0.1	0.0	0.02	0.03	0.052	0.001	0.2	0.4
2	Saskatoon, Sask.	Rural	1.4	0.1	0.2	0.6	0.074	0.001	0.2	0.6
3	Fort Clayton, C. Z.	Tropical jungle	2.8	0.1	0.4		0.266	0.018	0.8	
4	Rocky Point, B.C.	Marine	4.6	0.1	0.7	0.5	0.110	0.009	0.3	0.4
5	Potter County, Pa.	Rural	5.2	0.1	0.8		0.250	0.029	0.7	
6	Detroit, Mich. (R)‡	Urban	5.9	0.1	0.9		0.270	0.015	0.8	
7	Ottawa, Ont.	Rural	5.8	0.0	0.9		0.233	0.002	0.7	1.3
8	Morenei, Mich.	Rural	6.4	0.2	1.0		0.261	0.014	0.8	
9	State College, Pa.	Rural	6.4	0.1	1.0	1.0	0.334	0.035	1.0	1.0
10	York Redoubt, N.S.	Marine	7.4	0.3	1.2	1.5	0.371	0.008	1.1	2.1
11	Montreal, Que. (R)‡	Industrial	8.3	0.0	1.3	1.5	0.492	0.033	1.5	3.0
12	Middletown, Ohio	Semi-industrial	8.8	0.1	1.4	1.2	0.323	0.010	1.0	1.0
13	New Hampshire Coast, N.H.	Marine	9.4	0.3	1.5		0.458	0.032	1.4	
14	South Bend, Pa.	Semirural	10.5	0.1	1.6	1.5	0.396	0.014	1.2	1.0
15	Columbus, Ohio (R)‡	Urban	10.9	0.1	1.7		0.468	0.009	1.4	
16	New Cristobal, C.Z. (R)‡	Tropical marine	11.1	0.0	1.7		0.740	0.003	2.2	

17	Pittsburgh, Pa. (R)‡	Industrial	11.4	1.2	1.8		0.512	0.002	1.5	
18	London (Battersea), England (R)‡	Industrial	12.7	0.0	2.0		0.409	0.004	1.2	
19	Trail, B.C.	Industrial	13.2	0.1	2.1	1.4	0.501	0.004	1.5	
20	Mirafiores, C.Z.	Tropical urban	13.2	0.1	2.1		0.334	0.008	1.0	
21	Research Center, Pittsburg, Pa.	Semi-industrial	14.0	0.2	2.2		0.488	0.006	1.5	
22	Daytona Beach, Fla.	Marine	14.3	1.2	2.2	7.1	0.418	0.017	1.3	2.1
23	Bethlehem, Pa.	Industrial	15.2	0.2	2.4		0.262	0.003	0.8	
24	Cleveland, Ohio	Industrial	15.5	0.1	2.4		0.485	0.016	1.5	
25	Newark, N.J.	Industrial	16.8	0.2	2.6		0.779	0.002	2.3	
26	Brazos River, Tex.	Marine	17.1	0.3	2.7		0.466	0.001	1.4	
27	Bayone, N.J.	Industrial	21.6	0.5	3.4		1.020	0.026	3.1	
28	Kure Beach (800-ft site), N.C.	Marine	23.2	3.4	3.6	2.0	0.640	0.000	1.9	1.4
29	Pilsey Island, England	Marine	25.8	0.8	4.0		1.058	0.016	3.2	
30	East Chicago, Ind.	Industrial	33.5	0.0	5.2			STOLEN		
31	London (Stratford), England	Industrial	41.3	1.6	6.5		1.608	0.019	4.8	
32	Halifax, N.S. (Fed. Bldg.)	Marine industrial	46.9	2.7	7.3	3.8	2.913	0.096	8.7	22
33	Point Reyes, Calif.	Marine	61.0	0.6	9.5	1.8	0.338	0.017	1.0	1.4
34	Dungeness, England	Marine	97.2	5.1	15		0.768	0.001	2.3	
35	Galeta Point Beach, C.Z.	Tropical marine	117.3	§	18		2.577	§	7.7	
36	Widnes, England	Industrial	119.5	1.3	19		2.716	0.038	8.1	
37	Kure Beach (80-ft site), N.C.	Marine	210.1	12.4	33	13	2.114	0.071	6.4	5.7

*Plus or minus.

†Table XXII, Report of Committee B-3 (Appendix II), *Proc. ASTM*, **59**:200 (1959).

‡Exposure made on roof of building.

§Only one specimen exposed at this location.

Source: *ASTM Bull.*, Report of Committee B-3, Dec. 1961.

Oxygen is the primary cause of cracking, with as little as 0.01 ppm required. The mechanism is a slip step/film rupture process and not a hydrogen-induced brittle fracture, as previously suggested. SCC is inhibited by the addition of 0.2% water. Where water is undesirable (e.g., refrigeration) 0.025% hydrazine (oxygen scavenger) is effective.

An interesting, unusual, and very expensive SCC problem occurred in the southern states, where farmers use ammonia for fertilizer and liquid propane gas (LPG) for heating and cooking. Their LPG tanks have brass fittings (e.g., valves). Distributors who sell both ammonia and LPG did not *clean* the storage tanks, tank trucks, etc., before switching from ammonia to LPG. Some residual ammonia wound up in the LPG and all the fittings cracked. Investigation by the LPG producer and me showed that *one part per million* of ammonia would cause SCC of brass fittings. Replacement of the fittings cost about $80 million. This occurred several years ago when a million dollars was a real fortune!

8-3 Atmospheric Corrosion

Corrosion by various atmospheres accounts for more failures on a cost and tonnage basis than any other single environment, with an estimated cost of $2 billion in the United States. Atmospheres can be classified as industrial, marine, and rural. Corrosion is primarily due to moisture and oxygen but is accentuated by contaminants such as sulfur compounds and sodium chloride. Corrosion of steel on the seacoast is 400 to 500 times greater than in a desert area. Steel specimens 80 ft from the seashore corroded 12 times faster than those 800 ft away. Sodium chloride is the chief contaminant. This salt causes a large amount of corrosion of automobiles when it is used for deicing roads. Industrial atmospheres can be 50 to 100 times more corrosive than desert areas.

Industrial atmospheres are more corrosive than rural atmospheres, primarily because of sulfur gases generated by the burning of fuels. SO_2 in the presence of moisture forms sulfurous and sulfuric acids, which are both very corrosive. Table 8-3 illustrates the wide variation of corrosion in different parts of the world.

Small amounts of copper (tenths of one percent) increase resistance of steel to atmospheric corrosion because it forms a tighter, more protective rust film. Small amounts of nickel and chromium produce similar effects. Nickel and copper are helpful in industrial atmospheres because they form insoluble sulfates that do not wash away and thus afford some protection. For almost complete rust resistance in ferrous alloys, we must go to the stainless steels. Improving the corrosion resistance of steel with small alloy additions (low-alloy steels) is now commonly used in weight-saving applications and to increase durability of paint coatings. Some of these steels form a protective, pleasant-appearing rust, and painting is not required in rural and other inland locations. Copper, lead, aluminum, and galvanized steel are also widely used for atmospheric applications.

The low-alloy steels described in the previous paragraph are called *weathering steels*. They are used for buildings, statues, bridges, guard rails, etc. However they cannot be used for all conditions. Areas of high humidity and/or "polluted" atmospheres (such as at the seashore) result in a loose rust that washes off and causes unsightly stains on adjacent areas. Prolonged rainy seasons are harmful. To be successful these steels need alternate wetting and drying conditions for the rust to form and "set." They offer no advantage when submerged in water or in wet soil. The reader is referred to a book of 1028 pages for in-depth information.*

Sheltered corrosion causes much damage. This term is used to describe locations where moisture condenses or accumulates and does not dry out for long periods of time. A good example of this is the corrosion that occurs on the inside surface of an automobile door. Another is the condensation of moisture on the inside of a partially filled automobile gasoline tank when the temperature drops at night. Rapid corrosion on the inside of large storage tanks occurs when moisture-laden salt air is drawn in. This is often described as "breathing." Changes in ambient temperatures results in condensation of salt water on the inside walls of the tank. This problem is solved by blanketing the inside of the tank with nitrogen gas or dry air.

There are many types of protective coatings for steel that tend to inhibit corrosion.

8-4 Seawater

Seawater contains about 3.4% salt and is slightly alkaline, pH 8. It is a good electrolyte and can cause galvanic corrosion and crevice corrosion. Corrosion is affected by oxygen content, velocity, temperature, and biological organisms. Additional information is presented in Chap. 3.

Figure 8-1 shows typical corrosion rates for ordinary steel in the sea. Greatest attack occurs in the splash zone because of alternate wetting and drying and also aeration. Figure 8-2 shows corrosion rates and Fig. 8-3 pitting of metals and alloys in quiet seawater. Figure 8-4 shows the effect of velocity on corrosion of pipe and tubing. Figure 8-5 illustrates two-metal corrosion effects in pumps and valves. Corrosion rates in these figures may vary depending on local conditions, but the values are relatively accurate.

Table 8-4 indicates resistance to crevice corrosion and Table 8-5 relative resistance to fouling. Table 8-6 indicates cavitation resistance and Table 8-7 lists galvanic potentials. Corrosion by seawater at great depths (i.e.,

* *Atmospheric Corrosion*, W. H. Ailor, ed., Wiley-Interscience, New York (1982).

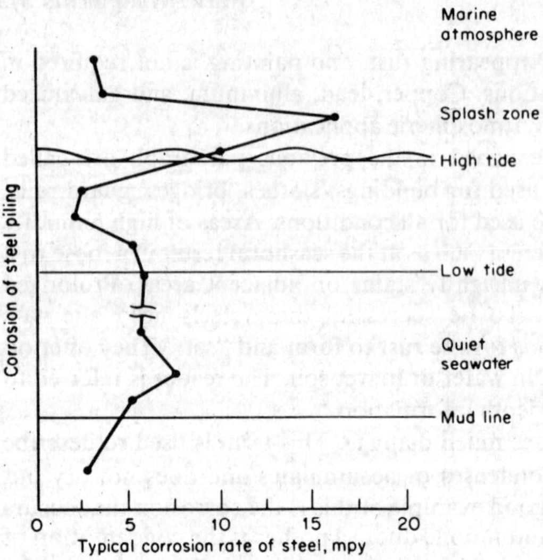

Figure 8-1 Corrosion of ordinary steel in the sea.

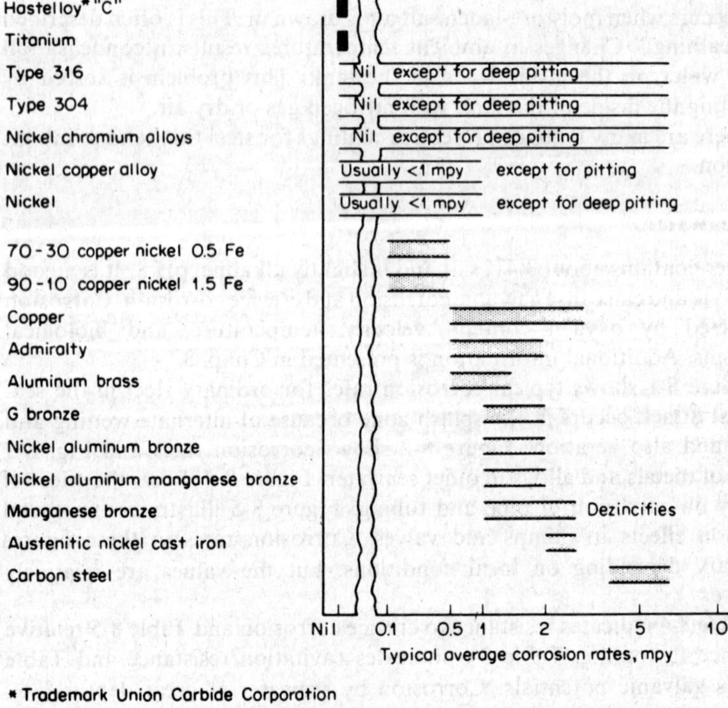

Figure 8-2 Corrosion of metals and alloys by quiet seawater—less than 2 ft/sec.

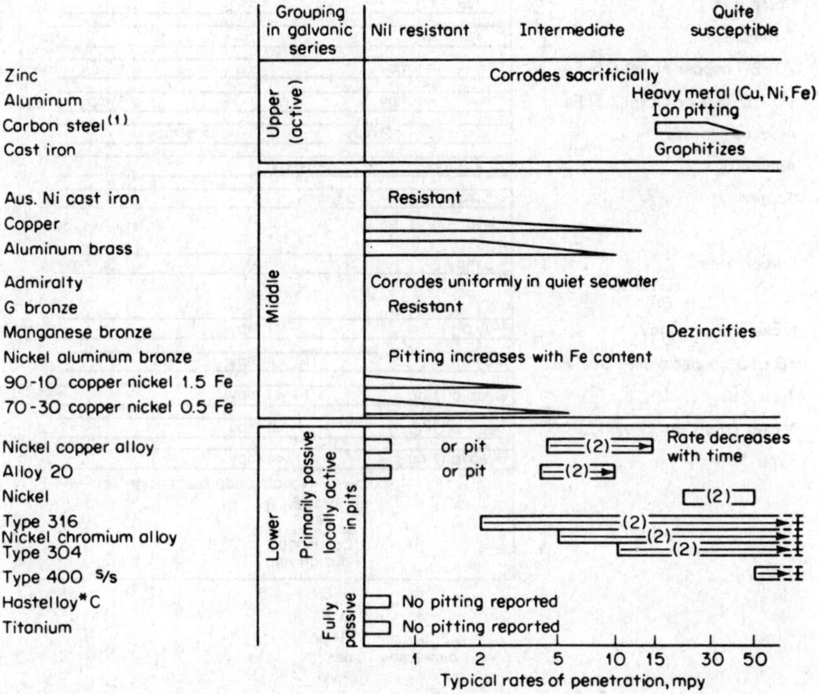

(1) Shallow round-bottom pits

(2) As velocity increases above 3 fps, pitting decreases. When continuously exposed to 5 fps and higher velocities these metals, except series 400 S/s, tend to remain passive without any pitting over the full surface in the absence of crevices.

* Trademark Union Carbide Corporation

Figure 8-3 Pitting in quiet seawater.

1 mile) is usually decreased because of the lower temperature (about 40°F).*

Brackish water is water contaminated with chlorides usually because of tidal action in rivers and bays near the ocean. Corrosion problems are less severe than in full-strength seawater, but guidelines for materials of construction are approximately the same.

For in-depth study the reader is referred to NACE Recommended Practice, RP-01-76 (1983 revision), *Corrosion Control of Steel, Fixed Offshore Platforms Associated with Petroleum Production*, which includes structural

*Figures 8-2 to 8-7 and Tables 8-4 to 8-7 are from A. H. Tuthill and C. M. Schillmoller (of The International Nickel Co., Inc.), Useful Guidelines in the Selection of Corrosion Resistant Materials for Marine Service, *Marine Technol. Trans.*, 3 (1965). Permission of the authors and Inco to use this material is gratefully acknowledged. The reader is referred to this paper for an excellent discussion of this subject.

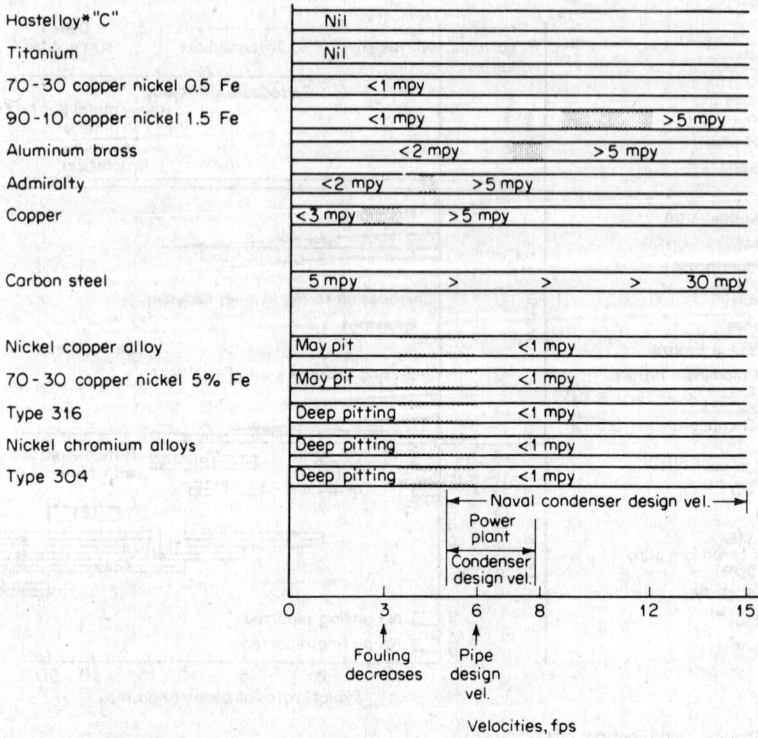

* Trademark Union Carbide Corporation

Figure 8-4 Effect of velocity on corrosion of piping by seawater.

Table 8-4 Tolerance for crevices immersed in quiet seawater

Inert	Best	Neutral	Less	Crevices tend to initiate deep pitting
	Useful resistance			
Hastelloy* "C"	90/10 copper nickel 1.5 Fe	Aus. nickel cast iron	Incoloy† alloy 825	Type 316
	70/30 copper nickel 0.5 Fe	Cast iron	Alloy 20	Nickel-chromium alloys
Titanium	Bronze	Carbon steel	Nickel-copper alloy	Type 304
	Brass		Copper	Series 400 S/S

*Trademark Union Carbide Corporation.
†INCO trademark.

Body material ↓	Trim: Brass or bronze	Trim: Nickel copper alloy	Trim: Type 316
Cast iron	Protected	Protected	Protected
Austenitic nickel Cast iron	Protected	Protected	Protected
M or G bronze 70/30 copper nickel	May vary (1)	Protected	Protected
Nickel copper alloy	Unsatisfactory	Neutral	May vary (2)
Alloy 20	Unsatisfactory	Neutral	May vary (2)

(1) Bronze trim commonly used. Trim may become anodic to body if velocity and turbulence keep stable protective film from forming on seat.

(2) Type 316 is so close to nickel-copper alloy in potential that it does not receive enough cathodic protection to protect it from pitting under low velocity and crevice conditions.

Figure 8-5 Galvanic compatibility—pump and valve trim.

Table 8-5 Fouling resistance—quiet seawater*

Arbitrary rating scale of fouling resistance	Materials
Best 90–100	Copper 90/10 copper-nickel
Good 70–90	Brass & bronze
Fair 50	70/30 copper-nickel, aluminum bronzes, zinc
Very slight 10	Nickel-copper alloy
Least 0	Carbon and low-alloy steels, stainless steels, Hastelloy† "C" Titanium

*Above 3 ft/sec continuous velocity (about 1.8 knots) fouling organisms have increasing difficulty in attaching themselves and clinging to the surface unless already attached securely.

†Trademark Union Carbide Corporation.

Table 8-6 Cavitation resistance

Ship propellers, pump impellers, hydrofoils

Based on field experience		Based on laboratory test
Stellite*	1	Stellite*
17-7 Cr-Ni stainless steel weld	2	Two layers 17-7 Cr-Ni stainless steel weld
18-8 Cr-Ni stainless steel weld	3	18-8 Cr-Ni stainless steel weld
Ampco† No. 10 weld	4	Ampco† No. 10 weld
25-20 Cr-Ni weld	5	Cast Ampco† No. 18 bronze
Eutectic-Xyron 2-24 weld	6	Nickel-aluminum bronze
Ampco† bronze castings	7	18-8 Cr-Ni cast stainless
18-8 Cr-Ni cast stainless	8	13% Cr, cast stainless
Nickel-aluminum bronze, cast	9	Manganese bronze, cast
13% Cr cast	10	Cast steel
Manganese bronze, cast	11	Bronze
18-8 stainless spray metallizing	12	Cast iron
Cast steel	13	Sprayed stainless 18-8 Cr-Ni
Bronze	14	Rubber
Rubber	15	Aluminum
Cast iron	16	
Aluminum	17	

*Trademark Union Carbide Corporation.

†Trademark Ampco Metals, Inc.

design for corrosion control, design criteria for cathodic protection systems, coatings, and a bibliography.

An excellent reference book by F. L. LaQue, *Marine Corrosion*, John Wiley & Sons, Inc., New York (1975) describes various forms of corrosion, desalination plants and equipment such as propellers, shafts, pumps, piping systems, heat exchangers, ship hulls, and ropes.

8-5 Fresh Water

Corrosivity in fresh water varies depending on oxygen content, hardness, chloride content, sulfur content, and many other factors. For example, a steel hot-water tank in a home may last 20 years in one area but only a year or two in other areas. Chloride contents may vary from a few parts per million (ppm) to several hundred within one county. Sulfur compounds in some localities in Ohio, for example, cause rapid corrosion of steel. For this reason, it is difficult to make general recommendations—it is a local problem.

Fresh water can be hard or soft, depending on minerals dissolved. In hard water, carbonates often deposit on the metal surface and protect it, but pitting may occur if the coating is not complete. Soft waters are usually more corrosive because protective deposits do not form.

Table 8-7 Galvanic potentials in flowing seawater
Velocity = 13 ft/sec except where noted

Metal or alloy	Temperature, °C	Volt* vs. saturated calomel
Zinc	26	−1.03
Mild steel	24	0.61
Gray cast iron	24	0.61
Austenitic cast iron†	14	0.47
Copper	24	0.36
Admiralty brass	24.6	0.36
Gunmetal	24	0.31
Aluminum brass	24.6	0.29
Admiralty brass	11.9	0.30‡
Lead-tin solder (50-50)	17	0.28
90/10 Cu/Ni (1.4 Fe)	6	0.24
90/10 Cu/Ni (1.4 Fe)	17	0.29
90/10 Cu/Ni (1.5 Fe)	24	0.22
70/30 Cu/Ni (0.51 Fe)	6	0.22
70/30 Cu/Ni (0.51 Fe)	17	0.24
70/30 Cu/Ni (0.51 Fe)	26.7	0.20
Monel § alloy 400	22	0.11
Nickel	25	0.10‡
Titanium	27	−0.10
Graphite	24	+0.25
Platinum	18	+0.26‡

*All values negative vs. saturated calomel reference electrode except those for graphite and platinum.

†Ni-Resist§ ductile cast iron type D-2 (3.0 C, 1.5–3 Si, 0.7–1.25 Mn, 18–22 Ni, 1.75–2.75 Cr).

‡Seawater velocity = 7.8 ft/sec.

§INCO trademark.

Low-alloy steels do not offer any advantage over ordinary steel in water applications (as compared with atmospheric corrosion). For example, most boiler tubes and boiler-water systems are made from low-carbon steel. As is the case in atmospheric corrosion, complete corrosion resistance would require the more expensive stainless steels. Wrought iron offers no particular advantage over ordinary steel.

Cast iron, steel, and galvanized steel are the most widely used materials for handling fresh water. Copper, brass, aluminum, some stainless steels, Monel, and cupronickel are also used where temperature, contamination, or longer life are factors. Table 8-8* lists design and materials for heat exchangers using water as the coolant.

*W. C. Ashbaugh, Corrosion Problems in Industrial Cooling Waters, *Proceedings of Short Course on Process Industry Corrosion*, Ohio State University and Committee T-5, NACE, Sept. 12–16, 1960.

Table 8-8 Heat exchanger design and materials

Material required by process	Materials						
	Tube sheets	Channel or head	Tubes	Shell	Baffles	Tie rods	Spacers
Condensers and Coolers with Cycle Water on Shell Side							
Steel	Steel	Steel	Steel	Steel	Steel	Steel	Steel
Ss*	Ss	Ss	Copper clad ss or ss	Copper	Ss, Everdur, Admiralty or naval brass	Ss, Everdur	Evedur, copper, or ss
Copper	Everdur	Copper	Copper	Copper	Everdur, Admiralty or naval brass	Ss, Everdur	Everdur, copper, or ss
Condensers and Coolers with Cycle Water on Tube Side							
Steel	Steel	Steel	Steel	Steel	Steel	Steel	Steel
Ss	Ss	Copper	Copper-lined ss	Ss	Ss	Ss	Ss
Copper	Everdur	Copper	Copper	Copper	Everdur, Admiralty or naval brass	Everdur	Copper

*Stainless steel.

Practically every plant and process involves transfer of heating and cooling operations. Shell and tube *heat exchangers* are the most popular. Good design and selection and control of materials can represent the difference between success and failure.

Smallwood* presents an excellent summary of tubing reliability. He describes corrosion-type defects as: (1) substitution of an inadequate alloy through error, (2) selective weld metal attack, (3) improper pickling, (4) corrosion during testing or handling, (5) residual stresses, (6) improper heat treatment, (7) imbedded tramp metal, (8) preferred grain orientation, (9) surface roughness, (10) dents, and (11) high-temperature contamination. An example of the latter is carbon pickup because of local or general carburization of the metal. (Carbon pickup of castings is discussed in Sec. 3-20.) Examples of all of these defects are discussed. Mechanical defects are also covered, and methods for defect detection, such as ultrasonic, are presented. A quality assurance program is outlined.

A detailed discussion of water treatment is beyond the scope of this

*R. E. Smallwood, Heat Exchanger Tubing Reliability, *Materials Performance*, **16**:27–34 (Feb. 1977).

book and is considered to be a *specialty* area. Water is not simply H_2O but can contain almost any impurity imaginable. Compositions vary not only in different parts of the country but also in local communities. Perhaps the best book on the subject is the well-known *Betz Handbook of Industrial Water Conditioning*. It is revised periodically. The 1980 (Eighth) Edition of the *Betz Handbook* consists of 437 pages. This book includes sections on external treatment, boiler water systems, cooling water systems, chemical treatment feeding and control, wastewater treatment and reuse, special industrial processes, and control water analyses and their interpretation. Other water treatment companies also provide valuable information.

In addition to the information presented in Sec. 6-5, much work has been done in recent years on *nontoxic inhibitors* because of the EPA ban on heavy metals. In the MTI Manual No. 4, *State-of-the-Art Survey of Non-chromate Cooling Water Corrosion Inhibitors* (1980) the discussion includes zinc-, zinc/phosphate-, molybdate-, silicate-, and organic- based inhibitors. A few other pertinent references are included in the following Suggested Reading list. It is interesting to note that all references except one on this list have 1982 dates, indicating a big "push" a year or two earlier.

Suggested Reading

Breske, T. C.: Application Experience with Organic Boiler Water Treatments, *Materials Performance*, 24–28 (Aug. 1982).

Krisher, A. S. Evaluation of Nonmetallic Cooling Water Inhibitors, *Materials Performance*, Phase 1:21–26 (Jan. 1979); Phase 2:9–13 (May 1982).

Puckorius, P., Initial Conditioning of Cooling Water Equipment, (NACE Standard RP-01-82), *Materials Performance*, 83–88 (Mar. 1982).

Robitaille, D. R.: Molybdate Inhibitors for Problem Cooling Waters, *Chem. Eng.*, 139–142 (Oct. 4, 1982).

Scott, W. J., Handling Cooling Water Systems, *Chem. Eng.*, 121–124 (Feb. 22, 1982).

8-6 High-Purity Water

When water is used as a heat-transfer medium and very little corrosion can be tolerated, high-purity water is required. Atomic power plants and more conventional high-pressure power units are examples. Corrosion decreases with increasing purity of the water because of less solids and gases and increasing electrical resistance. Ordinary distilled water exhibits resistance around 200,000 ohm-cm. Resistance is a measure of water purity. In some atomic applications 1 to 2 megohm water is utilized. At high temperatures (600 to 700°F) about 10 ppm O_2 and H_2 are formed because of radiolytic decomposition of the water. Overpressure with hydrogen reduces the O_2 formed. Intergranular attack and cracking of solution-quenched stainless steels and alloys have been observed in high-purity water containing oxygen.

Corrosion products cause contamination, "crud" deposits on heat-transfer surfaces, and plugging of thin annular spaces and control mechanisms. Zirconium, Zircalloys (Zr base), stainless steels, Inconel, and Incoloy are used in these services. In fact, the zirconium industry was born because of nuclear requirements.

Stress-corrosion cracking occurred in nuclear power plants almost from the very beginning and many years thereafter. Cracks occur in 304 stainless steels sensitized because of stress relief treatment or weld decay in the heat affected zones (HAZ). A local newspaper reported "cracked *pipes*" in 20 of a group of 21 similar plants. *Newer* plants and replacements utilize types 304L and 316L stainless steels (instead of 304) for piping.

I investigated failures in one of the first nuclear power plants. Type 304 superheater *tubes* were badly sensitized during stress relief of this Code heat exchanger. These tubes failed because of SCC (IGSCC) during "check-out" and warming operations long before scheduled plant start-up! Incoloy tubing was then installed. Replacement was more costly than need be because the building design did not allow for easy replacement of the superheater tubing.

The Fontana Corrosion Center at Ohio State University is conducting an extensive program (for several years) on SCC in pure water under sponsorship by the Electric Power Research Institute (EPRI). Figure 8-6, which appeared in an excellent review paper,* shows regions for no SCC, intergranular (IGSCC), and transgranular (TGSCC) cracking versus temperature and oxygen content of the water for *sensitized* 304 stainless steel. This paper also includes discussion of effects of pH.

Two additional papers of interest are by G. Cragnolino et al, "SCC of Sensitized Type 304 in Sulfate and Chloride Solutions at 250C and 100C" (*Corrosion* **37**:312–320, June 1981) and by Cragnolino and Macdonald, "Intergranular SCC of Austenitic Steel Below 100C—A Review" (*Corrosion* **38**:406–424, August 1982). The latter includes discussions of fluoride-containing environments, sulfur compounds, and IGSCC by polythionic acid.

The book *Corrosion in Nuclear Applications* by W. E. Berry (John Wiley & Sons, New York, 1971) is also highly recommended to the reader.

Stress-corrosion cracking of steam turbines is a source of high costs in power plants. SCC is largely due to the presence of caustic, chlorides, and sulfites in the boiler water or steam. Use of hydrazine as an oxygen scavenger, instead of sulfites, solves the sulfite problem. Part of these problems is due to the use of higher-strength steels for turbine rotors during recent years. For example, 4130 can crack in distilled water if hardness is over $R_C 40$ (hydrogen embrittlement).

Fontana served as a consultant in connection with the first nuclear-powered submarine—the *Nautilus*.

*S. Szklarska-Smialowska and G. Cragnolino, Stress Corrosion Cracking of Sensitized Type 304 Stainless Steel in Oxygenated Pure Water at Elevated Temperatures (Review), *Corrosion* **36**:653–665 (Dec. 1980).

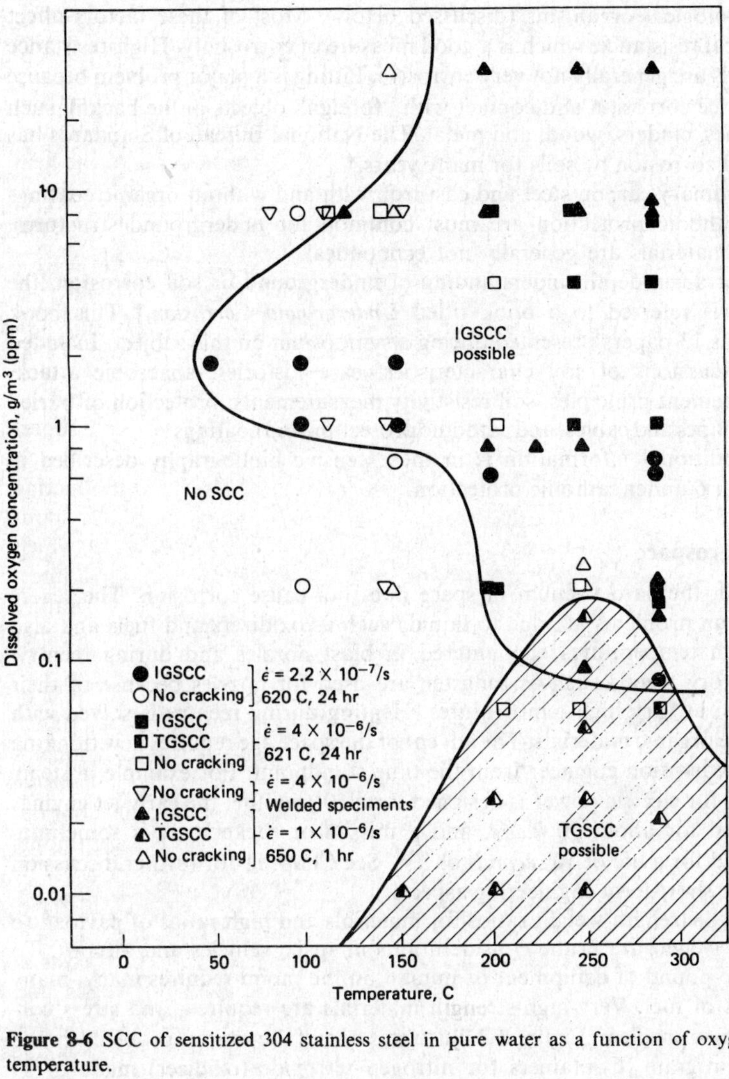

Figure 8-6 SCC of sensitized 304 stainless steel in pure water as a function of oxygen and temperature.

8-7 Soils

Corrosivity of soils varies over a wide range because of the variety of compositions. Tests in one location are generally applicable only to that location. Tests of several years' duration are needed to obtain reliable data. Factors affecting corrosiveness of soils are moisture, alkalinity, acidity, permeability of water and air (compactness or texture), oxygen, salts, stray currents, and biological organisms (discussed below). Most of these factors affect electrical resistance, which is a good measure of corrosivity. High-resistance dry soils are generally not very corrosive. Pitting is a major problem because of crevice corrosion and contact with "foreign" objects in the backfill such as stones, cinders, wood, and metal. The National Bureau of Standards has studied corrosion by soils for many years.*

Ordinary carbon steel and cast iron with and without organic coatings and cathodic protection are most common for underground structures. Other materials are generally not economical.

For an in-depth understanding of underground or soil corrosion, the reader is referred to a book titled *Underground Corrosion*.† This book contains 13 papers presented during a symposium on this subject. Included are discussions of soil characteristics, case histories, anaerobic attack, measurement principles, soil resistivity measurements, protection of buried tanks, pipes and cables, and cathodic protection and coatings.

Additional information is in the extensive bibliography described in Chapter 6 under cathodic protection.

8-8 Aerospace

In itself, the hard vacuum of space does not cause corrosion. The severe corrosion problems are due to liquids such as oxidizers and fuels and also the high temperatures encountered in blast nozzles and during reentry. Refractory metals such as tungsten are used for nozzles because of their strength at very high temperatures. Heating during reentry is solved with ablative organic materials. The advent of the space age requires new thinking by the corrosion engineer from the time standpoint. For example, a steam turbine for electric power is designed for a 50-*year* life, the early jet engines operated for about 50 *hours*, and a missile or rocket part is sometimes designed for a life of 50 *seconds* or less. See Chap. 11 for further discussion of these short-term engineering aspects.

High strength-weight ratios for materials and high ratios of payload to vehicle weight are prime considerations in space vehicles and aircraft. To put one pound of equipment or human on the moon requires many, many pounds of fuel. Very-high-strength materials are required, and stress corrosion is a problem. Figure 8-7 illustrates one of the serious problems in the lunar program. Containers for nitrogen tetroxide (oxidizer) made of Ti base, 6% Al, 4% V alloy ruptured in a matter of hours. Figure 8-7 shows a ruptured container, cracks in the metal, and the fracture surfaces. Another oxidizer, fuming nitric acid, caused stress corrosion of martensitic stainless steels.

*K. H. Logan, NBS Circular C450 (1945); and M. Romanoff, C579 (1957).

†*Underground Corrosion*, E. Escalante, ed., 212 pages. STP 741, ASTM, 1916 Race St., Philadelphia, Pa, 19103 (1980).

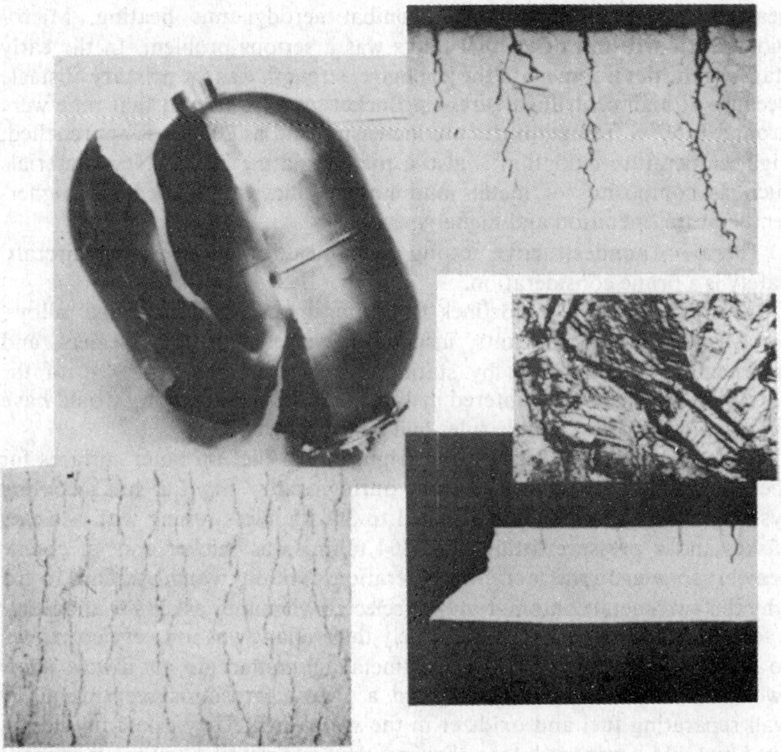

Figure 8-7 Stress-corrosion failure of N_2O_4 container. *(Textron's Bell Aerosystems Co.)*

Cleanliness and careful inspection of all parts during fabrication and assembly are paramount to avoid contamination. Rusting of high-strength steel due to resident water can cause stress corrosion and hydrogen embrittlement. Hydrogen charging of steel during electroplating can cause brittle fracture. Extensive corrosion of missiles stored in underground silos is caused by excessive moisture and inadequate coatings.*

Al, Mg, Ti, and Fe and their alloys are the primary metallic materials involved for aircraft. Stress corrosion, pitting, intergranular attack, crevice corrosion, and two-metal corrosion cause difficulties. Al and Mg alloys are anodic to most other metals. Protective systems include anodizing, cladding, and conversion coatings. Rivets are sources of trouble. Proper design and selection of materials are important. Supersonic aircraft require

* *Materials Performance*, **3** (Ap. 1964) contains several papers on corrosion of spacecraft and aircraft.

heat-resistant skin materials to combat aerodynamic heating. Microbiological corrosion of jet fuel tanks was a serious problem. In the early stages of the development of the jet engine, strength was the primary obstacle because of high centrifugal stresses. Bucket temperatures at that time were around 1450°F. Temperatures are increasing and as 2000°F is approached, high-temperature oxidation is also a predominating factor. New materials such as composites of metals and nonmetallics are needed for higher-temperature operation and higher speeds.

Periodic nondestructive testing is a must for all operating aircraft. Safety is a prime consideration.

Cohen* discusses guidelines for water ingress and drainage, alloys, adhesive bonding, lubricants, insulation, gaskets, coatings, sealers, and fasteners. He summarizes by stating, "It is estimated that 90% of the corrosion problems encountered in aerospace weapon handling would have been avoided if these simple rules had been followed."

The Space Shuttle has a layer of pure silica tiles on outer surfaces for thermal insulation to prevent burnup during reentry. Sky Lab had a cooling system with aluminum tubing joined to 304 stainless tubing with a nickel gasket and a pressure fitting. The 304 tubing was thicker and, of course heavier, so weight was not a consideration. Nobody would explain to me why this two-metal combination was selected when they asked me about this a few weeks before liftoff! Fortunately the coolant was not very corrosive, so failure did not occur. Dissimilar metals in contact do *not* always show two-metal corrosion. Vanguard had a stress-corrosion-susceptible alloy wall separating fuel and oxidizer in the *same* tank! They asked my advice but I was told it was too late to change—the first Sputnik had been launched!

8-9 Petroleum Industry

The petroleum industry contains a wide variety of corrosive environments. Some of these have been described elsewhere in the book, whereas many are unique to this industry. Thus it is convenient to group all these environments together. Corrosion problems occur in the petroleum industry in at least three general areas: (1) production, (2) transportation and storage, and (3) refinery operations.

Production Oil and gas fields consume a tremendous amount of iron and steel pipe, tubing, casings, pumps, valves, and sucker rods. Leaks cause loss of oil and gas and also permit infiltration of water and silt, thus increasing corrosion damage. Saline water and sulfides are often present in oil and gas wells. Corrosion in wells occurs inside and outside the casing. Surface

*Bennie Cohen, Designing to Prevent Aerospace Weapon Corrosion, *Materials Performance*, **15**:23–25 (Sept. 1976).

equipment is subject to atmospheric corrosion. In secondary recovery operations, water is pumped into the well to force up the oil.

Condensate wells Condensate wells handle fluids (gas containing dissolved hydrocarbons) at pressures up to 10,000 lb/in.^{2}. Depths run up to 15,000 ft.

Carbon dioxide is the chief corrosive agent, with organic acids contributing to the attack. Approximately 90% of the corrosive condensate wells encounter conditions as follows: (1) depth greater than 5000 ft, (2) bottom hole temperature above 160°F and pressure above 1500 lb/in.^{2}, (3) a carbon dioxide partial pressure above 15 lb/in.^{2}, and (4) a wellhead pH of less than 5.4.

Corrosion characteristics of a well are determined by (1) inspection of surface equipment, (2) analysis for carbon dioxide, organic acid, and iron, (3) coupon exposure tests, and (4) tubing-caliper surveys. Determination of iron content and tubing-caliper surveys are used to measure the effectiveness of inhibitor treatment.

Earlier practices involved addition of neutralizers such as ammonia, sodium carbonate, sodium hydroxide, and sodium silicate, but these were replaced in many cases by organic inhibitors, available in oil-soluble, water-dispersible, or water-soluble forms. See Chap. 6 for further information about inhibitors.

In some applications, alloy steels have replaced the medium-carbon manganese steels (J-55 and N-80) previously used. Table 8-9 shows the effect of chromium and nickel on corrosion of steel by condensate-well fluid. Straight chromium stainless steels, Stellite, Monel, and copper-base alloys are commonly used for valves and other wellhead parts. Galvanic corrosion is apparently not a factor because substantial amounts of high-conductivity water are not present.

Sweet oil wells It appears that corrosion in high-pressure flowing wells that produce pipeline oil has become almost commonplace in many areas. Three methods are used to combat this corrosion—coated tubing, inhibitors, and

Table 8-9 Effect of chromium and nickel additions on corrosion of steel by condensate-well fluid

% Ni	Corrosion, mpy	% Cr	Corrosion, mpy
0	36	0	36
3	4	2.25	50
5	3	5	21
9	2	9	1.5
		12	0

alloys. Coated tubing has found most favor, and until recently, baked-on phenolics have been used for almost all coating installations. Air-dried and baked epoxy resins are now being used in increasing amounts. Coatings have been discussed in Chap. 6.

Sour oil wells These wells handle oil with higher sulfur contents than sweet wells and represent a more corrosive environment. In high-H_2S wells there may be severe attack on the casing in the upper part of the well where the space is filled with gas. Water vapor condenses in this area and picks up H_2S and CO_2.

Corrosion is reduced by inhibitors which are injected continuously or periodically depending on the well corrosivity.

Offshore drilling Offshore drilling presents many interesting corrosion problems. Platforms are built over the water and supported by beam piles driven into the ocean floor. Each beam is surrounded by a pipe casing for protection. Similar platforms are used far out at sea for radar towers.

A variety of corrosion prevention methods are used in such structures. These include:

1. Adding inhibitors to the stagnant seawater between beams and casings.
2. Cathodic protection, with sacrificial anodes or impressed currents, of underwater structures.
3. Paints and other organic coatings to protect exposed structures above the splash zone.
4. Monel sheathing at the casing splash zone. This portion of offshore structures is the most susceptible to rapid corrosion. (See Fig. 8-1.)

Transportation and storage Petroleum products are transported by tankers, pipelines, railway tank cars, and tank trucks. The outside submerged surfaces of tanks and the outside surface of underground pipelines are protected with coatings and by using cathodic protection. Cathodic protection is also applied to the inside of tankers to prevent corrosion by seawater used for washing or ballast. Gasoline-carrying tankers present a more severe internal corrosion problem than oil tanks because the gasoline keeps the metal too clean. Oil leaves a film that affords some protection. Tank cars and tank trucks are coated on the outside for atmospheric corrosion.

Internal corrosion of storage tanks is due chiefly to water, which settles and remains on the bottom. Coatings and cathodic protection are used. Alkaline sodium chromate (or sodium nitrate) has been found to be an effective inhibitor for corrosion of domestic fuel oil tanks.

Internal corrosion of product pipelines can be controlled with coatings and inhibitors (a few parts per million) such as amines and nitrites. Ingenious methods for coating pipelines *in place* underground have also been developed.

Refinery operations Most of the corrosion difficulties in refineries are due to inorganics such as water, H_2S, CO_2, sulfuric acid, and sodium chloride, and not to the organics themselves. For this reason, the petroleum industry has much in common with the chemical industry.

Corrosive agents may be classified into two general categories: (1) those present in feedstock or crude oil, and (2) those associated with processes or control.

Water is usually present in crude oils, and complete removal is difficult. Water acts as an electrolyte and causes corrosion. It also tends to hydrolyze other materials, particularly chlorides, and thus forms an acidic environment.

Carbon dioxide has, in recent years, come to be recognized as one of the most important corrosive agents, especially in operations where gas is the feedstock, or raw material. Many gas wells produce large quantities of carbon dioxide.

Salt water is produced in most oil wells, and relatively large quantities of it get into the refinery, either in the water emulsified in the crude or in the crystalline form dispersed in the crude. The salts are calcium chloride, magnesium chloride, and sodium chloride. Desalting methods include washing and settling, addition of chemicals such as sulfonates to break the emulsion, centrifuging, and filtering. Salts and water are usually removed as quickly as possible, but the operations are frequently incomplete. If they are not removed, or only partially removed, hydrochloric acid often forms. Magnesium chloride is readily hydrolyzed. In this case, ammonia may be needed in amounts equivalent to three times the stoichiometric equivalent of sulfide and chloride ions (see the discussion of ammonia below).

Hydrogen sulfide, mercaptans, and other sulfide compounds are present in many of the crudes and gases processed by refineries. These are removed by reaction with sodium hydroxide, lime, iron oxide, or sodium carbonate, but for various reasons they are frequently not removed until the final operation is approached. Corrosion problems are associated with the refining process itself or with processes utilized to remove sulfur compounds.

Nitrogen is becoming an important consideration in some of the newer processes. Nitrogen is present in some crudes, but a more important source is the nitrogen in air. Large quantities of air are used in some of the burning operations associated with catalytic cracking processes. Ammonia and cyanides will form under certain conditions when nitrogen is present. The former can damage heat exchangers made of copper-bearing alloys. Cyanides are an important factor controlling the diffusion of hydrogen into steel. Hydrogen-metal interactions are discussed in Chaps. 3 and 11.

Oxygen (or *air*) is drawn into tanks and other equipment as they are emptied, or enters during shutdown periods. It could also be drawn into the system by pumps. Oxygen can also be present as a result of reactions of other compounds, such as water and carbon dioxide. The water used in the system often contains oxygen in solution.

Sulfuric acid is used in large quantities in many refinery operations such as alkylation and polymerization. The acid becomes contaminated and its corrosion characteristics may change. Utilization of this acid and its recovery or concentration presents corrosion problems that are extremely important to the refinery. For example, sludges often contain large quantities of carbon or carbonaceous material which make the acid strongly reducing in nature. These may attack stainless steels, and under the same conditions the copper-base alloys will give better performance.

Ammonia is used to control the pH of water and to reduce chloride acidity in the process streams. This procedure works well if the pH is 7, but is damaging to copper-bearing alloys if the pH is 8 or above. Ammonia is frequently misapplied or improperly controlled. Ammonia is added to vapors in the process and also to condensers to neutralize acid condensate. It is desirable to add ammonia just before the aqueous phase forms.

Hydrochloric acid forms because of hydrolysis as described earlier. Sometimes it is an intentional addition to the process stream. This is a fairly volatile acid so it is often present in distillation columns and also in the condensed petroleum fractions. (Hydrofluoric acid is used in one alkylation process.)

Caustic (sodium hydroxide) and *lime* are sometimes added for hydrogen sulfide removal and for neutralization. Lime and caustic additions to the crude reduces the amount of HCl present in the overhead vapors. These chemicals are dispersed in oil before adding to the stream for better mixing. Less than the theoretical additions are made to avoid an excess of alkali. Caustic sometimes causes deposits (and clogging) that are difficult to remove. It also causes stress corrosion.

Naphthenic acid,* when present in oils, can be quite corrosive at 430 to 750°F, and type 316 stainless is sometimes required as a constructional material. Substantial amounts of this acid are present in some oils. For less severe conditions 5% Cr steel is satisfactory. Monel is used when temperatures are below 500°F.

Polythionic acid causes rapid intergranular SCC of *sensitized* austenitic stainless steels in some refinery operations. Type 304 is susceptible. This attack is minimized if properly heat-treated (not sensitized) 304 or the low-carbon or stabilized types are used.

Refinery corrosion is sometimes separated into two classifications: (1) low-temperature corrosion and (2) high-temperature corrosion. The dividing point is usually 500°F. Presumably, water can exist below 500°F, and the mechanisms of aqueous corrosion apply. The high-temperature mechanisms take over above 500°F. Perhaps another reason for the division at 500°F is that ordinary carbon steel is economical for handling most crudes and naphthas up to this temperature, but alloy steels and other

*W. A. Derungs, Naphthenic Acid Corrosion—An Old Enemy of the Petroleum Industry, *Corrosion*, **12**:617t–622t (1956).

materials must be used at higher temperatures. This is a general classification and should not be regarded as a strict division.

Such a classification is not entirely satisfactory, even if it applies directly for actual operating conditions at temperature. For example, high-temperature equipment is generally affected by water and other condensates that form when the equipment is shut down, when it is purged with steam or water, or when it is started up again. Many fail to recognize the effect of the conditions that exist when the equipment is not in operation—not only in refineries but in many other process industries as well.

Alloys used in refinery operations Ordinary carbon steel is by far the most important alloy, since it accounts for over 98% of the construction materials used in the industry. As a general rule, every attempt should be made to use steel. This can be done by modifying the process in some manner such as lowering the temperature or adding inhibitors. Steel is the least expensive engineering metal aside from cast iron. In some cases, alloy steels are more economical because they have a longer service life, and they should be judiciously selected, where applicable.

Carbon steel is often unsuitable for heat-exchanger tubes because of corrosion by the cooling water. Brass, arsenical Admiralty Metal, red brass, and cupronickels are widely used. Austenitic stainless steels are expensive and may crack in chloride-containing waters. These steels, however, are used for tubing in stills and gas-cracking tubes. In some cases, a single tower is lined with two or three different materials to take care of the changing corrosiveness from the top to the bottom of the tower.

Corrosion by sour crudes increases with temperature (increases rapidly around 800°F) and with increasing sulfur content. Chromium is the most beneficial alloying element in steel for resistance to sulfur compounds. Accordingly, the chromium content of steel is increased with increasing sulfur and temperature starting as low as 1% Cr. Experience indicates that $2\frac{1}{4}$% Cr, 1% Mo steel is generally adequate for less than 0.2% H_2S in the gas stream. High sulfide contents require 5% Cr or higher. The Cr-Mo steel mentioned above and 4 to 6% Cr, 0.5% Mo steels are widely used in refineries. The corrosion characteristics of these materials are shown in Chap. 11.

In addition to the "naturally occurring" carbon dioxide described above, some severe problems have been encountered because of CO_2 injection or flooding to enhance recovery of oil. This problem is discussed by R. S. Treseder.* He states that two basic components of the mechanism are consistent with actual experience: (1) The main cathodic reaction is the reduction of undissociated carbonic acid or hydrogen ion, and (2) the expected high corrosion rates from the latter reaction are not achieved in

*R. S. Treseder, "Some Comments on Proposed Mechanisms for Carbon Dioxide Corrosion," NACE Paper 284, CORROSION/84, New Orleans Annual Meeting of NACE.

many systems because of the inhibiting effect of ferrous carbonate scale. He also states the severe corrosion in amine-gas-treating systems occurs because the cathodic reaction involves carbonic acid, which comes from thermal decomposition of bicarbonate ion on heating surfaces.

Kane and Boyd (referenced below) suggest use of high-chromium steels and also high alloys such as Incoloy 825 where severe corrosion of carbon steel is observed. They also recommend coatings and inhibitors.

T. Murata et al.* describe a criterion for evaluating and predicting the sour resistance of high alloys through multiple regression analysis of laboratory data with the aid of a computer. *Sour resistance* (SR) value consists of major alloy components and their coefficients, which are a function of given environmental parameters. Useful relationships with corrosion behavior in brine solutions containing H_2S and CO_2 up to 250°C are shown.

Suggested Readings

Graff, W. J.: *Introduction to Offshore Structures: Design Fabrication and Installation*, Gulf Publishing Company, Houston, Tex., 1981.

Kane, R. D., and W. K. Boyd: Materials Technology for Oil and Gas Production, *Metal Progr.*, 54–65 (Feb. 1981).

Corrosion Control of Steel, Fixed Offshore Platforms Associated with Petroleum Production (NACE Standard RP-01-76, (1983 Revision), *Materials Performance*, 9–36 (May 1983). This is a *very good* publication and includes discussion of design, coatings, and cathodic protection.

Corrosion Control in Petroleum Production, NACE Book.

Special Issue on Corrosion Control in Refinery Operations, *Materials Performance* (Sept. 1983).

Special Issue on Offshore Corrosion, *Materials Performance* (May 1983). Highlights coatings and cathodic protection.

Special Issue on Pipeline Corrosion Protection, *Materials Performance* (July 1983).

8-10 Biological Corrosion

Biological corrosion is not a type of corrosion; it is the deterioration of a metal by corrosion processes that occur directly or indirectly as a result of the activity of living organisms. These organisms include micro forms such as bacteria and macro types such as algae and barnacles. Microscopic and macroscopic organisms have been observed to live and reproduce in mediums with pH values between 0 and 11, a temperatures between 30 and 180°F, and under pressures up to 15,000 lb/in.^{2}. Thus biological activity may influence corrosion in a variety of environments including soil, natural water and seawater, natural petroleum products, and oil emulsion-cutting fluids.

*T. Murata, R. Matsuhashi, M. Sakakibara and M. Onoyama, NACE Paper 208, *Corrosion*/84.

Living organisms are sutained by chemical reactions. That is, organisms ingest a reactant or food and eliminate waste products. These processes can affect corrosion behavior in the following ways:

1. By directly influencing anodic and cathodic reactions
2. By influencing protective surface films
3. By creating corrosive conditions
4. By producing deposits

These effects may occur singly or in combination, depending on the environment and the organism involved.

Microorganisms Usually, microorganisms are classified according to their ability to grow in the presence or absence of oxygen. Organisms that require oxygen in their metabolic processes are termed *aerobic*; they grow only in nutrient mediums containing dissolved oxygen. Other organisms, called *anaerobic*, grow most favorably in environments containing little or no oxygen.

Although the acceleration of corrosion by microbiological organisms is quite widespread, there has been relatively little detailed research concerned with the identification of these species and the precise mechanism involved. Below, some of the more important and more completely studied microorganisms are discussed, together with other, lesser known types.

Anaerobic bacteria Probably the most important anaerobic bacteria that influence the corrosion behavior of buried steel structures are the sulfate-reducing types (*D. desulfuricans*). These reduce sulfate to sulfide according to the following schematic equation:

$$SO_4^{2-} + 4H_2 \rightarrow S^{2-} + 4H_2O \qquad (8.1)$$

The source of hydrogen shown in the above equation can be that evolved during the corrosion reaction or that derived from cellulose, sugars, or other organic products present in the soil.

Sulfate-reducing bacteria are most prevalent under anaerobic conditions, as in wet clay, boggy soils, and marshes. The presence of sulfide ion markedly influences both the cathodic and anodic reactions occurring on iron surfaces. Sulfide tends to retard cathodic reactions, particularly hydrogen evolution, and accelerates anodic dissolution. Under most conditions, the acceleration of dissolution is the most pronounced effect, which causes increased corrosion. As suggested by Eq. (8.1), the corrosion product in the presence of sulfate-reducing bacteria is iron sulfide, which precipitates when ferrous and sulfide ions are in contact.

Aerobic bacteria Aerobic sulfur-oxidizing bacteria, such as thiobaccillus thiooxidans, are capable of oxidizing elemental sulfur or sulfur-bearing compounds to sulfuric acid according to the following equation:

$$2S + 3O_2 + 2H_2O \rightarrow 2H_2SO_4 \tag{8.2}$$

These organisms thrive best in environments at low pH and can produce localized sulfuric acid concentrations up to 5% by weight. Thus, sulfur-oxidizing bacteria are capable of creating extremely corrosive conditions. These organisms require sulfur in either elemental or combined form for their existence and are therefore found frequently in sulfur fields, in oil fields, and in and about sewage disposal piping that contains sulfur-bearing organic waste products. In the case of sewage lines, sulfuric-oxidizing bacteria cause rapid acid attack of cement piping.

Sulfate-reducing and sulfur-oxidizing bacteria can operate in a cyclic fashion when soil conditions change. That is, sulfate-reducing bacteria grow rapidly during rainy seasons when the soil is wet and air is excluded, and sulfur-oxidizing bacteria grow rapidly during dry seasons when air permeates the soil. In certain areas, this cyclic effect causes extensive corrosion damage of buried steel pipelines. Also, it is evident that the presence of micro-organisms can accentuate conditions of differential aeration in soils.

Other microorganisms There are various other microorganisms that directly or indirectly influence the corrosion behavior of metals and but have not been studied in great detail. For example, several types of bacteria utilize hydro-carbons and can damage asphaltic pipe coatings. Iron bacteria are a group of microorganisms that assimilate ferrous iron from solution and precipitate it as ferrous or ferric hydroxide in sheets surrounding their cell walls. The growth of iron bacteria frequently results in tubercles on steel surfaces and tends to produce crevice attack. Certain bacteria are capable of oxidizing ammonia to nitric acid. Dilute nitric acid corrodes iron and most other metals. However, in most soils the amount of available ammonia is not high enough to cause an appreciable accumulation of nitric acid. However, these kinds of bacteria may be important where extensive use of synthetic ammonia fertilizers has been employed on cultivated fields above buried pipelines. Finally, most bacteria also produce carbon dioxide, which can contribute to the formation of carbonic acid and increase corrosivity.

Some of the more important microorganisms and their characteristics are listed in Table 8-10.

Prevention of microbiological corrosion It is important to correctly diagnose the presence of microbiological corrosion before applying corrective measures. The most direct and accurate method of identification is accom-plished by culturing samples of soil and examining them for evidence of microorganisms. In the case of sulfate-reducting bacteria, the existence

Table 8-10 Physiological properties of microorganisms

Group and type	Oxygen requirement	Soil components reduced or oxidized	Major end products	Habitat	Approximate optimum reaction pH	Temperature limits, °C
I. Sulfate reducing (Desulfovibriodesulfuricans)	Anaerobic	Sulfates, thiosulfates, sulfites, sulfur, hyposulfides	Hydrogen sulfide	Water, muds, sewage, oil wells, soils, bottom deposits, concrete	Optimum: 6.–7.5 Limits: 5.–9.0	Optimum: 25–30 Maximum: 55–65
II. Sulfur oxiding (Thiobacillus thioxidans)	Aerobic	Sulfur, sulfides, thiosulfates	Sulfuric acid	Soil composts, sulfur and rock phosphates, soils containing incompletely oxidized sulfur compounds	Optimum: 2.0–4.0 Limits: 0.5–6.0	Optimum: 28–30 18–37 slow growth
III. Thiosulfate oxidizing (Thiobacillus thioparus)	Aerobic	Thiosulfates, sulfur	Thiosulfate to sulfate and sulfur Sulfur to sulfate	Widely distributed, sea and river water, mud, sewage, soil	Optimum: Close to neutral Limits: 7.0–9.0	Optimum: 30
IV. Iron bacteria (Crenothrix and leptothrix)	Aerobic	Ferrous carbonate, ferrous bicarbonate, manganese bicarbonate	Ferric hydroxide	Stagnant and running water containing iron salts and organic matter		Optimum: 24 Limits: 5–40

Source: F. E. Kulman, *Corrosion*, **9**:11 (1953).

of sulfide corrosion product on buried steel structures is usually a strong indication of biological activity. However, the presence of sulfide corrosion product is not always the result of sulfate-reducing bacteria.

There are several general techniques for preventing microbiological corrosion. Coating the buried structure with asphalt, enamel, plastic tape, or concrete is frequently used to prevent contact between the steel structure and the environment. All of these techniques have been used with success. Concrete is less satisfactory in the presence of sulfur-oxidizing bacteria because it is also rapidly attacked by the sulfuric acid environment. Cathodic protection has also been used to prevent microbiological corrosion and is especially effective when used with coatings. In some cases it is possible to alter the environment and thereby reduce the effects of microbiological corrosion. For example, sulfur and sulfur-containing compounds can be frequently removed by aeration of sewage. Also, corrosion inhibitors can be added, and germicides such as chlorine and chlorinated compounds can be employed in recirculating systems. In some cases it is possible to avoid wet, boggy soils in constructing pipeline networks. The use of substitute materials such as asbestos and plastic pipe in place of steel pipe has also been used as an effective means of preventing the detrimental effects of microbiological activity in certain undesirable soil locations.

MACROORGANISMS

Fungus and mold Actually, fungus and mold are the same inasmuch as both terms refer to a group of plants characterized by their lack of chlorophyll. These species assimilate organic matter and produce considerable quantities of organic acids including oxalic, lactic, acetic, and citric acids. Fungi are capable of growing on a variety of substrates and are a particularly troublesome problem, especially in tropical areas. The most familiar type of attack of this kind is the mildewing of leather and other fabrics. In addition, fungi can attack rubber and bare and coated metal surfaces. In many instances the presence of fungi does not cause severe mechanical damage but affects the appearance of the product, which is undesirable. In addition to producing organic acids, fungi can also initiate crevice attack of metal surfaces.

Mold growth on coated and uncoated metal surfaces can be prevented or reduced by periodic cleaning. Also, reducing the relative humidity during storage and the employment of toxic organic agents (e.g., gentian violet) have also been found effective in reducing mold growth on metal surfaces. Mold growth on rubber is particularly troublesome in underground cables, since localized perforation of the rubber coating results in electrical leakage. The substitution of synthetic instead of natural rubber has been found to be an effective method for preventing this kind of failure.

Aqueous organisms Both fresh water and salt water can sustain thousands of types of animal and plant life, including barnacles, mussels, algae, and others. These animal and plant forms attach themselves to solid surfaces during their growth cycle. The accumulation of these organisms causes crevice corrosion and, probably more important, fouling of structures. Ship bottoms rapidly accumulate barnacles and other organisms that markedly affect the streamlining and increase power requirements. A severely fouled ship may require as much as 30% more power during its operation. Similarly, the accumulation of macroorganisms in heat exchangers and other such devices severely limits heat transfer and fluid flow and may result in complete obstruction.

The accumulation of aqueous macroorganisms is a function of environmental conditions. The most severe problem occurs in relatively shallow water, since in deeper water there are no surfaces to which the organisms may adhere. Thus, harbor conditions are specially conducive to the formation of deposits on ship hulls. In general, warm temperatures favor long breeding seasons and rapid multiplication of macroorganisms such as barnacles and mussels. In northern seawater, fouling generally occurs only in summer months, whereas in southern tropical waters, fouling is practically continuous. Relative motion between an object and water generally tends to inhibit the attachment of organisms. Thus, rapidly moving vessels accumulate only small quantities of organisms, and the major portion of fouling occurs when the vessel is docked. The same effect is also noted in heat exchangers employing seawater as a coolant. Rapid fluid flow tends to suppress fouling of heat exchangers, whereas rapid accumulation occurs at low fluid rates or during shutdown periods. Also, the nature of the surface strongly influences the attachment of macroorganisms. Smooth, hard surfaces offer an excellent point for adhesion, whereas rough, flaking surfaces tend to inhibit adhesion. For example, the fouling of stainless steel and iron occurs initially at about the same rate in seawater. However, after some exposure, the surface of the iron is covered by a loosely adhering iron oxide and fouling tends to be less on iron than on stainless steel after long exposure periods.

Fouling by organisms is most effectively inhibited by the use of antifouling paints. These paints contain toxic substances, usually copper compounds. They function by slowly releasing copper ions into the aqueous environment, which poisons the growth of barnacles and other creatures. A similar technique is used in closed systems where various toxic agents and algaecides such as chlorine and chlorine-containing compounds are added to the environment. These methods work more or less successfully depending on their application. However, under conditions conducive to the growth of aqueous organisms, periodic cleaning is almost always necessary to ensure unimpeded fluid flow and the absence of crevice attack.

A most comprehensive report is MTI Publication No. 13, titled *Microbiologically Influenced Corrosion: A State of the Art Review* (1984). The mission of this study was to survey the available literature on MIC for steel, stainless steels, copper alloys, aluminum and its alloys, nickel-chromium alloys and titanium—also mechanisms of MIC, methods of prevention, methods for detection in the laboratory and the field, and a list of priority areas for research. In many instances, particularly case histories, documentation is not good because a multidisciplinary study (i.e., microbiologist and corrosion engineer) was not involved.

This MTI publication contains an excellent glossary and list of organisms to bring the reader up to date, and also an excellent literature review (103 references). Mechanisms, methods for detection (including field kit), sampling, case histories (e.g., aluminum jet fuel tanks) and methods for prevention are discussed. Anyone interested in delving into MIC would save much time and effort by reading the MTI report.

One of my experiences with MIC involved stainless steel tanks that were hydrotested and then let stand for several months before operation. Tap water was used to test and the tanks were not drained. Severe pitting of the 304L occurred. This could have been prevented by flushing with good-quality water and then blowing dry.

Puckorius* describes perforation of a utility condenser during a six-week shutdown (28,000 tubes, 90 ft long). He concludes that failure of the 304 stainless steel resulted because of sulfide-producing bacteria and because the tubes were not completely drained. Iverson† describes "a new mechanism involving the production of a highly corrosive metabolite containing phosphorus."

8-11 Human Body†

For more than a hundred years, foreign materials have been routinely implanted in the mouth for dental treatment. These include silver amalgams, gold, cements, porcelain, and more recently, stainless steels and plastics. As the art and science of medicine has progressed, the applications of implants have increased rapidly. Today, screws, plates, and rods are used to repair severe fractures; cosmetic surgery utilizes liquid and solid polymers; pulse rate, blood pressure, and bladder function in diseased patients are controlled by internal electric devices; reliable contraception has been achieved by the implantation of objects in the uterus; and defective heart valves have been replaced by artificial ball-check valves. Experiments with animals and limited clinical tests with humans indicate that functioning artificial hearts (and other organs) are possible.

As would be expected, there are few corrosion problems associated with

*P. R. Puckorius, Massive Utility Condenser Failure Caused by Sulfide Producing Bacteria, *Materials Performance* **22**(12), 19–22 (Dec. 1983).

†W. P. Iverson, Mechanism of Anaerobic Corrosion of Steel by Sulfate Reducing Bacteria, *Materials Performance* **23**(3), 28–30 (Mar. 1984).

implanted plastics and ceramics. Teflon, Dacron, nylon, and silicone polymers show little degradation after long-term implantation and have been successfully used in many applications. There are other problems (e.g., blood clotting), but these are outside the realm of corrosion.

Metallic components are generally used for orthopedic applications, since high-strength components are needed for bone repair and replacement. The localized stresses encountered in human bone are surprising. For example, it is not uncommon to observe plastic deformation in bone plates possessing yield strengths of 40,000 lb/in.^{2}. In addition, most orthopedic implants are subjected to high cyclic stresses.

Body fluid consists of an aerated solution containing approximately 1% sodium chloride, together with minor amounts of other salts and organic compounds at 98 to 99°F. Thus, the corrosivity of the human body is similar to that exhibited by warm, aerated seawater. As discussed in Sec. 8-4, seawater tends to cause localized corrosion, including crevice attack, pitting, and galvanic corrosion.

Figures 8-8 and 8-9 show a nail-plate assembly after a four-year exposure in a human. This device is used to fix fractures of the femur neck (hip). In practice, the nail and plate are connected, with a locking bolt. As shown in these figures, intense crevice attack has occurred at this junction point. A plate from a similar device, removed after two years, shows evidence of pitting (Fig. 8-10). The pit has grown out of the viewing plane and reappeared at another point. As discussed in Chap. 3, pits grow in the direction of

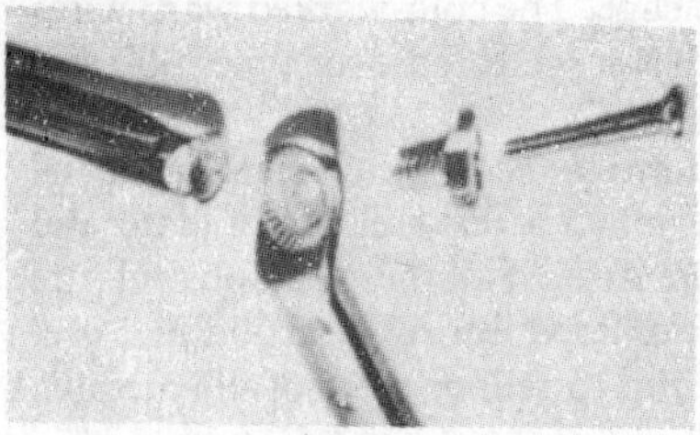

Figure 8-8 Thornton nail and plate disassembled after a four-year exposure. Nail on left: plate and locking bolt assembly on right. Material: type 316 stainless.

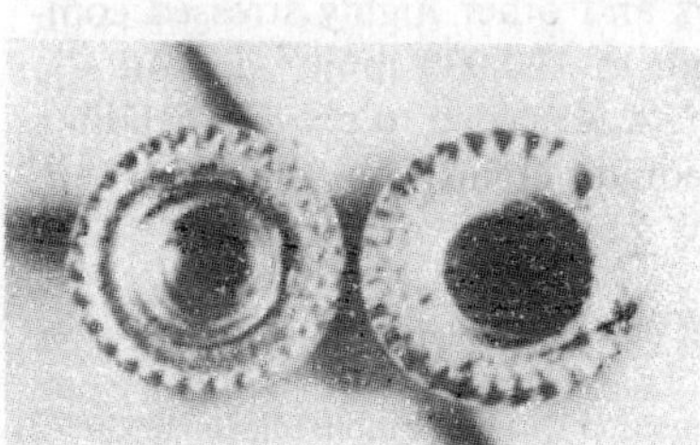

Figure 8-9 Enlarged view of Thornton nail shown in Fig. 8-8, together with an unused specimen (2 ×).

Figure 8-10 Cross section of a Thornton plate after a two-year exposure (30 ×). Material: type 316.

gravity, and this tunneling effect is probably due to changes in sample orientation (i.e., the horizontal-vertical sequence associated with sleeping-waking activities). The undercutting of this pit is characteristic of pits formed in seawater and other chloride-containing solutions.

None of the devices shown above failed mechanically due to corrosion damage. Rather, they were removed because of tissue irritation or infection produced by the products of corrosion. Compatibility, or the absence of body reaction, is one of the major problems associated with metal implants. Qualititive estimates suggest that implants must corrode at rates of 0.01 mpy or less to avoid the possibility of tissue reaction.

Another interesting example* of failure of a prosthetic device involves severe pitting of a type 304 wire-mesh rib cage. The cold-worked warp wires are badly attacked and ruptured. Type 316 is preferred over types 302 and 304.

Figure 8-11 shows some orthopedic implants. A skull plate, artificial elbows, a total hip replacement, a knee joint, a femoral screw plate and intermedullary rods (at bottom) are visible. Almost all these components are metallic because of the high stresses encountered during insertion and use. Some total hip joints use a polyethylene cup to reduce wear of the metallic ball.

New information about the performance of metallic implants has become available. Alloys for bone screws and other highly stressed components require a minimum yield strength of 120,000 lb/in.^{2} and 10+% elongation.† High yield (and tensile) strength is needed to prevent bending, breaking, and fatigue fracture, while adequate elongation is necessary to avoid brittleness. Two alloys, Ti-6A1-4V (see Table 5-10) and MP-35N

*E. H. Bucknall, Corrosion of a Stainless Steel Prosthetic Device, *Materials Protection*, 4:56–62 (1965).

†F. C. Sessa and N. D. Greene, Mechanical Evaluation Tests of Orthopaedic Bone Screws, *Eighth Annual International Biomaterials Symposium*, Philadelphia, Pa. (Apr. 1976).

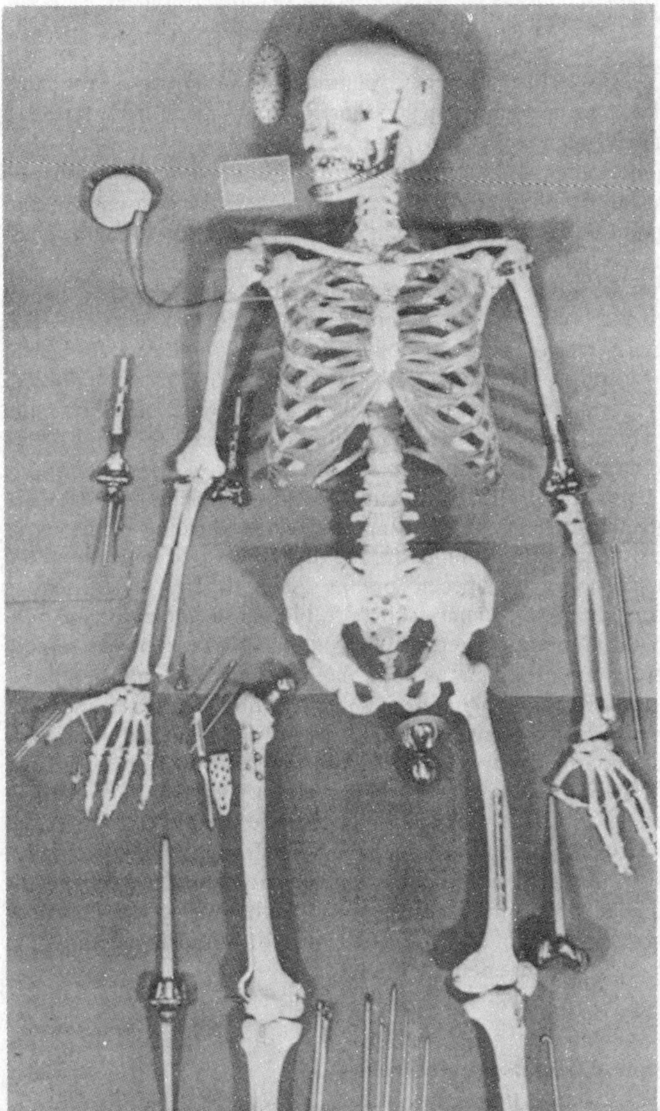

Figure 8-11 Orthopedic implants.

(20 Cr, 35 Ni, 35 Co, 10 Mo) meet these requirements. Bone screws fabricated from these alloys are mechanically superior to screws made from type 316 stainless steel or Vitallium (30 Cr, 5 Mo, 65 Co).

In addition to adequate mechanical properties, implant alloys must be corrosion-resistant in body fluids. The corrosion rates of most current and potential implant alloys are so low that loss of mechanical strength is not an important consideration. Corrosion rates have only biological significance. That is, corrosion releases foreign substances (e.g., corrosion products) into the body that may interfere with its normal functions. If a material and its corrosion products do not significantly influence biological processes it is said to be biologically compatible or "biocompatible." Compatibility can be described in terms of local and systemic effects—the influence on adjacent tissues vs. remote effects produced by transport of corrosion products away from the implant site.

Metallic corrosion products are capable of suppressing cells that are critical to infection-defense mechanisms and wound-healing reactions of local tissues. Also, several metal ions are rapidly transported throughout the body, and some collect in specific organs. Even though an element is eliminated from the body, it may still pose biological dangers. Nickel, a common implant alloying element, is an example. Nickel ions are rapidly and completely eliminated from the body via urine and feces, but the steady-state plasma level remains high. In view of the known carcinogenicity of nickel, this introduces a potential risk of cancer for patients with long-term implants. This and other properties of various implant alloys are compared in Table 8-11. It is obvious that extreme caution should be exercised in choosing an alloy for surgical-implant applications, especially those intended for long-term service. No alloy system appears free from potential biological hazards. Future research in this area will require careful assessment of the biological effects of various metallic ions together with a knowledge of the in vivo corrosion rates of alloys (see Sec. 4-30 for further discussion of in vivo corrosion rate measurement methods).

8-12 Corrosion of Metals by Halogens

Chlorine is discussed in Sec. 7-26 and fluorine in Sec. 7-29. The following paragraphs are quoted from P. L. Daniel and R. A. Rapp, Halogen Corrosion of Metals, in *Advances in Corrosion Science and Technology*, vol. 5, pp. 55–172, Plenum Press, New York, 1976. The interested reader is referred to this excellent paper for detailed in-depth information including 266 literature references, corrosion data on metals and their alloy systems, compounds formed, properties of halides, and thermodynamic and kinetic data.

Table 8-11 Comparison of alloy systems for surgical-implant applications

Alloy system	Mechanical properties	Corrosion resistance	Effect of corrosion products on wound healing*	Effect of corrosion products on local infection*	Systemic distribution of corrosion products*
Ni-containing	Several alloys in each category possess sufficient strength for application as screws, plates, etc.	Several alloys in each category possess adequate corrosion resistance for application as surgical implant components	Moderate interference	None	Eliminated, but high plasma levels: possible long-term carcinogenic effect
Co-containing	↓	↓	Strong interference	None	Eliminated, but high plasma, liver, and kidney levels: possible long term cardiovascular and carcinogenic effects
Cr-containing	↓	↓	No effect	Markedly increased probability	Accumulates in all organs, especially bone. Long-term effects unknown
(Noble metals Pt, Au, Rh, etc.)	↓	↓	Very strong interference	Markedly increased probability	Unknown
Reactive metals (Ti, Ta, etc.)	↓	↓	Unknown (corrosion products insoluble)	Unknown (corrosion products insoluble	Unknown (corrosion products insoluble)

*Predictions based on data from:

N. D. Greene, C. Onkelinx, L. J. Richelle, and P. A. Ward, Engineering and Biological Studies of Metallic Implant Materials, *Biomaterials*, NBS Spec. Publ. 415, pp. 45–54 (May 1975).

P. A. Ward, P. Goldschmidt, and N. D. Greene, *J. Retic. Soc.*, **18**:313 (1975).

C. Onkelinx, "Pharmacokinetics of Nickel (II), Cobalt (II), and Chromium (III), in the Rat," Orthopaedic Research Society, San Francisco (Feb. 1975) (to be published).

In summary, boron, carbon, silicon, germanium, titanium, vanadium, and many of the second and third row transition metals form low-melting volatile halides: hence, these metals are particularly subject to halogen attack, even at low temperatures. It should be noted that, although readily attacked by dry chlorine, titanium is extensively used in contact with wet chlorine and may be fairly resistant to dry fluorine at room temperature.

Silver zirconium, and probably hafnium are fairly resistant to halogen attack at 25°C but not at temperatures above 100–150°C. Substantive data on halogenation of many other second and third row transition metals are lacking. A few of the more noble of these metals, for example, iridium, platinum, and gold, may resist fluorine attack at temperatures up to 100°C or higher.

The halides of the alkali and alkaline earth metals exhibit melting points ranging from 445°C for LiI to 1450°C for SrF_2 and their vapor pressures attain 10^{-4} atm only at temperatures of 600°C or higher. Consequently, in chlorine, bromine, and iodine at low temperatures these metals may form protective films. However, in fluorine, the alkali and alkaline earth metals, excluding beryllium and magnesium, form only porous (non-protective) fluoride scales.

Beryllium, magnesium, and aluminum form protective fluoride films which keep the first-year metal loss to less than 5 μm at temperatures as high as 200°C. Magnesium may also be resistant to other halogens, since its halides have high melting points and low vapor pressures. However, beryllium is likely to be less resistant to other halogens, since the vapor pressures of its halides attain values of 10^{-4} atm at only 225–270°C. Aluminum reacts readily with both chlorine and bromine.

At least at room temperature and a little above, zinc, cadmium, indium, thallium, tin, and lead are also potentially resistant to fluorine attack, since their fluorides have relatively high melting points and low vapor pressures. However, other halides of these metals have significantly lower melting points and higher vapor pressures; consequently, these metals are likely to be less resistant to attack by other halogens (chlorine, bromine, and iodine).

The first row transition metals, except titanium and vanadium, are generally resistant to halogen attack. Of these metals, iron and chromium are the least resistant while nickel and copper are the most resistant. In fluorine, potentially protective iron fluoride scale tends to spall from the metal; in other halogens, the iron halide scales have relatively high vapor pressures. Although there are some indications that manganese, chromium, and cobalt, at least at low temperatures, are also fairly resistant to halogen attack, there is no substantive evidence to this effect. Nickel and iron fluorination is said to occur by inward diffusion of fluorine anions. Such diffusion may also control the fluorination of cobalt, magnesium, manganese, and zinc, which form fluorides with the same tetragonal SnO_2-TiI_2 structure as that of NiF_2.

Little reliable information is available on the halogen resistance of alloys. The alloys most commonly used for handling halogens are those of the above-mentioned halogen-resistant metals, such as nickel, copper, iron, chromium, magnesium, aluminum, and silver. For example, Monel (Ni-Cu) is used for handling all the halogens, aluminum-magnesium alloys are commonly used for fluorine, and stainless steel (Fe-Ni-Cr) for chlorine.

A knowledge of the above-discussed metal and metal-halide scale properties and an understanding of corrosion mechanisms provide the basis for the designing and engineering of corrosion-resistant alloys. This idea of alloys design (used, for example, to develop alloys resistant to high-temperature oxidation and hot corrosion) rather than simple alloy evaluation has not generally been applied to develop halogen-resistant alloys. Such alloy design could be very productive.

Support information needed for this alloy development includes kinetic, morphological, and mechanistic data for halogenation of pure metals. Such data are also needed to give the materials engineer (1) a broader selection of pure metals for use in halogen environments, (2) the ability to forecast the long-term effect of changes in the halogen environment (for example, industrial process gas composition) on industrial processes and process equipment. Metals of particular interest for which data are particularly scarce include chromium, manganese, cobalt, aluminum, zinc, tin, and lead.

8-13 Corrosion of Automobiles

This is a severe economic problem, particularly in marine atmospheres (seacoasts) and northern climates where deicing salts (chlorides) are used on roadways. A badly rusted car loses much of its value. One simple procedure I use that is very effective involves obtaining touch-up paint when the car is brand new. Look for damaged coating (i.e., nicks) and touch up starting at day one. Do *not wait* until an actual rust spot appears. Frequent washing (to remove salt) and waxing the "chrome" is very helpful. "Rustproofing" when the car is brand new is a must to prevent inside-out corrosion (i.e., doors and rocker panels). In 1983 I traded in a seven-year old steel body car without a rust spot on it!

In December 1983, an article in the local press stated "Six hundred new pickup trucks worth nearly $5 million have been junked because they rusted while parked at a Tampa, Florida, port awaiting shipment for up to a year." These trucks (made in Japan) were crushed and sold for scrap (about $50 each).

Another way of avoiding car body corrosion is to use a plastic body. I have been driving Corvettes (glass fiber-polyester) since 1960. The rationale is that a rusty car would ruin my reputation!

A paper* in 1984 describes methods used that year by manufacturers. These include galvanized steel, electroplated steels, and complete immersion of the body in a rust-preventing primer coat.

Other preventive procedures are (1) use inhibitor in engine cooling system, (2) avoid very short runs, like moving the car from the driveway to the garage (water condenses in the exhaust system), and (3) keep the gasoline tank at least half full. This minimizes condensation of water when the temperature drops (at night). Water in the tank corrodes the steel, and corrosion products clog the fuel filter and carburetor jets.

8-14 Nuclear Waste Isolation

During December 1982, Congress passed landmark legislation that established the first national policy for burying nuclear wastes that have been accumulating for many years. Construction of the first actual facility will begin about 1990 and completed in 1996. Natural and engineered inner

*H. E. Chandler, "Update on Corrosion Control Coatings for Autos," *Metal Progr.*, 39–43 (Feb. 1984).

and outer barriers must be provided to contain radioactive materials and other contaminates from the environment (i.e., groundwater and the atmosphere) for 10,000 *years*. High-level wastes (i.e., from nuclear reactors) are of primary concern. This time period requires corrosion and materials engineers to "reverse" their thinking. A steam power turbine is designed to last 50 *years*, the early jet engines lasted 50 *hours*, and in some rocket applications 5 *seconds* was sufficient life. Corrosion is, of course, an important consideration.

Nuclear waste would be placed in repositories mined in salt, volcanic ash rock, granite, or in natural or man-made caverns. The package or isolation system must protect the waste from intrusions. A system under consideration consists of various barriers* as follows (looking at a cross section from above, it looks like a series of concentric circles):

1. *Waste form*. In the center, radioactive material encapsulated in glass to make the waste less accessible to intrusion and corrosion by water. This water would be hot (about 200°C) and would dissolve ordinary glasses. A titanate ceramic† is proposed because it displays a "high degree of stability under these conditions."
2. A metallic *canister* surrounds the waste and a *filler* (stabilizer) is packed in between the encapsulated waste and the canister. Titanium, or a titanium alloy, is a front-runner as the material of construction for the cannister.
3. Moving outward, there is a gap surrounded by an *overpack*, a second cannister, which also could be titanium.
4. Next is a gap and then the hole *sleeve*, which provides structural support and another corrosion barrier.
5. Next is backfill placed between the hole sleeve and the hole in the geologic medium (i.e., granite).

The canister, overpack, and hole sleeve are structural components and also provide corrosion resistance.

Not only technical but also political considerations are involved. The technical problems can be solved, but politics is another matter—nobody wants the waste in his or her "backyard."

"Backyard" has a different twist in another scheme‡ that seems quite feasible—namely, ocean burial. This consists of a projectile-type canister, containing vitrified high-level waste, that is dropped into the ocean; it penetrates the seabed and is covered with sediment. In an early experiment,

*A simplified diagram is shown in C. Klingsberg and J. Duguid, Isolating Radioactive Wastes, *Amer. Scient.*, 185 (Mar.–Apr. 1982).

†T. Ringwood, Immobilization of Radioactive Wastes in SYNROC, *Amer. Scient.*, 201–207 (Mar.–Apr. 1982).

‡Herb Short, Sea Burial of Radwaste: Still Drowned in Debate, *Chem. Eng.*, 14–18 (Mar. 5, 1984).

an ordinary steel drum was recovered after 15 years from the Atlantic Ocean (2800 meters depth). The steel was in relatively good condition, and a life of 25–37 years was predicted (thickness of steel not stated). Present thinking is to use a titanium alloy, TiCode 12, for the canister, which should last a very long time—100 years or much, much more.

8-15 Liquid Metals and Fused Salts

These materials are important in technology because of their superior heat-transfer characteristics. Historically, molten lead and fused salts have been used for heat treatment of steels. More recently, these materials have become important as heat-transfer mediums for power generation, particularly for nuclear power plants.

The efficiency of a heat engine or power plant is increased by operating at higher temperatures. Although water and water vapor are the most common heat-transfer fluids, they become rather difficult to handle at the extremely high temperatures and pressures encountered in modern conventional and nuclear power plants. High-pressure equipment is expensive and hazardous. Liquid metals and fused salts, because of their low boiling points and very high thermal conductivities, are much more useful in these circumstances. Some of the pertinent properties of fused salts and liquid metals for heat-transfer applications are:

1. *Low vapor pressure.* Mediums with high boiling points permit the use of thin heat-exchanger tubes, which aids thermal energy transfer. The higher the boiling point, the greater the ultimate temperatures that may be employed.
2. *Low melting point.* Salts and metals that melt above room temperature require some type of preheating prior to the start of a power plant. The higher the melting temperature, the greater the difficulty of plant operation and the greater the possibility of freeze-up during low-temperature operations.
3. *Low pumping power.* Pumping power is the energy required to pump a given volume of fused metal or salt. Hence, pumping power is a direct function of density; mercury requires very high pumping power, whereas sodium requires low pumping power.
4. *Low neutron cross section.* In nuclear applications it is important that the liquid metal or fused salt have a low absorption coefficient for neutrons, since these materials are circulated in the center of the reactor. Thus, in nuclear applications both the thermal and nuclear properties are important. The following liquid metals and fused salts have sufficiently low neutron absorption cross sections to be successfully used in nuclear power reactors: aluminum, bismuth, lead, magnesium, sodium, tin, zinc, sodium-potassium alloys, and fused sodium hydroxide.

Below, the corrosion characteristics of both fused salts and liquid metals are considered. It will be noted that there is less information available on fused salts, probably because their thermal properties are slightly inferior to those of liquid metals.

Liquid metals Liquid metals, like aqueous solutions, cause different *types* of corrosive attack. However, the interaction that occurs is not electrochemical in nature. That is, liquid-metal corrosion is primarily a physical rather than a chemical effect. It is the result of direct solution or solid-state interactions. The types of liquid-metal corrosion that have been observed are briefly tabulated below:*

1. Solution of the structural metal
   a. Uniform solution
   b. Intergranular solution
   c. Selective solution of a phase or component
2. Diffusion of liquid into solid metal
   a. Uniform diffusion
   b. Intergranular solution
3. Intermetallic compound formation
   a. Surface reaction
   b. Subsurface precipitation
4. Mass transfer
   a. Composition-gradient mass transfer
   b. Thermal-gradient mass transfer

Although all forms of liquid-metal attack are undesirable, there are several that require special attention. Mass transfer by thermal gradients (4 above) is very prevalent in heat-exchange systems because of thermal gradients. Solubilities of 1 or 2 ppm in the liquid metal are sufficient to cause deposition in cool areas, with subsequent restriction of fluid flow. Mass transfer is the most troublesome form of liquid-metal attack since it occurs in virtually all liquid-metal systems. The most dangerous form of attack is the formation of intermetallic compounds on or under the surface of structural material (3). In heat exchangers employing liquid metals, most of the resistance to thermal transfer occurs in the tube wall. As a consequence, large thermal gradients and large stresses exist in these areas. The existence of a brittle intermetallic compound on or in a structural component can cause cracking under these conditions and therefore should be avoided.

The physical nature of liquid-metal corrosion is most clearly demonstrated by comparing the solubility of the structural material with its corrosion resistance. In Fig. 8-12 such a comparison is made between a

*A. deS. Brasunas, *Corrosion*, **9**:78 (1953).

variety of pure metals and mercury as a solvent. As shown, there is a direct relationship between corrosion rate and solubility of the structural component in liquid mercury. Although Fig. 8-12 shows that iron is quite resistant to liquid mercury, in actual practice with moving mercury, it is necessary to have small amounts of titanium or magnesium present in the solute to prevent excessive attack.

Resistance of various metals and nonmetals to liquid-metal systems are summarized in Fig. 8-13. Consistent with the format used in this book, some additional comments on corrosion resistance are listed below and are organized under the corrosive rather than the structural material.

Sodium, potassium, and sodium-potassium alloys As shown in Fig. 8-13, a large number of materials can be used to successfully contain this group of

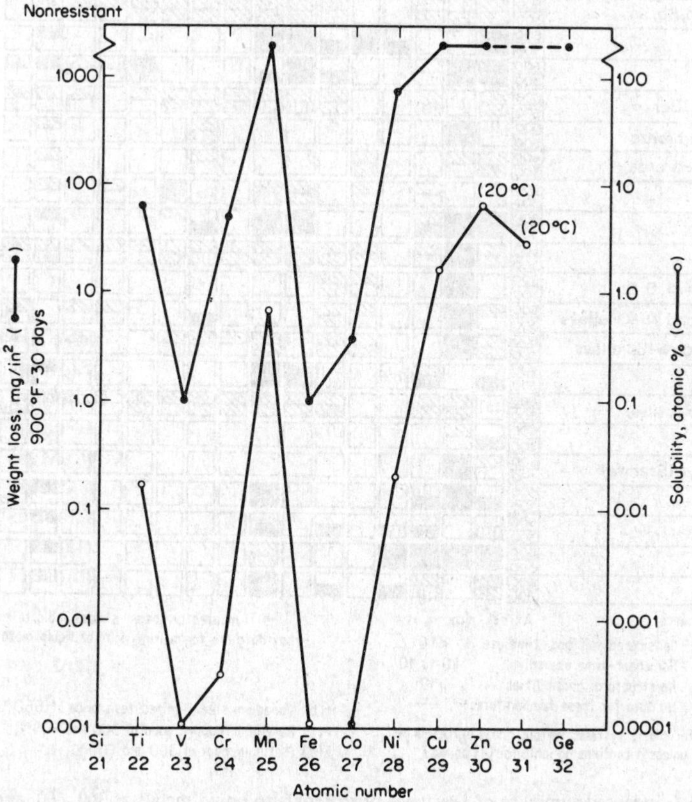

Figure 8-12 Comparison of corrosion resistance with solubility. [*J. F. Nejedlik and E. J. Vargo, Electrochem. Technol.*, **3**:*250 (1965)*.]

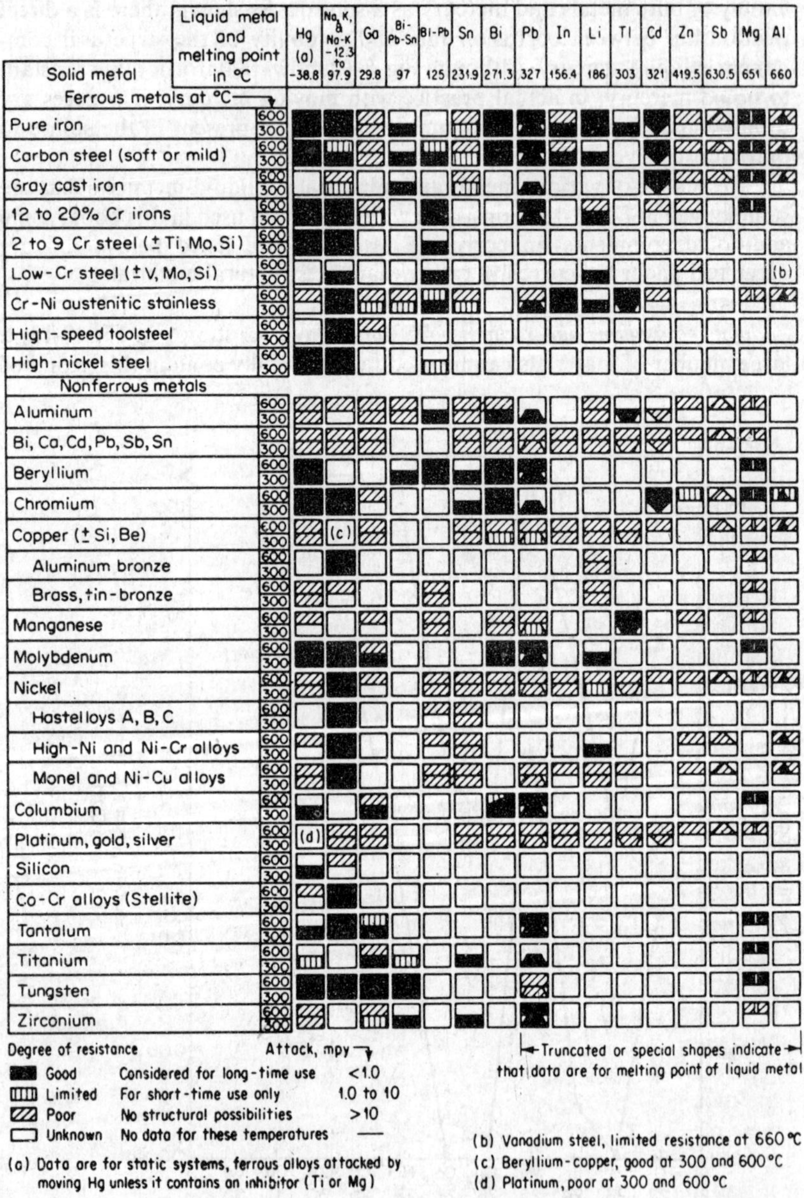

Figure 8-13 Condensed summary of resistance of materials to liquid metals at 300, 600, and 800°C. *(E. C. Miller, Liquid Metals Handbook, 2d ed., p. 144, Government Printing Office, Washington, D.C., 1952.)*

liquid metals and alloys. 18Cr-8Ni and higher alloy stainless steels afford good resistance to both liquid-metal and high-temperature oxidation. Nickel and nickel-base alloys such as Monel and the Hastelloys also show favorable resistance. Liquid sodium has no effect on the stress rupture strength, tensile strength, or fatigue life of 18-8 steels at 500°C. Also, liquid sodium does not produce stress-corrosion cracking of the stainless steels, nickel, and Monel at this temperature. These liquid metals and alloys rapidly oxidize on exposure to air, and the presence of oxides makes these metals very corrosive at high temperatures. Therefore, special precautions have to be taken to keep the oxygen level low. Inert-gas atmospheres are frequently employed for this purpose.

Lithium Lithum is, in general, more aggresive in its corrosive attack than either sodium or sodium-potassium alloys, as shown in Fig. 8-13.

Magnesium Steel and cast iron have been successfully used to contain magnesium at high temperatures in die-casting machines. High-chromium steels have similar resistance, while austenitic stainless steels are attacked by magnesium at its melting point.

Mercury Carbon steel has been successfully used to contain mercury in temperatures of 540°C. As noted previously, the presence of traces of titanium and magnesium in the mercury inhibits the corrosion of iron in ferrous-base materials. In recent years, 5% chromium steel and Si-Cr-Mo steels have replaced carbon steel because of their superior corrosion resistance and higher rupture strength.

Aluminum There are no known metals or alloys totally immune to attack by liquid aluminum. Above its melting point almost every metal is severely attacked, as shown in Fig. 8-13.

Gallium Although gallium has the widest liquidus range of any known metal, it is one of the most corrosive liquid metals. Only the refractory metals are capable of resisting attack by liquid gallium.

Lead, bismuth, tin, and their alloys Stress-corrosion cracking is common in materials that are exposed to these molten metals. Nickel-chromium steels and most other metals suffer rupture failures when stressed in contact with these mediums. Nickel, Monel, and cupronickel alloys possess the greatest resistance to stress cracking in these liquid metals. The cracking tendency of lead and other materials has been known for some time, since most steel and cast iron containers for lead baths usually fail due to cracking.

Fused salts Corrosion in fused salts has been studied less extensively than liquid metals. Fused salts apparently occupy a position somewhere between liquid metals and aqueous solutions. That is, physical solution of many metals in fused salts is possible, and also fused salts can act as a conducting electrolyte and thus sustain electrochemical reactions. Mass transport due to thermal gradients has been observed in numerous fused-salts systems, and electrochemical interactions with the formation of corrosion products has also been observed.

Sodium hydroxide is one of the more common fused salts used for heat-transfer applications. Nickel and nickel-base alloys have been observed to be the most resistant to this fused salt. Sodium hydroxide and also fused carbonates and chlorides have been employed in high-temperature fuel-cell applications, and again, nickel and nickel-base alloys have been found to be generally the most useful and economical. Some corrosion data for various fused salts in contact with nickel and other materials are presented in Table 8-12.

Testing Numerous methods have been employed for evaluating the suitability of structural materials for use in liquid-metal and fused-salt systems. Usually a static test is used for preliminary screening. In its simplest form, this test consists of sealing a liquid metal in a tube of the structural material to be investigated and heating it for several days at constant temperature. The material is then evaluated by visual examination, metal-

Table 8-12 Corrosion in sodium hydroxide at 800–815°C after 100-hr exposure (He atmosphere)

Composition				
Ni	Mo	Fe	Cr	Maximum depth of porosity,* mils
90	10	—		0.5
85	15	—		3.2
80	20	—		2.2
75	25	—		6.0
80	—	20		5.2
67	—	33		4.0
80	10	10		1.8
70	15	15		5.8
60	20	20		5.0
78	—	7	14 (Inconel)	9.0

*Corrosion selectively removes iron, molybdenum, and chromium, leaving a porous surface.

Source: G. P. Smith, M. E. Steidlitz, and E. E. Hoffman, *Corrosion*, **13**:561*t*, 627*t* (1957) and **13**:47*t* (1958).

lographic studies, weight change, and analysis of the liquid metal. In general, metallographic examination is the best method for evaluating liquid-metal attack since weight change does not have as much significance as it does in aqueous corrosion testing (e.g., intermetallic compound formation).

There are several types of dynamic tests commonly used for evaluating liquid-metal corrosion. The one which most closely approximates actual operating conditions is a convection-loop test. In this test, a piece of continuous tubing in the shape of a heart or the letter D is used, and heat is applied to one vertical leg and removed from the other by means of cooling coils. Thermal convection causes metal flow throughout the loop, and maximum velocities of 6 to 10 ft/min are possible. This type of test is most useful for determining the nature and extent of mass transport.

8-16 Solar Energy

The rays of the sun have been used throughout history for warmth and drying commodities. However, since the oil embargo of 1973, considerable attention is being given to the use of the sun to decrease dependence on petroleum. Systems vary from simple to sophisticated (i.e., electric power generation). There are thousands of small installations for heating domestic hot water. The container is usually galvanized steel, stainless steel, or glass-lined steel with a sacrificial anode mounted on a rooftop. Black plastic pipe or black iron pipe is laid on a roof and water pumped through it to a domestic water tank or, for example, a swimming pool.

The United States leads the world in photovoltaic power plant installations using single crystal technology.* Most of the cost of the plant is for solar cells. Total cost is $8000–12,000/kW for plants above 100 kW in size.

Solar ponds contain salts with depths of several (about 10) feet deep. Three distinct layers form. The top (upper convective zone), middle (nonconvective zone), and the bottom storage (lower convective) zone. Concentration of the salt is low at the surface and high at the bottom. Temperatures in the bottom zone vary from 120°F in winter to 200°F in summer. Heat is trapped in the bottom zone. The salt container could be pit-resistant stainless or a plastic-sheet liner for large ponds (i.e., one acre).

Another system under study consists of a large tower containing heliostats, which produces steam that is stored or piped to a turbine generator.† Another uses a molten salt (60% potassium nitrate, 40% sodium nitrate) producing steam at 1050°F. Inconel 800 is the preferred choice of material of construction.

Chem. Eng., 26 (Apr. 16, 1984).
†*Chem. Eng.*, 41 (May 31, 1982).

8-17 Geothermal Energy

In some areas of the world such as New Zealand, Italy, and California, high temperatures exist not far below the earth's surface and hot water is present. In Wairakei, New Zealand, the water is 400°F and above. This is tapped (like drilling for ordinary groundwater or oil); the wet steam is flashed and the dry steam pumped to an electric power plant. This has been done for more than 25 years at this site. I visited Wairakei in 1972 and experienced an eerie feeling standing on ground that shook *continuously*.

Corrosiveness of the geothermal fluids varies considerably. Most contain a variety of salts (i.e., chlorides), solids, and some contain substantial amounts of H_2S (sour). The latter often causes stress corrosion.* The usual corrosion problems such as those described in Chapter 3 and elsewhere in this book are encountered.

The Heber geothermal power plant under construction in Southern California will install 1.6 million ft (490 km) of 0.75 in. (19 mm) diameter, 29 Cr–4 Mo ferritic stainless steel in heat exchangers that extract heat from brine for conversion to electricity.†

Ellis et al. (Radian Corp., Austin, Texas) prepared three *comprehensive* reports on corrosion in geothermal systems. These are *Materials Selection Guidelines for Geothermal Energy Utilization Systems* in 1981, an addendum to this one in 1983, and *Corrosion Reference for Geothermal Downhole Materials Selection* in 1983. Descriptions of systems and a tremendous amount of corrosion data are presented for 20 geothermal plants in seven different countries in the free world.

8-18 Sewage and Plant-Waste Treatment

The effects of chloride ions on stress corrosion and pitting are discussed in Chaps. 3 and 6. Chlorides can cause failures of austenitic stainless steels but in some situations these materials tolerate high chloride levels. An example is treatment of municipal sewage sludge using the wet air-oxidation process. Briefly this involves adding air to the sludge under pressure and heating the mixture at high temperatures in the range of 350 to 600°F.

Figure 8-14 is based on over twelve years of successful actual plant experience with heat exchangers and reactors handling sewage by this process.‡ In recent years 316L has been used instead of 316 and 304. Over 100 municipalities use the wet air-oxidation process. Locations of some of

*A. R. Troiano and R. F. Hehemann. Hydrogen Sulfide Cracking in Materials for Geothermal Power. Materials Performance, **18**:31–39 (1979) (Contains a number of references on materials problems with geothermal wells.)

†Metal Progr., 18 (July 1984).

‡Up-dated chart, old chart appeared in T. P. Oettinger and M. G. Fontana, *Materials Performance*, **15**:29–35 (Nov. 1976).

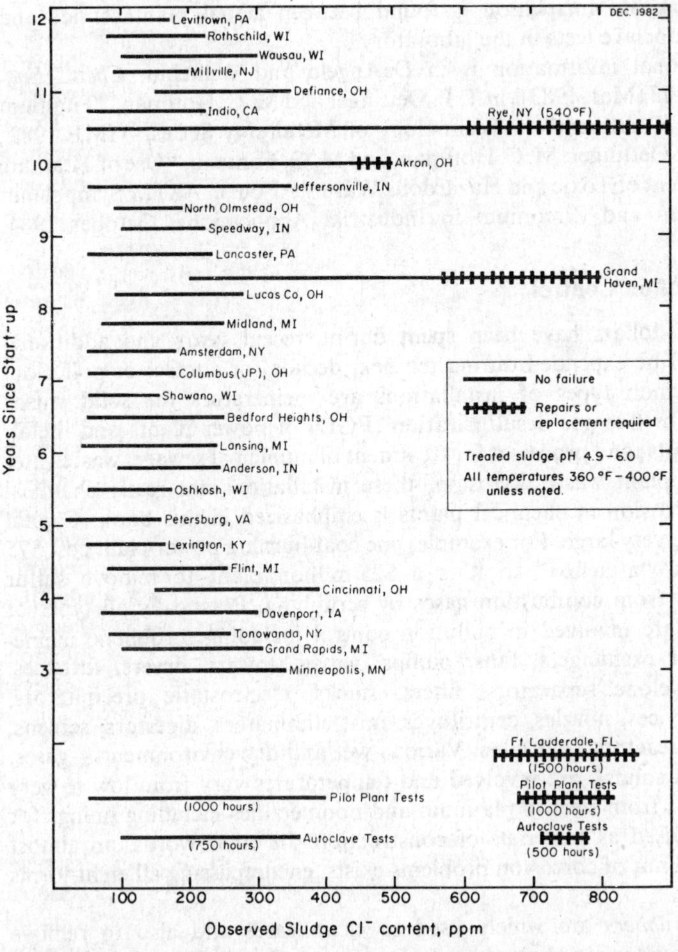

Figure 8-14 Performance data from several municipal sewage plants constructed of types 304, 316 and 316L stainless steels. Pilot plant and shaking autoclave corrosion test results are also indicated. [*This chart is an update (Dec. 1982) of the chart in T. P. Oettinger and M. G. Fontana, Materials Performance* **15**:*29–35 (Nov. 1976). Courtesy Zimpro, Inc.*]

these plants are shown in the figure. Below about 300 to 400 ppm chlorides, these sustenitic stainless steels show excellent performance without pitting and cracking. Higher chloride contents cause attack but titanium exhibits excellent performance. One titanium plant in Japan has been operating for more than eight years with over 5000 ppm chlorides. Alloy C-276 (Table 5-9) also shows good resistance under high chloride and temperature conditions.

Remarkable correlation is found between actual plant service and shaking autoclave tests in the laboratory.

Additional information is in DeAngelo and Wilhelmi, *Chem. Eng. Progr*., 68–73 (Mar. 1983); in T. P. Oettinger and M. C. Hoffman, "Titanium For Waste Treatment" in Technology of Metallurgy Series, AIME 1982; and in T. P. Oettinger, M. C. Hoffman, and M. G. Fontana, "Use of Titanium for Treatment of Toxic and Hazardous Wastes," Fourth ASTM Symposium on Titanium and Zirconium in Industrial Applications, October 1984.

8-19 Pollution Control

Billions of dollars have been spent during recent years and additional billions will be expended during the next decade for control of pollution. Three common types of installations are incinerators for solid trash; scrubbers for flue gas desulfurization (FGD) of power plants and metallurgical plants; and processes for treatment of municipal sewage, wastewater and liquid plant wastes. Most of these installations are really chemical plants. Corrosion in chemical plants is emphasized in this book. Capital costs can be very large. For example, one coal-burning power plant cost $75 million and "attached" to it is a $25-million plant to remove sulfur compounds from combustion gases by scrubbing. Practically all types of equipment are involved in pollution control including scrubbers, incinerators, heat exchangers, fans, pumps, valves, towers, dryers, furnaces, burners, cyclone separators, filters, stacks, electrostatic precipitators, catalytic devices, nozzles, centrifuges, mist eliminators, digestors, screens, conveyors, reactors, and kilns. Various wet and dry environments—gases, liquids, and solids—are involved and temperatures vary from low to very high. Metals from steel to platinum and nonmetallics including linings for metals are used as materials of construction. In other words, an almost complete gamut of corrosion problems exists, encompassing all eight forms of corrosion.

Wet *scrubbers* are widely used to clean gases and also to remove particulate matter in pollution control systems.* Gases of principal concern are SO_2, SO_3, and HCl. When wetted these form sulfurous, sulfuric, and hydrochloric acids. Halogens may also be present. Chlorides are usually present in sufficient abundance to cause pitting, crevice corrosion, and stress corrosion. Presence of fly ash accentuates erosion-corrosion problems. Before selecting materials of construction† the environment should be carefully defined, including (1) composition and temperature of the gas stream and scrubbing liquor, (2) pH of liquors, (3) velocity of gas and liquid

*T. G. Gleason, How To Avoid Scrubber Corrosion, *Chem. Eng. Progr*., **71**:43–47 (Mar. 1975).

†N. T. Mistry, Materials Selection for Gas Scrubbers, *Corrosion*, **15**:27–33 (Apr. 1976).

streams, (4) abrasiveness of particles in gas and liquid streams, (5) inhibitors in liquids, and (6) continuous or intermittent operation of equipment. Water is often the chief scrubbing agent, but wet lime/limestone processes and others are also used. Dry collection devices, especially for removal of solid particles, use fabric filters and bag houses and are not subject to aqueous corrosion.

Corrosion problems can be minimized by maintaining pH as high as is feasible, avoiding deposition of solids, reducing velocity of streams, keeping chlorides as low as possible, staying above dew point (using reheaters if necessary), and minimizing temperatures. Proper selection of alloys and other materials for the most corrosive portions of the system is also important.

Equipment in scrubbing systems can vary widely according to the specific functions to be performed. Figure 8-15 (from Mistry) is a schematic flow diagram for a generalized scrubbing system with two options. Reheaters are sometimes used to raise the effluent gases above the dew point to minimize corrosion of the stack. Water is usually recycled unless cheap water is available. Table 8-13 (from Mistry) lists materials of construction for some of the equipment for different process scrubbing systems. Where Carpenter 20 is listed this means alloy 20 for castings (i.e., Durimet 20) and also wrought Carpenter 20Cb-3.

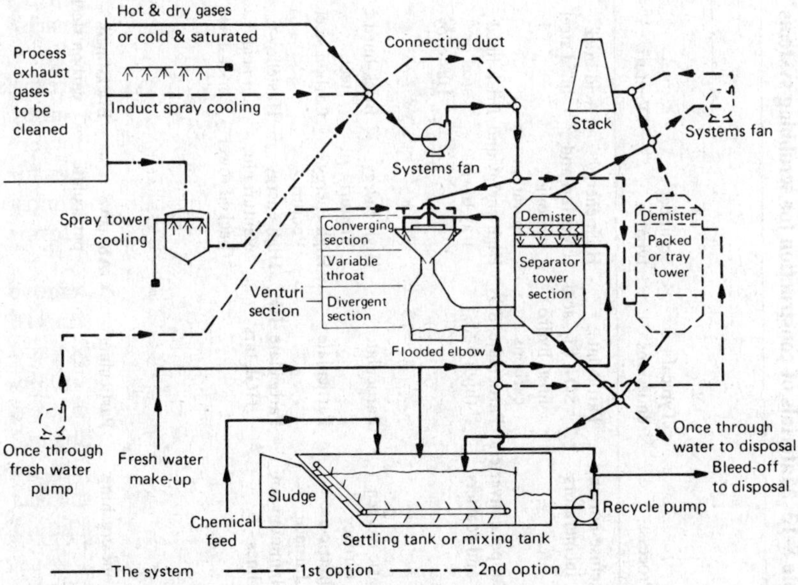

Figure 8-15 Schematic flow diagram of generalized gas-scrubbing system. [*From N. T. Mistry, Materials Selection for Gas Scrubbers, Corrosion* **15**:*27–33 (Apr. 1976).*]

Table 8-13 Materials of construction for scrubbing systems*

No.	Process	Typical emissions	Scrubber type	Materials of construction: Venturi	Separator tower	Wet fan at outlet: Casing	Wheel and shaft	Demister or packing	Remarks
1	Refuse-burning incinerators	Particulate, SO_2, Cl_2, acid mist, hydrocarbons	High-energy Venturi and cyclonic separation	Acid brick-lined steel	FRP- or rubber-lined steel	Rubber-lined	Incoloy-825 or Hastelloy-C	—	Highly corrosive environments (organics are also present)
2	Copper converters and roasters	Particulate, SO_2 (high 3-5%)	High-energy and tray tower	Brick-lined steel or Hastelloy-C	FRP-lined carbon steel	Rubber-lined	Carpenter-20	Carpenter-20 Hastelloy G	Exhaust gases used to make sulfuric acid for electrolytic refining of copper
3	Basic oxygen furnace	Particulate	High-energy Venturi	Brick-lined steel	Rubber- or FRP-lined steel	Rubber-lined	Hastelloy-C	Polypropylene	Contains abrasive particulate
4	Electric-arc furnace	Particulate	High-energy Venturi	Carbon steel	Carbon steel	Carbon steel	Carbon steel	Carbon steel	
5	Aluminum pot lines	Particulate, HF, HGl, tars	High-energy Venturi plus packed tower	Hastelloy-C or acid-brick-lined	Rubber-lined carbon steel	Rubber-lined	Hastelloy-C or Inconel 625	Polypropylene	Very corrosive and tacky environment because of tars. Electrostatic precipitators have been used successfully
6	Stone crushing	Particulate	Abrasive particulate	Brick-lined carbon steel	Carbon steel with wear plate	Carbon steel	Carbon steel	Carbon steel	Metal should be extra thick because of abrasive particles

7	Fertilizer industry (superphosphate)	Silica, SiF_4, HF, CaF_2	Venturi with tray or packed tower	Hastelloy-C or Carbon steel with rubber or brick lining	FRP with Dynel lining or carbon steel with rubber lining	Rubber-lined	Hastelloy-C or Inconel 625	Polypropylene	FRP not recommended because of fluorides
8	Gray iron foundry	Abrasive particulate	High-energy Venturi	Carbon steel refractory- or rubber-lined	Carbon steel	Carbon steel	Carbon steel	Carbon steel	Dust is very abrasive
9	Black liquor recovery boiler	Fumes, SO_2, H_2S, Mercaptans	Venturi and packed or tray tower	316L or rubber-lined steel	316L or FRP	Rubber-lined	Allegheny Ludlum 216 Armco 22-13-5	Polypropylene Inconel 625	Exhaust gases contains tars and are odorous
10	Bark boiler	Very abrasive particulate and SO_2	High-energy Venturi	Ceramic- or refractory lined-steel	Rubber-lined steel or 316L stainless	Rubber-lined	316L or Allegheny Ludlum 216	316L or Ploypropylene	Silica is present in particulate which is very abrasive
11	Catalytic cracker	Particulate and SO_2	High-energy scrubber	Ceramic-lined	Ceramic-lined	—	—	—	Very high energy required to recover expensive catalyst
12	Titanium pigment plant	Particulate (TiO_2), HCl, $TiCl_4$, etc.	High-energy scrubber	Rubber-lined steel Hastelloy-C	FRP- or rubber-lined steel	Rubber-lined	Hastelloy-C Inconel 625	Hastelloy-C Polypropylene	Titanium particulate (fume) is very fine (submicron). Hence, very high energy required.
13	Acid pickling	HCl, H_2SO_4, HNO_3, (Cl_2, SO_2)	Packed-tower scrubber	—	FRP	FRP	FRP	Polypropylene	No particulate, therefore low energy, hence material strength not as critical
14	Sulfuric acid plant	Acid mist	Packed-tower scrubber	—	Carpenter-20 Teflon-lined steel	Rubber-lined	Carpenter-20	Carpenter-20	No particulate emission but large process

(continued)

Table 8-13 Materials of construction for scrubbing systems* (continued)

No.	Process	Typical emissions	Scrubber type	Materials of construction: Venturi	Separator tower	Wet fan at outlet: Casing	Wet fan at outlet: Wheel and shaft	Demister or packing	Remarks
									equipment requiring good strength
15	Steam generators	Particulate and SO_2	High-energy Venturi	316L, refractory-lined steel	316L, FRP-lined steel	Rubber-lined or 316L	316L, Allegheny Ludlum 216, Hastelloy-C Armco 22-13-5	Hastelloy-G Incoloy 825 Polypropylene	Many types of systems in operation using various materials of construction

*N. T. Mistry, Materials Selection for Gas Scrubbers, *Corrosion*, **15**:27–33 (Apr. 1976).

Corrosion problems can vary widely between once-through and recycle systems. Once-through processes could use anything from fresh water to seawater. Recycling usually involves increasing chloride and acid contents and more aggressive conditions. As chloride and acid increase, better and more expensive materials are required (alloy 20, Incoloys 625 and 825, Hastelloy C, and titanium). Table 8-14 (from Gleason) indicates materials of construction for recycle systems vs. chloride and sulfuric acid concentrations. Titanium is good for chlorides (i.e., seawater) but is not suitable for sulfuric acid.

In most cases, if corrosion resistance is adequate, metals and alloys are preferred. They often cost less (e.g., than brick lining), are easier to maintain and can better handle temperature upsets and scrubber liquor failure (e.g., than rubber or plastics).

Many different materials are used for *pumps* including alloys, rubber-lined steel, and fiber-reinforced plastics (FRP). It is important to select, whenever feasible, standard materials because of lower cost of mass-produced items.* For example, pumps made of more resistant alloys such

Table 8-14 Materials of construction for sulfur dioxide gas scrubbers (recycle systems)*

Chloride concentration in scrubbing liquor (as NaCl)	Sulfuric acid concentration	Material of construction
Below 100–150 ppm	less than 2%	Humidifier—mild steel, lead, and brick Upper tower—316 ELC stainless steel
Below 100–150 ppm	2–20%	Humidifier—mild steel, lead, and brick Upper tower—Alloy 20 Upper tower—Hastelloy C (+70°C)
Below 100–150 ppm	20–30%	Humidifier—mild steel, lead, and brick Upper tower—mild steel, lead, or rubber. Hastelloy C trays
Seawater	0.25%	Humidifier—mild steel, rubber, and brick Upper tower—mild steel, rubber, and Hastelloy C
150–1000 ppm	2–5%	Humidifier—mild steel, rubber, and brick Upper tower—Alloy 20
1000–5000 ppm	2–5%	Humidifier—mild steel, rubber, and brick Upper tower—mild steel, rubber-lined, Hastelloy C trays and piping

*T. G. Gleason, How To Avoid Scrubber Corrosion, *Chem. Eng. Progr.*, **71**:43–47 (Mar. 1975).

*W. G. Ashbaugh, in *Resolving Corrosion Problems in Air Pollution Control Equipment*, p. 29, NACE Book (1976). Other references in this book, for sake of brevity, will list the author and page number in: NACE Book (1976).

as CD4MCu and Durimet 20 often cost less than CF-8M. Gleason states that CD4MCu and alloy 20 pumps give good service except when solids contents are too high. Bailey* reports selection of Durimet 20 over CF-8M because of high chloride content in the recovery solution and also the use of Hastelloy-C (Chlorimet 3) pumps for very aggressive conditions. Mockridge† prefers building up alloy 20 recycle pumps with weld metal over larger, belt-driven rubber-lined pumps. However, soft-rubber-lined pumps are used under severe erosion conditions. The amount of fly ash varies with the type of coal being burned. Precipitators preceding the SO_2 absorbers lower fly ash content and thereby reduce erosion corrosion in the scrubber system.‡

Spray *nozzles* are made of PVC, FRP, ceramics, and high alloys depending on temperatures, solids content, and acid conditions. *Venturi* throats involve high velocities and erosion corrosion. Rubber and plastics are not suitable for high temperatures. Polyester linings wear away quickly (Mockridge). *Piping* and *ducts* can be metal, plastic, or lined. In one case, FRP eroded within four years under abrasive conditions. *Valves* present similar problems. Materials for *fans* and *towers* are listed in Table 8-13. *Elbows* and *tees* can be protected from impingement (erosion) by "wear plates" and hard facing.

Reheaters are installed to protect ducts, fans, and stacks from corrosion by heating the gases above the dew point. Materials selection depends upon acid conditions, amounts of chlorides, and solids deposition on the metal surface. Soot blowers help to keep metal surfaces clean. Materials can vary from carbon steel to Hastelloy C. This alloy and Incoloy 625 show good corrosion resistance. Type 316 heater tubes have failed because of stress corrosion and pitting. A good discussion of this problem was presented by Crow.¶

One of the first early pushes for clean air (Clean Air Act of 1970) was to control or decrease emissions from automobiles. Then came the output of fossil fuel (mostly coal) power plants. Much has been written in newspapers on SO_2 emissions, causing the so-called "acid rain." The EPA requires reduction of sulfur dioxide emissions by 70–90% for domestic coals. About 80% of the utility *flue gas desulfurization* (FGD) systems operational or planned in the United States are wet lime or limestone processes.§ Wet scrubbing removes SO_2 by contacting the combustion gases with a lime or

*E. E. Bailey and R. W. Heinz, SO_2 Recovery Plants—Materials of Construction, *Chem. Eng. Progr.*, **71**:64–68 (Mar. 1975).

†P. C. Mockridge and A. Saleem, p. 83, NACE Book (1976).

§J. F. Stewart, p. 61, NACE Book (1976).

¶G. L. Crow, Reheaters in Systems for Scrubbing Stack Gases, *Materials Performance*, **14**:9–14 (July 1975).

§G. W. Driggers, P. C. Rader, and J. E. Johnson, Applied Field and Laboratory Experience to Alloy and Coating Selection in Combustion Engineering FGD Systems, *Materials Performance*, 34–39 (Dec. 1983).

limestone slurry. The reaction products, calcium sulfite and sulfate, are precipitated. This scrubbing system is really a chemical plant and the usual considerations apply.

An article by R. W. Kirchner* is of great interest with regard to use of alloys, and Table 8-15 and Fig. 8-16 are extracted from this work. Pitt and Anderson† discuss FGD for smelters. Froelich and Ware‡ describe corrosion problems with a closed loop limestone FGD system. Magnesium sulfite to remove SO_2 is in operation in a power plant in Montana. Sales of new FGD systems will reach \$3 billion by 1990.¶ In an excellent paper, Rosenberg et al.§ present actual operating experience.

Incinerators differ from conventional power and plant processes because of the variability and diverse nature of the materials being burned. For example, municipal trash and garbage could contain almost anything. PVC plastics are a common item. These introduce hydrochloric acid and high chlorides into the gases to be cleaned. Plastics containing nitrogen can form hydrogen cyanide. Hydrofluoric acid, organics, organic acids, and

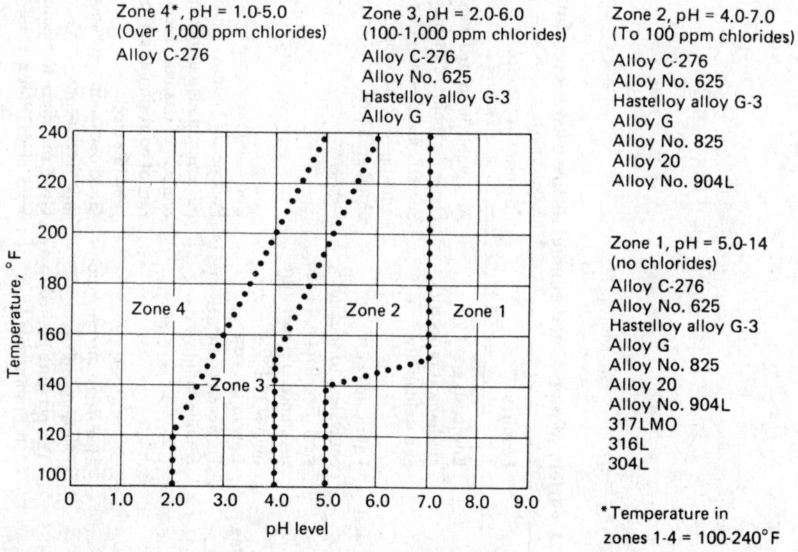

Figure 8-16 A general guide for selecting alloys to use in flue-gas-desulfurization systems.

*R. W. Kirchner, Materials of Construction for Flue Gas Desulfurization Systems, *Chem. Eng.*, 81–86 (Sept. 19, 1983).

†W. G. Pitt and T. N. Anderson, *Materials Performance*, 26–29 (May 1982).

‡D. A. Froelich and M. Ware, *Materials Performance*, 40–43 (Oct. 1982).

¶*Chem. Eng.*, 36 (July 9, 1984).

§H. S. Rosenberg, C. W. Kistler, L. O. Nilsson, L. J. Nowacki, J. A. Beavers, G. H. Koch, H. H. Krause, Jr., and C. E. Dene, "Recent Operating Experience with Construction Materials for Wet Flue Gas Scrubbers," NACE Paper 182, CORROSION/83.

Table 8-15 A variety of metals and nonmetals are used successfully in different parts of FGD systems

Material	Features	Corrosion resistance	Uses
Steel	Low cost, good mechanical properties and weldability.	Acceptable in treated water and above dewpoint of fluegases. Severely attacked in other FGD conditions—corrosion of > 1 in./yr.	General-purpose in noncorrosive applications, such as dry-inlet ducting. Normally used for structural support.
Rubber-lined steel	Fabricated cost equivalent to stainless steel. Excellent abrasion resistance. Temperature limited to 180°F.	Good acid erosion-corrosion resistance.	Rubber-lined pipe, pumps and absorber sections in high-erosion conditions.
Organic coating over steel	Low initial cost. However, coatings can contain holidays.	Corrosion resistance unacceptable. Presence of holidays causes blistering, undercutting and removal of coating, due to high rate of corrosion attack on substrate steel.	Various coatings have been evaluated in ducting with limited success. Service life is 3 to 4 yr, maximum.
Refractories	Heat-resistant, corrosion-resistant. Bulky.	Good corrosion resistance; "weak link" is cement.	Stack linings with acidproof cement.
FRP plastics (Polyester or	Temperature limited to 250°F. Fabricated	Good resistance to alkalies and weak acids. Polyester dissolves	Linings in flooded areas of absorber, piping, etc.

epoxy)	costs cheaper than stainless steel.	in strong sulfuric acid. Glass fibers attacked by HF. Temperature-limited, and can be destroyed (charred) by temperature excursion. Epoxy not resistant to strong acids, but acceptable in strong alkalies.	
Polypropylene	Temperature limited to 225°F.	Good acid and alkaline resistance.	Mist eliminators.
Stainless steels	Good mechanical properties, high temperature resistance, weldability, and fabricability.	Good resistance to mild media. However, they suffer from pitting, crevice corrosion and SCC in various chlorides environments. High corrosion attack under strongly acid chloride conditions.	Scrubber vessels, nozzles, ducts—areas washed by alkaline reagents and chemicals with pH levels of 5.5–7.0.
High-performance alloys	Excellent mechanical properties, high temperature resistance, and good weldability and fabricability.	Good corrosion resistance to strongly acidic chloride conditions. Higher alloy content provides resistance to SCC, and improved pitting and crevice corrosion resistance.	Quench and scrubber units, induced-draft fans, guillotine valves, louvers, venturis, stack liners, reheaters and ductwork.

high sulfur (auto tires) can be present. Burners, furnaces, and kilns are subject to the variability of the charge. The reader is referred to papers by Vaughn,* Velsey,† Conybear,‡ and Dunn§ for more detailed discussions. Tables 8-16 and 8-17 are from Velsey.

Table 8-16 Summary of corrosion studies*

1. At scrubber inlet, samples were suspended in gas stream which had been conditioned or cooled with sprays.
   Temperature 250 to 600°F (122 to 315°C). 60–90 day tests.
   Best Materials: Titanium, Hastelloy C-275, Inconel 625, Hastelloy G, Incoloy 825, Carpenter 20Cb-3
   Questionable: 316, 216, Hastelloy B, 304, Incoloy 800, Inconel 600
   Unsatisfactory: (avg corrosion rate > 10 mpy)—Carbon steel, weathering steel, Ni-Resist II, Monel 400
2. In scrubber, after gases initially saturated and before demister.
   Temperature 140 to 190°F (60 to 88°C). 85–104 day tests.
   Best Materials: Titanium, Hastelloy C-276, Inconel 625, Hastelloy G, Incoloy 825
   Questionable: Carpenter 20Cb-3, 216, 317
   Unsatisfactory: (avg corrosion rate > 10 mpy)—Carbon steel, weathering steel, Ni-Resist II, Monel 400, Hastelloy B, 304, Inconel 600, Incoloy 800, 316
3. In scrubber above demister.
   Temperature 140 to 200°F (60 to 93°C). 42–98 day tests.
   Best Materials: Titanium, Inconel 625, Hastelloy C and C-276, Hastelloy G, Incoloy 825, 216, Carpenter 20Cb-3
   Questionable: 316, 317, 304, Incoloy 800, Inconel 600, 310, 430
   Unsatisfactory: (avg corrosion rate > 10 mpy)—Carbon steel, weathering steel, cast iron, Ni-Resist II, Monel 400, Hastelloy
4. ID fans and stacks.
   Temperature 160 to 200°F (72 to 93°C). 42-92 day tests.
   Best Materials: Titanium, Inconel 625, Hastelloy C, Incoloy 825
   Questionable: Hastelloy B, Incoloy 800
   Unsatisfactory: (avg corrosion rate > 10 mpy)—Carbon steel, weathering steel, Monel 400, 430, Inconel 600, 304, 329, 317, Carpenter 20Cb-3, 316
5. In scrubber water.
   Temperature 105 to 212°F (41 to 100°C). 23-137 day tests.
   Best Materials: Inconel 625, Hastelloy C-276 + C, Hastelloy G, titanium, Incoloy 825, 216, Incoloy 800
   Questionable: Carpenter 20Cb-3, 317, 316, 304, 310, 309, Inconel 600
   Unsatisfactory: (avg corrosion rate > 10 mpy)—Carbon steel, weathering steel, Ni-Resist II, Hastelloy B, Monel 400, Nickel 200

*C. O. Velsey, Materials of Construction for Wet Scrubbers for Incinerator Applications, pp. 34–45, NACE Book (1976).

*D. A. Vaughn, H. H. Krause, and W. K. Boyd, Fireside Corrosion in Municipal Incinerators Versus Refuse Composition, *Materials Performance*, **14**:16–24 (May 1975).

†C. O. Velsey, Materials of Construction for Wet Scrubbers for Incinerator Applications, pp. 35–45, NACE Book (1976).

‡J. G. Conybear, Corrosion Concerns in Waste Incinerators, pp. 109–114, NACE Book (1976).

§K. S. Dunn, Incineration's Role in Ultimate Disposal of Process Wastes in "Chemical Engineering Deskbook," *Environmental Eng.*, **82**:141–150 (Oct. 5, 1975).

Table 8-17 Corrosion experience from operation, Plants A—E upstream of scrubber entrance* gas temperatures from 2000 to 500°F (1093 to 260°C)

Materials used	Experience	Corrective action (if necessary)
Plant A		
Scrubber vessel—316 ELC SS	Failed after approx. 15,000 to 20,000 operating hours	Replaced with FRP lining covered by firebrick
Plant B		
Gas cooling chamber—refractory	Problems at wet/dry interface	
Plant C		
Venturi section—castable refractory lining	Some wear	Patched with flake-filled polyester
Plant D		
Breeching—silicon carbide refractory	Failed after 2000 operating hours	Replaced with acid brick
Plant E		
Flue—steel, polyurethane lining, castable refractory	No reported problems	

*C. O. Velsey, Materials of Construction for Wet Scrubbers for Incinerator Applications, pp. 34–45, NACE Book (1976).

8-20 Coal Conversion

The conversion of coal to gas and oil has opened a new era in the field of corrosion and materials of construction. The problems are challenging and tough. Examples are: valves that close at 2500°F; erosion corrosion failure of a type 347 stainless steel pipe within 4 hours; reactor vessels weighing up to 4000 tons, 250 ft high, 22 ft ID, with walls 12 in. thick. Present designs, technologies, and materials are inadequate in many cases and new ones must be developed. Reliability is important not only because of the expensive equipment involved but also because downtime is very costly. A troublesome and unique feature of coal conversion processes is severe erosion because of the presence of solids (fly ash). Another difference is that high-temperature alloys in the past were developed for oxidizing conditions (i.e., gas turbines and heat-treating furnaces), whereas highly reducing conditions often exist in coal conversion processes. The chemistry of the environment is also important because changes in composition (i.e., hydrogen sulfide) can alter the aggressiveness of attack tremendously. Coal conversion technology has been neglected in this country since the advent of cheap natural gas and oil decades ago. Gas and oil from coal will not be cheap. Energy shortages will be with us for a long time.

One of the first extensive workshops was held at The Ohio State University in April 1974 under the direction of Professor R. W. Staehle, published as *Materials Problems and Research Opportunities in Coal Conversion*, vols. 1 and 2, Fontana Corrosion Center, Ohio State University (Apr. 1974). Materials, design, and process persons interacted and discussed mutual problems. These people must work closely together in order to define and solve the problems involved. For example, see A. J. McNab, The Materials/Design Interface in Coal Conversion Technology, vol. 2, pp. 34–47.

It seems that reserves of oil and gas may be depleted within the next 30 to 50 years, and we will have to depend on our largest reserve of fossil fuel—namely, coal. Conversion of coal to oil and gas must be done with consideration of environmental issues such as emission of sulfur and nitrogen oxides, dust, hazardous solid waste and contamination of water. Much of the following was supplied to me by Dr. John Stringer. Different approaches to the problem are as follows:

1. Advanced combustion processes
   a. Staged combustion to limit the formation of oxides of nitrogen.
   b. Atmospheric fluidized bed combustion.
   c. Pressurized fluidized bed combustion.
2. Coal conversion systems
   a. Gasification of coal.
   b. Hydrogenation of coal to produce either a liquid fuel or a clean solid fuel.
   c. Beneficiation of coal, usually coupled with the production of a coal-water or coal-oil slurry.

According to Stringer, the materials issues in these systems can be listed generically as follows:

1. The availability of materials of construction having adequate mechanical properties, typically strength and toughness, at the required temperatures and in the appropriate environments.
2. High-temperature corrosion in atmospheres having low oxygen and relatively high sulfur and carbon activities; the presence of chlorine, probably as hydrogen chloride, may be an additional complicating factor.
3. High-temperature corrosion induced by molten deposits, including alkali sulfates, oxides of lead or vanadium, and so forth.
4. Erosion by particles of ash or of the uncombusted coal minerals; this erosion may be in gas or liquid streams.
5. Wear by abrasion.
6. Low-temperature corrosion by aqueous liquors containing a wide variety of impurities.
7. The interaction of two or more of the above effects to produce damage greater than the simple sum of the two acting separately.

It is apparent from this list that most of Chapter 11 (High-Temperature Corrosion) is applicable to this discussion—for example, Sec. 11-12 (Corrosion of Metals by Sulfur Compounds at High Temperatures) and Sec. 11-13 (Hot Corrosion of Alloys).

Schad and Hafke* discuss recent developments in coal gasification. In addition, "Materials, Equipment for Conversion of Coal" is a News Feature article in *Chemical Engineering*, 35–38, June 1, 1983.

Jones et al.† report on a 250-ton-per-day coal liquefaction plant. In response to a question I asked after presentation of this excellent paper, one author stated that this precess was *economically viable* for the Illinois coal involved.

Lochman and Howell‡ described some of the problems in coal conversion and the need for interdisciplinary action. Examples of environmental conditions and problems are listed in Table 8-18. Reliability of operation (90% or better) is a primary factor and requires spare critical equipment, better materials for longer life, and improved quality control including nondestructive testing. For large heavy-walled gasifiers and liquefaction reactors, brittle fracture and quality inspection are problems. For use at temperatures up to 3000°F reliable refractory linings are needed to withstand corrosion and erosion from slags and solid ash. Piping up to 72 in. in diameter with walls up to 5 in. can be subject to severe erosion corrosion. In one case right-angle turns are used with sintered carbide faceplates to take the impingement. Valves are a difficult problem, and adequate designs and materials are not available at this writing. Erosion corrosion of pumps now requires positive displacement pumps in some cases, but an adequate centrifugal pump could replace five reciprocating pumps at 20% of the cost of the latter. Gas turbines using coal-converted "dirty" gas require improved materials.

Lochman further states that heat-exchanger problems include high temperature and pressure closures, erosion, corrosion, plugging, prediction of heat-transfer rate, and uniform flow. He cites an example of design where the inlet tubes were extended 4 in. and their life doubled. This technique is described in Chapter 3.

In another paper, O'Hara, Lochman, and Jentz discuss materials for coal liquefaction.¶ Table 8-19 summarizes operating conditions. An added factor here as compared to gas production is hydrogenation which can

*M. K. Schad and Hafke, *Chem. Eng. Progr.*, 45–51 (May 1983).

†J. P. Jones, G. M. Buchheim, E. Lendvai-Litner, and G. Sorell, "Materials Performance in the EDS Coal Liquefaction Pilot Plant," NACE Paper 80, CORROSION/84.

‡W. J. Lochman and R. D. Howell, "Corrosion Engineering—Design Interface for Coal Conversion," presented at CORROSION/77, March 15, 1977, San Francisco.

¶J. B. O'Hara, W. J. Lochman, and N. E. Jentz, Materials Considerations in Coal Liquefaction," *Metal Progr.*, **110**:33–38 (Nov. 1976).

Table 8-18 Gasification processes and environmental conditions

Area	Methods	Constituents	Problems
I Coal preparation	A. Dry grinding	Coal and low O_2 air	Moisture and SO_2 in stream of flue gas, sulfur in coal, erosion Temperatures for dust removal are limited by bag materials (can use Venturis for wet scrubbing)
	B. Wet grinding	Coal and water	Corrosion: sulfur in coal attacking metals Erosion: solids
II Coal feed injection	*A.* Lock hoppers Lurgi CO_2 acceptor	Coal, pressurizing gas	Valve wear due to erosion
	B. Slurry feeding and vaporizing IGT Hy-Gas BCR Bi-Gas	Coal with either oil or water Heating medium is process gas containing CO, H_2, H_2S, CO_2, H_2O, CH_4, and possibly oils and tars	Pump wear due to erosion Piping wear due to erosion Flow metering: erosion Gas velocity: 35 to 100 ft/sec Liquid velocity: 6 to 10 ft/sec
	C. Mechanical inspection	Pulverized coal	Erosion within injector
III Gasification	*A.* Moving bed Lurgi Wellman General Electric (GE) gas	Coal, char, CH_4, CO, CO_2, H_2S, COS, tars, oil, H_2O, NH_3, Ash	Erosion Corrosion Temperatures: 1700 to 2600°F Pressures: atoms to 450 psig
	B. Fluidized bed IGT Hy-gas Agglomerating ash CO_2 Acceptor Synthane Winkler Exxon	Same as above plus inert solids for CO_2 acceptor (calcined dolomite)	Erosion Corrosion Temperatures: 1500 to 2100°F Pressures: atmos to 1200 psig
	C. Entrained flow BCR Bi-Gas Koppers-Totzek Texaco B and W-DuPont Combustion Engineering Foster-Wheeler	Same as above	Erosion due to char and ash particles Corrosion Also possible slag attack Temperatures: 2500 to over 3000°F Pressures: Atmos to 1200 psig

(continued)

Table 8-18 Gasification processes and environmental conditions (continued)

Area	Methods	Constituents	Problems
IV Ash removal	*A*. Lock hopper (dry)	Ash, inert gas	Valve wear due to erosion Temperatures: to 1200°F Pressures: to 450 psig
	B. Lock hopper (slurry)	Ash, Cl, water, H_2S, CO_2, CO	Valve wear Temperatures: to 500°F Pressures: to 1200 psig
	C. Slurry with letdown down valves	Same as *B*, above	Valve wear: Flashing across valves Temperatures: to 500°F Pressures: to 1200 psig
V Dust removal	*A*. Dry: Electrostatic Dust bag Cyclones Sand filters	High solids loadings same gas as in gasifier	Erosion Inlet temperatures: to 1700°F Pressures: to 1200 psig
	B. Wet: Scrubbing columns Venturi	Same as *A*	Corrosion: CO_2 in H_2O, H_2S, H_2, chlorides High velocity through Venturi throats
VI Sulfur removal	*A*. Absorption Physical Chemical At low to moderate temperatures	CH_4, CO, CO_2, H_2S, COS, H_2O, HCN, plus solvent	Corrosion: These processes are in commercial use, satisfactory materials of construction have been established
	B. High-temperature sulfur-removal processes (1200 to 1800°F)	Same gases as above with a strong alkali-based reactive agent either as a solid or a liquid	Corrosion: Caused by alkalis and sulfur compounds Erosion: By solids Temperature

cause hydrogen embrittlement and carburization. Corrosion and erosion may be tougher for liquefaction than gasification. A catastrophic failure of a 316 preheater because of stress corrosion occured in a lignite-processing pilot plant.

Nateson discusses erosion of materials in coal gasification, and Fig. 8-17 is from his paper. Ductile mode means cutting of the metal and brittle mode means cracking away of particles of metal.

Table 8-19 Summary of operating conditions for liquefaction processes*

Process	Process steps	Temperature, F (C)	Pressure, psig (MPa)	Environment; flow-phase conditions	Constituents
Hydroliquefaction Noncatalytic—SRC	Coal dissolving	800–900 (425–480)	1000–2000 (7–14)	Three-phase (liquid, gas, and solid) up-flow; low linear velocity	Coal fines; coal-derived aromatic solvent of high sulfur and nitrogen content. Gases with H_2, CO, CO_2, H_2S, NH_3, HCN, and H_2O
Catalytic	Coal dissolving Catalytic hydrogenation	800–900 (425–480)	3000–4000 (21–28)	Three-phase (liquid gas, and solid) up-flow through packed catalyst bed	Coal fines; coal-derived solvent with high aromatic, sulfur, and nitrogen content
Extraction—CSF process	Donor solvent Coal dissolving	800–900 (425–480)	400 (2.7)	Oil-coal slurry-agitated	Ground coal; coal derived solvent which has been enriched with H_2
	Solvent hydrogenation	800 + (425 +)	3000–4000 (21–28)	Three-phase (liquid, hydrogen, and catalyst particles)	Hydrogen-rich gas with H_2S, NH_3, and H_2O. Coal-derived aromatic

				low velocity, upflow	liquid of high sulfur and nitrogen content
Pyrolysis—COED	Pyrolysis	800–1150 (425–620)	10–15 (0.06–0.10)	Fluid bed; coal and/or char gas from pyrolysis	Ground cola and/or char. Unsaturated pyrolysis vapors composed of tar, H_2O, unsaturated gas. H_2S, CO_2, NH_3 COS, HCN, and char fines
	Char conversion	1800 (980)	5–10 (0.03–0.06)	Fluid bed; char syn-gas product; steam and oxygen feed	Char fines; syngas composed of H_2, CO, CO_2, H_2O, H_2S, NH_3, COS, CS_2, HCN, and entrained char
Indirect—Fischer-Tropsch	Liquefaction (synthesis)	600 (315)	400 (2.7)	Gas phase	Syngas of high purity plus products; paraffin hydrocarbons, oxygenated compounds, CO_2, and H_2O

*J. B. O'Hara, W. J. Lochman, and N. E. Jentz, Materials Considerations in Coal Liquefaction, *Metal Progr.*, **110**:33–38 (Nov. 1976).

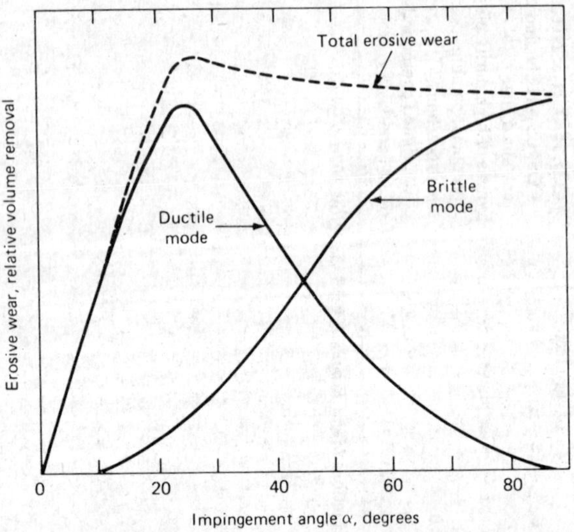

Figure 8-17 Degree of erosion as influenced by the particle-impingement angle. [*From K. Nateson, Corrosion,* **32**:*364–369 (Sept. 1976).*]

Suggested Reading

Bryers, R. W., ed.: *Ash Deposits and Corrosion Due to Combustion Gases*, Hemisphere Publishing Corp., Washington, D.C., 1977.

Hart, A. B., and A. J. B. Cutler, eds.: *Deposition and Corrosion in Gas Turbines*, Applied Science Publishers, London, 1973.

Hawbolt, E. B., and A. Mitchel, eds.: *Materials to Supply the Energy Demand*, American Society of Metals, Metals Park, Ohio, 1981.

Hill, V. L., and H. L. Black, eds.: *The Properties and Performance of Materials in Coal Gasification*, American Society of Metals, Metals Park, Ohio, 1981.

Levy, A. V.: *Corrosion-Erosion-Wear of Materials in Emerging Fossil Energy Systems*, NACE, Houston, Tex., 1982.

Meadowcroft, D. B., and M. I. Manning, eds.: *Corrosion Resistant Materials for Coal Conversion Systems*, Applied Science Publishers, London, 1983.

Rapp, R. A., ed.: *High Temperature Corrosion*, NACE, Houston, Tex., 1983.

Construction Materials for Coal Conversion—Performance and Properties Data, NBS Special Publication 642, Superintendent of Documents, Washington, D.C., 1982. ($16). This contains much "raw data" on a large variety of aspects.

8-21 Pulp and Paper Industry

An integrated paper mill represents a large cross section of industrial processes. Many of the corrosion problems and considerations for corrosion-resistant construction materials, described in previous chapters, are found in the paper industry.

The most common method for converting wood to fibers for papermaking is the kraft pulping process (Fig. 8-18). Both batch and continuous digesters cook the wood chips in a caustic/sulfide liquor (pH 11). The *batch* digester is a welded, thick wall, carbon steel pressure vessel that is *fully* stress-relieved. A rimmed steel of low silicon content (0.02 wt %) is used with sufficient corrosion allowance to resist uniform attack as it provides average lifetimes of 25 to 30 years. Type 316L stainless steel overlay is commonly used in regions subject to erosion corrosion (e.g., the chip/liquor inlet and discharge zones). Increased life of the vessel is also achieved, via partial or complete 310 stainless steel overlay. Most *continuous* digesters are mild steel construction—typically, ASTM-A516 Grade 70 at the top and A285 Grade C in the lower portions. The more uniform conditions—e.g., temperature, pressure, wood/liquor flow—help keep corrosion rates low. However, stress corrosion cracking of the weldments in the top of the digester has been revealed recently. Although proper corrective measures are still under study, many of the SCC affected welds have been overlaid with Inconel 82 or AISI Type 309.

Indirect heating of the digester liquor is accomplished in vertical, shell-tube type heat exchangers, called liquor heaters. Heavy-wall, mild-steel

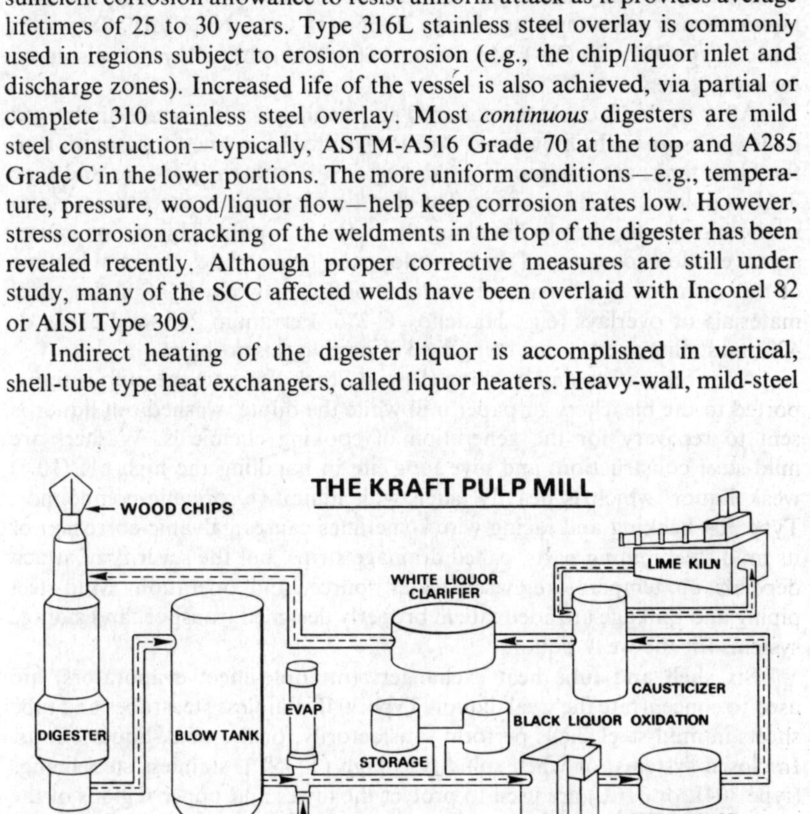

Figure 8-18 Plant layout for kraft pulping processes and chemical recovery in the manufacture of paper products. *(Courtesy The Institute of Paper Chemistry, Appleton, Wis.)*

piping is used for circulation, and the most common combination of heater materials includes stainless steel tubes and tube sheets in a mild-steel shell. A variety of stainless steel tube alloys are used because, dependent on the system, SCC has occurred. The tubes are frequently acid cleaned with muriatic acid, and type 304L is inadequate. The annealed, seamless 300 series stainless alloy tubes provide satisfactory life (> 10 years) in milder liquor/cleaner environments. Alloys 3RE60 (18 Cr, 5 Ni, 3 Mo), Carpenter 7 Mo (Type 329), 26 Cr–1 Mo, Inconel 600, and other alloys are also used for replacement.

After the kraft cook, the softened chips and spent liquor are discharged under pressure to the blow tank. Pipelines, called blow lines, for transport to the blow tank, are commonly mild steel in the straight portions and type 316L stainless steel els or bends. The chip/liquor slurry impacts target plates in the blow tank, thereby breaking the chips into pulp. This action causes severe erosion corrosion of the low-alloy-steel target plates. Normal lifetime of this common plate material is nine months to two years. Stainless steel materials or overlays (e.g., Hastelloy C-276, Ferralium 255, and 316L, or 409 and stellite overlays are being used to combat this problem.

After separation and washing (Brown Stock Washer) the pulp is transported to the bleachery or paper mill while the dilute, washed-out liquor is sent to recovery for the generation of cooking chemicals. Washers are mild-steel construction, and give long life in handling the high-pH (10.5) weak liquor, which is heavily laden with inhibitive, organic compounds. Type 304 backing and facing wire sometimes causes galvanic corrosion of its mild-steel mating parts, called drainage strips, but the severity of attack depends on temperature, wash water source, and operation. Mild-steel piping and tankage are adequate in properly designed transport and storage systems for the weak liquor.

Six shell and tube heat exchangers (multiple-effect evaporators) are used to concentrate the weak liquor. Type 304L stainless steel tubes and tube sheets in mild-steel shells perform satisfactorily for the weak-liquor effects. In closed systems, or when sulfidity is high (> 35%), stainless steel linings (type 304L or 316L) are used to protect the lower and upper regions of the mild-steel shell. Type 316L tubes and sheets are used in the heavy liquor effects. Fouling of the tubes, via calcium carbonate precipitation, necessitates frequent cleaning by boiling, hydroblast, and/or acid cleaning. Tube failures by stress corrosion and corrosion fatigue have occurred in some systems.

The recovery boiler is the heart of the kraft chemical recovery cycle. Heavy black liquor (60–65% solids) is burned to produce heat for steam and subsequent power generation. As it burns, the residue, called char, is reduced at high temperature in the lower furnace to form a molten smelt (Na_2S and Na_2CO_3), which flows through spout openings into water sprays and the dissolving tank. The furnace is composed (wall to wall and floor to ceiling) of boiler tubes. Fireside protection against the reducing conditions and sulfur gases in the lower furnace is accomplished by pinstudding (small studs, about 3/8 in. dia. by 5/8 in. long, attached to the tube o.d.) and/or controlled temperature and combustion conditions. Today, with higher-pressure boilers and/or more aggressive liquor chemistry, thermal spray (metallized type 316, nickel, and aluminum tricoats or plasma applied high-chromium stainless steel) and type 304L stainless-clad tubes are employed in this region for protection. Carbon steel or chrome-moly steel are used in the upper regions—i.e., the superheater, generating, and economizer sections. In many instances, high-pressure ($\geqslant$1000-psig) recovery boilers, either clad or solid type 304L stainless steel, are used in the lower bends (highest-temperature regions) of the superheater.

A refractory (brick or gunite) lined, carbon steel vessel is used to dissolve the smelt in water and form green liquor (pH 10–11). The fume from the recovery boiler is cleaned of fly ash (Na_2SO_4) by electrostatic precipitators. These carbon steel units are subject to severe corrosion when fume temperature is below the *dew point* for formation of sulfuric acid. Proper temperature control, heavy insulation, and coating inside wall surfaces (vinyl ester) are all used to combat this costly corrosion problem.

Solids (Na-Ca carbonates) are removed from the green liquor, and the dissolved sodium carbonate is converted to caustic (NaOH) via the addition of lime from the kiln. The resulting white liquor, after clarification, is used for cooking in the digester, thus completing the chemical recovery cycle. The high-pH, low-temperature ($\geqslant$100°C) conditions permit the use of mild steel in piping and equipment for the green and white liquor systems. Type 316L is used where conditions for erosion corrosion persist. Equipment life varies from 10 to 30 years, depending on process control and the degree of contamination. Many of the older (10 years or more) mills have reverted to type 304L replacements, particularly for components handling the more aggressive white liquor. Anodic protection is also under study as a means for prolonging the life of mild-steel equipment—for example, white liquor clarifiers, storage tanks, and continuous digesters.

There are other chemical pulping methods. The process systems are similar to kraft, but the construction materials are stainless steel. For example, sulfite pulping, using sodium, calcium, magnesium, or ammonium base, is an acid process (pH 1–3) that requires the alloy content of type 316L as a minimum. A so-called papermaker's grade of stainless steel, type 317 (type 316 with 2.75% Mo minimum), is often specified for sulfite mill equipment. For neutral sulfite pulping (pH 7–9), type 304L and 316L stainless steels are common construction materials. These same materials find widespread use in thermomechanical pulp mills, where fiber is produced by grinding the wood chips in pressurized refiners, with or without chemicals.

The pulp from the Brown Stock Washers may be bleached. The low-pH, highly oxidizing bleach environment associated with chlorine and chlorine dioxide chemicals requires titanium or nonmetallics, such as (tile or fiberglass-reinforced plastic (FRP), for construction materials in the bleach tower. A caustic stage follows each bleach to remove or extract the contaminants (lignin). After each bleach stage, the pulp is washed and again titanium is used in washing drums. Alternatively, type 316L and 317L stainless steels are commonly used in this application, and these materials give good service provided process conditions (e.g., residual chlorine/chlorine dioxide) are closely controlled. Recently, anodic protection has found use for prolonging the life of stainless steel washers, even under occasional adverse conditions or upsets. The extraction stage washers are type 316L. The piping and seal tanks beneath all washers are made of stainless steel (316L minimum) or FRP. The use of FRP in the bleach plant has grown to such an extent in the past few years that complete washing systems—e.g. drums, piping, and tanks—are of this construction. Also today, the former stainless steel facing and backing wires on the washer drums are being replaced with Kynar* material.

There are many other corrosion-resistant alloys finding increased application in the bleach plant. For example, 254SMO, 904L, and Hastelloy C-276 are among many alloys used in various locations. Bleaching agents other than chlorine are also used to brighten fiber, including oxygen, peroxide, hydrosulfite, and ozone. In these systems, type 316L stainless steel is the most common material.

Large amounts of water are used in washing the bleached or unbleached pulp, and in transporting it to the paper mill. Stainless steel (types 304L and 316L) pumps and piping conduct the fiber/water slurry, called stock, through several stages of preparation for the paper machines. In the refining stages, fiber surfaces are prepared for paper formation by passing the stock between serrated plates, which rotate at high speeds. The refiner plates are commonly made of Ni-Hard or white cast iron, although precipitation-hardened stainless steels (such as 17–4 PH, AM355) are also used to combat erosion-corrosion (cavitation) damage. Refiner casing materials may be alloyed cast stainless iron (CA 15) or 300 series stainless steel. The stock is centrifugally cleaned in stainless steel, ceramic or high-density polyethylene conical-shaped cleaners prior to the paper machine. A closely controlled flow of stock is conducted through a polished stainless steel (type 316L or higher alloy) headbox onto a fine-mesh plastic fourdrinier wire. Water removal through the wire is accomplished with high-density polyethylene or ceramic foils and suction boxes. Suction couch and press rolls provide further dewatering when the fiber mat moves off the end of the fourdrinier section and into the press section. The rolls are perforated and equipped with internal suction boxes connected to vacuum systems to remove the process water, called white water (pH 4–7). The speed of rotation (around 600 rpm), and accompanying stress, provides conditions for corrosion fatigue, which is a

*Kynar is a trade name of Pennwalt Corporation.

failure mode for the rolls. At least 21 alloys are used for suction roll manufacture by casting or forging processes. Bronze alloys (e.g., 85 Cu, 5 Sn, 5 Pb, 5 Zn), and nickel aluminum bronze (84 Cu, 9 Al, 4 Ni), are used for lower stress conditions. Martensitic stainless steel (CA-15) or duplex alloy (26 Cr, 6 Ni) are used for conditions of higher stress. Austenitic alloys, such as CF-3M, are also available.

More detailed accounts of construction materials and their performance in the paper industry can be found in the four volumes of *Pulp and Paper Industry Corrosion Problems* (NACE, Houston, Texas, 1974, 1977, 1980, and 1983), and four papers in the January 1984 issue of *Materials Performance*.

8-22 Dew Point Corrosion

Section 1-6 classifies corrosion as wet or dry. Mechanisms and degree of attack can vary widely. When the temperature of the environment drops below the dew point, water, acid, or some other liquid condenses on exposed surfaces. Storage tanks, your auto gasoline tank, and other hollow structures are examples. Fossil fuel power plants often have long ducts connecting the combustion chambers and the brick-lined fume stacks. Sulfuric acid tends to condense, particularly during cold weather, before it reaches the stack and corrodes the ducts. FGD plants often reheat the treated gases to avoid condensation.

Water (dew) that condenses (during the night) on metal surfaces is usually saturated with oxygen and corrodes steel. Shaded or undersurfaces not exposed to sunlight could stay wet for a long time (sheltered corrosion).

Two of my experiences illustrate these points. A flagpole made of steel pipe at a gas station broke when a painter was on a ladder; he fell and was severely injured. A lawsuit resulted. Corrosion occurred on the inside of the pipe. The other example involved rusting of steel files (cutting) during manufacture and storage. The company air-conditioned the plant to 72°F to decrease humidity but the rusting problem became worse! When trucks entered the plant on a hot day, outside air rushed in and water condensed on the steel files. My "remuneration" for this investigation was a file to sharpen my lawn mover. This was OK because the company was an Ohio taxpayer!

8-23 Corrosion Under Insulation

This is a costly and difficult problem because untold number of vessels and miles of piping are insulated and the corrosion is hidden. An example of external stress-corrosion cracking (SCC) is shown in Fig. 3-67, and the attendant text describes the mechanism. The attack under insulation also includes general corrosion, pitting, and crevice corrosion. Such failures are distressing because they are not related to known process or cooling-water conditions. Like internal cracking, ESCC causes shutdowns.

Visible wet spots are a sign of trouble, and these areas can be investigated. Otherwise, inspection is difficult and not effective. A nondestructive method to determine corrosion under insulation would be a great boon. The Materials Technology Institute is sponsoring work on electromagnetic methods, but a viable method is not available at this writing. The method is difficult because the insulation is often covered with sheet metal (i.e., aluminum) to keep out water. Pipe configurations are sometimes complicated.

The best known way to control corrosion under insulation is to paint the metal with a good coating. For example, Simmons et al.* used an epoxy primer to protect tanks in the Space Shuttle program. Further information is in MTI Manual No. 15 (see Sec. 3-44).

8-24 Electronic Equipment

This equipment often contains a conglomeration of different materials. Gold-plated parts may be in contact with aluminum or solder. The parts are usually small in size, and a little corrosion can cause shorts and dysfunction. Many large computers are in air-conditioned and clean rooms. It would be nice if they could operate in ordinary rooms. Unfortunately, electronic equipment is often exposed to the atmosphere and pollutants that cause corrosion.

The American Society for Metals held a symposium in August 1984 and one session was on corrosion. Information should be published later by ASM. The July 1984 issue of the *Journal of Metals* contains several papers on electronic materials, but they discuss their metallurgy and nothing on corrosion.

Dobbs and Slenski† investigated a number of failures and state, "Experience has shown that about 20% of failed devices submitted for examination are caused by *corrosion*." Poor cleaning of solder flux, moisture penetration of coatings, lack of cleanliness, poor choice of materials, and cleaning with halogen-containing solutions are cited as causes of failure.

Actual service conditions are difficult to simulate in laboratory tests, and they may yield misleading results. Much work needs to be done in the area of electronic materials.

*J. R. Simmons, C. Johnson, A. Daech, and R. Mershell, Epoxy Primer/Foam Insulation on Aluminum Tanks for Liquid Oxygen and Nitrogen, *Materials Performance*, 14–18 (June 1982).

†B. Dobbs and G. Slenski, Investigation of Corrosion Related Failure in Electronic Systems, *Materials Performance*, 35–38 (Mar. 1984).

8-25 Liquid-Metal Embrittlement or Cracking

Liquid-metal-assisted cracking (LMC) usually does not involve corrosion (the electrochemical reduction of mercurous salts in aqueous solution is an exception). However, liquid-metal penetration along the grain boundaries results in cracking, and constitutes a very real problem. Many metal systems, including copper, aluminum, and titanium as well as the stainless and nickel alloys, are subject to LMC, usually by molten metals that react specifically with the alloy system.

Solutions of mercurous nitrate have long been used to detect residual stresses in yellow or cartridge brass, for example, but it is the elemental mercury that usually poses a problem in plant. Mercury is a liquid at room temperature, but it has a relatively high vapor pressure, enabling it to travel freely about a process stream. Mercury from broken thermometers or blown manometers is a real hazard for Alloy 400 components even at room temperature, and for Alloy 200 at elevated temperature. Traces of mercury have been reported to accelerate SCC of austenitic stainless steels in chloride solutions.

Copper, lead, cadmium, aluminum, and zinc can crack stainless steels and chromium-bearing nickel alloys at moderately elevated temperatures, the molten metal penetrating along the grain boundaries in a manner analogous to the penetration of steel by molten copper, or of high-strength copper alloys by molten tin or lead-tin solder. Lead-containing lubricants or antigalling compounds have reportedly caused cracking of Alloy 600 in nuclear applications. The higher alloys (e.g., Alloys G, B, C, etc.) are reported to resist molten zinc when stressed as U-bends.

Zinc is the most common offender, because of the widespread use of galvanized steel and zinc-pigmented primers in plant, but cadmium-plated hardware has also caused problem. Contamination of stainless steel by contact with galvanized wire during heat treatment is an established cause of LMC. Galvanized grating and handrails, as well as miscellaneous hardware (e.g., bolts, cables, chains, pipe hangers, etc.) are commonly employed to resist atmospheric corrosion. Zinc embrittlement from walkways during a plant fire has been reported. "Overspray" from priming of structural steel with zinc-pigmented primers or paints can contaminate stainless steels or Alloy 600, for example, and render the alloy unweldable until every trace of the zinc has been removed.

Special care must be taken in removing zinc contamination by grinding, because the grinding wheels can then spread zinc contamination by smearing. Zinc-pigmented paint must be removed by sandblasting followed by an acid wash, while metallic zinc is removed by the wash alone. Dilute nitric or phosphoric acid (the latter is safer for personnel) may be employed; usually a 10% solution suffices. After washing, a spot test for residual zinc may be employed using an ammonium acetate/ammonium fluoride solution mixed with diphenylthio-carbazone.

Welding of galvanized steel to austenitic stainless steels or Alloy 600 is not recommended. Many companies specifically prohibit such weldments. Attempts to accomplish this bimetallic weld have been made by grinding off the zinc coating, "buttering" the steel with a high-alloy weld deposit (e.g., type 309), and then welding the buttered end to the stainless equipment. Reportedly this has sometimes been successful in shop fabrication. In the field, attempts to employ this technique on fire safety spray rings, ladder and platform supports, etc., have led to cracking of stainless equipment. Also, cracks can occur under the weld, where they are not amenable to visual or dye-penetrant inspection, and in the carbon steel part as well, leading to a lack of structural integrity.

Zinc is not harmful in the ionic state (e.g., as a zinc chromate primer) unless subsequently reduced to the metallic state. It is conceivable that this could occur on exposure to fire under reducing conditions.

The vapors of molten zinc are readily oxidized in contact with hot air, which could render them harmless. It is certain, however, that one could not rely on this favorable condition in practice. Molten zinc and/or its unoxidized vapors pose a real hazard to stainless steel and chromium-bearing nickel alloy equipment.

For cracking to occur, certain combinations of circumstances must exist:

1. A temperature above 400°C (752°F).
2. Direct contact with the molten metal (e.g., zinc, cadmium-zinc eutectic, aluminum).
3. A highly stressed alloy surface. (*Note*: Residual stress in fabricated equipment is more than adequate to initiate LMC.)

The resulting cracking can constitute anything from a hazard to a potential disaster, depending upon the possibility of releasing a flammable, explosive, or lethal material. Obviously, stainless or nickel-alloy equipment should be shielded from possible contact with molten zinc, cadmium, or aluminum, if possible, and should be carefully inspected after fire exposure.

Carbon steel structural components on stainless or nickel-alloy equipment should be protected from atmospheric rusting by a zinc-free, heavy duty coating system, such as the high-build epoxies. All pipe fabrication and process vessel design criteria should specifically prohibit galvanizing or zinc-pigmented organic or inorganic paints on members to be welded to the equipment. Zinc-coated parts may be mechanically attached to such equipment if the service temperature is limited to 350°C (662°F), although they are a possible source of contamination in case of fire.

Nickel-base alloys can suffer intercrystalline cracking during heat-treatment or welding, due to sulfur contamination. Sulfur-bearing lubricants or cutting oils, or sulfate or other sulfurous contaminants, can cause cracking, especially during welding. Proper cleaning or decontamination before welding or heat-treating is an effective remedy.

An AIME book, *Embrittlement of Liquid and Solid Metals*, M. H. Kamdar, ed., 1983, includes liquid-metal embrittlement, solid-metal-induced embrittlement and embrittlement of nuclear reactor metals and alloys.

8-26 Hydrogen Peroxide

The problem here is not corrosion of metal but decomposition of the H_2O_2. For example, molybdenum in stainless alloys and other than high-purity aluminum alloys can cause decomposition. A paper by G. Ayling and H. M. Castrantas in *Chemical Engineering*, 80–82, Nov. 30, 1981, covers this subject nicely.

8-27 Rebar Corrosion

Corrosion of steel reinforcing bars is a problem in many structures (i.e., piers and buildings), but it is a *major* problem (because of deicing salts) in highway bridges. Costs are mentioned in Chapter 1. Thousands of bridges are in need of repair. Kilareski* considers the application of anodic protection, cathodic protection, and water-resistant coatings. Vrable† discusses galvanic corrosion.

A cornice on a building fell on a busy sidewalk in Columbus and injured several people in 1984. Examination showed that the supportive metal bars had corroded.

The theme of the July 1985 issue of *Materials Performance* is "Solving Rebar Corrosion in Concrete." Papers on bridges, offshore platforms, and other items are included.

8-28 Bolting

Many failures of bolts occur in flanges, pumps, valves, and other structures probably because not enough attention is paid to bolts when specifying materials. MTI Publication No. 14, *Closure Bolting for the Chemical Process Industries* (1984) is a comprehenisve review of this problem; for example, it lists "Basic Bolting Dos and Don'ts," such as *do* review vendors' bolting specifications and *don't* use austenitic stainless steels where there is a potential

*W. P. Kilareski, Corrosion Induced Deterioration of Reinforced Concrete—An Overview, *Materials Performance*, 48–50 (Mar. 1980).

†J. B. Vrable, Galvanic Corrosion of Reinforcing Steel Exposed to Concrete and Water, *Materials Performance*, 51–52 (Mar. 1982).

of exposure to chloride ions. Since bolting is subject to stresses, SCC should be considered.

In one serious case, involving an oxidizing environment, the valves were stainless steel but the bolts were a copper alloy!

8-29 Statue of Liberty

No discussion of corrosion would be complete without mention of the restoration of the Statue of Liberty, which is under way and should be completed in 1986. The Grand Old Lady has suffered the ravages of corrosion. The structure consists of an iron framework, with a $\frac{3}{32}$-in. copper sheet shell or robe. The copper in the torch is thinner because of its intricate design. The wrought iron ribs making up the strap work have lost up to two-thirds of their original cross-sectional area because of rusting. Copper saddles hold the copper sheet to the iron. Voluminous iron rust caused failure of the saddles because of the wedging action of corrosion products. Rain enters the structure and, in addition, the marine environment is naturally corrosive. One suggestion is that acid rain during recent years has resulted in a soluble patina on the copper, which has replaced the protective insoluble patina. Restoration involves better design; sealing, ventilation, and air conditioning; and the use of stainless steels and Ferallium* (donated by producers). N. A. Nielson describes the situation in "Keeping the Torch Lit," *Materials Performance*, 78–83, April 1984.

*Ferallium is a slight modification of CD4MCu, which was originally patented by Ohio State University (Fontana, Beck, and Bidwell).

CHAPTER

NINE

MODERN THEORY—PRINCIPLES

9-1 Introduction

In this chapter the modern electrochemical theory of aqueous corrosion is introduced and described. Although the presentation is necessarily simplified and brief, all generalizations and simplifications are clearly indicated. Equation derivations have been omitted for clarity and brevity. The principles described in this chapter differ from classic corrosion theories in that no mention is made of local anodes and cathodes. These concepts are not in conflict with classic corrosion theories, but represent a different approach to the problem of understanding the electrochemical behavior of corrosion reactions. We have chosen to utilize this particular approach since it has proved to be an unusually useful method for understanding and controlling corrosion as evidenced by the recent numerous advances in the corrosion field. Anodic protection, noble metal alloying, and methods for rapidly measuring corrosion rates by electrochemical measurements have been developed in the past 20 years by application of modern electrode kinetic principles.

This chapter is divided into two main sections: thermodynamic principles and electrode kinetics. Applications of these principles are described in Chap. 10.

THERMODYNAMICS

Thermodynamics, the science of energy changes, has been widely applied to corrosion studies for many years. Below, the principles applicable to corrosion phenomena and their limitations are reviewed.

9-2 Free Energy

The change in free energy ΔG is a direct measure of the work capacity or maximum electric energy available from a system. If the change in free energy accompanying the transition of a system from one state to another is negative, this indicates a loss in free energy and also the spontaneous reaction direction of the system. That is, if no external forces act on the system, the system will tend to transform to its lowest energy state. If the change in free energy is positive, this indicates that the transition represents an increase in energy, and this requires that additional energy be added to the system. These principles are illustrated in Fig. 9-1 by a mechanical analogy. If the ball moves from position 1 to position 2, this represents a decrease in free energy. The transition from position 1 to position 2 is the spontaneous direction for this particular system. The reverse transformation—from position 2 to position 1—is not a spontaneous direction and requires the application of energy.

The change in free energy is a state function and is independent of the reaction path. This is illustrated in Fig. 9-2, which is similar to Fig. 9-1 except that there are two possible reaction paths *A* and *B*. For either path, free-energy change for the transition from state 1 to state 2*A* or 2*B* is exactly the same. It is obvious, however, that the transformation along path *B* will require more time and will be slower than along path *A*. Chemical and corrosion reactions behave in exactly the same fashion. It is not possible to accurately predict the velocity of a reaction from the change in free energy. This parameter reflects *only* the direction of reaction by its sign, and any predictions of velocity based on the magnitude of the change in free energy may be erroneous, as illustrated in Fig. 9-2.

The free-energy change accompanying an electrochemical reaction can be calculated by the following equation:

$$\Delta G = -nFE \tag{9.1}$$

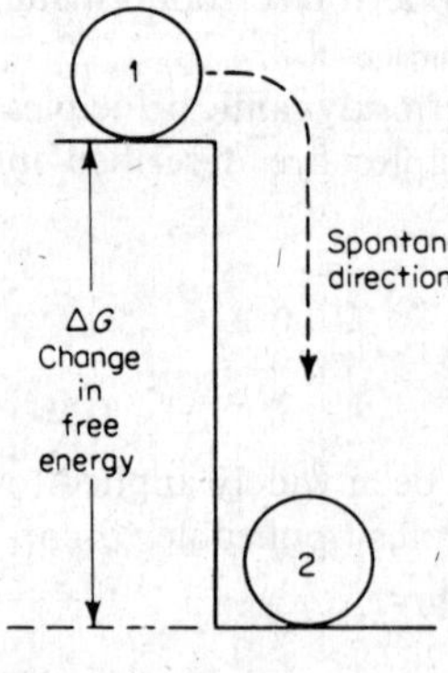

Figure 9-1 Mechanical analogy of free-energy change.

where ΔG is the free-energy change, n is the number of electrons involved in the reaction, F is the Faraday constant, and E equals the cell potential. Below we shall discuss how cell potential E is determined.

9-3 Cell Potentials and the EMF Series

The change in free energy accompanying an electrochemical or corrosion reaction can be calculated from a knowledge of the cell potential of the reaction. Since there has been considerable confusion regarding sign conventions in electrochemistry, we shall use actual electrode potentials in all discussions.

To illustrate the principle of a reversible cell potential, consider the replacement reaction between copper and zinc occurring at equilibrium:

$$Cu + Zn^{2+} = Cu^{2+} + Zn \tag{9.2}$$

The above reaction is written with an equal sign to indicate an equilibrium reaction. Considering zinc and copper electrodes in equilibrium with their ions, it is apparent that Eq. (9.2) represents their summation. Thus

$$Cu = Cu^{2+} + 2e \tag{9.3}$$

$$+Zn^{2+} + 2e = Zn \tag{9.4}$$

$$\overline{Cu + Zn^{2+} = Cu^{2+} + Zn} \tag{9.2}$$

To study the free-energy change associated with the above reaction, we can construct an electrochemical cell containing copper and zinc electrodes in equilibrium with their ions separated by a porous membrane to retard mixing, as illustrated in Fig. 9-3. For purposes of simplicity, the concentrations of metal ions are maintained at unit activity; each solution contains approximately 1 gram-atomic weight of metal ion per liter. It is necessary that both electrodes be at equilibrium. That is, the reactions in each compartment are represented by Eqs. (9.3) and (9.4), and the rates of metal dissolution and deposition must be the same; there is no net change in the system. This is illustrated in Fig. 9-4. At certain points on the metal surface,

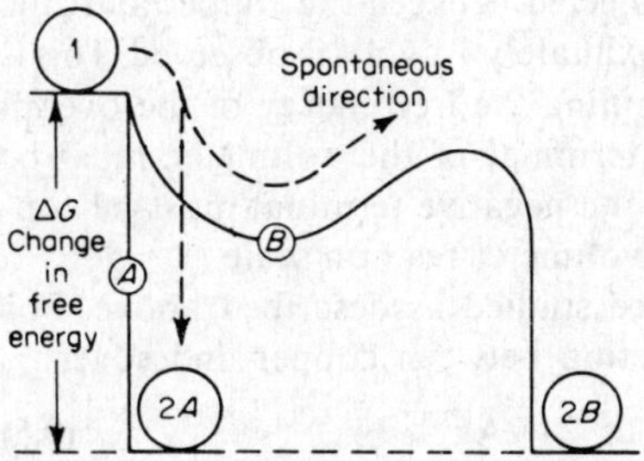

Figure 9-2 Effect of reaction path on reaction rate.

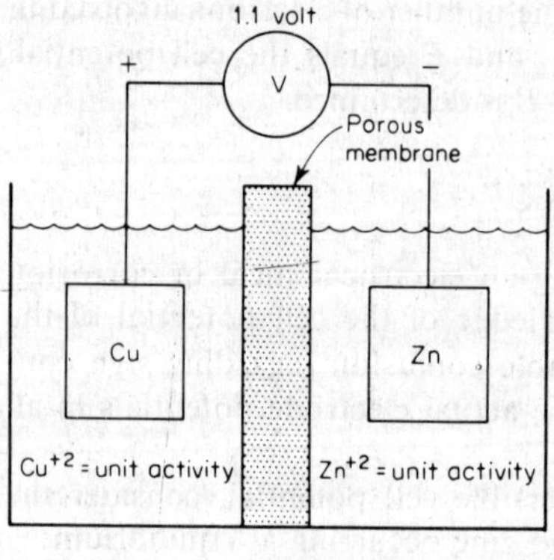

Figure 9-3 Reversible cell containing copper and zinc in equilibrium with their ions.

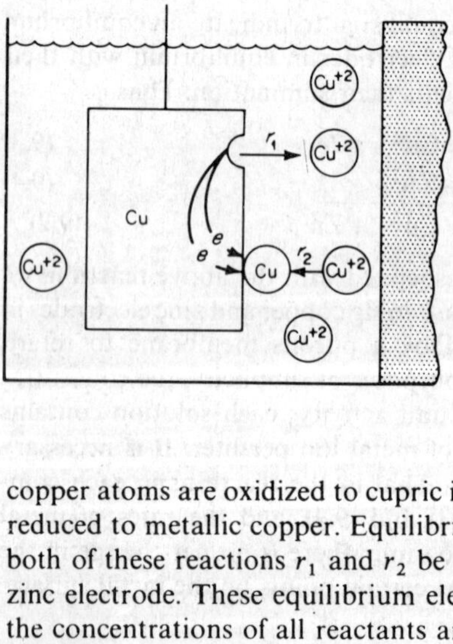

Figure 9-4 Reversible copper electrode.

copper atoms are oxidized to cupric ions, and at other points cupric ions are reduced to metallic copper. Equilibrium conditions dictate that the rates of both of these reactions r_1 and r_2 be equal. Similar restrictions apply to the zinc electrode. These equilibrium electrodes are called *half-cells*, and when the concentrations of all reactants are maintained at unit activity, they are termed *standard half-cells*.

If a high-resistance voltmeter is connected between the copper and zinc electrodes, a potential difference of approximately 1.1 volts is observed. This is the cell potential that is used in determining the free energy of the overall electrochemical reaction. The positive terminal of the voltmeter must be connected to the copper electrode, and the negative terminal must be connected to the zinc electrode to have the voltmeter read on scale.

Any electrochemical reaction can be studied as described above. For example, consider the replacement reaction between copper and silver:

$$Cu + 2Ag^{+} = Cu^{2+} + 2Ag \tag{9.5}$$

To study this reaction, reversible electrodes of copper and silver can be established as shown in Fig. 9-5, and the potential difference between the two electrodes is 0.45 volt. In this cell, copper is negative with respect to silver.

Although it is usually possible to establish a reversible electrochemical cell for any given reaction, the fact that there is an infinite number of combinations makes both the experimental measurements and tabulations of such data almost impossible. To simplify the representation and calculations of cell potentials, the concept of half-cell potentials has been developed. An arbitrary half-cell reaction is used as a reference by defining its potential as zero, and all other half-cell potentials are calculated with respect to this zero reference. Although any half-cell reaction can be chosen for this standard reference point, the hydrogen-hydrogen ion reaction ($2H^+ + 2e = H_2$) is universally accepted, since it is relatively easy to establish a reversible hydrogen electrode. Consider for example the reaction between zinc and hydrogen ions:

$$Zn + 2H^+ = Zn^{2+} + H_2 \tag{9.6}$$

This reaction can be divided into two half-cell reactions:

$$Zn = Zn^{2+} + 2e \tag{9.7}$$

$$2H^+ + 2e = H_2 \tag{9.8}$$

Since it is not possible to make an electrode from hydrogen gas, an inert electrode is used. As shown in Fig. 9-6, a reversible divided cell between zinc and its ions and hydrogen gas and hydrogen ions is established by using zinc and platinum electrodes. The platinum electrode acts as an inert substrate for the electrochemical reaction as shown in Fig. 9-7. At different points on the platinum electrode, hydrogen ions are reduced to hydrogen gas and hydrogen gas is oxidized to hydrogen ions, with electron transfer occurring between these points. It is important to note that the platinum

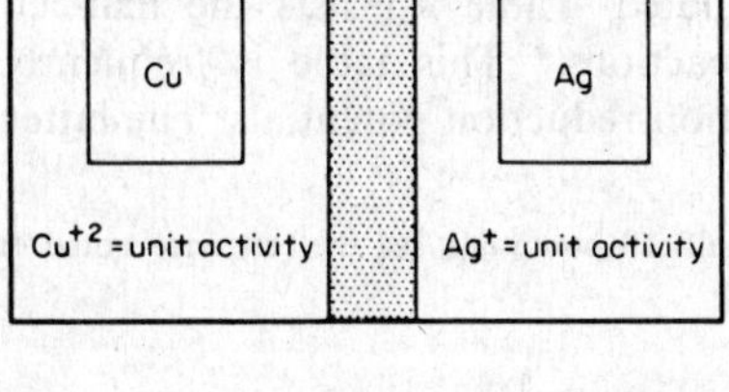

Figure 9-5 Reversible cell containing copper and silver in equilibrium with their ions.

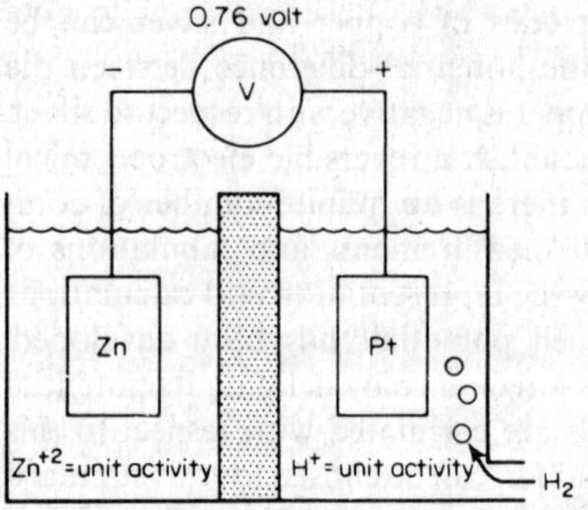

Figure 9-6 Cell containing reversible zinc and hydrogen electrodes.

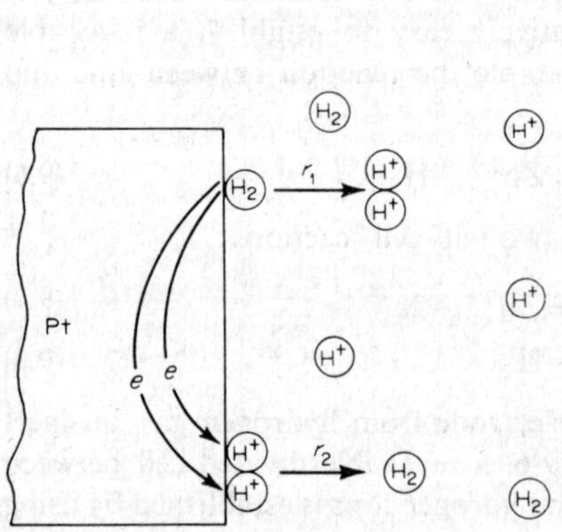

Figure 9-7 Hydrogen electrode on platinum.

electrode does not take part in this reaction but merely serves as a solid interface at which this reaction can occur. Many metals function as reversible hydrogen electrodes; platinum is usually preferred due to its inertness and the ease with which electron transfer occurs on its surface. As with other half-cell electrodes, the concentration of hydrogen ions is maintained at unit activity, and the solution is saturated with hydrogen gas at 1 atmosphere pressure.

The cell as shown in Fig. 9-6 has a potential of 0.763 and zinc is negative with respect to the hydrogen electrode. Using the above convention, the hydrogen electrode is defined as having a potential of zero, and it follows that the potential of the zinc half-cell is −0.76 volt. In a similar fashion, other half-cell potentials can be calculated. Table 9-1 lists the half-cell potentials for some electrochemical reactions.* This table is frequently called the emf series, half-cell, or oxidation-reduction potentials. The latter

*For a very complete and accurate list of half-cell potentials, see the reference noted in Table 9-1.

term is frequently abbreviated as *redox* potentials. It is important to note that in all cases these potentials refer to electrodes in which all reactants are at unit activity and at 25°C.

From the data presented in Table 9-1, it is possible to calculate the cell potential for numerous electrochemical reactions. Note that thc absolute potential difference between the copper electrode and zinc electrode is approximately 1.1 volts and that the copper electrode is positive with respect

Table 9-1 Standard oxidation-reduction (redox) potentials

25°C, volts vs. normal hydrogen electrode*

$Au = Au^{3+} + 3e$	+1.498
$O_2 + 4H^+ + 4e = 2H_2O$	+1.229
$Pt = Pt^{2+} + 2e$	+1.2
$Pd = Pd^{2+} + 2e$	+0.987
$Ag = Ag^+ + e$	+0.799
$2Hg = Hg_2^{2+} + 2e$	+0.788
$Fe^{3+} + e = Fe^{2+}$	+0.771
$O_2 + 2H_2O + 4e = 4\ OH$	+0.401
$Cu = Cu^{2+} + 2e$	+0.337
$Sn^{4+} + 2e = Sn^{2+}$	+0.15
$2H^+ + 2e = H_2$	0.000
$Pb = Pb^{2+} + 2e$	−0.126
$Sn = Sn^{2+} + 2e$	−0.136
$Ni = Ni^{2+} + 2e$	−0.250
$Co = Co^{2+} + 2e$	−0.277
$Cd = Cd^{2+} + 2e$	−0.403
$Fe = Fe^{2+} + 2e$	−0.440
$Cr = Cr^{3+} + 3e$	−0.744
$Zn = Zn^{2+} + 2e$	−0.763
$Al = Al^{3+} + 3e$	−1.662
$Mg = Mg^{2+} + 2e$	−2.363
$Na = Na^+ + e$	−2.714
$K = K^+ + e$	−2.925

*Electrode potential values are given and are invariant (e.g., $Zn = Zn^{2+} + 2e$, and $Zn^{2+} + 2e = Zn$, are identical and represent zinc in equilibrium with its ions with a potential of −0.763 volts vs. normal hydrogen electrode).

Source: A. J. de Bethune and N. A. S. Loud, *Standard Aqueous Electrode Potentials and Temperature Coefficients at* 25°C, Clifford A. Hampel, Skokie, Ill., 1964.

to the zinc. Likewise, the potential difference between the copper and silver electrode is 0.45 volt and copper is negative with respect to silver. Thus, the use of redox potentials greatly simplifies the calculation of cell potentials.

To determine the potential of a system in which the reactants are not at unit activity, the familiar Nernst equation can be employed; that is,

$$E = E_0 + 2.3\frac{RT}{nF}\log\frac{a_{oxid}}{a_{red}} \tag{9.9}$$

where E is the half-cell potential, E_0 the standard half-cell potential, R is the gas constant, T is absolute temperature, n is the number of electrons transferred, F is the Faraday constant, a_{oxid} and a_{red} are the activities (concentrations) of oxidized and reduced species. As indicated in the above equation, half-cell potential becomes more positive as the amount of oxidized species increases. For each tenfold increase in oxidized reactant, the half-cell potential increases by 59 mV for a single electron reaction.

9-4 Applications of Thermodynamics to Corrosion

As discussed above, there is a definite relation between the free-energy change and the cell potential of an electrochemical reaction. In most instances, the actual magnitude of the free-energy change is relatively unimportant in corrosion applications. The most important factor is the sign of the free-energy change for a given reaction, since this indicates whether or not the reaction is spontaneous. Hence, although Eq. (9.1) forms the basis for thermodynamic calculations, it is rarely used in studying corrosion phenomena. However, a simple rule derived from Eq. (9.1) is used to predict the spontaneous direction of any electrochemical reaction. This rule can be simply stated as: *In any electrochemical reaction, the most negative or active half-cell tends to be oxidized, and the most positive or noble half-cell tends to be reduced.* Considering Table 9-1 and applying this rule, the spontaneous direction of the zinc-copper replacement reaction given in Eq. (9.2) is toward the oxidation of zinc and the deposition of copper. Similarly, in the copper-silver replacement reaction, Eq. (9.5), copper tends to be oxidized and the silver ions tend to be reduced, indicating that the spontaneous direction of this reaction is to the right, as written.

Redox potentials are very useful in predicting corrosion behavior. From the above rule it follows that all metals with reversible potentials more active (negative) than hydrogen will tend to be corroded by acid solutions. Copper and silver, which have more noble potentials, are not corroded in acid solutions. Although copper and silver are not corroded by acid solutions, if dissolved oxygen is present there is a possibility of oxygen reduction. Table 9-1 indicates that in the presence of oxygen, copper and silver tend to corrode spontaneously. For example:

$$Cu + H_2SO_4 = \text{no reaction} \qquad (Cu/Cu^{2+} \text{ more} + \text{than } H_2/H^+)$$
$$2Cu + 2H_2SO_4 + O_2 \rightleftharpoons 2CuSO_4 + 2H_2O \qquad (O_2/H_2O \text{ more} + \text{than } Cu/Cu^{2+})$$

As the reversible potential of a metal becomes more noble, its tendency to corrode in the presence of oxidizing agents decreases. Hence, the metals at the uppermost part of the redox series, such as platinum and gold, are very inert, since there will be no tendency to corrode except in the presence of extremely powerful oxidizing agents. It is important to note that all of the above discussions refer to systems at unit activity. Since half-cell potentials change with concentration, Nernst calculations must be made before making predictions about spontaneous direction at concentrations other than unit activity.

Thermodynamics or, more specifically, half-cell potentials can be used to state a criterion for corrosion. *Corrosion will not occur unless the spontaneous direction of the reaction indicates metal oxidation.* As mentioned above, it is important to remember that although the spontaneous direction of a reaction may be in the direction of metal corrosion, this does not necessarily indicate that corrosion will occur. If the reaction proceeds at a negligible rate, then the metal will be essentially inert. Hence, the major use of thermodynamic calculations is a negative one. That is, thermodynamics can indicate unambiguously that corrosion will not occur. However, corrosion may or may not occur if the reaction direction indicates metal oxidation.

The applications of thermodynamics to corrosion phenomena have been further generalized by means of potential-pH plots. These are frequently called Pourbaix diagrams, after Dr. M. Pourbaix who first suggested their use. The potential-pH diagram for iron is shown in Fig. 9-8. Such diagrams are constructed from calculations based on the Nernst equation (9.9 and

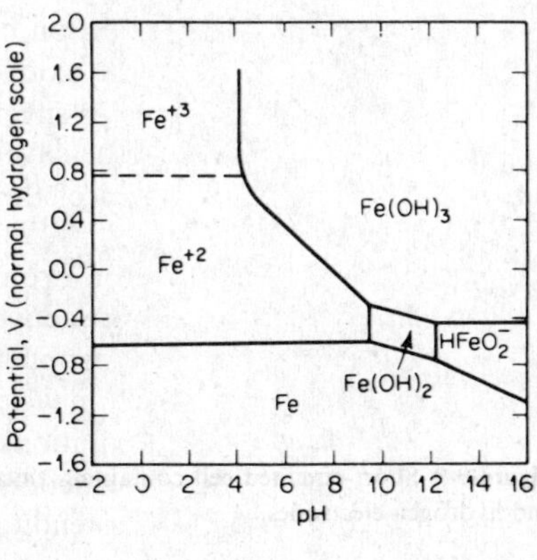

Figure 9-8 Simplified potential-pH diagram for the Fe-H_2O system. (*M. Pourbaix, Atlas of Electrochemical Equilibria in Aqueous Solutions*, pp. 307–321, Pergamon Press, New York, 1966.)

solubility data for various metal compounds. As shown, it is possible to delineate areas in which iron, iron hydroxide, ferrous ions, etc., are thermodynamically stable. That is, these forms represent states of lowest free energy.

The main uses of these diagrams are (1) predicting the spontaneous direction of reactions, (2) estimating the composition of corrosion products, and (3) predicting environmental changes that will prevent or reduce corrosive attack. For example, the large region in Fig. 9-8 labeled Fe indicates that iron is inert under these conditions of potential and pH.

Potential-pH diagrams are subject to the same limitations as any thermodynamic calculation. They represent equilibrium conditions and should never be used to predict the velocity of a reaction. Unfortunately, many investigators have ignored this limitation, which has been repeatedly stated by Pourbaix, and have attempted to use these plots to predict reaction kinetics. Similar diagrams have been prepared for other metals (see Suggested Reading at the end of Sec. 9-11).

ELECTRODE KINETICS

From an engineering standpoint, the major interest is in the kinetics or rate of corrosion. Corroding systems are not at equilibrium, and therefore thermodynamic calculations cannot be applied. Essentially, we are interested in what happens when cells such as have been described above are short-circuited, as illustrated in Fig. 9-9. In this instance a vigorous reaction occurs: the zinc electrode rapidly dissolves in the solution and simultaneously a rapid evolution of hydrogen is observed at the platinum electrode. Electrons released from the zinc-dissolution reaction are transferred through the connecting wire to the platinum electrode where they are consumed in the hydrogen-reduction reaction. The process that occurs in Fig. 9-9 is exactly the same process that occurs when zinc metal is immersed in a hydrogen-

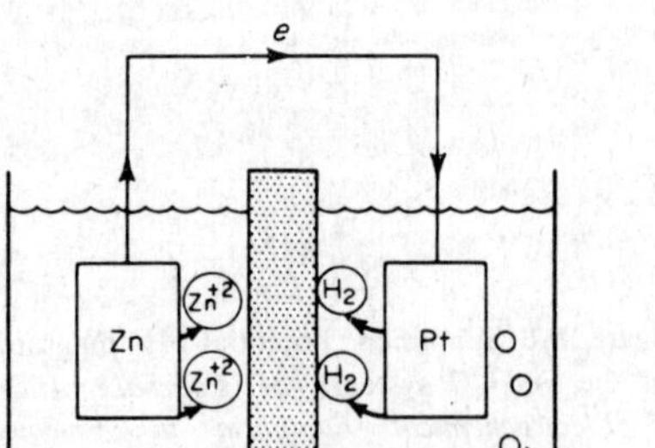

Figure 9-9 Short-circuited cell containing zinc and hydrogen electrodes.

saturated acid solution containing zinc ions, as shown in Fig. 9-10. In both instances the overall reaction is the dissolution of zinc and the evolution of hydrogen. In the divided cell shown in Fig. 9-9, the reactions occur on separate electrodes, whereas in Fig. 9-10 these reactions occur on the same metal surface. In both instances, the free-energy change for the reaction is exactly the same, since the platinum metal does not participate in the reaction.

Before proceeding with discussions of electrode-kinetics principles, several useful terms should be defined. *Anode* refers to an electrode at which a net oxidation process occurs, and *cathode* refers to an electrode at which a net reduction reaction occurs. Referring to Fig. 9-9, the zinc electrode is the anode, and the platinum or hydrogen electrode is the cathode in this particular cell when the terminals are short-circuited. Similarly, *anodic reaction* is synonymous with an oxidation reaction and *cathodic reaction* is synonymous with a reduction reaction. When a cell such as is shown in Fig. 9-9 is short-circuited, and net oxidation and reduction processes occur at the electrode interfaces, the potentials of these electrodes will no longer be at their equilibrium potential. This deviation from equilibrium potential is called *polarization*. Polarization can be defined as the displacement of electrode potential resulting from a net current. The magnitude of polarization is frequently measured in terms of *overvoltage*. Overvoltage, usually abbreviated as η, is a measure of polarization with respect to the equilibrium potential of an electrode. That is, the equilibrium potential of an electrode is considered as zero, and the overvoltage is stated in terms of volts or millivolts plus or minus with respect to this zero reference.

To illustrate overvoltage, consider that the zinc electrode which is coupled to the platinum electrode in Fig. 9-8 has a potential after coupling of -0.66 volt. Thus, its overvoltage is $+100$ mV or $+0.10$ volt. The terminology described above will be used consistently throughout the rest of this chapter and in Chap. 10.

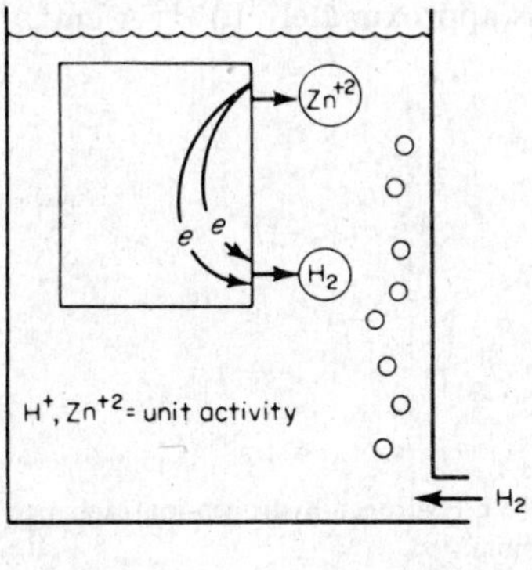

Figure 9-10 Corroding zinc, shown schematically.

9-5 Exchange Current Density

Consider the reversible hydrogen electrode established on platinum as shown in Fig. 9-7. Equilibrium conditions dictate that the rate of oxidation and reduction, r_1 and r_2 respectively, must be equal. Hence, at an equilibrium hydrogen electrode there is a finite rate of interchange between hydrogen molecules and hydrogen ions in solution. By plotting electrode potential versus reaction rate as shown in Fig. 9-11, it is possible to establish a point corresponding to the platinum-hydrogen electrode. This point represents the particular exchange reaction rate of the electrode expressed in terms of moles reacting per square centimeter per second. Note that there is no net reaction, since both the oxidation and reduction rates are equal; the exchange reaction rate is equal to the rate of oxidation and reduction. The exchange reaction rate can be more conveniently expressed in terms of current density. Since two electrons are consumed during the reduction of the two hydrogen ions, and since two electrons are released during the oxidation of the single hydrogen molecule, the reaction rate can be expressed in terms of current density. More precisely, the relationship between exchange reaction rate and current density can be directly derived from Faraday's law:

$$r_{oxid} = r_{red} = \frac{i_0}{nF} \tag{9.10}$$

where r_{oxid} and r_{red} are the equilibrium oxidation and reduction rates and i_0 is the exchange current density; n and F have been defined previously.

Exchange current density i_0 is the rate of oxidation and reduction reactions at an equilibrium electrode expressed in terms of current density. Exchange current density is a misnomer since there is no net current. It is merely a convenient way of representing the rates of oxidation and reduction at equilibrium. It is observed that the exchange current density varies depending on the metal electrode as shown in Fig. 9-12. In this figure, a logarithmic current-density scale is used to permit a wide range of points to be plotted. Note that the exchange current density for platinum is approximately 1 mA/cm², and that for mercury it is approximately 10^{-12} A/cm².

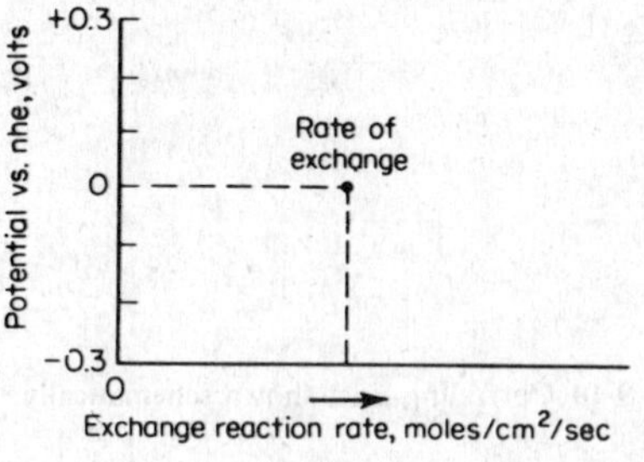

Figure 9-11 Hydrogen-hydrogen-ion exchange on platinum.

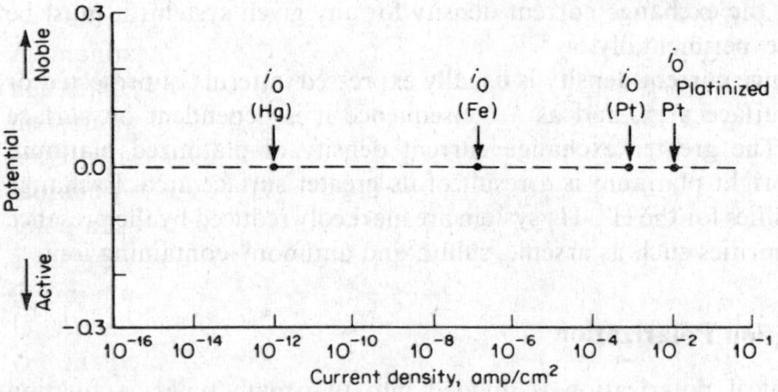

Figure 9-12 Hydrogen-hydrogen-ion exchange current densities.

Table 9-2 lists some experimentally determined exchange current densities. The magnitude of exchange current density is a function of several variables. First, it is a specific function of the particular redox reaction. Further, as shown in Fig. 9-12 and Table 9-2, it is also related to electrode composition. Like reversible potentials, exchange current densities are influenced by the ratio of oxidized and reduced species that are present and by the temperature of the system. There is no thereotical way of precisely

Table 9-2 Exchange current densities

Reaction	Electrode	Solution	i_0, A/cm^2
$2H^+ + 2e = H_2$	Al	$2N\ H_2SO_4$	10^{-10}
$2H^+ + 2e = H_2$	Au	$1N$ HCl	10^{-6}
$2H^+ + 2e = H_2$	Cu	$0.1N$ HCl	2×10^{-7}
$2H^+ + 2e = H_2$	Fe	$2N\ H_2SO_4$	10^{-6}
$2H^+ + 2e = H_2$	Hg	$1N$ HCl	2×10^{-12}
$2H^+ + 2e = H_2$	Hg	$5N$ HCl	4×10^{-11}
$2H^+ + 2e = H_2$	Ni	$1N$ HCl	4×10^{-6}
$2H^+ + 2e = H_2$	Pb	$1N$ HCl	2×10^{-13}
$2H^+ + 2e = H_2$	Pt	$1N$ HCl	10^{-3}
$2H^+ + 2e = H_2$	Pd	$0.6N$ HCl	2×10^{-4}
$2H^+ + 2e = H_2$	Sn	$1N$ HCl	10^{-8}
$O_2 + 4H^+ + 4e = 2H_2O$	Au	$0.1N$ NaOH	5×10^{-13}
$O_2 + 4H^+ + 4e = 2H_2O$	Pt	$0.1N$ NaOH	4×10^{-13}
$Fe^{3+} + e = Fe^{2+}$	Pt		2×10^{-3}
$Ni = Ni^{2+} + 2e$	Ni	$0.5N\ NiSO_4$	10^{-6}

Source: J. O'M. Bockris, Parameters of Electrode Kinetics, *Electrochemical Constants*, NBS Circular 524, U.S. Government Printing Office, Washington, D.C., pp. 243–262, 1953.

determining the exchange current density for any given system; it must be determined experimentally.

Exchange current density is usually expressed in terms of projected or geometric surface area, and as a consequence it is dependent on surface roughness. The greater exchange current density of platinized platinum relative to bright platinum is a result of its greater surface area. Exchange current densities for the H^+-H_2 system are markedly reduced by the presence of trace impurities such as arsenic, sulfur, and antimony-containing ions.

9-6 Activation Polarization

Electrochemical polarization is divided into two main types—activation and concentration polarization. Activation polarization refers to electrochemical reactions that are controlled by a slow step in the reaction sequence. As discussed in Chap. 2, this slow step during hydrogen evolution might be the electron-transfer step or the formation of hydrogen molecules: The relationship between reaction rate and overvoltage for activation polarization is:

$$\eta_a = \pm\beta \log \frac{i}{i_0} \tag{9.11}$$

where η_a is overvoltage, β is a constant, and i is the rate of oxidation or reduction in terms of current density. Equation (9.11) is called the Tafel equation, and β is frequently termed "β slope" or Tafel constant.* Equation (9.11) is graphically illustrated in Fig. 9-13. If a logarithmic current scale is

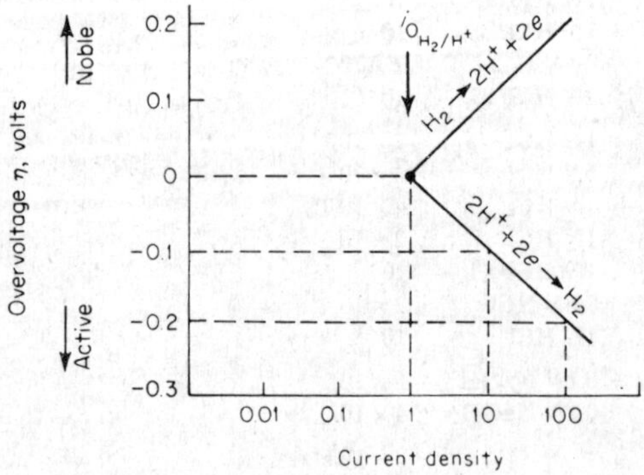

Figure 9-13 Activation-polarization curve of a hydrogen electrode.

*β represents the expression $2.3RT/\alpha nF$, where R, T, n, and F are as before, and α is the symmetry coefficient, which describes the shape of the rate-controlling energy barrier. See Bockris for a detailed derivation of the Tafel equation (Suggested Reading at end of Sec. 9-11).

used, the relationship between overvoltage or potential and current density is a linear function.* The value of β for electrochemical reactions ranges between 0.05 and 0.15 volt. In general, the value of β is usually 0.1 volt. The significance of this parameter can be seen upon examination of Fig. 9-13. Here the oxidation and reduction reactions corresponding to a hydrogen electrode are plotted with a beta value of 0.1 volt. Note that the reaction rate changes by one order of magnitude for each 100-mV, or 0.1-volt, change in overvoltage. This illustration shows that the reaction rate of an electrochemical reaction is very sensitive to small changes in electrode potential. Further, it can be seen that at all potentials more noble than the reversible potential a net oxidation process occurs, and that at all potentials more active or more negative than the reversible potential a net reduction occurs. At the reversible potential, or at zero overvoltage, there is no net rate of oxidation or reduction since both rates are equal at this intersection point.

9-7 Concentration Polarization

To illustrate the phenomenon of concentration polarization, consider the hydrogen-evolution reaction. As shown schematically in Fig. 9-14, at low reduction rates the distribution of hydrogen ions in the solution adjacent to the electrode surface is relatively uniform. At very high reduction rates the region adjacent to the electrode surface will become depleted of hydrogen ions. If the reduction rate is increased further, a limiting rate will be reached that is determined by the diffusion rate of hydrogen ions to the electrode

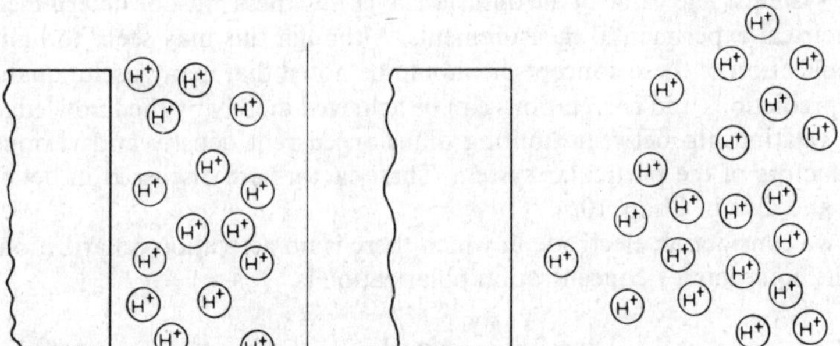

Figure 9-14 Concentration gradients during hydrogen evolution, shown schematically.

*This applies only at overvoltages greater than approximately ± 50 mV. However, since polarization during corrosion reactions is usually greater than ± 50 mV, a linear relationship can be generally assumed.

surface. This limiting rate is the *limiting diffusion current density* i_L. It represents the maximum rate of reduction possible for a given system; the equation expressing this parameter is

$$i_L = \frac{DnFC_B}{x} \tag{9.12}$$

where i_L is the limiting diffusion current density, D is the diffusion coefficient of the reacting ions, C_B is the concentration of the reacting ions in the bulk solution, and x is the thickness of the diffusion layer.*

Equation (9.12) shows that limiting diffusion current is a function of the diffusion coefficient, the concentration of reacting ions in solution, and the thickness of the diffusion layer. Changes that affect these parameters influence the limiting diffusion current. It is generally observed that there is a linear relationship between the concentration of reactive ions in solution and the limiting diffusion current density, as indicated in Eq. (9.12). The diffusion layer thickness is influenced by the shape of the particular electrode, the geometry of the system, and by agitation. Agitation tends to decrease the diffusion layer thickness because of convection currents and consequently increases the limiting diffusion current density. Limiting diffusion current density is usually only significant during reduction processes and is usually negligible during metal-dissolution reactions. Hence, limiting diffusion current density can be ignored during most metal-dissolution reactions. The reason for this is, simply, that there is an almost unlimited supply of metal atoms for dissolution. Although the limiting diffusion current density of a particular system is precisely defined by Eq. (9.12), the magnitude of the diffusion layer thickness is extremely difficult to calculate except for very simple systems. The value of the diffusion layer thickness must be determined by empirical experimental measurements. Although this may seem to limit the application of these concepts, it should be noted that many useful qualitative predictions and correlations can be achieved merely by the knowledge of the relationship between limiting diffusion current density and various other factors of the particular system. These factors are discussed in detail below and also in Chap. 10.

If we consider an electrode in which there is no activation polarization, then the equation for concentration polarization is

$$\eta_c = 2.3\frac{RT}{nF}\log\left(1 - \frac{i}{i_L}\right) \tag{9.13}$$

where the terms are as defined in Eqs. (9.9) and (9.12). A graphical representation of the Eq. (9.13) is shown in Fig. 9-15. Concentration polarization does not become apparent until the net reduction current density approaches the

*See Glasstone and Potter for the derivation of this equation (Suggested Reading at end of Sec. 9-11).

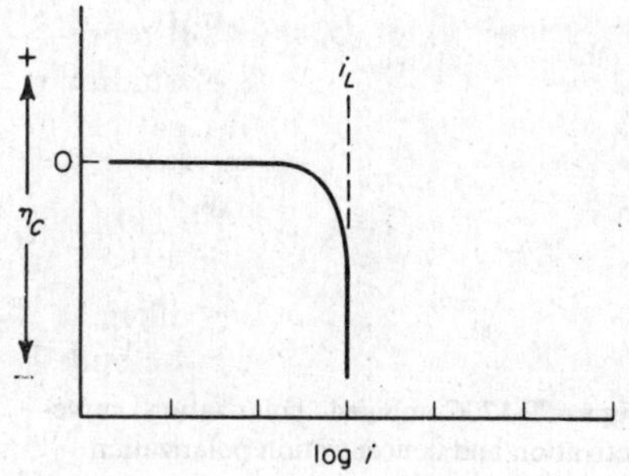

Figure 9-15 Concentration polarization curve (reduction process).

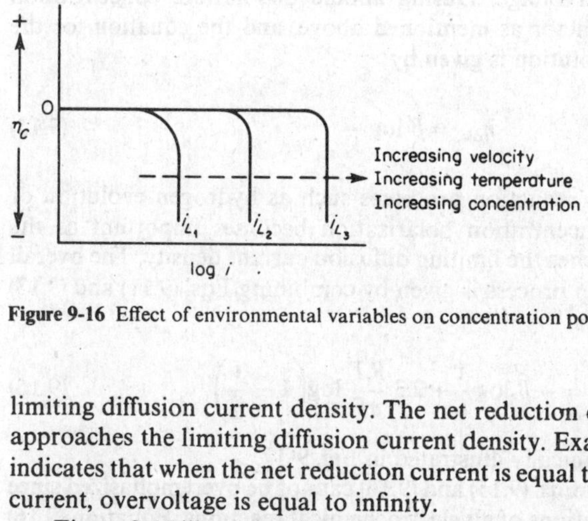

Figure 9-16 Effect of environmental variables on concentration polarization curve.

limiting diffusion current density. The net reduction current asymptotically approaches the limiting diffusion current density. Examination of Eq. (9.13) indicates that when the net reduction current is equal to the limiting diffusion current, overvoltage is equal to infinity.

Figure 9-16 illustrates the effects of changing limiting diffusion current on the shape of the polarization curve encountered during concentration polarization. As the solution velocity, concentration, or temperature are increased, limiting diffusion current increases since all of these factors exert an influence, as indicated in Eq. (9.12).

9-8 Combined Polarization

Both activation and concentration polarization usually occur at an electrode. At low reaction rates, activation polarization usually controls, whereas at higher reaction rates concentration polarization becomes controlling. The total polarization of an electrode is the sum of the contributions of activation polarization and concentration polarization:

$$\eta_T = \eta_a + \eta_c \tag{9.14}$$

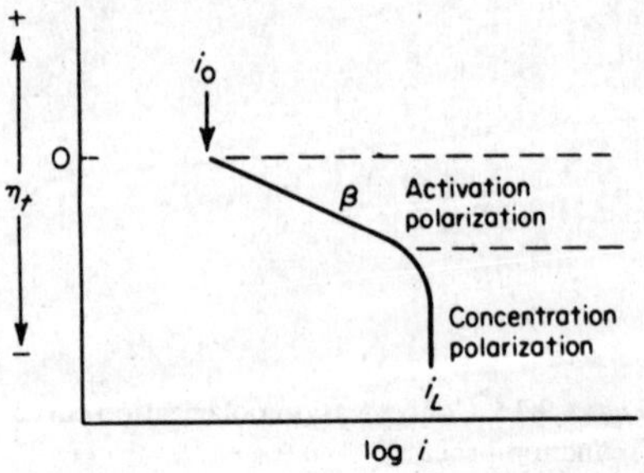

Figure 9-17 Combined polarization curve—activation and concentration polarization.

where η_T is total overvoltage. During anodic dissolution, concentration polarization is not a factor as mentioned above, and the equation for the kinetics of anodic dissolution is given by:

$$\eta_{\text{diss}} = \beta \log \frac{i}{i_o} \tag{9.15}$$

See Eq. (9.11). During reduction processes such as hydrogen evolution or oxygen reduction, concentration polarization becomes important as the reduction rate approaches the limiting diffusion current density. The overall reaction for a reduction process is given by combining Eqs. (9.11) and (9.13) with appropriate signs:

$$\eta_{\text{red}} = -\beta \log \frac{i}{i_o} + 2.3 \frac{RT}{nF} \log\left(1 - \frac{i}{i_L}\right) \tag{9.16}$$

Equation (9.16) is graphically illustrated in Fig. 9-17.

The importance of Eqs. (9.15) and (9.16) cannot be overemphasized since they are the basic equations of all electrochemical reactions. Equation (9.16) applies to any reduction reaction, and Eq. (9.15) applies to almost all anodic dissolution reactions. Exceptions to Eq. (9.15) are metals which demonstrate active-passive behavior; these discussed in detail below. Using only three basic parameters, namely, β, i_0 and i_L, the kinetics of virtually every corrosion reaction can be precisely described. Equations (9.15) and (9.16) represent an outstanding simplification of the complex phenomena observed during corrosion reactions. The use and application of these two equations are described below and also in greater detail in Chap. 10.

9-9 Mixed-Potential Theory

Although the concepts utilized in the mixed-potential theory were known before 1900, the first formal presentation of this theory is usually attributed to Wagner and Traud in 1938. The mixed-potential theory consists of two simple hypotheses:

1. Any electrochemical reaction can be divided into two or more partial oxidation and reduction reactions.
2. There can be no net accumulation of electric charge during an electrochemical reaction.

The first hypothesis is quite obvious, and it can be experimentally demonstrated that electrochemical reactions are composed of two or more partial oxidation or reduction reactions. The second hypothesis is merely a restatement of the law of conservation of charge. That is, a metal immersed in an electrolyte cannot spontaneously accumulate electric charge. From this it follows that *during the corrosion of an electrically isolated metal sample, the total rate of oxidation must equal the total rate of reduction.*

The mixed-potential theory, together with the kinetic equations described above, constitute the basis of modern electrode-kinetics theory.

9-10 Mixed Electrodes

The utilization of the mixed-potential theory can be best demonstrated by considering mixed electrodes. A mixed electrode is an electrode or metal sample that is in contact with two or more oxidation-reduction systems. To illustrate, let us consider the case of zinc immersed in hydrochloric acid. Under these conditions the zinc is rapidly corroded by the hydrochloric acid, and the electrochemical reactions occurring can be represented as in Fig. 9-18. If we consider a zinc electrode in equilibrium with its ions, it would

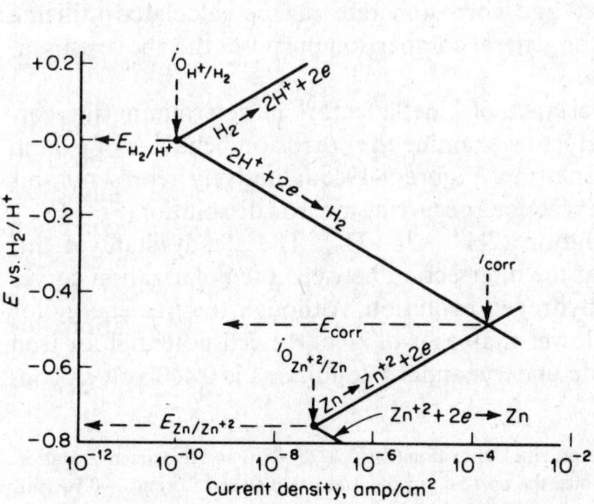

Figure 9-18 Electrode kinetic behavior of pure zinc in acid solution, shown schematically.

be represented by a reversible potential corresponding to the zinc–zinc-ion electrode reaction, and a corresponding exchange current density. Likewise, if we consider the hydrogen-electrode reaction occurring on a zinc surface under equilibrium conditions, then this particular equilibrium state would be represented by the reversible potential of the hydrogen electrode and the corresponding exchange current density for this reaction on a zinc metal surface. However, if a piece of zinc is inserted in hydrochloric acid containing some zinc ions, the electrode cannot remain at either of these two reversible potentials but must lie at some other potential. Zinc, since it is metallic, is an excellent conductor, and its entire surface must be at a constant potential. This potential is achieved when the second hypothesis of the mixed-potential theory is satisfied; that is, the total rate of oxidation must equal the total rate of reduction.

Examination of Fig. 9-18 illustrates this principle graphically. The only point in this system where the total rates of oxidation and reduction are equal is at the intersection represented by a "mixed" or corrosion potential E_{corr}. At this point the rate of zinc dissolution is equal to the rate of hydrogen evolution expressed in terms of current density.* For every zinc ion released, two electrons are utilized in forming a hydrogen molecule. Only at this point is charge conservation maintained. The current density corresponding to this point is usually called corrosion current density, i_{corr}, since it represents the rate of zinc dissolution. It should be noted that i_{corr} also corresponds to the rate at which hydrogen gas is evolved. If the β values and exchange current densities for this system are known, it is possible to predict the corrosion rate of zinc in hydrochloric acid from electrochemical data. For purposes of comparison, a current density of 1 $\mu A/cm^2$ roughly corresponds to a corrosion rate of 1 mpy for most metals. The exact relationship between dissolution current density and corrosion rate can be calculated utilizing Faraday's law. However, for general comparison purposes the above estimation is reasonably accurate.

To illustrate the importance of kinetic factors in determining the corrosion behavior of a metal let us examine the corrosion behavior of iron in dilute hydrochloric acid solution. Figure 9-19 qualitatively represents this particular system. The two reactions occurring are iron dissolution ($Fe \rightarrow Fe^{2+} + 2e$) and hydrogen evolution ($2H^+ + 2e \rightarrow H_2$). The steady state of this particular system occurs at the intersection between the polarization curves for iron dissolution and hydrogen evolution. Although the free energy for the dissolution of iron is lower than that of zinc (the cell potential for iron and the hydrogen electrode under standard conditions is 0.440 volt as con-

*Note that the reverse reactions, zinc deposition ($Zn^{2+} + 2e \rightarrow Zn$) and hydrogen oxidation ($H_2 \rightarrow 2H^+ + 2e$), do not occur since the corrosion potential lies between 0.0 and −0.76 volt. Zinc deposition can only occur at potentials more negative than −0.76, and hydrogen oxidation only occurs at potentials more positive than 0.00 volt.

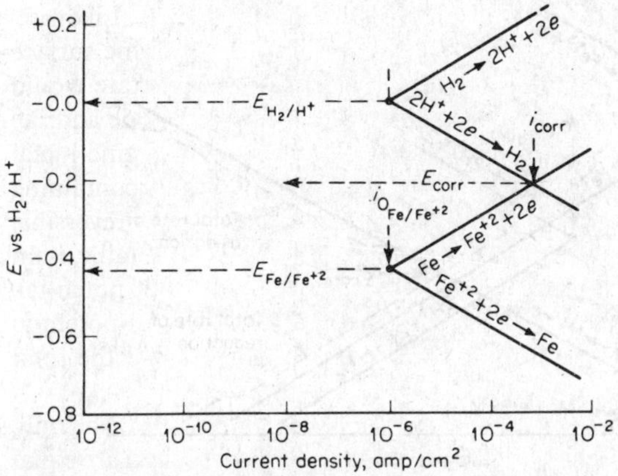

Figure 9-19 Schematic representation of electrode kinetic behavior of pure iron in acid solution.

trasted to 0.76 volt for zinc), the corrosion rate of iron is greater than that of pure zinc when exposed to identical concentrations of hydrochloric acid. This is due to the very low exchange current density for the hydrogen-evolution reaction on zinc surfaces. Thus, although the free-energy change for the corrosion of zinc is negative and greater than that for iron, the corrosion rate of zinc is less than that of iron. This illustrates the error that may be introduced by assuming that free-energy change and corrosion rate are proportional.

The system illustrated in Fig. 9-18 represents one of the simplest corrosion systems, a metal in contact with a single redox system. Under many actual corrosion conditions the environment is more complicated. Consider the corrosion behavior of a metal M in an acid containing ferric salts. This system is schematically illustrated in Fig. 9-20. Note that reversible potentials are indicated for the three redox systems, metal-metal ion, hydrogen ion–hydrogen gas, and ferric-ferrous ions. These can be assumed since all ferric salts contain traces of ferrous ions, and trace amounts of hydrogen gas and metal ion are present in this medium. The basic principles of the mixed-potential theory also apply to this more complex system. At steady state, the total rate of oxidation must equal the total rate of reduction. To determine steady-state conditions, the total rate of oxidation is determined by summing the individual oxidation currents corresponding to metal dissolution, hydrogen-gas oxidation, and ferrous-ion oxidations at constant potentials. As shown in Fig. 9-20, the total rate of oxidation follows the metal-dissolution rate until the reversible hydrogen potential is reached and an increase is noted because of the addition of hydrogen-oxidation currents. In a similar fashion, the total rate of reduction is determined by summing the total reduction currents corresponding to ferric-ion reduction, hydrogen-ion

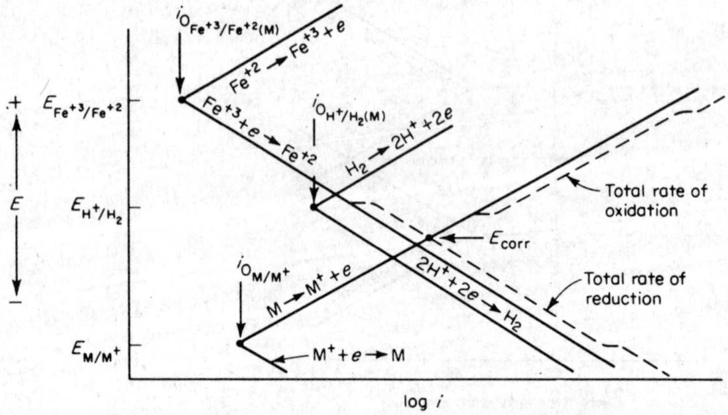

Figure 9-20 Behavior of metal M in acid solution containing ferric salts showing determination of E_{corr}.

reduction, and metal-ion reduction, as shown in Fig. 9-20. The point at which the total rate of oxidation equals the total rate of reduction is the mixed or corrosion potential of this system. The rates of the individual processes which are occurring in this system are illustrated in Fig. 9-21. A horizontal line is drawn at E_{corr} since the metal is equipotential. The rate of metal dissolution or the corrosion current is given by i_{corr}, the rate of ferric-ion reduction is equal to $i_{(Fe^{3+} \rightarrow Fe^{2+})}$, and the rate of hydrogen evolution is given by $i_{(H^+ \rightarrow H_2)}$. Note that this graphical construction leads to the equation

$$i_{corr} = i_{(Fe^{3+} \rightarrow Fe^{2+})} + i_{(H^+ \rightarrow H_2)} \tag{9.17}$$

Equation (9.17) satisfies the charge conservation principle of the mixed-potential theory.

Figure 9-21 illustrates some interesting principles concerning the corrosion of a metal in acid solutions containing oxidizers. In the absence of oxidizers, the corrosion rate of metal M is given by the intersection of the hydrogen-reduction and metal-dissolution polarization curves. The addition of an oxidizer, such as ferric ions, shifts the corrosion potential to E_{corr} and consequently increases corrosion rate from i'_{corr} to i_{corr} and decreases hydrogen evolution from i'_{corr} to $i_{(H^+ \rightarrow H_2)}$. Note that in oxidizer-free acids i_{corr} = rate of H_2 evolution. Experimentally, it is usually observed that the rate of hydrogen evolution is decreased markedly by the addition of oxygen or oxidizing agents to acid solutions. This phenomenon has often been termed *depolarization* and is assumed to be the result of interactions between the oxidizing agents and hydrogen gas on the surface. Figure 9-21 indicates that this is not the case; the reduction in hydrogen-evolution rate is a direct result of the shift in corrosion potential and is completely independent of the chemical character of the oxidizing agent.

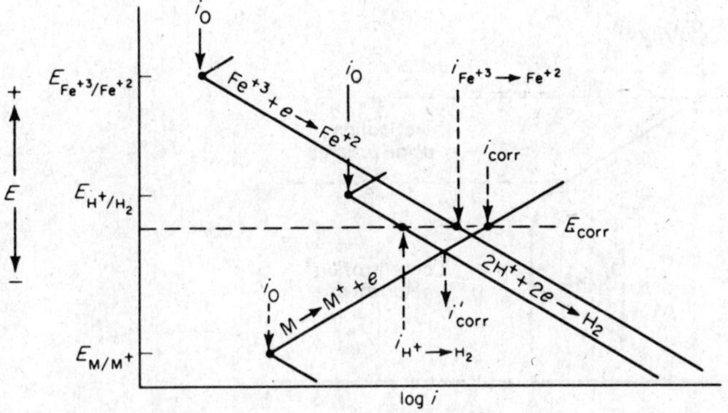

Figure 9-21 Behavior of metal M in acid solution containing ferric salts showing calculations of reaction rates.

The effect of an oxidizing agent is dependent both on its redox potential and its particular reduction kinetics. In the example shown above, the addition of ferric ions causes a pronounced change because of the relatively noble redox potential of the ferric-ferrous half-cell and its relatively high exchange current density of the surface of metal M. Figure 9-22 illustrates the importance of kinetic factors in determining the effect of an oxidizing-agent condition. In this figure, the exchange current density for the ferric-ferrous half-cell on metal M is considered to be very small. Reduction currents influence the total reduction rate only if they are within one order of magnitude of the major reduction process. As shown in Fig. 9-22, the addition of ferric ions produces no change in the corrosion potential and corrosion rate

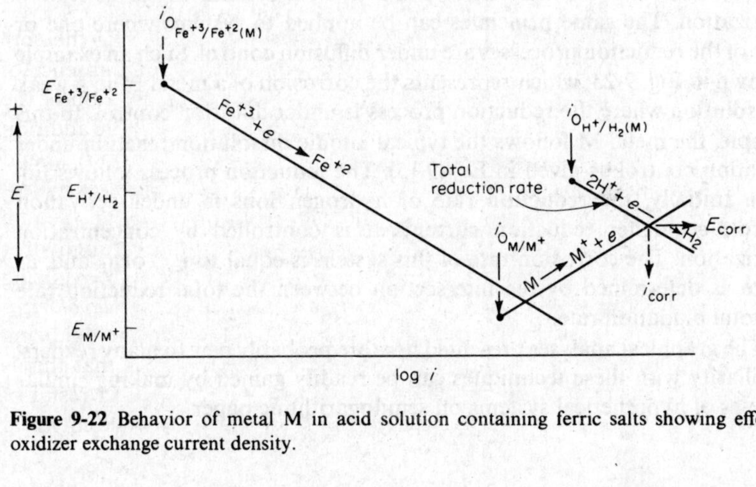

Figure 9-22 Behavior of metal M in acid solution containing ferric salts showing effect of oxidizer exchange current density.

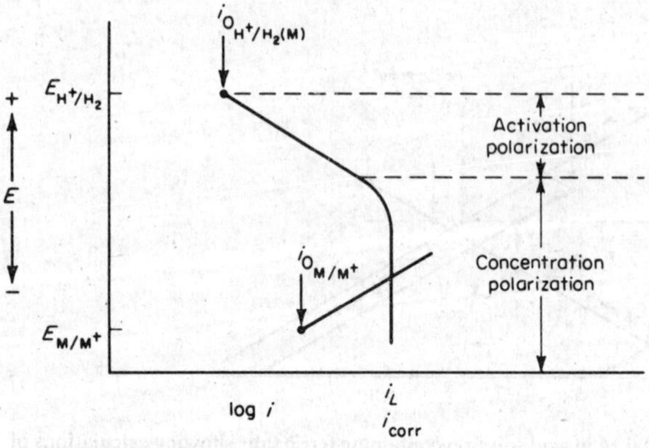

Figure 9-23 Corrosion of metal M under reduction-diffusion control.

of the system. This example serves to illustrate that not only is the redox potential of a particular oxidizing agent important, but also its exchange current density on the particular metal surface involved. This same kind of analysis can be applied to more complex systems; for example, a metal immersed in an acid solution containing several oxidizers such as ferric ion, cupric ion, and oxygen. Further, as we will see later, it can also be applied to the corrosion of two different metals immersed in contact in the same electrolyte. In all cases the analysis is exactly the same. The total rates of oxidation and reduction are determined, and the resulting corrosion potential of the system is graphically located. From this, the dissolution rate and the rates of the individual processes can be determined.

All of the above examples were shown with systems under activation polarization. The same principles can be applied to systems where one or more of the reduction processes are under diffusion control. Such an example is shown in Fig. 9-23, which represents the corrosion of a metal M in a weak acid solution where the reduction process is under diffusion control. In this example, the metal M follows the typical anodic dissolution reaction under activation control as given in Eq. (9.15). The reduction process follows Eq. (9.16). Initially, the reduction rate of hydrogen ions is under activation control; at higher reduction currents it is controlled by concentration polarization. The corrosion rate of this system is equal to i_{corr} or i_L and, as before, is determined by the intersection between the total reduction rate and total oxidation rate.

The graphical analyses presented here are probably new to many readers. Familiarity with these techniques can be readily gained by making similar analyses of hypothetical systems on semilogarithmic paper.

9-11 Passivity

Passivity is an unusual phenomenon observed during the corrosion of certain metals and alloys. Simply, it can be defined as a loss of chemical reactivity under certain environmental conditions. This is a very simple statement—we shall see later that it is possible to define this phenomenon more precisely by electrochemical parameters. To illustrate the nature of passivity, let us consider some of the very earliest experiments conducted by Faraday in the 1840s with the corrosion of iron in nitric acid. This particular series of experiments can be readily and simply accomplished in the laboratory and illustrates both the spectacular and unusual behavior of the passive state. If a small piece of iron or steel is immersed in nitric acid of approximately 70% concentration at room temperature, no reaction is observed. Weight loss determinations indicate that the corrosion rate of the iron in this system is extremely small, approaching zero. This experiment is illustrated in Fig. 9-24*A*. If water is now added, diluting the nitric acid approximately 1 to 1, no change occurs. The iron sample remains inert as shown in Fig. 9-24*B*. However, if the iron is scratched with a glass rod or if the beaker is shaken violently so that the sample strikes the sides, a violent reaction occurs. The iron rapidly goes into solution, and large volumes of nitrogen oxide gases are released (Fig. 9-24*C*). A similar effect occurs if the iron is directly introduced into diluted nitric acid.

Figures 9-24*B* and *C* illustrate the phenomenon of passivity. Both systems are identical. In one, the iron is almost inert, while in the other it corrodes at an extremely rapid rate. The difference in rates is of the order of about 100,000 to 1. In Fig. 9-24*B* the iron is considered to be in the passive state; in Fig. 9-24*C* the iron is in the so-called active state.

The above simple experiments demonstrate several important points concerning passivity. First, in the passive state, the corrosion rate of a metal is very low. Frequently the reduction in corrosion rate accompanying the transition from active to passive state will be of the order of 10^4 to 10^6. Secondly, the passive state often is relatively unstable and subject to damage

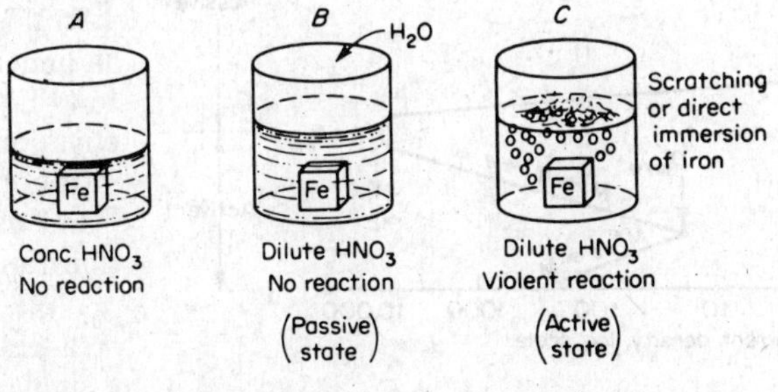

Figure 9-24 Schematic illustration of Faraday's passivity experiments with iron.

as shown by the effect of scratching in the above experiments. Hence, from an engineering viewpoint, passivity offers a unique possibility for reducing corrosion, but it must be used with caution because of the possibility of a transition from the passive to active state. The unusual characteristics of passivity, together with the possibility of utilizing this during engineering applications, explain why this subject has been studied extensively since its first observation more than 120 years ago.

The experiments described above, and numerous others, indicate that passivity is the result of a surface film. It is estimated that the film is only approximately 30 angstroms or less in thickness, contains considerable water of hydration and, as noted above, is extremely delicate and subject to changes when removed from a metal surface or when the metal is removed from the corrosive environment. Hence, the nature of the passive film, and consequently the basic nature of passivity, still remains an unsolved problem. Electrochemical studies conducted during the past 10 years have led to numerous and significant developments in the field of corrosion based on an increased knowledge of passivity. These studies do not give information concerning the nature of the passive film, but rather, are completely independent of the mechanism.

Iron, chromium, nickel, titanium and alloys containing major amounts of these elements demonstrate active-passive transitions. There are several other metals and alloys which also demonstrate passivity, but for our purposes we are primarily interested in the above metals since they are important materials of construction. Figure 9-25 schematically illustrates the typical behavior of an active-passive metal. The metal initially demonstrates behavior similar to nonpassivating metals. That is, as electrode potential is

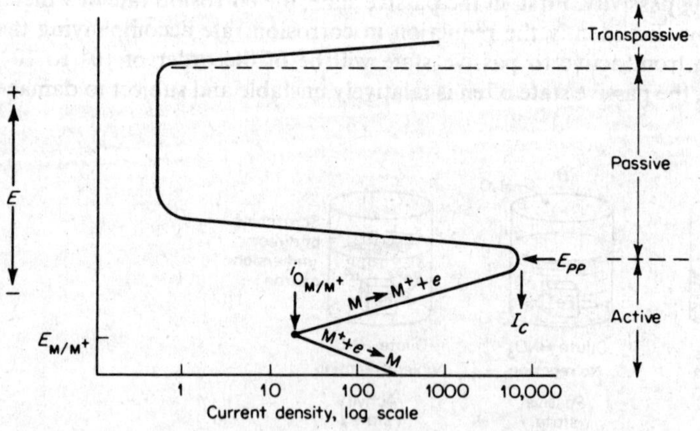

Figure 9-25 Typical anodic dissolution behavior of an active-passive metal.

made more positive, the metal follows typical Tafel behavior, and dissolution rate increases exponentially. This is the active region. At more noble potentials, dissolution rate decreases to a very small value and remains essentially independent of potential over a considerable potential region. This is termed the passive region. Finally, at very noble potentials, dissolution rate again increases with increasing potential in the transpassive region. As discussed below, one of the important characteristics of an active-passive metal is the position of its anodic current density maximum characterized by the primary passive potential E_{pp} and the critical anodic current density for passivity I_c. Electrode potential can be crudely equated to the oxidizing power of a medium, and hence a similarity between Fig. 9-25 and those shown previously in Chap. 2 is evident.

Figure 9-25 presents very useful information. First, it provides a method for defining passivity. A passive metal or alloy is one which demonstrates the typical S-shaped dissolution curve shown.* Further, it also illustrates the decrease in dissolution rate accompanying the active-to-passive transition. This decrease in dissolution rate just above the primary passive potential is the result of film formation at this point. The transpassive region where dissolution rate again increases with increasing potential is apparently due to the destruction of the passive film at very positive potentials.

Figure 9-26 illustrates the effect of increasing temperature and acid concentration on the behavior of an active-passive metal. Both temperature and hydrogen-ion concentration tend to increase the critical anodic current density and usually have relatively little effect on the primary passive potential and passive dissolution rate. A similar effect is noticed upon increasing

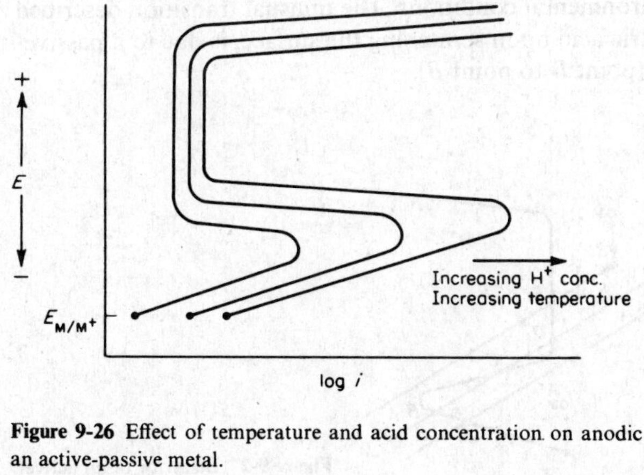

Figure 9-26 Effect of temperature and acid concentration on anodic dissolution behavior of an active-passive metal.

*Titanium is an exception; it does not possess a transpassive region.

chloride additions in the case of stainless steels and other ferrous-base alloys. Additional information concerning the effect of environmental variables on the characteristic of active-passive metals can be found in the sources listed at the end of this section.

When considering mixed electrodes involving an active-passive metal, the peculiar S-shaped anodic polarization curves of these metals often leads to unusual results. Figure 9-27 illustrates three possible cases that may occur when an active-passive metal is exposed to a corrosive environment such as an acid solution. Reduction processes under activation polarization control are shown in Fig. 9-27. However, it should be noted that the cases shown here are general and apply regardless of the shape of the reduction polarization curve. Figure 9-27 shows a single reduction process such as hydrogen evolution with three different possible exchange current densities. In case 1 there is only one stable intersection point, point *A*, which is in the active region, and a high corrosion rate is observed. Case 1 is characteristic of titanium in dilute, air-free sulfuric or hydrochloric acid. Under these conditions, titanium corrodes rapidly and cannot passivate. Case 2 is particularly interesting since there are three possible intersection points at which the total rate of oxidation and total rate of reduction are equal. These are points *B*, *C*, and *D*. Although all three of these points meet the basic requirements of the mixed-potential theory (rates of oxidation and reduction are equal), point *C* is electrically unstable and, as a consequence, the system cannot exist at this point. Hence, both points *B* and *D* are stable; *B* is in the active region corresponding to a high corrosion rate, while *D* is in the passive region with a low corrosion rate. This system may exist in either the active or passive state. Chromium in air-free sulfuric acid and iron in dilute nitric acid are typical of this behavior. That is, both active and passive states are stable under identical environmental conditions. The unusual transition described for iron in dilute nitric acid upon scratching the surface, is due to a passive-to-active transition (point *D* to point *B*).

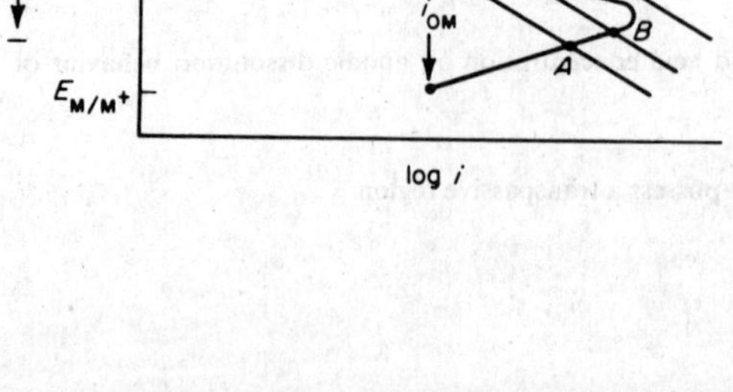

Figure 9-27 Behavior of an active-passive metal under corrosive conditions.

In case 3 there is only one stable point, in the passive region at point *E*. For such a system, the metal or alloy will spontaneously passivate and remain passive. This system cannot be made active and always demonstrates a very low corrosion rate. The system represented by case 3 is typical of stainless steels and titanium in acid solutions containing oxidizers such as ferric salts or dissolved oxygen and also iron in concentrated nitric acid (Fig. 9-24*A*).

From an engineering viewpoint, case 3 is the most desirable. This system will spontaneously passivate and corrode very slowly. Although in case 2 the passive state is possible, this particular situation is not desirable. In fact, it is the least desirable of the three cases, since an unexpected transition from the passive to active state as a result of surface damage or similar factors could lead to rapid attack. Case histories of such failure are quite common, the most common being stainless steels exposed to aerated acid solutions or acid solutions containing traces of oxidizers. During preliminary corrosion testing where specimens are handled carefully, it is frequently observed that all specimens exist in the passive state. However, when a chemical plant or component is fabricated from the material and installed, damage to the surface may result in a transition from the passive to active states. Case 1 is not particularly desirable, since the corrosion rate is usually quite high. However, since the corrosion rate is constant, predictions based on laboratory and pilot-plant tests are more accurate than those obtained with a system similar to case 2.

Considering Fig. 9-27, it is readily apparent why the position of the current maximum or "nose" of the anodic polarization curve is important. Spontaneous passivation only occurs if the cathodic reduction process clears the tip of the nose of the anodic dissolution curve as shown in case 3. More precisely, at the primary passive potential, the cathodic reduction rate must be equal to or greater than the anodic dissolution rate for spontaneous passivation to occur. It follows that a metal or alloy will be more readily passivated if it has a small critical anodic current density and an active primary passive potential. From this it is easy to see that a knowledge of the anodic dissolution behavior of a metal or alloy can be used to quantitatively determine its ease of passivation and consequently its ultimate corrosion resistance. This method of evaluating the corrosion resistance of a metal or alloy is discussed in greater detail in Chap. 10.

Suggested Reading

Bockris, J. O'M., and A. K. N. Reddy: *Modern Electrochemistry*, Plenum Press, New York, 1970.

Glasstone, S.: *Introduction to Electrochemistry*, D. Van Nostrand Company, Inc., Princeton, N.J., 1942.

Pourbaix, M. J. N.: *Atlas of Electrochemical Equilibria in Aqueous Solutions*, Pergamon Press, New York, 1966.

Rapp, R. A.: *Metallic Corrosion*, NACE, 1985. Translation of German book *Die Korrosion* by Helmut Kalsche.
Stern, M., and A. L. Geary: *J. Electrochem. Soc.*, **104**:56 (1957).
———: *J. Electrochem. Soc.*, **105**: 638 (1958).
Wagner, C., and W. Traud: *Z. Electrochem.*, **44**:391 (1938).

9-12 Mechanisms of the Growth and Breakdown of Passive Films

Because of the importance of passivity in protecting industrially important metals and alloys from corrosion, extensive work has been carried out over the past hundred years to elucidate the mechanism(s) of the growth and breakdown of passive films. A vast literature exists on this subject, and no attempt is made here to provide a comprehensive review. Instead, only some of the more recent models are discussed, and the reader is referred to the references given at the end of this section for more comprehensive accounts of the subject.

In spite of the intensive effort that has been invested, relatively little is known of the electronic and physical properties of passive films. However, it is known that many, if not all, films grow according to either a logarithmic-rate law

$$L = A + B \ln t \tag{9.18}$$

or an inverse logarithmic law

$$\frac{1}{L} = C - D \ln t \tag{9.19}$$

where L is the film thickness, and A, B, C, and D are constants. Furthermore, many passive films exhibit photoeffects that can be explained only by assigning semiconductor properties to the film. Indeed, a picture is emerging of passive films being highly doped semiconductors existing under the influence of a strong electric field.

One of the earliest models for the growth of passive films is that due to Cabrera and Mott. This model assumed that: (1) The film grows by the transport of cations across the film to the film-solution interface, where they react with electrolyte; (2) the penetration of cations through the film is assisted by the high electric field; (3) the electric field strength is constant throughout the film and is equal to ϕ_f/L, where ϕ_f is the total potential drop across the film. For potentiostatic conditions ϕ_f is constant; and (4) the rate-limiting step for film growth is the emission of metal cations from the metal into the film. The energy-distance diagram corresponding to this last case is shown in Fig. 9-28; note that the energy barrier for the initial entry of cations into the film is greater than that for subsequent movement through the film. Also note that the potential-energy surface is tilted downward because of the electric field. The assumptions noted above yield the

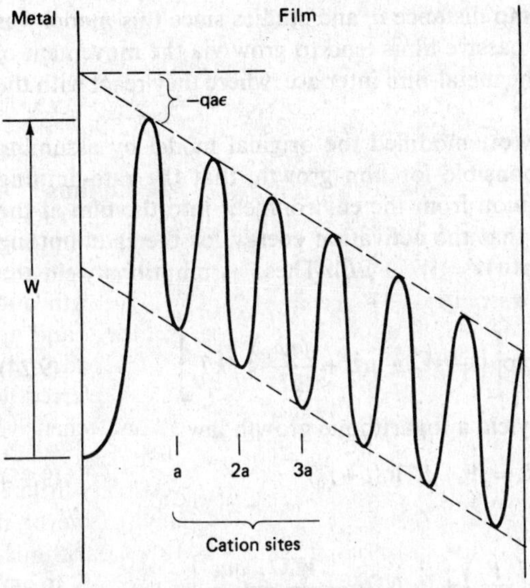

Figure 9-28 Potential energy diagram for the movement of a cation through an oxide film assuming that cation injection from the metal is rate-limiting.

following expression for the differential film growth rate law:

$$\frac{dL}{dt} = N\Omega v \exp\left[\left(-W + \frac{qa\phi_f}{2L}\right)\Big/ kT\right] \tag{9.20}$$

where N is the number of mobile cations per unit surface area, Ω is the molecular volume per cation, v is the vibrational frequency, W is the activation energy for the entry of cations into the film, q is the charge on the cation, a is the jump distance, and kT is the thermal energy. This equation can be integrated approximately to yield an inverse logarithmic rate law:

$$\frac{1}{L} = C - D \ln t \tag{9.21}$$

$$C = \frac{2}{qa\phi_f}\left\{W - kt \ln\left[\frac{2(W - 39kT)^2 N\Omega v}{\phi_f aqkT}\right]\right. \tag{9.22}$$

$$D = \frac{2kT}{qa\phi_f} \tag{9.23}$$

The Cabrera-Mott model has been applied extensively for interpreting the growth of oxide films on metal surfaces. However, the model tends to yield unrealistic values for the jump distance a, and studies since this model was first proposed indicate that passive films tend to grow via the movement of anions through the film to the metal-film interface, where they react with the metal.

In 1970, Fehlner and Mott modified the original model by assuming that anion diffusion is responsible for film growth, that the rate-limiting step is the emission of an anion from the environment into the film at the film-solution interface, and that the activation energy for the rate limiting step increases with thickness ($W = W^\circ + \mu L$). These assumptions yield the following differential-rate law

$$\frac{dL}{dt} = N\Omega v \exp\left[\left(-W^\circ - \mu L + \frac{qa\phi_f}{2L}\right)\Big/ kT\right] \tag{9.24}$$

which may be integrated to yield a logarithmic growth law

$$L = A' + B' \ln(t + t_0) \tag{9.25}$$

where

$$A' = \frac{kT}{\mu}\left[\ln\left(\frac{\mu}{kT}\right) + \ln N\Omega v - \frac{W^\circ}{kT} + \frac{qa\varepsilon}{2kT}\right] \tag{9.26}$$

$$B' = \frac{kT}{\mu}$$

and $\varepsilon = \phi_f / L$. The principal objection to this model is that experimental evidence indicates that the interfacial reaction

$$O_{(\text{oxide})} + 2H^+ + 2e^- \rightleftharpoons H_2O \tag{9.27}$$

is fast (i.e., equilibrium is established rapidly) so that emission of an anion from the environment into the film is unlikely to be the rate-limiting step.

On the basis of their experimental work on the growth of passive films on iron in borate buffer solutions, Sato and Cohen proposed the place exchange mechanism (Fig. 9-29) to explain the empirical rate law:

$$i = k' \exp(mV_{\text{app}} - Q_t/n) \tag{9.28}$$

where i is the current, V_{app} the applied voltage, Q_t is the accumulated charge (proportional to L) and k', m, and n are constants. This equation can be integrated to yield a logarithmic law of the form

$$L = A + B \ln(t + t_0) \tag{9.29}$$

in which A and B are functions of various fundamental parameters (see Sato and Notoya*). This model requires the cooperative rotation of MO pairs; an event which becomes less and less probable as the film thickens.

*N. Sato and T. Notoya, *J. Electrochem. Soc.*, **114**:585 (1967).

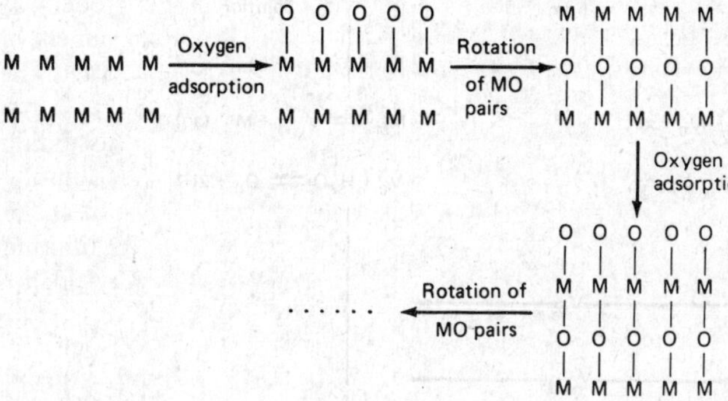

Figure 9-29 Schematic diagram of place exchange mechanism for the growth of oxide films on metal surfaces.

However, Fromhold has developed this model further, and he argues that the place exchange mechanism is viable even for relatively thick films.

More recently Chao, Lin, and Macdonald* have developed the point defect model for the growth and breakdown of passive films. The basis for this model is that films are considered to be highly defect structures with the principal defects being cation vacancies ($V_{M^{\chi\prime}}$), anion vacancies ($V_{\ddot{O}}$), electrons (e') and holes (h°). The various reactions involving these species at the metal-film and film-solution interfaces are assumed to be in quasi-equilibrium (Fig. 9-30), such that during film growth anion vacancies are created and consumed at the metal-film and film-solution interfaces, respectively. Similarly, cation vacancies are created at the film-solution interface but are consumed at the metal-film interface. The net effect of the movement of anion vacancies from the metal-film interface to the film-solution interface (or anions in the reverse direction)

$$(2/\chi)\text{m} + H_2O \rightarrow [(2/\chi)M_M + O_O] + 2H^+ + 2e' \tag{9.30}$$

is film (square brackets) growth, whereas movement of cations from the film-solution interface inwards towards the metal leads only to metal dissolution:

$$\text{m} \rightarrow M^{\chi+}(aq) + \chi e' \tag{9.31}$$

Thus, we conclude that film growth occurs inwards into the metal phase via the movement of anions through the film. This conclusion, which is independent of any mathematical assumptions, is in agreement with the substantial amount of tracer work that has been carried out in recent years.

*C. Y. Chao, L. F. Lin, and D. D. Macdonald, *J. Electrochem. Soc.*, **128**:1187 (1981); **129**:1874 (1982).

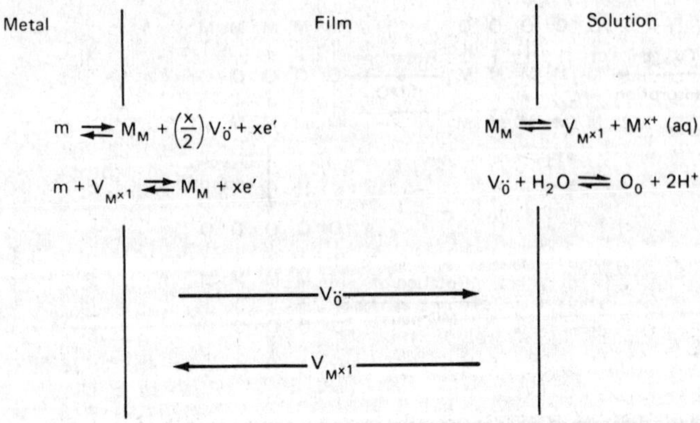

Figure 9-30 Processes assumed to occur during the anodic growth of passive films according to the point defect model. m = metal atom, M_M = metal cation in cation site, O_0 = oxygen ion in anion site, $V_{M\chi 1}$ = cation vacancy, $V_{\ddot{O}}$ = anion vacancy.

Perhaps the most controversial assumption of this model is that the electric field strength within the film is constant, and is independent of film thickness and applied voltage, even under potentiostatic conditions. This was assumed to be the case since the field strength ($\sim 10^6$ V/cm) is sufficiently high that the film probably exists on the verge of dielectric breakdown. Thus, the field is buffered by the cogeneration of electron-hole pairs, with each species moving in the appropriate direction to counter the field. This "constant-field" assumption appears to be supported by recent ac impedance studies on passive films on nickel (Chao et al., 1982; see preceding footnote). The one other assumption that is different from those employed in previous models is that the potential drop across the film-solution interface is a linear function of the applied voltage (V_{ext}) and pH; that is,

$$\phi_{F/s} = \alpha V_{ext} + \beta \text{pH} + \phi^{\circ}_{f/s} \tag{9.32}$$

Evaluation of α for passive films on iron and nickel in borate and phosphate solutions indicates that more than 70% of the applied voltage appears across the film-solution interface rather than across the film. This is an important finding, since previous models have assumed that all of the potential drop occurs across the film itself.

Mathematical development of the point defect model yields the following integral-rate law:

$$\exp(2KL) - 2KL - 1 = 2KA(B-1)t \tag{9.33}$$

where $K = F\varepsilon/RT$, ε is the electric field strength, and A and B are constants that depend upon the fundamental electrochemical and thermodynamic parameters for the system. For sufficiently thick films ($L > 5$ Å) this equation converts to the logarithmic expression

$$L = \frac{\ln 2KA(B-1) + \ln t}{2K} \tag{9.34}$$

This model has been subjected to extensive evaluation for passive films formed on iron and nickel in borate and phosphate solutions, and it has been found to provide a quantitative description of the film growth phenomena. The model is now being extended to account for the semiconductor properties of passive films, including the observed photocurrents.

A considerable amount of work has been reported on the breakdown of passive films, particularly in halide solutions, and excellent reviews are available.* The numerous models that have been proposed to explain passivity breakdown, and hence the initiation of pitting corrosion, include the competitive-ion-adsorption theory, complex-ion theory, ion penetration theory, chemicomechanical theory, and the point defect model. In this latter case (Lin et al., 1981), the initiation of pitting is due to the accumulation of cation vacancies to form a cation vacancy condensate at the metal-film interface (Fig. 9-31). Once the condensate reaches a critical size, mechanical instability occurs and pitting attack ensues. The enhanced flux of cation vacancies across the film is envisaged to result from the absorption of halide ions into oxygen ion vacancies at the film-solution interface:

$$X^- + V_{\ddot{O}} \rightarrow X_{\dot{O}} \tag{9.35}$$

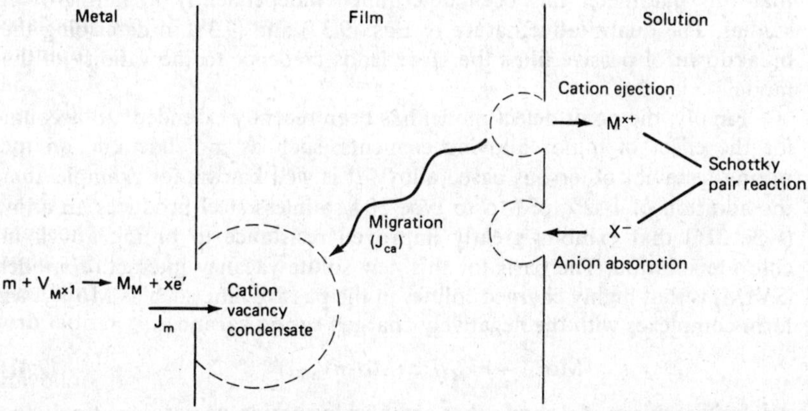

Figure 9-31 Processes leading to the breakdown of passive films according to the point defect model.

*L. F. Lin, C. Y. Chao, and D. D. Macdonald, *J. Electrochem. Soc.*, **128**:1195 (1981); M. Janik-Czachor, *J. Electrochem. Soc.*, **128**:513c (1981).

followed by the Schottky pair reaction:

$$\text{Null} \rightarrow V_{M^{\chi\prime}} + (\chi/2)V_{\ddot{O}} \tag{9.36}$$

to generate cation vacancies. Mathematical development of this model leads to the following expressions for the critical breakdown potential (V_c) and the induction time (t_{ind}):

$$V_c = \frac{4.606RT}{\chi F\alpha}\log\left(\frac{J_m}{J^\circ u^{-\chi/2}}\right) - \frac{2.303RT}{\alpha F}\log a_x \tag{9.37}$$

$$t_{\text{ind}} = \xi'\left[\exp\left(\frac{xF\alpha\Delta V}{2RT}\right) - 1\right]^{-1} + \tau \tag{9.38}$$

$$V = V - V_c \tag{9.39}$$

$$\xi' = \xi/J^\circ u^{-\chi/2}[c\bar{1}]^{\chi/2}\exp\left(\frac{\chi F\alpha V_c}{2RT}\right) \tag{9.40}$$

where ξ, J_m, J°, and u are constants, α is defined in Eq. (9-32), and a_x is the activity of the halide ion X^- in the solution. These equations have been found to provide a quantitative account for the breakdown of passive films on iron and nickel in chloride-containing solutions (Lin et al., 1981). Of particular importance is the fact that the dependence of V_c on a_x, Eq. (9.32), is determined solely by the parameter α (the others being constants), and that this parameter has been determined independently in film growth studies. The quantitative nature of Eqs. (9.37) and (9.38) in describing the breakdown of passive films therefore lends credence to the validity of the model.

Finally, the point defect model has been recently extended* to account for the effect of minor alloying elements, such as molybdenum, on the pitting behavior of ferrous based alloys. It is well known, for example, that the addition of 1–2% of Mo to type 304 stainless steel produces an alloy (type 316) that exhibits greatly improved resistance to pitting attack in chloride solution. The basis for this new solute-vacancy interaction model (SVIM) is that highly charged solutes in the passive film, such as Mo^{6+}, can form complexes with the negatively-charged cation vacancies:

$$Mo^{6+} + nV_{M^{\chi\prime}} \rightleftharpoons (Mo \cdot nV_{M^{\chi\prime}})^{(6-n\chi)+} \tag{9.41}$$

Analysis of this model shows that complex formation decreases not only the concentration of free cation vacancies in the film but also the diffusivity of these species. Both of these phenomena lead to an increase in the breakdown

*M. Urquidi and D. D. Macdonald, *J. Electrochem. Soc.*, **132**, 555–558 (1985).

potential and to an increase in the induction time. The changes in these two parameters have been found to be in quantitative accord with experiment.

References

Cabrera, N., and N. F. Mott: *Rep. Prog. Phys.*, **12**:163 (1948–49).
Fehlner, F. P., and N. F. Mott: *Oxid. Met.*, **2**:59 (1970).
Fromhold, A. T.: *J. Electrochem. Soc.*, **127**:422 (1980).
Sato, N., and M. Cohen: *J. Electrochem. Soc.*, **111**:512(1964).
Vetter, K. J.: *Electrochemical Kinetics*, Academic Press, New York, 1967.
Young, L.: *Anodic Oxide Films*, Academic Press, New York, 1961.

CHAPTER
TEN

MODERN THEORY—APPLICATIONS

10-1 Introduction

Mixed-potential theory has proved to be useful in corrosion studies because (1) it permits prediction of complex corrosion behavior, (2) it has been used to develop new corrosion prevention methods, and (3) it has been used as a basis for new rapid corrosion rate measurement techniques. In this chapter, examples of these three different kinds of applications are discussed.

The purpose of this chapter is to illustrate the value of electrochemical analyses of corrosion phenomena. For this reason, only the more important applications are briefly discussed. For additional information concerning the applications of mixed-potential theory to corrosion, consult the list of Suggested Reading at the end of this chapter. As discussed in Chap. 9, the active-passive transition provides a unique and useful means for reducing or preventing corrosion, and many of the applications covered in this chapter are based on this unusual transition.

PREDICTING CORROSION BEHAVIOR

Frequently, corrosion engineers are asked questions such as: What effect will increasing temperature have on the life of the equipment? Will increasing velocity cause an increase or decrease in corrosion rate? If dissimilar metals are used, will this affect corrosion rate? Below, the effect of several different environmental variables are considered from an electrochemical point of view. The examples discussed here do not cover all possible variables but should serve to show how mixed-potential theory is applied to corrosion problems.

10-2 Effect of Oxidizers

The effects of oxidizing agents have been discussed in Chap. 2, and the electrochemical nature of oxidizers has been briefly covered in the latter part of Chap. 9. When an oxidizing agent is added to a corroding system containing a nonpassivating metal, corrosion rate is increased and corrosion potential shifts in the noble direction. The magnitude of these effects is dependent on the redox potential of the oxidizer and its kinetic parameters (exchange current density and beta slope). The peculiar anodic dissolution behavior of active-passive metals in the presence of oxidizers is illustrated in Fig. 10-1, which shows a typical active-passive metal M immersed in an electrolyte containing a redox agent R. This redox system could be any typical oxidizing agent such as ferric, cupric, or chromate ions. Let us consider the behavior of this system as the amount of oxidizer is increased, which is equivalent to increasing the ratio of oxidized to reduced species. For simplicity, it is assumed that the exchange current density for this redox system remains constant and that the only effect produced by increasing the amount of oxidizer is to shift reversible potential in the positive direction according to the Nernst equation (9.9). The increase in oxidizing agent from concentration 1 to 7 is represented by curves 1 to 7.

Initially, metal M corrodes in the active state at a rate corresponding to point *A*. As the concentration of oxidizer is increased from 1 to 3, corrosion rate continuously increases from *A* to *C*. In this particular range of oxidizer concentrations, metal M acts like a nonpassivating metal; its corrosion rate increases with oxidizer concentration. At a concentration corresponding to curve 4, there is a rapid transition in corrosion potential from point *D* in the active state to *G* in the passive state. This follows directly from mixed-potential theory; the steady state is determined by the intersection of total oxidation and reduction rates. As oxidizer concentration is increased from 4 to 5, the metal remains in the passive state and its corrosion rate remains

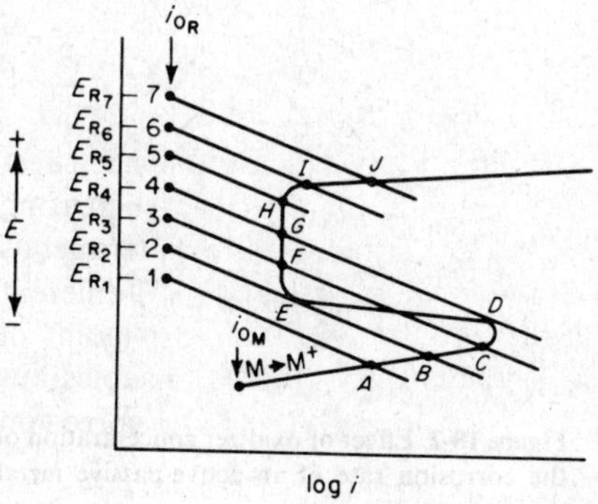

Figure 10-1 Effect of oxidizer concentration on the electrochemical behavior of an active-passive metal.

low and constant. As the concentration of oxidizer is increased further, the transpassive region is intersected and corrosion rate increases rapidly with increasing oxidizer concentration, as shown by curves 6 and 7 in Fig. 10-1.

Let us consider the effects of reducing oxidizer concentration in an active-passive system. Assume that the system is composed of metal M in an electrolyte containing oxidizer R at concentration 7. At this point the metal will be corroding in the transpassive region at point *J*. If the concentration of oxidizer is now reduced by dilution or replacement of the electrolyte, the corrosion rate will decrease from point *J* to *I*. Further reductions in the concentration of oxidizer will cause the system to shift into the passive region following points *H*, *G*, and *F*. As the concentration of oxidizer is reduced to a concentration corresponding to curve 2, a transition from point *E* to *B* occurs. As the concentration is reduced from 2 to 1, the corrosion rate will decrease from point *B* to point *A*.

The effects of increasing and decreasing oxidizer concentration illustrated in Fig. 10-1 are plotted in terms of the logarithm of corrosion rate versus oxidizer concentration in Fig. 10-2. As discussed above, the path during oxidizer additions follows the sequence *A*, *B*, *C*, *D*, *G*, *H*, *I*, *J*. As oxidizer concentration is reduced, the path is *J*, *I*, *H*, *G*, *F*, *E*, *B*, *A*. The hysteresis evident in Fig. 10-2 results from the shape and interaction of the individual electrochemical processes illustrated in Fig. 10-1. Simply, this hysteresis indicates that the amount of oxidizer necessary to cause passivation is greater than that required to maintain passivity. Thus, in Fig. 10-2, an oxidizer concentration of 4 is required to produce passivity, whereas a concentration of oxidizer as low as 2 can maintain passivity.

Figures 10-1 and 10-2 may be used to explain the unusual active-passive behavior observed during immersion of iron or steel in nitric acid (see Sec. 9-11). When iron is immersed in concentrated nitric acid, the system is characterized by a very noble redox potential and an intersection point corresponding to approximately curve 5 and point *H* in Fig. 10-1. Under these conditions, the system spontaneously passivates and remains passive even though the surface is scratched. If the concentration of nitric acid is diluted

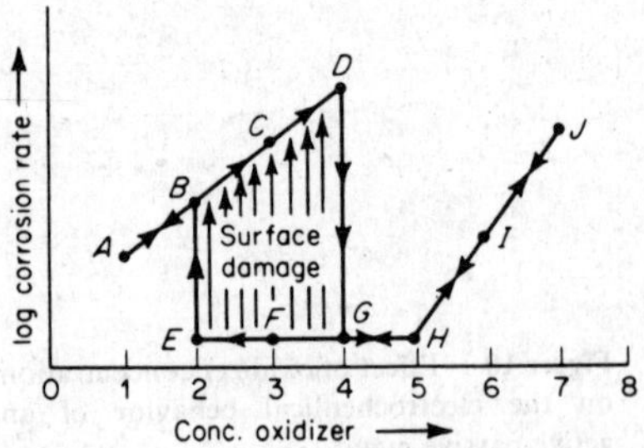

Figure 10-2 Effect of oxidizer concentration on the corrosion rate of an active-passive metal.

so that both active and passive states are stable (oxidizer concentration between 2 and 4), then scratching or damaging the surface causes a transition from the passive to active state. This analysis leads to a very useful, practical rule. *To safely maintain passivity, oxidizer concentration should be equal to or greater than the minimum amount necessary to produce spontaneous passivation.* In the example shown here, this would correspond to a concentration equal to 4 or greater. Although lower concentrations of oxidizers can maintain the passive state, surface damage may produce a transition from the passive to active as shown in Fig. 10-2. The above rule is general since it applies to all active-passive metals and alloys and is independent of the shapes of the anodic dissolution and specific reduction curves involved.

10-3 Velocity Effects

In Chap. 9 it was shown that limiting diffusion current density increases with increasing electrolyte velocity. Let us consider a system containing a metal M immersed in a corrosive system in which the reduction process is under diffusion control as illustrated in Fig. 10-3. Curves 1 through 6 correspond to limiting diffusion current densities at relative solution velocities ranging from 1 to 6. Velocity effects may be determined by plotting corrosion rate versus the relative velocity of the electrolyte, as shown in Fig. 10-4. As velocity is increased from 1 to 3, corrosion rate increases continuously from A to C. However, as velocity is increased further, the reduction reaction becomes activation-controlled. As a consequence, the corrosion rate becomes independent of velocity at very high velocities. As the velocity is increased from 4 to 6, the corrosion rate remains fixed at point D. Corrosion rate is often observed to be initially dependent on velocity, whereas at higher

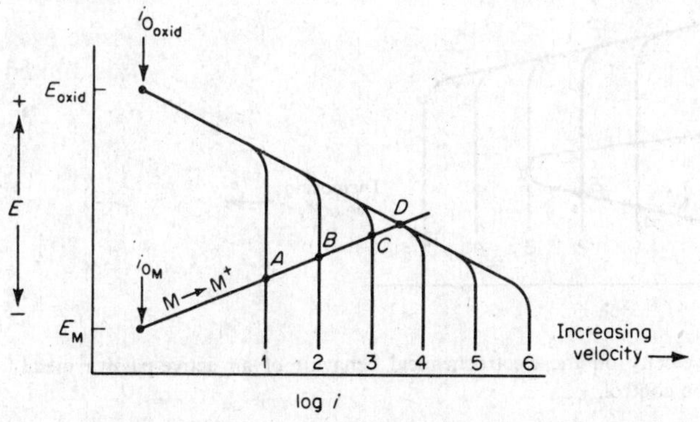

Figure 10-3 Effect of velocity on the electrochemical behavior of a normal metal corroding with a diffusion-controlled cathodic process.

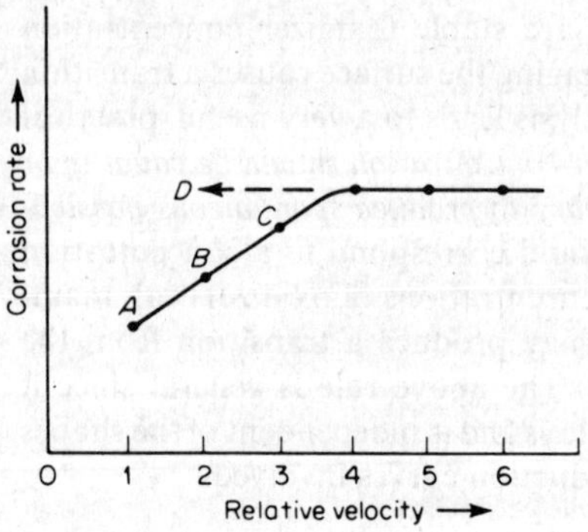

Figure 10-4 Effect of velocity on the corrosion rate of a normal metal corroding under diffusion control.

velocities it becomes independent of this variable, as shown in Fig. 10-3 and 10-4.

From the above discussion, two general rules concerning velocity effects are evident:

1. Solution velocity affects the corrosion rate of a diffusion-controlled system. Velocity has no effect on activation-controlled systems.
2. The corrosion rate of a metal in a diffusion-controlled system becomes independent of solution velocity at very high velocities.*

The effects of velocity on an active-passive metal are illustrated in Figs. 10-5 and 10-6. In this example, an active-passive metal M is corroding in an

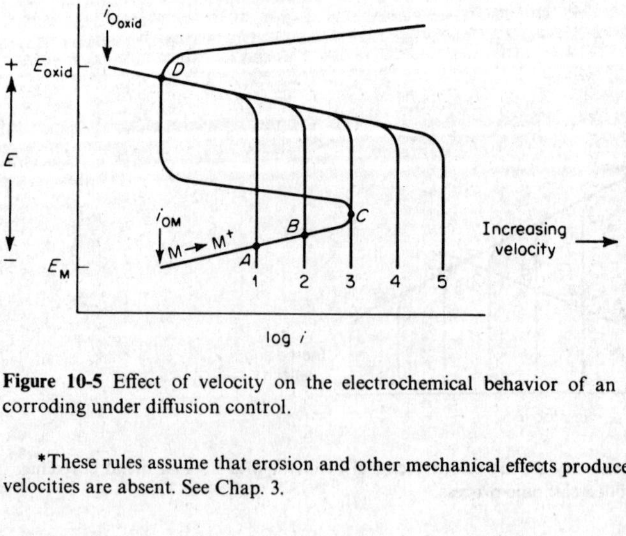

Figure 10-5 Effect of velocity on the electrochemical behavior of an active-passive metal corroding under diffusion control.

*These rules assume that erosion and other mechanical effects produced by extremely high velocities are absent. See Chap. 3.

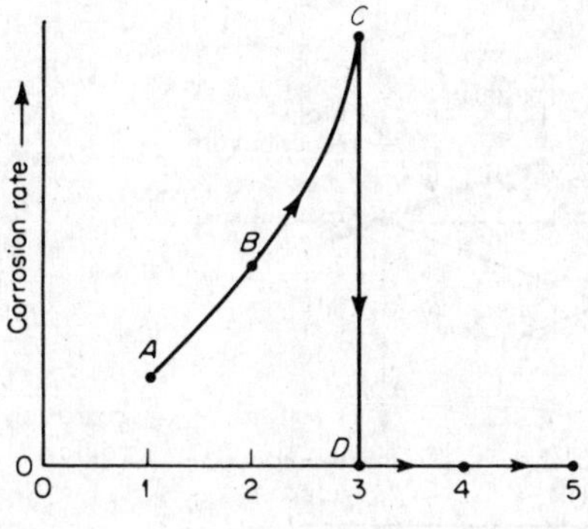

Figure 10-6 Effect of velocity on the corrosion rate of an active-passive metal corroding under diffusion control.

electrolyte under complete diffusion control, and curves 1 through 5 correspond to increases in limiting diffusion current density with increasing velocity. As velocity is increased, the corrosion rate increases along the path *ABC*. When velocity is increased beyond 3, there is a rapid transition from point *C* in the active region to point *D* in the passive state. These results are shown in terms of velocity versus corrosion rate in Fig. 10-6, which is identical to those shown in Chap. 2. The difference in velocity dependence between an ordinary metal and one demonstrating active-passive behavior is the result of the unusual dissolution behavior of active-passive metals. This behavior is typical of all active-passive metals that are corroding under diffusion control. Further, it is obvious that the ease with which a given active metal is passivated by increasing velocity is determined by the critical anodic current density (point *C* in Fig. 10-5). The smaller the critical anodic current density, the easier a metal will be passivated by an increase in velocity. This aspect of active-passive metals is discussed further in Sec. 10-5.

10-4 Galvanic Coupling

Due to mechanical and economic considerations, engineering structures frequently contain several different metals and alloys. As shown in Chap. 3, electric contact between dissimilar metals frequently leads to severe corrosion problems. The galvanic couple between dissimilar metals can be treated by application of the mixed-potential theory. Let us first, for simplicity, consider a galvanic couple between a corroding and an inert metal. If a piece of platinum is coupled to zinc corroding in an air-free acid solution, vigorous hydrogen evolution occurs on the platinum surface and the rate of hydrogen evolution on the zinc sample is decreased. Also, the corrosion rate of zinc is greater when coupled to platinum. The electrochemical characteristics of this system are schematically illustrated in Fig. 10-7. The corrosion rate of zinc in an air-free acid is determined by the intersection between the

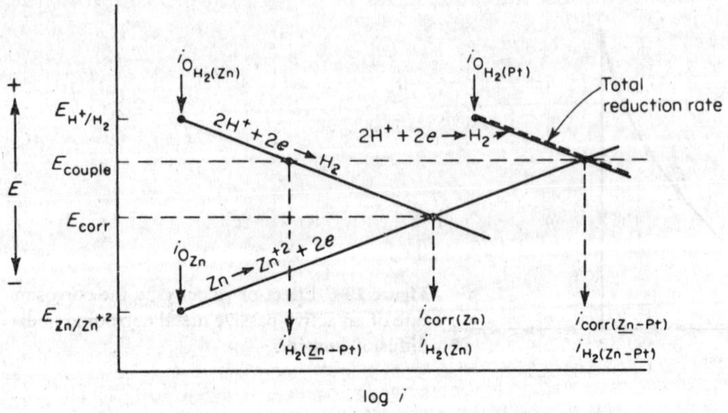

Figure 10-7 Effect of galvanically coupling zinc to platinum.

polarization curves corresponding to the hydrogen-evolution and zinc-dissolution reactions, yielding a corrosion rate equal to $i_{corr(Zn)}$. When equal areas of platinum and zinc are coupled, the total rate of hydrogen evolution is equal to the sum of the rates of this reaction on both the zinc and platinum surfaces. Since the hydrogen-hydrogen ion exchange current density is very high on platinum and very low on zinc, the total rate of hydrogen evolution is effectively equal to the rate of hydrogen evolution on the platinum surface, as shown in Fig. 10-7. Figure 10-7 shows that coupling zinc to platinum shifts the mixed potential from E_{corr} to E_{couple}, increases corrosion rate from $i_{corr(Zn)}$ to $i_{corr(\underline{Zn}-Pt)}$ and decreases the rate of hydrogen evolution on the zinc from $i_{H_2(Zn)}$ to $i_{H_2(\underline{Zn}-Pt)}$. The rate of hydrogen-ion reduction on the platinum is $i_{H_2(Zn-\underline{Pt})}$. These conclusions are consistent with the previously noted experimental observations. The graphical calculations used in Fig. 10-7 are identical to those described in Chap. 9.

In Chap. 3, the galvanic and emf series were presented, and the similarity between them was briefly discussed. This similarity has led many to believe that they are roughly equivalent, and that the emf series can be used to determine the severity of galvanic corrosion. From previous discussions, it is easy to see that this is erroneous, since the emf series refers to reversible thermodynamic conditions and cannot be used to predict corrosion rate. As mentioned above, the increase in corrosion rate of zinc observed when this metal is coupled to platinum is the result of the higher exchange current density for hydrogen evolution on platinum surfaces. It is not due to the noble reversible potential of the platinum–platinum-ion electrode, as frequently stated in the literature. To illustrate this point, consider the relative positions of platinum and gold in the emf and galvanic series. The reversible potential of the gold electrode is more positive than that of platinum in the emf series, whereas in most galvanic series tabulations the position of the gold is below platinum. In Fig. 10-8 the effect of coupling zinc to gold and to platinum is compared. As mentioned before, the exchange current density for the hydrogen reaction on the zinc metal surface is very low, and as a consequence the rate of hydrogen evolved in a galvanic couple can be assumed to be almost equal to the rate of hydrogen evolution on either gold or platinum. The table of exchange current densities (Chap. 9) indicates that the hydrogen–hydrogen-ion exchange current density is lower on gold than on platinum. Thus, if equal areas of gold and zinc are coupled, the corrosion rate increase is less than that observed if equal areas of zinc and platinum are coupled. Uncoupled zinc corrodes at a rate equal to $i_{corr(Zn)}$; coupled to gold, its rate increases to $i_{corr(\underline{Zn}-Au)}$; and coupled to platinum its rate is $i_{corr(\underline{Zn}-Pt)}$. The reason why gold produces a less severe galvanic effect is not related to its reversible potential but rather to the fact that it has a lower hydrogen exchange current density than platinum.

A couple between a corroding and an inert metal represents the simplest example of galvanic corrosion. A couple between two corroding metals may also be examined by application of mixed-potential principles, as shown in Fig. 10-9. This figure shows the corrosion rate of two metals before and after coupling. Metal M has a relatively noble corrosion potential and a low corrosion rate $i_{corr(M)}$, while metal N corrodes at a high rate $i_{corr(N)}$ at an active corrosion potential. If equal areas of these two metals are coupled, the resultant mixed potential of this system occurs at the point where the total oxidation rate equals total reduction rate. The rates of the individual partial processes are determined by this mixed potential. As shown in Fig. 10-9, coupling equal areas of these two metals decreases the corrosion rate of metal M to $i_{corr(\underline{M}-N)}$ and increases the corrosion rate of metal N to $i_{corr(M-\underline{N})}$.

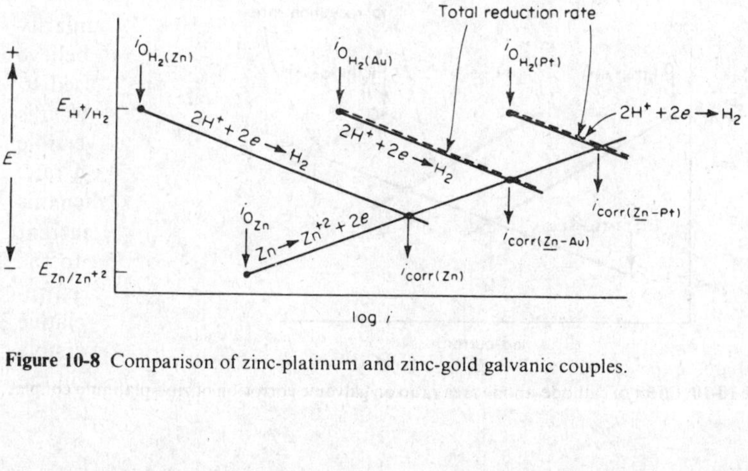

Figure 10-8 Comparison of zinc-platinum and zinc-gold galvanic couples.

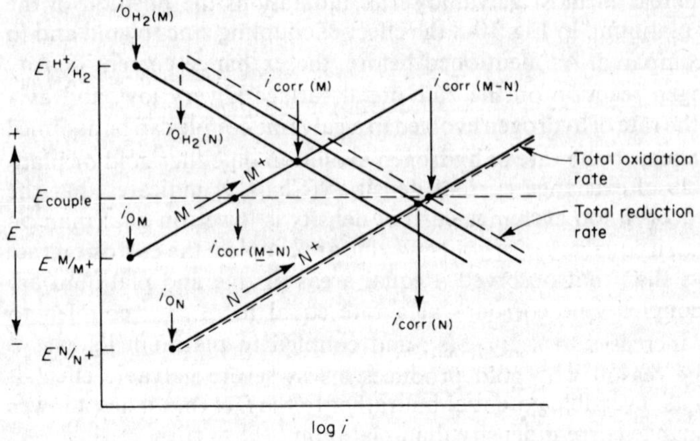

Figure 10-9 Galvanic couple between two corroding metals.

The relative areas of the two electrodes in a galvanic couple also influences galvanic behavior. Figure 10-10 illustrates the effect of cathode area on the behavior of a galvanic couple composed of zinc and platinum. Current rather than current density is used in this figure. If a piece of zinc 1 cm^2 in area is exposed to the acid solution it will corrode with a rate equal to i_A. Note that since 1 cm^2 of zinc is considered, current and current density i_A are equal. If this zinc specimen is coupled to a platinum electrode of 1-cm^2 area, the zinc corrosion rate is equal to i_B. Again, since electrodes with 1-cm^2 areas are used, current and current density are equal. However, if we consider a platinum electrode with an area of 10 cm^2 and plot its behavior in terms of

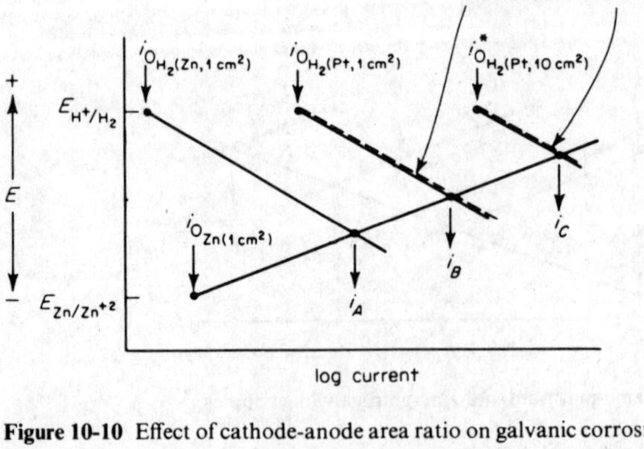

Figure 10-10 Effect of cathode-anode area ratio on galvanic corrosion of zinc-platinum couples.

current, it has an exchange current i_0^*, which is 10 times greater than a 1-cm^2 electrode. Thus, increasing the area of an electrode increases its exchange *current*, which is directly proportional to specimen area. This is illustrated in Fig. 10-10. As shown, the corrosion rate of the couple is increased as the area of platinum is increased. These results are consistent with the observations reported in Chap. 3. As the size of the cathode in a galvanic couple is increased, the corrosion rate of the anode is increased. If the relative area of the anode electrode in a galvanic couple is increased, its overall corrosion rate is reduced.

From the discussion above, and Figs. 10-7 to 10-10, it is possible to draw some general conclusions regarding galvanic corrosion.

1. If two corroding metals are galvanically coupled, the corrosion rate of the metal with the most active corrosion potential is accelerated and that of the other metal is retarded. The metal with the active corrosion potential becomes the anode and the one with the noble corrosion potential, the cathode. These conclusions apply to all galvanic couples and are independent of the absolute corrosion rates of the two metals involved. Hence, in determining the effect of galvanic coupling, the anode should be defined on the basis of corrosion potential rather than corrosion rate.
2. The corrosion behavior of a galvanic couple is determined by the reversible electrode potential of the actual processes involved, their exchange current densities and Tafel slopes, and the relative areas of the two metals. Galvanic-corrosion behavior cannot be predicted accurately on the basis of emf potentials.

Galvanic couples containing active-passive metals produce unusual effects under certain conditions. Consider the system shown in Fig. 10-11,

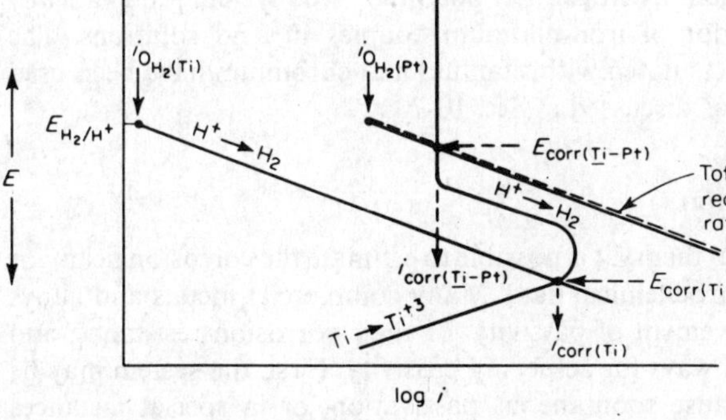

Figure 10-11 Spontaneous passivation of titanium by galvanically coupling to platinum.

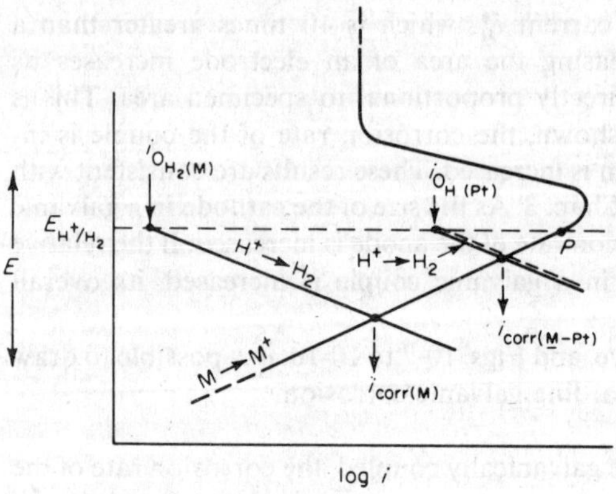

Figure 10-12 Galvanic couple between an active-passive metal and platinum in air-free acid solution.

which illustrates the behavior of titanium in acid solution before and after coupling to a platinum electrode. An isolated titanium electrode will corrode in the active state with a corrosion rate of $i_{corr(Ti)}$. Upon coupling to an equal area of platinum, the titanium spontaneously passivates and its corrosion rate decreases to $i_{corr(\underline{Ti}-Pt)}$. This particular system represents an exception to the general rule that the metal with the most active corrosion potential is accelerated when galvanically coupled. This unusual behavior only occurs if the passive region of the metal begins at a potential more active than the reversible potential of the redox system. In the case shown in Fig. 10-11, titanium can exist in the passive state at potentials more active than the reversible hydrogen potential. Coupling with platinum produces spontaneous passivation of titanium in the absence of oxygen or oxidizers. Only two metals, titanium and chromium, demonstrate the effect shown in Fig. 10-11. If the passive range of a metal begins at potentials more noble than the reversible hydrogen potential, coupling to platinum in the absence of oxidizers increases the corrosion rate as shown in Fig. 10-12. If a very large platinum electrode is coupled to metal M, the corrosion is increased to P at the reversible hydrogen–hydrogen-ion potential. This system schematically illustrates the behavior of iron-platinum couples in acid solutions. The unusual galvanic effects noted with titanium and chromium have been used to reduce corrosion, as discussed in Sec. 10-7.

10-5 Alloy Evaluation

Using mixed-potential theory it is possible to estimate the corrosion behavior of an alloy from electrochemical data. Many commercial metals and alloys depend on the achievement of passivity for their corrosion resistance, and there are two general ways for achieving passivity. First, the system may be made oxidizing to cause spontaneous passivation, or in special instances galvanic coupling can be used to achieve passivity as in the case of titanium. One of the most common oxidizers encountered in practice is oxygen. Oxygen is very slightly soluble in water and aqueous solutions, and, as a result, its reduction is usually under diffusion control. In air-saturated non-agitated solutions, the limiting diffusion current for oxygen reduction is approximately 100 $\mu A/cm^2$. If an active-passive metal is exposed to an aerated corrosive medium, it spontaneously passivates if its critical anodic current density is equal to or less than approximately 100 $\mu A/cm^2$. Thus, if the anodic dissolution behavior of a metal or alloy is known, it is possible to predict whether or not it will passivate in an aerated solution.

To illustrate some of these concepts, let us consider the anodic polarization behavior of several commercial alloys shown in Figs. 10-13 to 10-16. These figures represent experimental data measured in dilute sulfuric acid, and although they are applied-current polarization curves, they very closely approximate the actual anodic-dissolution curves of these materials. The anodic-dissolution curve of iron (Fig. 10-13), because of its very large critical anodic density and relatively noble primary passive potential, is difficult to passivate in acid solutions. It is obvious that a limiting oxygen diffusion current density of 100 μA is insufficient to cause its passivation. The addition of chromium to iron increases the ease of passivation by reducing critical anodic current density, as shown in Fig. 10-14. Although the critical anodic current density of this alloy is too high to be passivated by dissolved oxygen, it is more easily passivated by oxidizing agents than is pure iron. The addition of both chromium and nickel to iron markedly increases the ease of passivation, as indicated in Fig. 10-15. This is the typical 18Cr-8Ni austenitic

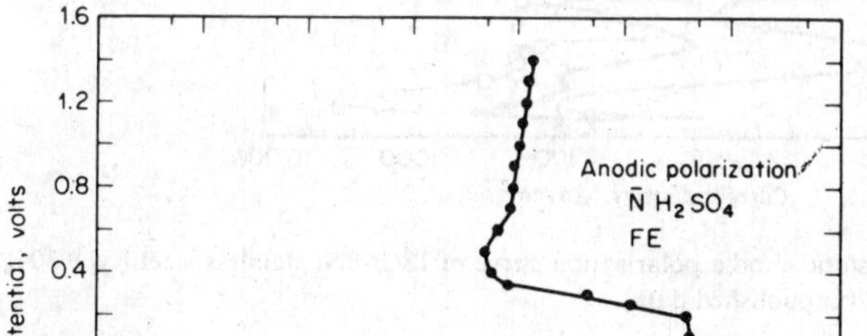

Figure 10-13 Potentiostatic anodic polarization curve of iron in normal sulfuric acid. [*R. F. Steigerwald and N. D. Greene, J. Electrochem. Soc.*, **109**:*1026 (1962)*]

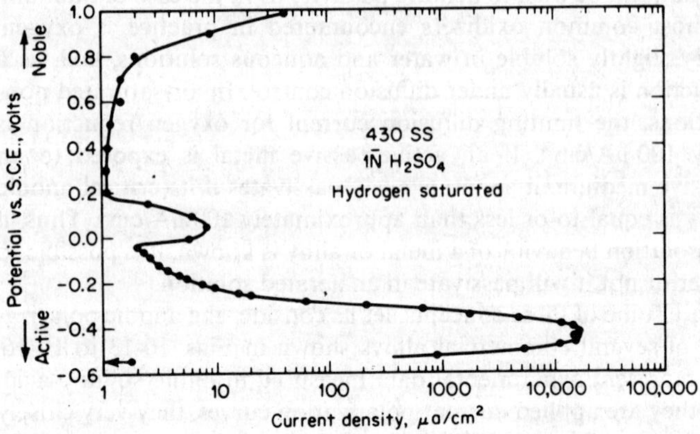

Figure 10-14 Potentiostatic anodic polarization curve of 18Cr stainless steel (type 430) in normal sulfuric acid. [*N. D. Greene, Corrosion, 18:136t (1962)*]

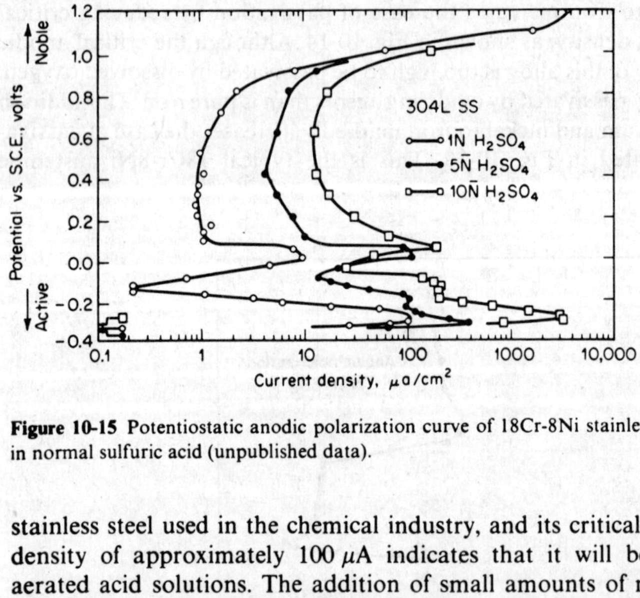

Figure 10-15 Potentiostatic anodic polarization curve of 18Cr-8Ni stainless steel (type 304L) in normal sulfuric acid (unpublished data).

stainless steel used in the chemical industry, and its critical anodic current density of approximately 100 μA indicates that it will be passivated in aerated acid solutions. The addition of small amounts of molybdenum to austenitic stainless steels (Fig. 10-16) produces a further decrease in the critical anodic current density, and this material is passivated by solutions that are not completely air-saturated.

Measurement of experimental data such as Figs. 10-13 to 10-16 requires approximately 2 to 3 hr per alloy and, as discussed above, may be used to

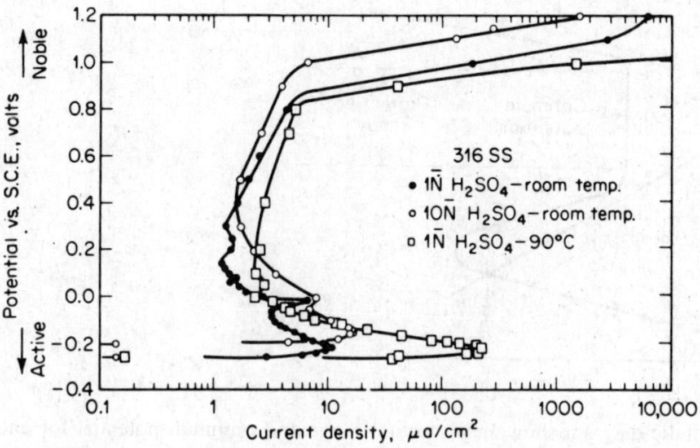

Figure 10-16 Potentiostatic anodic polarization curve of 18Cr-8Ni-2Mo stainless steel (type 316) in normal sulfuric acid (unpublished data).

compare quantitatively the corrosion characteristics of metals and alloys and to determine the effect of alloying additions. These electrochemical techniques are faster and more sensitive than conventional corrosion tests. For further details regarding these techniques and their uses in evaluating and developing corrosion-resistant alloys, consult the sources listed at the end of this chapter.

CORROSION PREVENTION

10-6 Anodic Protection

Anodic protection, or the prevention of corrosion by impressed anodic current, is a very recent innovation, and its development has been based entirely on electrode-kinetics principles. To illustrate the principles of anodic protection, consider the system illustrated in Fig. 10-17. An active-passive metal M is corroding in the active state at a rate equal to $i_{corr(active)}$. In this particular system, only the active state is stable and the metal cannot spontaneously transform to the passive condition without application of an external current. If this metal were in the passive state, its corrosion rate would be reduced to $i_{corr(passive)}$. The potential range in which anodic protection can be achieved is called the protection range, which corresponds to the passive region. The optimum potential for anodic protection is midway in the passive region, since this permits slight variations in the controlled potential without detrimentally affecting corrosion rate. As discussed in Chap. 6, the potential of an electrode or structure can be maintained constant by a constant-potential device called a potentiostat.

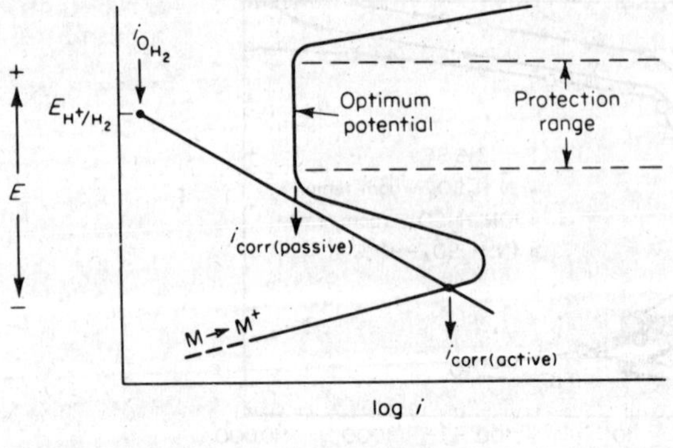

Figure 10-17 Schematic diagram showing protection range and optimum potential for anodically protecting an active-passive metal.

Consider the sequence of the events involved in anodically protecting a system such as shown in Fig. 10-18. Since mixed-potential theory requires charge conservation, the applied anodic current density is equal to the difference between the total oxidation and reduction rates of the system. Hence

$$i_{\text{app(anodic)}} = i_{\text{oxid}} - i_{\text{red}} \tag{10.1}$$

where $i_{\text{app(anodic)}}$ is the applied anodic current density. Figure 10-18 shows applied current densities as the potential of the system is shifted from the corrosion potential to the optimum corrosion potential E_4. At the corrosion potential, applied anodic current density is zero since $i_{\text{oxid}} = i_{\text{red}}$ and the

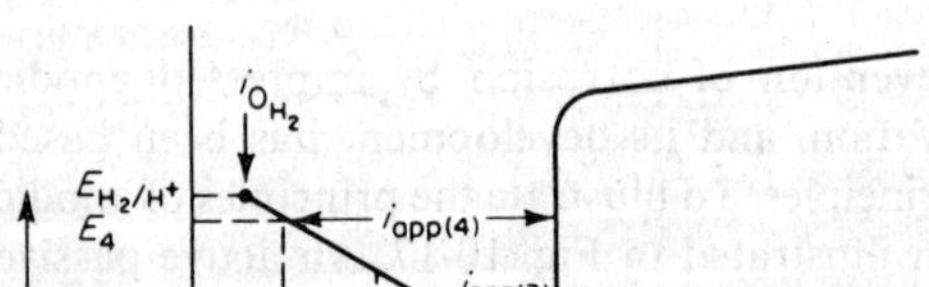

Figure 10-18 Effect of applied anodic and cathodic currents on the behavior of an active-passive system.

corrosion rate is 100 $\mu A/cm^2$. If potential is increased to E_1 with a potentiostat, an applied current density of $1000 - 10$ or 990 $\mu A/cm^2$ is required. At E_2, corresponding to the maximum of the active-passive curve, an applied anodic current of approximately 10,000 $\mu A/cm^2$ is required to maintain this potential, while at E_3, applied current decreases to approximately 0.9 $\mu A/cm^2$. At the optimum potential E_4, the applied anodic current is approximately 1 $\mu A/cm^2$, which is equal to the corrosion rate at this potential.

The system illustrated in Fig. 10-18 can be cathodically protected by applied currents. Applied cathodic current density is equal to the difference between the total reduction rate and total oxidation rate:

$$i_{\text{app(cathodic)}} = i_{\text{red}} - i_{\text{oxid}} \tag{10.2}$$

where $i_{\text{app(cathodic)}}$ is the applied cathodic current density. Equation (10.2) is similar to (10.1) and results from the charge-conservation principle of the mixed-potential theory. Applying Eq. (10.2) to the system shown in Fig. 10-18, it is apparent that if potential is shifted to E_c, an applied cathodic current density of $10{,}000 - 1$ or approximately 10,000 $\mu A/cm^2$ is required. The corrosion rate of M at E_c is 1 $\mu A/cm^2$ as shown. Thus, 10,000 $\mu A/cm^2$ of cathodic current, or 1 $\mu A/cm^2$ of anodic current produce the same reduction in corrosion rate. This example demonstrates that anodic protection is much more efficient than cathodic protection in acid solutions.

10-7 Noble-Metal Alloying

As discussed in Sec. 10-4, galvanic coupling of titanium and chromium produces spontaneous passivation in the complete absence of oxygen and oxidizers. Both titanium and chromium can be protected from corrosion in hot concentrated oxidizer-free acid by coupling them to a suitable area of platinum or other noble metals with a high exchange current density for the hydrogen-evolution reaction. Although galvanic coupling of such metals could be used to prevent corrosion, such a technique would be both cumbersome and unreliable, since the connection between the metals might be broken under industrial exposure conditions. These disadvantages can be avoided by alloying a low-hydrogen-overvoltage metal, such as platinum, with titanium or chromium. If a small amount of platinum is added to titanium, the resultant alloy will tend to produce an "automatic" galvanic couple on exposure to a corrosive medium.

As illustrated in Fig. 10-19, the original alloy is a homogeneous solution of platinum in titanium, and during exposure to a corrosive medium the titanium is preferentially dissolved from this alloy, since platinum is virtually inert in all electrolyte solutions. As a result, platinum enriches on the surface of the alloy and a galvanic couple of titanium and platinum is produced as shown. Such an alloy, when exposed to a corrosive medium, will continue to corrode in the active state until sufficient platinum accumulates on its surface

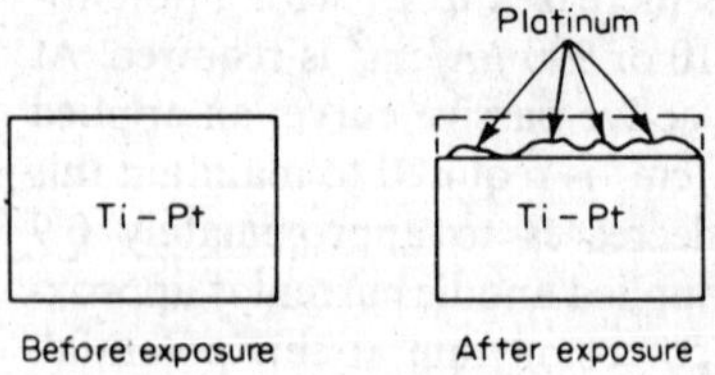

Figure 10-19 Surface enrichment of platinum during corrosion of a titanium-platinum alloy.

to cause spontaneous passivation. If the platinum is accidentally removed by abrasion, the process repeats, and additional platinum enrichment occurs.

Tables 10-1 and 10-2 present the results of corrosion tests performed with titanium and chromium containing various alloy additions. To be successful; a noble-metal alloying element must have a very high exchange current density for the hydrogen-evolution reaction and be essentially inert in the medium to which it is exposed. Note that the noble metals with low hydrogen overvoltage, such as platinum, palladium, and rhodium, markedly improve the corrosion resistance of both chromium and titanium. Other elements such as gold and silver, although inert, have lower exchange current densities for the hydrogen-evolution reaction and are less effective in reducing corrosion. The results achieved with chromium are more pronounced than those with titanium, since chromium corrodes more rapidly than titanium in acid mediums. As shown, the reduction in corrosion rate may exceed 5 orders of magnitude. Similar effects are also achieved by adding noble metals to alloys containing large amounts of chromium and titanium. Only chromium and titanium and their alloys can be protected in air-free acid solutions by this technique. The corrosion resistance of other active-passive metals in oxidizing

Table 10-1 Effect of alloy additions on the corrosion resistance of titanium

Corrosion rate, mpy

Nominal composition	Boiling H_2SO_4 1%	Boiling H_2SO_4 10%	Boiling HCl 3%	Boiling HCl 10%
Ti	460	3950	242	4500
Ti + 0.5% Pt	2	48	3	120
Ti + 0.4% Pd	2	45	2	67
Ti + 0.5% Rh	3	48	2	55
Ti + 0.6% Ir	2	45	3	88
Ti + 0.5% Au	3		9	146
Ti + 0.3% Ag				4850
Ti + 0.4% Cu	660		550	

Source: M. Stern and H. Wissenberg, *J. Electrochem. Soc.*, **106**:759 (1959).

Table 10-2 Effect of alloy additions on the corrosion resistance of chromium
Corrosion rate, mpy

Nominal composition	Boiling H_2SO_4 10%	Boiling H_2SO_4 20%	Boiling HCl 5%	Boiling HCl 10%
Cr	100,000	*D**	250,000	*D*
Cr + 0.5% Pt	3	16	1	25
Cr + 0.5% Pd	2	14	56	*D*
Cr + 0.5% Rh		3	11	45
Cr + 0.5% Ir	1	2	1	20
Cr + 0.5% Au	600	1900	*D*	
Cr + 0.5% Ag	2600		*D*	
Cr + 2% Cu	780	2700	*D*	*D*

**D* = dissolved during test.

Source: N. D. Greene, C. R. Bishop and M. Stern, *J. Electrochem. Soc.*, **108**:836 (1961).

mediums can also be improved by the addition of noble metals. That is, stainless steels containing small quantities of platinum passivate more readily in the presence of oxygen, ferric salts, and other oxidizing agents. However, in air-free solutions the addition of noble metals to other active-passive metals increases their corrosion rate. It is important to note that the noble-metal alloying concept is completely contrary to classic corrosion theory. This concept of alloying was a direct result of the application of modern mixed-potential theory to corrosion reactions.

CORROSION RATE MEASUREMENTS

Mixed-potential theory forms the basis for two electrochemical methods used to determine corrosion rate. These are the Tafel extrapolation and linear-polarization techniques. Both of these methods are briefly described below; additional information is contained in the sources listed at the end of this chapter.

10-8 Tafel Extrapolation

The Tafel extrapolation method for determining corrosion rate was used by Wagner and Traud to verify the mixed-potential theory. This technique uses data obtained from cathodic or anodic polarization measurements. Cathodic polarization data are preferred, since these are easier to measure experimentally. A schematic diagram for conducting a cathodic polarization measurement is shown in Fig. 10-20. The metal sample is termed the working electrode, and cathodic current is supplied to it by means of an auxiliary electrode composed of some inert material such as platinum. Current is measured by means of an ammeter *A*, and the potential of the working electrode is measured with respect to a reference electrode by a potentiometer-electrometer circuit. In practice, current is increased by reducing the value of the variable resistance *R*; the potential and current at various settings are simultaneously measured. It is important to note that Fig. 10-20 is schematic diagram and that polarization measurements cannot be conducted in the simple fashion shown here. Numerous precautions are required, and the actual experimental arrangement is much more complex than indicated. (See Suggested Reading at the end of the chapter.)

Let us consider the results obtained during the cathodic polarization of a metal M immersed in an air-free acid solution. Prior to the application of cathodic current, the voltmeter indicates the corrosion potential of the specimen with respect to the reference electrode. If the potential of the electrode is plotted against the logarithm of applied current, a figure similar to that shown in Fig. 10-21 is obtained. The applied-current polarization curve is indicated by points and a solid line. The curve is nonlinear at low currents, but at higher currents it becomes linear on a semilogarithmic plot. Applied cathodic current is equal to the difference between the current corresponding to the reduction process and that corresponding to the oxidation or dissolution process. Referring to both Eq. (10.2) and Fig. 10-21, it is apparent that at relatively high applied current densities, the applied current begins to approach total actual cathodic current, since the corresponding total anodic current becomes negligible. In actual practice, an applied polarization curve becomes linear on a semilogarithmic plot at

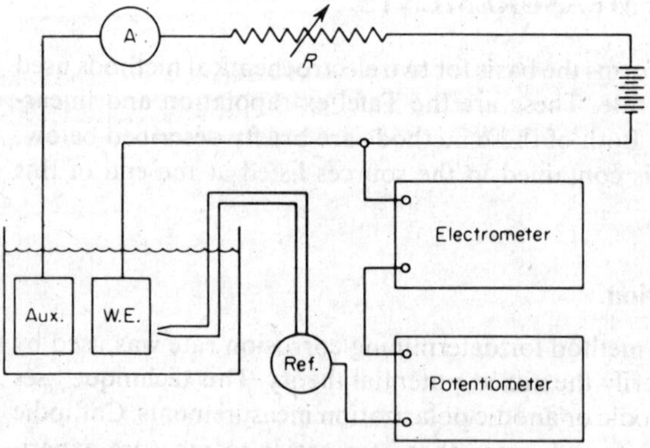

Figure 10-20 Electric circuit for cathodic polarization measurements.

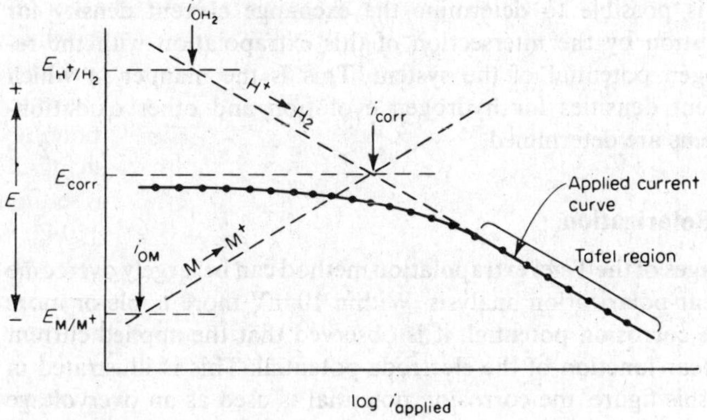

Figure 10-21 Applied current cathodic polarization curve of a corroding metal showing Tafel extrapolation.

approximately 50 mV more active than the corrosion potential. This region of linearity is referred to as the Tafel region. In Fig. 10-21 the total anodic and cathodic polarization curves corresponding to hydrogen evolution and metal dissolution are superimposed as dotted lines. It can be seen that at relatively high applied current densities the applied current density and that corresponding to hydrogen evolution have become virtually identical. To determine the corrosion rate from such polarization measurements, the Tafel region is extrapolated to the corrosion potential. At the corrosion potential, the rate of hydrogen evolution is equal to the rate of metal dissolution, and this point corresponds to the corrosion rate of the system expressed in terms of current density.

Under ideal conditions, the accuracy of the Tafel extrapolation method is equal to or greater than conventional weight loss methods. With this technique it is possible to measure extremely low corrosion rates, and it can be used for continuously monitoring the corrosion rate of a system (a polarization curve can be measured in 10 min or less). Although it can be performed rapidly with high accuracy, there are numerous restrictions that must be met before this method can be used successfully. To ensure reasonable accuracy, the Tafel region must extend over a current range of at least one order of magnitude. In many systems this cannot be achieved because of interference from concentration polarization and other extraneous effects. Further, the method can be applied only to systems containing one reduction process since the Tafel region is usually distorted if more than one reduction process occurs. In summary, the Tafel method is very useful and can be used in certain circumstances to measure corrosion rates rapidly.

Note that by continuing the extrapolation of the Tafel region shown in

Fig. 10-21, it is possible to determine the exchange current density for hydrogen evolution by the intersection of this extrapolation with the reversible hydrogen potential of the system. This is the manner in which exchange current densities for hydrogen evolution and other oxidation-reduction systems are determined.

10-9 Linear Polarization

The disadvantages of the Tafel extrapolation method can be largely overcome by using a linear-polarization analysis. Within 10 mV more noble or more active than the corrosion potential, it is observed that the applied current density is a linear function of the electrode potential. This is illustrated in Fig. 10-22. In this figure, the corrosion potential is used as an overvoltage reference point and a plot of overvoltage versus applied anodic and cathodic current is shown on a linear scale. This plot represents the first 20-mV polarization of the curve shown in Fig. 10-21. The slope of this linear-polarization curve is related to the kinetic parameters of the system as follows:

$$\frac{\Delta E}{\Delta i_{\text{app}}} = \frac{\beta_a \beta_c}{2.3(i_{\text{corr}})(\beta_a + \beta_c)} \tag{10.3}$$

where β_a and β_c are the Tafel slopes of the anodic and cathodic reactions respectively. The term $\Delta E/\Delta i_{\text{app}}$ is given in ohms (volts/amperes or millivolts/

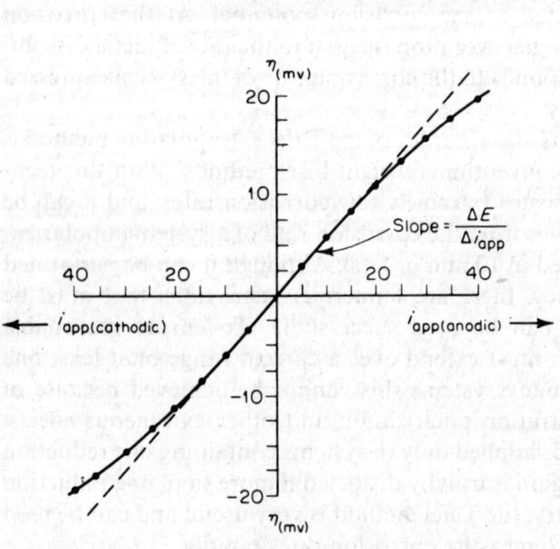

Figure 10-22 Applied-current linear-polarization curve.

milliamperes). If the beta values for the reactions are known, corrosion rate may be calculated by substitution into Eq. (10.3).

The slope of a linear-polarization curve $\Delta E/\Delta i_{app}$ is controlled mainly by i_{corr} and is relatively insensitive to changes in beta values as shown in Eq. (10.3). Hence, it is possible to formulate a reasonably accurate approximation of Eq. (10.3). Assuming that anodic and cathodic beta values of 0.12 volt represent the average of all corrosion systems, Eq. (10.3) reduces to

$$\frac{\Delta E}{\Delta i_{app}} = \frac{0.026}{i_{corr}} \tag{10.4}$$

Equation (10.4) may be used to calculate the corrosion rate of a system without knowledge of its electrode-kinetic parameters. It can be applied to systems with activation- or diffusion-controlled ($\beta_c = \infty$) reduction reactions and yields corrosion rates differing by no more than a factor of 3 from the actual rates (see Stern and Weisert reference at the end of this chapter). Although the accuracy of this approximation may not always be sufficient, Eq. (10.4) provides a unique basis for rapidly measuring relative corrosion rates or changes in corrosion rate.

The advantages of electrochemical corrosion-rate measurements, particularly linear-polarization techniques, are as follows:

1. They permit rapid corrosion rate measurements and can be used to monitor corrosion rate in various process streams.
2. These techniques may be used for accurately measuring very low corrosion rates (less than 0.1 mpy), which are both difficult and tedious to perform with conventional weight loss or chemical analytical techniques. The measurement of low corrosion rates is especially important in nuclear, pharmaceutical, and food processing industries, where trace impurities and contamination are problems.
3. Electrochemical corrosion rate measurements may be used to measure the corrosion rate of structures that cannot be visually inspected or subjected to weight loss tests. Underground pipes and tanks and large chemical plant components are examples.

Suggested Reading

Banks, W. P., and J. D. Sudbury: *Corrosion*, **19**:300*t* (1963). Anodic protection.

Edeleanu, C.: *Metallurgia*, **50**:113 (1954). Anodic protection.

———: *J. Iron Steel Inst.*, **188**:122 (1958). Alloy evaluation, potentiostatic polarization methods.

Evans, S., and E. L. Koehler: *J. Electrochem. Soc.*, **108**:509 (1961). Tafel extrapolation, linear polarization.

Fisher, A. O., and J. F. Brady: *Corrosion*, **19**:37*t* (1963). Anodic protection, potentiostatic polarization methods.

Greene, N. D.: *Corrosion*, **15**:369*t* (1959). Potentiostatic polarization methods.

———, C. R. Bishop, and M. Stern: *J. Electrochem. Soc.*, **108**:836 (1961). Noble-metal alloying.
———: *Corrosion*, **18**:136*t* (1962). Alloy evaluation.
———: *Experimental Electrode Kinetics*, Rensselaer Polytechnic Institute, Troy, N.Y., 1965. Experimental methods.
LeGault, R. A., and M. S. Walker: *Corrosion*, **19**:222*t* (1963). Linear polarization.
Mueller, W. A.: *Corrosion*, **18**:73*t* (1962). Alloy evaluation.
———: *Can. J. Technol.*, **34**:162 (1956). Anodic protection.
Prazak, M.: *Corrosion*, **19**:75*t* (1963). Alloy evaluation.
Riggs, O. L., and C. E. Locke: *Anodic Protection—Theory and Practice in the Prevention of Corrosion*, Plenum Press, New York, 1981.
Stern, M. and H. Wissenberg: *J. Electrochem. Soc.*, **106**:755, 759 (1959). Noble-metal alloying, potentiostatic polarization methods.
———, and A. L. Geary: *J. Electrochem. Soc.*, **105**:638 (1958). Tafel extrapolation, linear polarization.
———, and E. D. Weisert: *Proc. ASTM*, **59**:1280 (1959). Linear polarization.
Tomashov, N. D.: *Corrosion*, **14**:229*t* (1958). Noble-metal alloying.

CHAPTER

ELEVEN

HIGH-TEMPERATURE CORROSION

11-1 Introduction

Although oxidation generally refers to an electron-producing reaction, this term is also employed to designate the reaction between a metal and air or oxygen in the absence of water or an aqueous phase. Scaling, tarnishing, and dry corrosion are also sometimes used to describe this phenomenon. Since virtually every metal and alloy reacts with air at elevated temperatures, oxidation resistance must be considered in most metallurgical engineering applications. As the temperature is increased, the importance of metal oxidation also increases as in applications for gas turbines, rocket engines, furnaces, and high-temperature petrochemical systems.

A comprehensive conference on the subject of high-temperature corrosion has been published.* The main purposes of this chapter are to introduce briefly the basic mechanisms of high-temperature corrosion and to provide generalized engineering information regarding the resistance of commercially important metals and alloys to various corrosive high-temperature environments. For further details concerning metal-gas reactions, the reader is referred to the aforementioned publication and the journal *Oxidation of Metals*.

MECHANISMS AND KINETICS

11-2 Pilling-Bedworth Ratio

In one of the earliest scientific studies of oxidation, Pilling and Bedworth† proposed that oxidation resistance should be related to the volume ratio of

***High Temperature Corrosion*, R. A. Rapp, ed., NACE Publication, Houston, Tex. (1983).
†N. B. Pilling and R. E. Bedworth, *J. Inst. Metals*, **29**:529 (1923).

oxide and metal per gm atom of metal. According to Pilling and Bedworth, a volume ratio of less than 1 produces insufficient oxide to cover the metal and is unprotective. Similarly, it was argued that a ratio of much greater than 1 tends to introduce large compressive stresses in the oxide, which also causes poor oxidation resistance due to cracking and spalling. The ideal ratio, according to these investigators, would be close to 1.

The volume ratios for some metals are listed in Table 11-1. As shown, this ratio does not accurately predict oxidation resistance, although there is some qualitative agreement. In general, metals with volume ratios of less than 1 form nonprotective oxides, as do those with very high volume ratios (2 to 3), but this ratio was intended only as an emperical criterion and does not encompass other properties more important in determining oxidation resistance. To be protective to oxygen reaction an oxide must possess good adherence, a high melting point, a low vapor pressure, good high-temperature plasticity to resist fracture, and low electrical conductivity or low diffusion coefficients for metal ions and oxygen. For cyclic temperatures the metal and oxide should possess similar coefficients of expansion. Thus, oxidation resistance of a metal or alloy depends on a number of complex factors. When gaseous reactions involve several oxidants (oxygen plus sulfur or carbon or nitrogen, etc.), special considerations of competitive product stabilities and growth rates must be considered. Finally, if a fused salt (alkali sulfate, carbonate, nitrate, etc.) should contact a metal at high temperatures,

Table 11-1 Oxide-metal volume ratios

Protective oxides	Nonprotective oxides
Be—1.59	Li—0.57
Cu—1.68	Na—0.57
Al—1.28	K—0.45
Si—2.27	Ag—1.59
Cr—1.99	Cd—1.21
Mn—1.79	Ti—1.95
Fe—1.77	Mo—3.40
Co—1.99	Cb—2.61
Ni—1.52	Sb—2.35
Pd—1.60	W—3.40
Pb—1.40	Ta—2.33
Ce—1.16	U—3.05
	V—3.18

Source: B. Chalmers, *Physical Metallurgy*, p. 445, John Wiley & Sons, Inc., New York, 1959.

then electrochemistry and oxide/salt chemistry become important. These considerations will be discussed briefly in this chapter.

11-3 Electrochemical and Morphological Aspects of Oxidation

Oxidation by gaseous oxygen, like aqueous corrosion, is an electrochemical process. It is not simply the chemical combination of metal and oxygen on a molecular scale, $M + \frac{1}{2}O_2 \rightarrow MO$, but consists of two partial, spatially separated processes:

$$M \rightarrow M^{2} + 2e^{-} \qquad \text{(metal oxidation at metal-scale interface)} \qquad (11.1)$$

$$\frac{1}{2}O_2 + 2e^{-} \rightarrow O_2 \qquad \text{(oxidant reduction at scale-gas interface)} \qquad (11.2)$$

$$M + \frac{1}{2}O_2 \rightarrow MO \qquad \text{(overall)} \qquad (11.3)$$

with new MO lattice sites produced either at the metal-scale interface or at the scale-gas interface. To examine the electrochemical nature of gaseous oxidation, it is helpful to compare this process with aqueous galvanic corrosion, where the metal M is immersed in an oxygen-saturated aqueous solution and is electrically connected to an inert electrode at which oxygen reduction occurs readily (high exchange current density). Under these conditions, M dissolves to form metal ions and oxygen is reduced to form hydroxide ions at the inert electrode, while electrons pass through an electronically conducting connection. This galvanic effect occurs only if there is an electronic conductor between the two electrodes. In the aqueous electrolyte, current is carried by positive and negative ions.

Gaseous oxidation may be considered as analogous to aqueous galvanic corrosion. Metal ions are formed at the metal-scale interface and oxygen is reduced to oxygen ions at the scale-gas interface. Because all metal oxides conduct both ions and electrons to some extent, this electrochemical reaction occurs without the necessity of an external electronic conductor between the local anode and the local cathode. Figure 11-1 shows that the oxide layer serves simultaneously as (1) an ionic conductor (electrolyte), (2) an electronic conductor, (3) an electrode at which oxygen is reduced, and (4) a diffusion barrier through which electrons pass and ions must migrate over defect lattice sites (V_M and V_O).

As discussed in Chap. 3, aqueous galvanic corrosion can be retarded by either increasing the electrolyte resistance or avoiding metallic contact between dissimilar metals. Of these methods, the latter is usually the more practical for aqueous corrosion because the solution composition is often invariable due to process considerations. The alternative possibility prevails in gaseous oxidation. It is impossible to eliminate electronic contact between the metal and the oxide-gas interface since this is an inherent property of metal-oxide systems. The electronic conductivities of oxides are usually several orders of magnitude greater than their ionic conductivities, so the diffusion of either cations or oxygen ions usually controls the reaction rate.

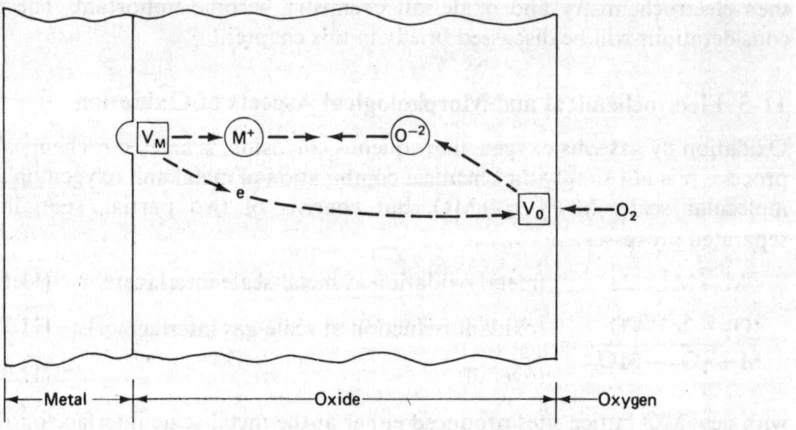

Figure 11-1 Schematic illustration of electrochemical processes occurring during gaseous oxidation.

Almost without exception, cations and oxygen ions do not diffuse with comparable ease in a given oxide because lattice defects for one species generally predominates. Thus simple diffusion control should result in the growth of the scale at either the metal-scale or else the scale-gas interface. For oxidation controlled by lattice diffusion, the oxidation rate is most effectively retarded in practice by reducing the flux of ions diffusing through the scale. This is equivalent to reducing the electrolyte conductivity in the case of galvanic corrosion. Although it is sometimes possible to reduce the ionic diffusivity by the "doping" of a given oxide lattice (as described later), more usually the base-metal composition is changed by alloying or else the alloy is coated so that a different and inherently more protective binary or ternary compound is formed as the scale.

Many metal-oxygen phase diagrams indicate several stable binary oxides. For example, iron may form the compounds FeO, Fe_3O_4, and Fe_2O_3; copper may form Cu_2O and CuO; etc. In the oxide-scale formation on pure metals, generally all of the potentially stable oxide phases from the metal-oxygen phase diagram are formed in sequence. The most oxygen-rich compound is found at the scale-gas interface and the most metal-rich compound at the metal-scale interface; thus for Fe above about 560°C the phase sequence is $Fe/FeO/Fe_3O_4/Fe_2O_3/O_2$, as shown in Fig. 11-2. The relative thickness of each phase is determined by the rate of ionic diffusion through that phase. Morphological or nucleation complications sometimes arise in the oxidation of some metals so that stable phases may be absent from the scale or metastable phases may be formed.

Both oxide and sulfide scales formed on the common base metals (Fe,

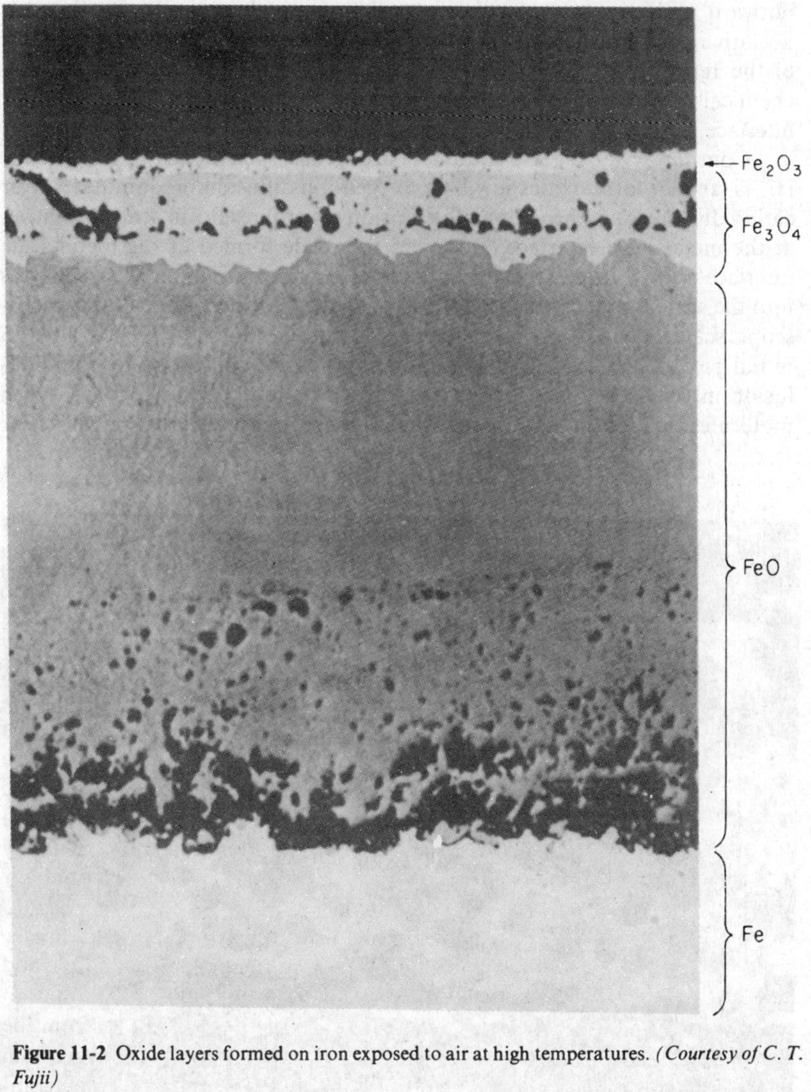

Figure 11-2 Oxide layers formed on iron exposed to air at high temperatures. *(Courtesy of C. T. Fujii)*

Ni, Cu, Cr, Co, and others) grow principally at the scale-gas interface by outward cation diffusion. However, because of condensation of cation vacancies at the metal-scale interface, a significant volume of voids often appears in the inner part of the scale, as shown in Fig. 11-2. It has been proposed that some of the oxide in the middle of the scale "dissociates," sending cations outward and oxygen molecules inward through these voids. By this dissociative mechanism, such scales are believed to grow on both sides because of the reaction of cations and oxygen at the scale-gas interface and the chemical reaction of oxygen molecules with the metal near the metal-scale interface.

Contrary to the more traditional base metals, such metals as Ta, Cb(Nb), Hf, Ti, and Zr form oxides in which oxygen-ion diffusion predominates over cation diffusion, and hence simple diffusion control results in scale formation at the metal-scale interface. However, the oxide formed at the metal-scale interface (with a large increase in volume) induces such severe tensile stress into the scale that fracture occurs and the scale becomes porous on a microscopic scale and cracked on a macroscopic scale (see Fig. 11-3). Thus, after an initial period, the oxidation of these metals is not controlled by ionic diffusion in the scale. These scales are said to be nonprotective, and oxygen molecules can diffuse in the gas phase filling the voids to a location very near

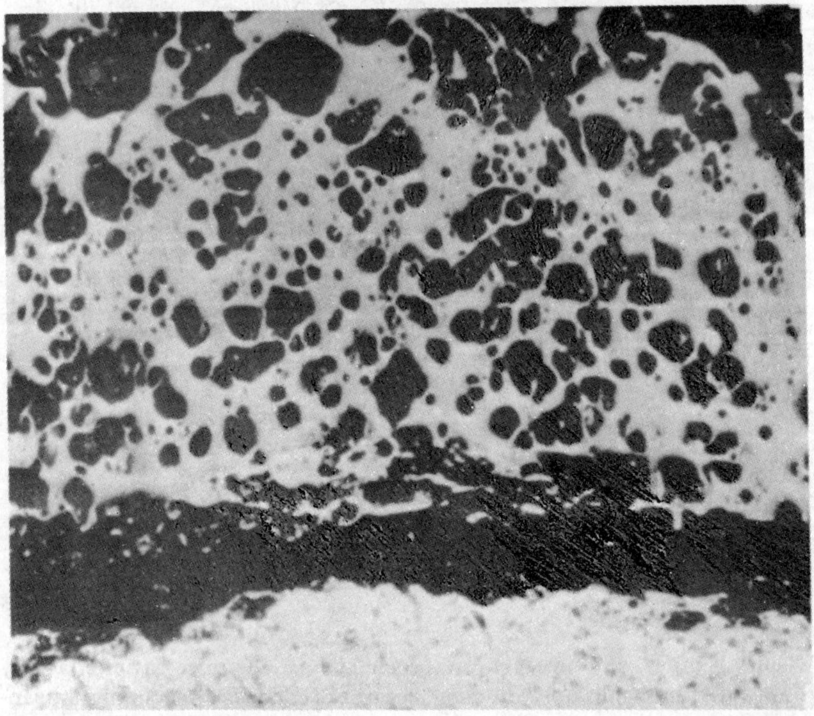

Figure 11-3 Nonprotective scale formed on columbium. Exposure for 64 hr at 1100°C (100 ×). *(Courtest R. A. Rapp)*

the metal-scale interface where the reduction reaction can occur. Then for these metals, the ideal electrochemical oxidation model with oxygen reduction at the oxide-gas interface is replaced by a mechanism offering much less resistance and much more rapid kinetics.

These particular examples serve to point out that many different morphological occurrences often cause the oxidation mechanism to deviate from the simple ideal electrochemical, lattice diffusion model that was described. In addition, the significant dissolution of oxygen and sulfur atoms into some metals, the high volatility of some oxides, halides, and metals, the low melting points of some oxides and sulfides, and grain boundaries in the scale and in the metal often complicate the oxidation mechanisms of pure metals. Despite these complications, much can be learned by an examination of the structures and the diffusion mechanisms in these reaction product compounds.

11-4 Oxide Defect Structure

In general, all oxides are nonstoichiometric compounds; i.e., their actual compositions are variable and deviate from their ideal molecular formulas. Some oxides have an excess and others a deficiency of metallic ions, or equivalently a deficiency or excess of oxygen ions. Figure 11-4*a* schematically shows a metal-excess zirconium oxide, ZrO_2. In this representation, two vacant anion sites (represented by squares) are compensated by four excess electrons for electroneutrality. In this structure, electronic current is carried by the excess electrons and ionic transport by the oxide ions. Zirconium oxide is termed an *n*-type semiconductor since it contains an excess of negatively charged electronic current carriers (electrons). Other *n*-type semiconducting oxides include CdO, TiO_2, Ta_2O_5, Al_2O_3, SiO_2, Cb_2O_5, and PbO_2. However, of this group, at least Ta_2O_5 and Cb_2O_5 are metal-deficient, with oxygen vacancies as the predominant ionic defect. Rutile TiO_2 is believed to exhibit both titanium interstitial ions and oxygen vacancies.

In Figs. 11-4*b* and *c*, substitutional ions of valence lower (Ca^{2+}) and higher (Ta^{5+}) than zirconium (Zr^{4+}) have been introduced into the ZrO_2 lattice. For CaO substitution, anion vacancies are formed to constitute charge compensation. For Ta_2O_5 dissolution into ZrO_2, an increase in excess electron concentration should provide electrical neutrality. Through an equilibrium among defects, a corresponding decrease in oxygen vacancies is expected. Since the "native" defect concentration in pure ZrO_2 is very low (in the ppm range), the substitution of even minor concentrations of "aliovalent" impurities leads to large changes in the ionic diffusivity and the electrical conductivity.

Nickel oxide (NiO), shown in illustration Fig. 11-5*a*, is a metal-deficient oxide. For each nickel ion vacancy there are two trivalent nickel ions in

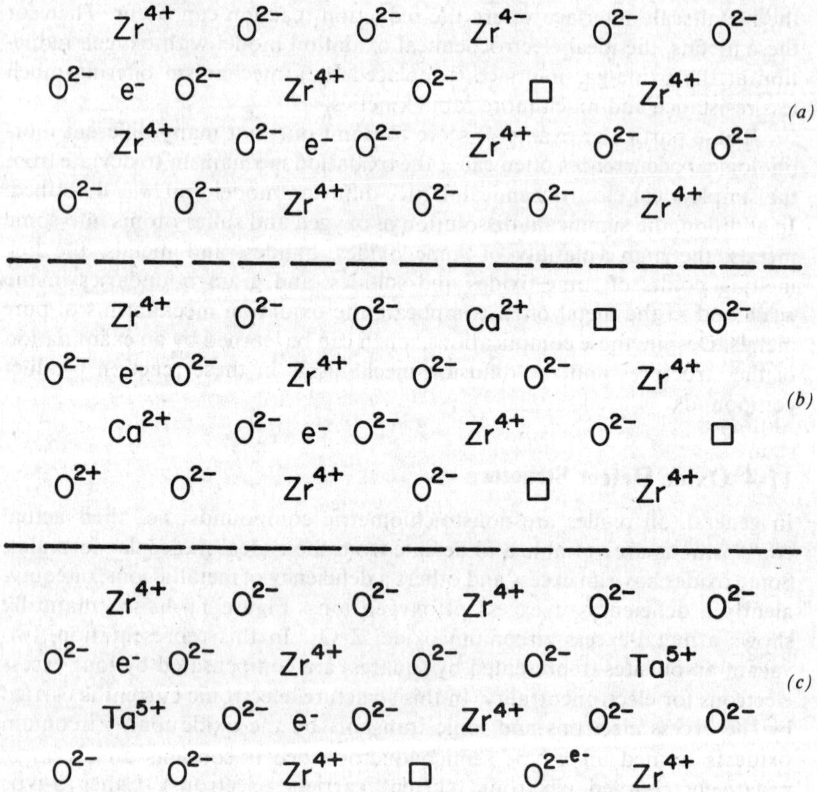

Figure 11-4 Idealized lattice structure of zirconium oxide, an *n*-type semiconductor: (a) pure ZrO; (b) effect of Ca^{2+} addition; (c) effect of Ta^{5+} addition.

normal lattice positions. The trivalent ions can be considered as a divalent ion and an associated "electron hole" (or absence of an electron). Electronic conduction occurs by the diffusion of these positively charged electron holes and hence, this oxide is termed a *p*-type semiconductor. Ionic transport occurs by the diffusion of nickel vacancies. Other *p*-type oxides are FeO, Cu_2O, MnO, and CoO.

In Figs. 11-5(*b*) and (*c*), monovalent (Li^+) and trivalent (Cr^{3+}) impurities have been substituted into the NiO structure. These impurities are compensated electrically by an increase in positive holes (Ni^{3+}) for Li_2O substitution, and by an increase in cation vacancies for Cr_2O_3 doping. These substitutions make (Ni, Li)O a good *p*-type electronic semiconductor (and catalyst), and (Ni, Cr)O has a higher cation diffusivity than pure NiO. For this reason, a nickel alloy containing a few percent of Cr oxidizes more rapidly

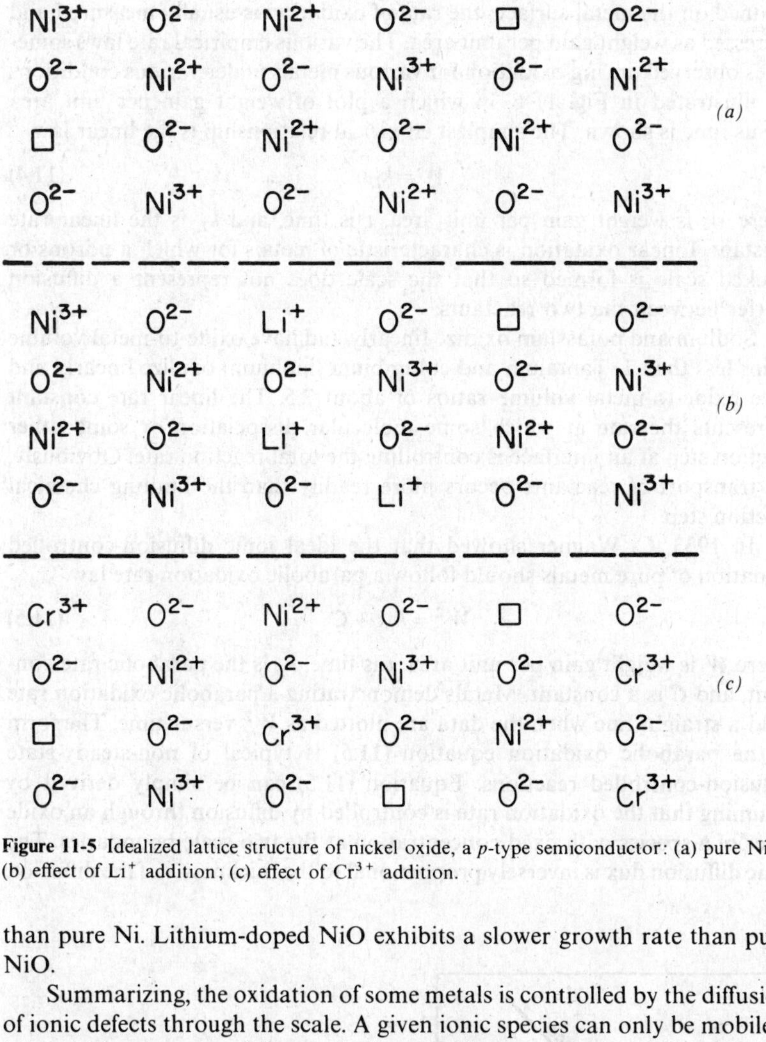

Figure 11-5 Idealized lattice structure of nickel oxide, a *p*-type semiconductor: (a) pure NiO; (b) effect of Li^+ addition; (c) effect of Cr^{3+} addition.

than pure Ni. Lithium-doped NiO exhibits a slower growth rate than pure NiO.

Summarizing, the oxidation of some metals is controlled by the diffusion of ionic defects through the scale. A given ionic species can only be mobile if interstitial or vacancy defects exist in its sublattice; likewise the interface at which new oxide scale is formed is determined by the ionic species which is mobile. In principle, a diffusion-controlled oxidation may be retarded by decreasing the concentration of ionic defects in the scale.

11-5 Oxidation Kinetics

The most important parameter of metal oxidation from an engineering viewpoint is the reaction rate. Since the oxide reaction product is generally retained on the metal surface, the rate of oxidation is usually measured and expressed as weight gain per unit area. The various empirical rate laws sometimes observed during oxidation for various metals under various conditions are illustrated in Fig. 11-6, in which a plot of weight gain per unit area versus time is shown. The simplest empirical relationship is the linear law

$$W = k_L t \tag{11.4}$$

where W is weight gain per unit area, t is time, and k_L is the linear rate constant. Linear oxidation is characteristic of metals for which a porous or cracked scale is formed so that the scale does not represent a diffusion barrier between the two reactants.

Sodium and potassium oxidize linearly and have oxide-to-metal volume ratios less than 1. Tantalum and columbium (niobium) oxidize linearly and have oxide-to-metal volume ratios of about 2.5. The linear rate constant represents the rate at which some molecular dissociation or some other reaction step at an interface is controlling the total reaction rate. Obviously, the transport of reactants occurs more readily than the limiting chemical reaction step.

In 1933, C. Wagner showed that the ideal ionic diffusion-controlled oxidation of pure metals should follow a parabolic oxidation rate law,

$$W^2 = k_p t + C \tag{11.5}$$

where W is weight gain per unit area, t is time, k_p is the parabolic rate constant, and C is a constant. Metals demonstrating a parabolic oxidation rate yield a straight line when the data are plotted as W^2 versus time. The form of the parabolic oxidation equation (11.5) is typical of non-steady-state diffusion-controlled reactions. Equation (11.5) can be simply derived by assuming that the oxidation rate is controlled by diffusion through an oxide layer of a species with fixed concentration at the two scale boundaries. The ionic diffusion flux is inversely proportional to the thickness x of the diffusion

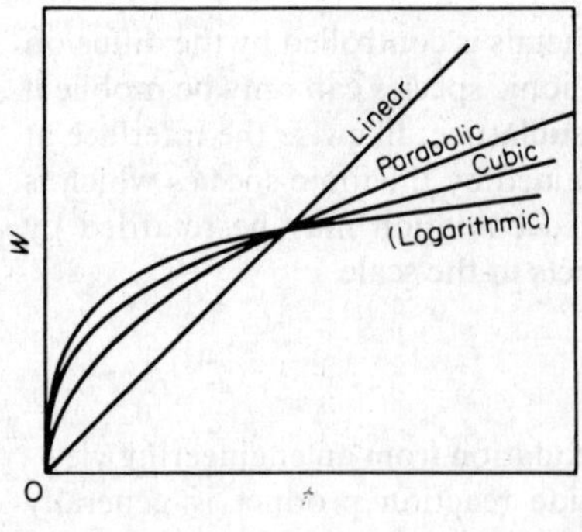

Figure 11-6 Oxidation rate laws.

barrier, and the change in scale thickness or weight is likewise proportional to the ionic diffusion flux. Wagner has derived expressions from which the absolute magnitude of the parabolic rate constant k_p for the ideal diffusion-controlled oxidation of a pure metal to form a single-phase scale may be calculated from a knowledge of either the partial ionic conductivities or the ionic self-diffusion coefficients in the scale.

For the oxidation of Fe, Co, Ni and Cu, where the required conditions are satisfied, the calculated and observed parabolic rate constants are in good agreement. (Rates controlled by slow electronic transport are experienced only in metal-halogen reactions.) In general, rate laws of a nearly parabolic nature are quite common and are usually associated with thick, coherent oxides. However, the failure of experimental data to satisfy exactly a parabolic rate dependence for a thick, coherent scale probably only indicates that morphological complications (such as voids in the scale) are preventing the retention of ideal parabolic conditions; ionic diffusion in the scale may still principally control or limit the oxidation.*

The logarithmic empirical reaction rate law

$$W = k_e \log(Ct + A) \tag{11.6}$$

where k_e, C, and A are constants, and the related inverse logarithmic oxidation rate law

$$\frac{1}{w} = C - k_i \log t \tag{11.7}$$

where k_i and C are constants, do not differ greatly because experimental data may fit both equations equally satisfactorily. Logarithmic oxidation behavior is generally observed with thin oxide layers (e.g., less than 1000 Å) at low temperatures. Aluminum, copper, iron, and some other metals oxidize in this manner at ambient or slightly elevated temperatures. The exact mechanism is not completely understood, but it is generally agreed that logarithmic (or inverse logarithmic) oxidation results from the effect of an electrical potential gradient within very thin oxide layers in assisting ionic transport across the scale.

Under specific conditions, some metals appear to oxidize according to a cubic law

$$W^3 = k_c t + C \tag{11.8}$$

where k_c and C are constants, or according to some other rate law with exponents different from 3 (e.g., 2.5, 3.4, etc.). Usually such behavior is restricted to short exposure periods. For the oxidation of zirconium and hafnium apparent cubic rate laws have been explained as a combination of diffusion-limited scale formation and oxygen dissolution into the metal. In

*R. A. Rapp, *Metall. Trans.* **15A**:765 (1984).

other such cases as well, such irrational rate laws can probably be explained by the superposition of a morphological complication and ionic diffusion through the scale. In general, then, such rate laws probably do not represent any new or significant mechanism.

Examination of Fig. 11-6 and Eqs. (11.4) to (11.8) indicates that a linear oxidation rate is the least desirable, since weight gain increases at a constant rate with time. Parabolic and logarithmic oxidation rates are the more desirable for alloys used in high-temperature oxidizing environments. Materials that oxidize logarithmically (or according to an inverse logarithmic relationship) reach an apparent limiting film thickness. Aluminum oxidizes in air at ambient temperature according to the logarithmic rate law, and as a consequence, film growth essentially stops after a few days' exposure. This is one of the reasons that aluminum possesses superior atmospheric oxidation resistance.

11-6 Effect of Alloying

As indicated in Figs 11-4 and 11-5, the concentration of ionic defects (anion vacancies and excess electrons, or metal ion vacancies and electron holes) may be influenced by the presence of foreign ions in the lattice (the doping effect). From the consideration of electroneutrality in compounds, the following rules describing these effects have been proposed by both Hauffe and Wagner (see references in Tables 11-2 and 11-3).

1. For *n*-type oxides (metal excess—e.g., ZrO_2, ZnO)
   a. Substitution of lower-valence cations into the lattice increases the concentration of interstitial cation or anion vacancies, and decreases the concentration of excess electrons. *A diffusion-controlled oxidation rate would be increased.*
   b. Substitution of cations possessing higher valence decreases the concentration of interstitial cations or anion vacancies and increases the concentration of excess electrons. *A diffusion-controlled oxidation rate would be decreased.*
2. For *p*-type oxides (metal deficient—e.g., NiO, CoO)
   a. The incorporation of lower valency cations decreases the concentration of cation vacancies or interstitial anions and increases the number of electron holes. *A diffusion-controlled oxidation rate would be decreased.*
   b. The addition of higher valency cations increases the cation vacancy or interstitial anion concentration and decreases the electron hole concentration. *A diffusion-controlled oxidation rate would be decreased.*

The experimental data in Table 11-2 have been presented as evidence for the validity of the doping effect in metal oxidation. Foreign metal ions

Table 11-2 Oxidation of zinc and zinc alloys 390°C, 1 atm O_2

Material	Parabolic oxidation constant K_p, g^2/cm^2-hr
Zn	8×10^{-10}
Zn + 1.0 atomic % Al	1×10^{-11}
Zn + 0.4 atomic % Li	2×10^{-7}

Source: C. Gensch and K. Hauffe, *Z. Physik. Chem.*, **196**:427 (1950).

Table 11-3 Oxidation of nickel and nickel alloys

Nickel and chromium-nickel alloys at 1000°C in pure oxygen*

Wt. % Cr	Parabolic oxidation constant K_p, g^2/cm^4-sec
0	3.8×10^{-10}
0.3	15×10^{-10}
1.0	28×10^{-10}
3.0	36×10^{-10}
10.0	5.0×10^{-10}

Effect of lithium oxide vapor on the oxidation of nickel at 1000°C in oxygen†

Atmosphere	Parabolic rate constant K_p, g^2/cm^4-sec
O_2	2.5×10^{-10}
$O_2 + Li_2O$	5.8×10^{-11}

*Data from C. Wagner and K. E. Zimens, *Acta Chem. Scand.*, **1**:547 (1947).

†Data from H. Pfeiffer and K. Hauffe, *Z. Metalkunde*, **43**:364 (1952).

have been presumably substituted into the *n*-type ZnO lattice by alloy additions to the matrix metal. The addition of aluminum to zinc decreased its oxidation rate, whereas alloying with lithium increased the oxidation rate.

The effects of foreign ions on the defect structure of nickel oxide, a *p*-type semiconductor, are shown in Fig. 11-5. The influence of substituted lithium ions is opposite of that observed in ZnO as predicted by the above rules. Again, electroneutrality is maintained. The preceding rules and Fig.

11-5 suggest most unusual conclusions regarding the oxidation rate of nickel. Lithium, a metal with virtually no resistance to oxidation (rapid linear law), should improve the oxidation characteristics of nickel when added in small amounts. On the other hand, small additions of chromium, which is invariably present in oxidation-resistant alloys (see Chap. 5), are predicted to be detrimental on the basis of this approach. These conclusions are confirmed by experiment, as shown in Table 11-3. The oxidation rate of nickel is reduced in the presence of lithium oxide vapor. The oxidation rate of Ni-Li alloys have provided comparable evidence. The oxidation rates of dilute nickel-chromium alloys increase continuously up to about 5% chromium. Above this level, $NiCr_2O_4$ tends to form preferentially at the alloy surface and the above interpretation no longer applies.

11-7 Catastrophic Oxidation

Catastrophic oxidation refers to metal-oxygen reactions that occur at continuously increasing rates or that break away from protective behavior and thereafter react very rapidly. Metals that follow linear oxidation kinetics, Eq. (11.4), at lower temperatures tend to oxidize catastrophically at high temperatures because of the rapid, exothermic reaction at their surfaces. If the rate of heat transfer to the metal and surroundings is less than the rate of heat evolution by the reaction, the surface temperature increases. This causes a chain reaction characteristic—the temperature and reaction rate increase. In the extreme case, for example, for magnesium or columbium (niobium), ignition of the metal occurs. Metals like molybdenum, tungsten, osmium, rhenium, and vanadium, which have volatile oxides, may oxidize catastrophically, but this does not result from surface ignition. Also, alloys containing molybdenum and vanadium may oxidize catastrophically as first postulated in 1948 by W. C. Leslie (masters thesis.* In this case, the formation of low-melting eutectic oxide mixtures produces a liquid beneath the scale, which is less protective. Catastrophic oxidation can also occur if vanadium oxide or lead oxide compounds are present in the gas phase. This is particularly important during the combustion of fuel oils with high vanadium contents. Figure 11-7 shows the relative resistances of various chromium-nickel-iron alloys exposed to the combustion products of a high-vanadium oil. As shown, alloys with high Cr and Ni contents retard the effect of catastrophic oxidation. The accelerated reaction of alloys in high-vanadium combustion product gases is a type of hot corrosion,† treated in Sec. 11-13. It is important to note that high-nickel alloys are not satisfactory in oxidizing environments containing sulfurous gases (e.g., high-sulfur fuel

*W. C. Leslie and M. G. Fontana, *Trans. ASM*, **41**:1213–1247 (1949); also S. S. Brenner, *J. Electrochem. Soc.*, **102**:16 (1955).

†K. L. Luthra and H. S. Spacil, *J. Electrochem. Soc.*, **129**:649 (1982).

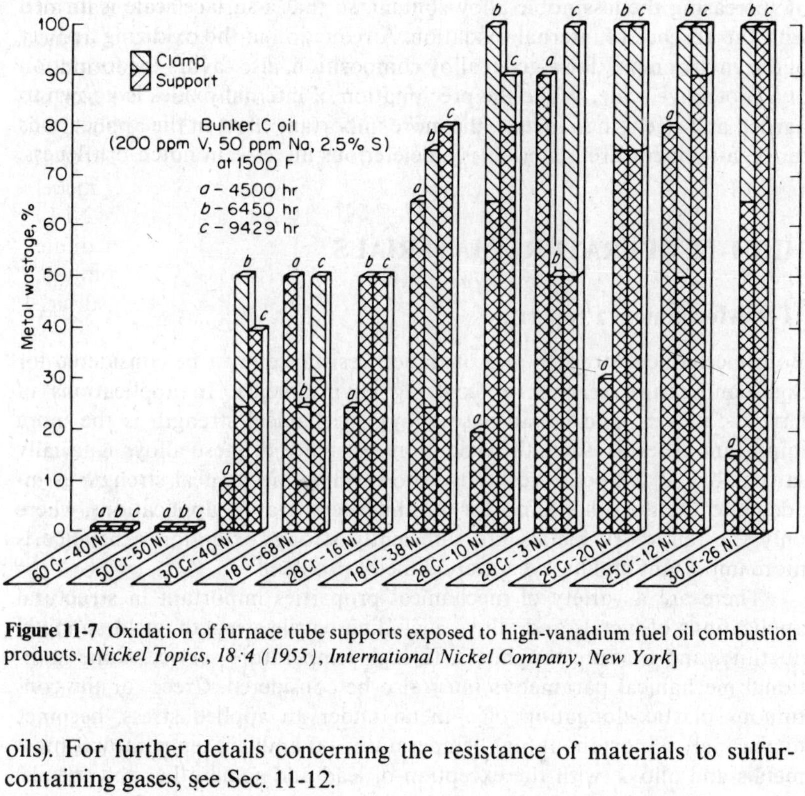

Figure 11-7 Oxidation of furnace tube supports exposed to high-vanadium fuel oil combustion products. [*Nickel Topics, 18:4 (1955), International Nickel Company, New York*]

oils). For further details concerning the resistance of materials to sulfur-containing gases, see Sec. 11-12.

11-8 Internal Oxidation

In certain alloy systems, one or more dilute components that may form more stable oxides than the base metal may oxidize preferentially below the external surface of the metal, or below the metal-scale interface. This is called internal oxidation because the oxide precipitate is formed within the metal matrix, rather than at the external surface. Dilute alloys of base metals exhibiting a relatively high solubility and diffusivity for atomic oxygen are subject to internal oxidation. Dilute copper- and silver-base alloys containing Al, Zn, Cd, Be, etc. show this kind of oxidation. Internal oxidation also occurs frequently in the iron, nickel, and cobalt alloys commonly used in high-temperature applications. Here it is particularly deleterious because the less noble alloying additions such as Cr and Al have been intended for the formation of a protective external scale. Upon forming a Cr_2O_3 or Al_2O_3 internal oxide precipitates, the preferential oxidation of these elements to form the surface scale is precluded. Internal oxidation may be prevented by increasing the less noble alloy content, so that a surface scale is formed with the absence of internal oxidation. A reduction in the oxidizing activity of the environment, for a given alloy composition, also favors the formation of a protective scale. While the precipitation of internal oxides is known to harden and strengthen metals, the more important effect in the applications for high-temperature materials is a deleterious increase in notch brittleness.

HIGH-TEMPERATURE MATERIALS

11-9 Mechanical Properties

Both mechanical strength and oxidation resistance must be considered for high-temperature service in oxidizing environments. In applications of ferrous-, nickel-, and cobalt-base alloys, mechanical strength is the more important criterion since the oxidation resistance of these alloys is usually satisfactory at temperatures above those where mechanical strength is inadequate. On the other hand, for ultrahigh-temperature applications, where only the refractory metals are sufficiently strong, oxidation resistance is more important and an even more difficult problem.

There are a variety of mechanical properties important in structural applications of metals and alloys, including tensile strength, yield strength, ductility, and fracture toughness. In high-temperature service, some additional mechanical parameters must also be considered. Creep, or the continuous plastic elongation of a metal under an applied stress, becomes increasingly important at high temperatures. At ambient temperatures most metals and alloys (with the exception of lead and some other soft metals) experience a rapid, but limited, plastic deformation immediately after application of a sufficiently large stress. However, at higher temperatures, plastic deformation continues for extended time periods. Creep is particularly important in applications in jet turbines, where close tolerance must be maintained, and also in applications such as furnace support rods, where extremely long-term exposure to high temperatures occurs. Creep rate or creep resistance is usually expressed as percent plastic deformation for a given time period at a constant applied load.

The introduction of gas turbines and rocket engines as a means of aircraft propulsion necessitated the revision and reconsideration of mechanical-strength properties. Gas turbines are also used to propel ships and trains, to pump fluids in pipelines, and to generate electricity for utilities. These devices, which operate for relatively short periods of time, pose problems quite different from stream turbines which are required to operate for periods of up to 20 years. Stress rupture, or the stress necessary to cause rupture in a predetermined time period at constant temperature, is far more important in short-term high-temperature materials applications. Alternatively, stress rupture life is reported as the time to cause failure for a constant applied stress. In summary, the mechanical and physical properties necessary or desired for a good high-temperature metal or alloy are:

1. Good room-temperature mechanical properties such as tensile strength, ductility, and toughness to permit cold-rolling and other fabrication operations. Adequate aqueous corrosion resistance to resist corrosion prior to installation and during downtime.
2. Adequate high-temperature mechanical properties including low creep rate, high stress-rupture strength, and toughness to avoid fracture are required. Strength at high temperatures is often achieved by solid-solution strengthening, precipitation hardening, or dispersion hardening. The high-temperature materials must also resist cyclic stresses and repeated heating and cooling cycles. Techniques have been developed recently for the introduction of several percent of stable submicron particles of ThO_2 or Y_2O_3 into pure Ni and Co and into alloys of these metals, which exhibit better oxidation resistance. The dislocation-particle reactions and the stabilized elongated grain structure in these extruded materials provide increased strength at temperatures required for service in jet engines.*
3. Structural stability at high temperatures is required to avoid graphitization, grain growth, phase changes, precipitate dissolution, precipitate overaging, new precipitation, and other solid-state metallurgical changes.

The compositions of some of the more common high-temperature materials (superalloys) used for oxidation resistance and especially in gas-turbine applications, are listed in Table 11-4. Nickel-base alloys are commonly hardened by the precipitation of the so-called gamma prime [Ni_3(Al, Ti)] phase. About 15 to 20% Cr is included in the alloys to provide oxidation resistance through the formation of protective Cr_2O_3 and $NiCr_2O_4$ in the oxide scale with the exclusion of NiO. Because higher Cr contents interfere with the γ' strengthening mechanism, the compositions of most superalloys represent a compromise to provide adequate high-temperature oxidation resistance in combination with high-temperature strength. Molybdenum and tungsten serve as the best substitutional solid-solution strengtheners, and columbium (niobium) and titanium are effective carbide formers. While interstitial carbon (and boron) is an effective solid-solution strengthener, excessive carbon is deleterious to oxidation resistance. Nickel is generally included in cobalt-base alloys to stabilize the face-centered cubic crystal lattice.

Figure 11-8 illustrates the stress-rupture properties of these classes of

*B. A. Wilcox and A. H. Clauser, Dispersion Strengthening, in C. T. Sims and W. C. Hagel, eds., *The Superalloys*, John Wiley & Sons, Inc., New York (1972).

Table 11-4 Composition of high-temperature alloys

	C	Cr	Ni	Co	Mo	W	Cb	Ti	Al	Fe	B	Other
Combustion Can Materials												
Hastelloy X	0.1	22.0	Bal	1.5	9.0	0.6	—	—	—	18.5	—	—
C 263	0.03	20.0	Bal	20.0	5.9	—	—	2.15	.45	—	—	—
Inconel 617	0.07	22.0	Bal	12.5	9.0	—	—	—	1.0	—	—	—
Haynes 188	0.10	22.0	22.0	39	—	14.0	—	—	—	3Max	—	.08La
Turbine Disk Materials												
Discaloy	0.04	13.5	26.0	—	2.9	—	—	2.15	0.2	Bal	0.003	.3V
Inconel 901	0.05	13.5	42.7	—	6.2	—	—	2.5	0.2	34	—	—
Inconel 718	0.04	19.0	52.5	—	3.0	—	9.2	0.8	.5	18	—	—
Waspalloy	0.07	19.5	Bal	13.5	4.3	—	—	3.0	1.4	—	0.006	0.09Zr
Rene 95	0.15	14.0	61	8.0	3.5	3.5	—	2.5	3.5	—	0.01	.05Zr
IN 100	0.18	10.0	Bal	15.0	3.0	3.0	—	4.7	—	—	0.014	.06Zr
Udimet 700	0.15	15.0	Bal	18.5	5.2	—	—	3.5	4.2	—	0.05	—
Nickel Alloys												
IN 738	0.17	16.0	61.0	8.5	1.7	2.6	0.9	3.4	3.4	—	0.01	0.1Zr

IN 792	0.12	12.4	61.0	9.0	1.9	3.8	—	4.5	3.5	—	0.02	0.1Zr
IN 100	0.18	10.0	60.0	15.0	3.0	—	—	4.7	5.5	—	0.014	.06Zr 1V
Cabot 214	0.04	16.0	75.5	—	—	—	—	—	4.5	4.0	—	0.01Y
B-1900	0.1	8.0	—	10.0	6.0	0.1	0.1	1.0	6.0	0.35	0.015	4.3Ta .08Zr
Rene 80	0.17	14.0	—	9.5	4.0	4.0	—	5.0	3.0	—	0.015	—
Alloy 713C	0.12	12.5	—	—	4.2	—	2.0	0.8	6.1	—	0.012	—
PWA 1480 monocrystal	0.15	10.0	Bal	5.0	—	4.0	—	1.5	5	—	—	12Ta
PWA 1422 direct solidification	0.14	10.0	Bal	10.0	—	12.5	1	2	5	—	—	2Hf
						Cobalt Alloys						
Haynes 188	0.10	22.0	22.0	39	—	14.0	—	—	—	3Max	—	.08La
X 40	0.50	25.5	10.5	54	—	7.5	—	—	—	.75	—	.75Si 3.5Ta
Mar-M 509	0.6	23.5	10.0	55	—	7.0	—	0.2	—	—	—	—
WI 52	0.45	21.0	1.0	—	—	11.0	—	—	—	—	—	—

Sources: Metal Progr., **126**:82–85 (June 1984) and private communication. For an extensive listing of superalloys see "Guide to Selection of Super-alloys," *Metal Progr*. (mid-June 1984). Also the book *High Temperature Materials in Gas Turbines*, P. R. Sahm and M. O. Speidel, ed., American Elsevier Publishing Co., New York, 1974. Also of interest is "Protection of Superalloys for Turbine Application," D. Chatterji, R. C. DeVries and G. Romeo, chap. 1 in *Advances in Corrosion Science and Technology*, Vol. 6, Plenum Press, New York, 1976.

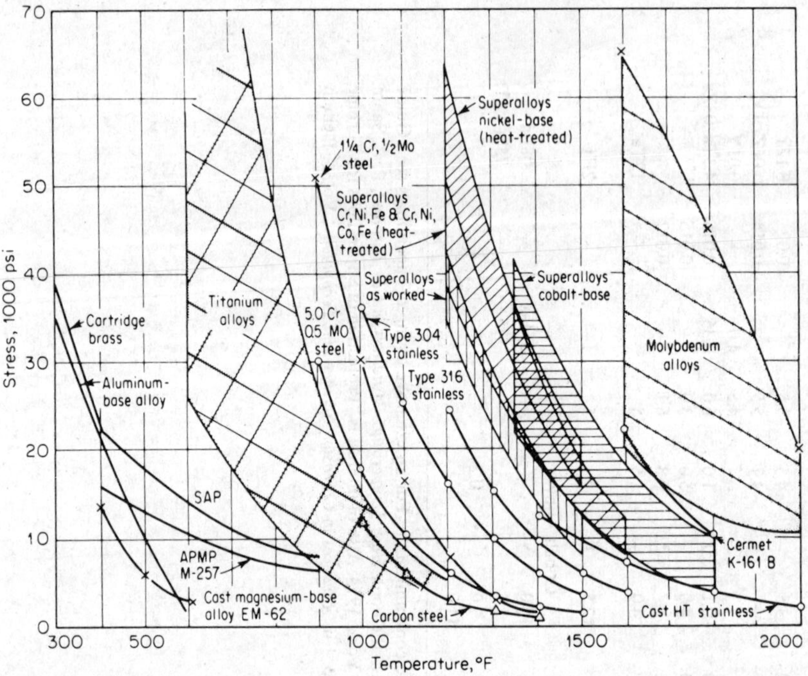

Figure 11-8 Stress to rupture in 1000 hr vs. temperature. *(H. C. Cross and W. F. Simmons, ASM Preprint, 1954; see also "Super Alloys," Universal-Cyclops Steel Corp., Bridgeville, Pa.)*

high-temperature alloys. Here, the stress necessary to cause rupture or fracture in 1000 hr is given over a range of temperature. Areas are outlined in Fig. 11-8 for the various alloy types, showing how composition affects mechanical properties. As noted above, stress rupture is usually used as a design criterion, since tensile strength becomes meaningless at elevated temperatures.

11-10 Oxidation Resistance

As mentioned before, there are numerous properties an oxide layer must possess to be protective. When most of these conditions are attained, a high degree of resistance to oxidation is achieved. Nickel, cobalt, and iron exhibit only a moderate degree of oxidation resistance; the alloying additions of chromium, silicon, and aluminum allow the formation of relatively protective spinel and rhombohedral oxide phases (Cr_2O_3 and Al_2O_3). However, if these alloying elements are too dilute and therefore internally oxidized, then their effectiveness at forming a protective scale will be seriously reduced.

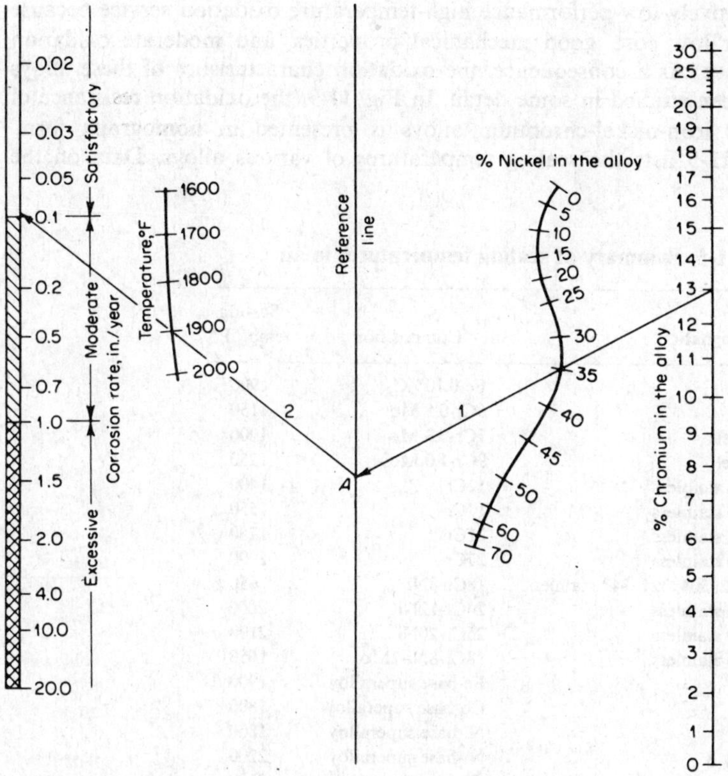

Figure 11-9 Air corrosion of Fe-Ni-Cr alloys. To use the nomograph: (1) Knowing the composition and temperature: If an alloy of 13% Cr and 35% Ni were contemplated for use in air at 1900°F, estimate the corrosion rate as follows: Draw a straight line (line 1) connecting the % Ni and % Cr to interest reference line A. Then draw another line (line 2) through the temperature selected, 1900°F, and read the corrosion rate, in this case, about 0.09 in./year. This is an average rate for a 100-hr test and would be a conservative estimate for a longer exposure. (2) Knowing the temperature and tolerable corrosion: By reversing the above sequence one may select a suitable composition from a series of alloys having equal corrosion resistance by a series of lines radiating from the intersection point on the reference line. This nomographic chart is based on the 100-hr test data of Brasunas, Gow, and Harder [*Proc. ASTM., 46:870 (1946)*] and presents air-corrosion data in concise and convenient form which would otherwise require pages of tables or numerous curves. Loss in accuracy is to be expected in condensing data and, should more precise data be required, the original article may be consulted. *(Metal Progr. Data Sheet, Dec. 1964, p. 114B.)*

Iron-nickel-chromium alloys are the most commonly employed materials for relatively low-performance high-temperature oxidation service because of their low cost, good mechanical properties, and moderate oxidation resistance. As a consequence, the oxidation characteristics of these alloys have been studied in some detail. In Fig. 11-9, the oxidation resistance of various iron-nickel-chromium alloys is presented in nomograph form. Table 11-5 lists the scaling temperatures of various alloys. Data on the

Table 11-5 Summary of scaling temperatures in air

Alloy designation	Composition	Scaling temp, °F*
1010 steel	Fe-0.10%C	900
502 steel	5Cr-0.5 Mo	1150
Alloy steel	7Cr-0.3 Mo	1200
Alloy steel	9Cr-1.0 Mo	1250
Type 410 stainless	12Cr	1400
Type 430 stainless	17Cr	1550
Type 442 stainless	21Cr	1750
Type 446 stainless	27Cr	1900
Types 302, 304, 321, 347 stainless	18Cr-8Ni	1650
Type 309 stainless	24Cr-12Ni	2000
Type 310 stainless	25Cr-20Ni	2100
Type 316 stainless	18Cr-8Ni-2Mo	1650
N-155	Fe-base superalloy	1900
S-816	Co-base superalloy	1800
M-252	Ni-base superalloy	1800
Hastelloy X	Ni-base superalloy	2200
HS-21	Co-base superalloy	2100
	Cr	1650
	Ni	1450
	Cu	850
Brass	70 Cu-30 Zn	1300
Hastelloy B	—	1400
Hastelloy C	—	2100
HW	12Cr-60Ni-bal Fe	2050
HT	15Cr-66Ni-bal Fe	2100
HX	17Cr-66Ni-bal Fe	2100

*Temperature below which oxidation rate is negligible. "Negligible" is often defined as less than about 0.002 g weight gain per square inch per hour.

Sources: F. H. Clark, *Metals at High Temperatures*, p. 348, Reinhold Publishing Corporation, New York, 1950; *Super Alloys*, Universal-Cyclops Steel Corp., Bridgeville, Pa; and *Metals Handbook*, vol. 1, pp. 405, 448, 449, 482, 486, 598, American Society for Metals, Metals Park, Ohio, 1961.

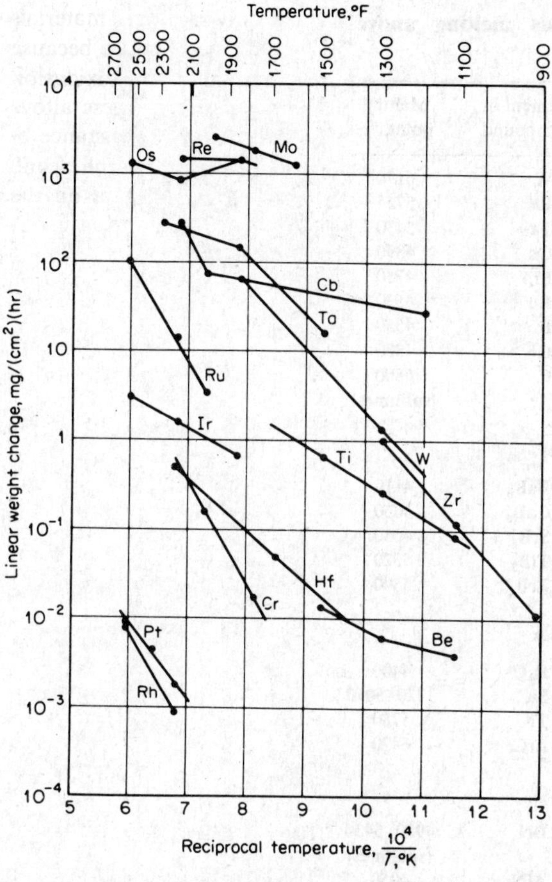

Figure 11-10 Oxidation of refractory metals. *(J. E. Campbell, DMIC Report 160. Oct. 21, 1961, p. 66, Battelle Memorial Institute, Columbus, Ohio)*

oxidation behavior of the refractory metals are shown in Fig. 11-10.* Comparing Table 11-5 and Fig. 11-8, it would appear that iron-, cobalt-, and nickel-base alloys possess satisfactory oxidation resistance at temperatures above those where mechanical properties become inadequate.

On the other hand, preferential grain-boundary oxidation, and exaggerated attack resulting from thermal cycling, applied stresses, and the presence of moisture and chloride- or sulfur-containing gas results in service

*Recent data on ruthenium indicate that information for this element in Fig. 11-10 is not good and should not be used.

Table 11-6 Some solids melting above 4000°F

	Element or compound	Melting point, °F
Tungsten	W	6170
Rhenium	Re	5755
Tantalum	Ta	5430
Osmium	Os	4890
Molybdenum	Mo	4750
Ruthenium	Ru	4530
Iridium	Ir	4550
Columbium	Cb	4380
Graphite (carbon)	C	6800 (sublimes)
	Borides	
Barium hexaboride	BaB_6	4110
Calcium hexaboride	CaB_6	4050
Strontium hexaboride	SrB_6	4050
Titanium diboride	TiB_2	5320
Zirconium diboride	ZrB_2	5500
	Carbides	
Boron carbide	B_4C	4400
Silicon carbide	SiC	3720–5070
Titanium carbide	TiC	5780
Zirconium carbide	ZrC	6420
	Nitrides	
Boron nitride	BN	4950–5430 (sublimes)
Aluminum nitride	AlN	4050
Titanium nitride	TiN	5340
Zirconium nitride	ZrN	5400
Hafnium nitride	HfN	5990
	Oxides	
Magnesia	MgO	5120
Calcia	CaO	4690
Strontia	SrO	4390, 4600
Zirconia	ZrO_2	5010
Chromia	Cr_2O_3	4230

Source: Data compiled by S. Sklarew and M. J. Albom, *Metal Progr.*, 106 (Feb. 1965).

lives far shorter than those predicted from isothermal oxidation experiments in air.

A comparison of Figs. 11-8 and 11-10 indicates that for very-high-temperature service, molybdenum and other high-melting metals such as tungsten, columbium, and tantalum must be used to achieve sufficient mechanical strength, and then oxidation resistance is the controlling factor. Because Mo and W form the very volatile oxides MoO_3 and WO_3, and Cb and Ta the nonprotective oxides Cb_2O_5 (see Fig. 11-3) and Ta_2O_5, all of these metals must be protectively coated before their use in an oxidizing atmosphere. At temperatures above 3000°F the number of metallic materials that possess adequate mechanical strength and oxidation resistance is very limited. As a consequence, nonmetals such as oxide, boride, carbide, and nitride compounds must be employed for the leading edges and nose cone of the space shuttle vehicle, a carbon/carbon composite is protected by a SiC-base pack cementation coating to provide oxidation resistance upon reentry. Table 11-6 presents the properties of some solids that melt above 4000°F. Examination shows that only a few metals are still solid at 4000°F, and considering the oxidation rates of these metals as shown in Fig. 11-10, these metals are useless in the uncoated condition in an oxidizing atmosphere.

Metal compounds such as borides, carbides, and nitrides possess moderate oxidation resistance, and because of their strong bonding forces possess extremely high melting points. Borides, carbides, and oxides are used primarily for extremely high temperatures. The major limitations of nonmetallic materials are their lack of ductility, low resistance to thermal shock, inability to be joined, and incompatibility with other materials.*

OTHER METAL-GAS REACTIONS

11-11 Decarburization and Hydrogen Attack

In Chap. 3, mechanical damage due to hydrogen absorption at ambient temperatures was discussed. At elevated temperatures, hydrogen can influence the mechanical properties of metals in a variety of ways. Since most high-temperature gas streams are mixtures of gases, it is necessary to consider the effect of hydrogen in the presence of other gases. One effect of hydrogen at high temperatures is that of decarburization, or removal of carbon from an alloy. If the alloy is strengthened by interstitial carbon or by carbide precipitates, decarburization results in a reduction of tensile strength and an increase in ductility and creep rate. Thus, after long-term exposure to hydrogen at elevated temperatures, steels tend to lose their strength. The reverse process, carburization, can also occur in hydrogen-hydrocarbon gas mixtures that are frequently encountered in petroleum-refining operations. For low-alloy steels, carburization is usually less detrimental than de-

*E. L. Hall, Y. M. Kouh, M. R. Jackson and R. L. Mehan, *Metall. Trans.*, **14A**:781 (1983).

carburization, although the addition of carbon to an alloy tends to decrease its ductility and remove certain solid-solution elements through carbide precipitation. Also, as noted previously in Chap. 3, hydrogen penetration into a metal can cause scale spallation and metal fracture or rupture resulting from the formation of molecular hydrogen in internal voids upon cooling.

Hydrogen and hydrocarbon gases If steel is exposed to hydrogen at high temperatures, the following reaction can occur:

$$\underline{C}(Fe) + 4\underline{H}(Fe) = CH_4 \tag{11.9}$$

Carbides or dissolved carbon [indicated as $\underline{C}$(Fe)], react with the dissolved hydrogen [$\underline{H}$(Fe)] to form methane. The rate and direction of the reaction depends on the amounts of hydrogen and methane present in the gas phase and the carbon content of the alloy. The equilibrium between carbon steel and hydrogen-methane gas mixtures of 1 atm total pressure can be obtained from thermodynamic data as shown in Fig. 11-11, which can be used to estimate the possibility of decarburization of carbon steels in methane-hydrogen mixtures. It has been previously noted that thermodynamics provides little information concerning the rate of a reaction at ambient temperatures. However, at high temperatures local equilibrium at interfaces is usually rapidly achieved, and thermodynamic data may be used to predict accurately the interfacial concentrations which serve as boundary conditions for diffusion-limited reactions.

Because atomic hydrogen diffuses readily in steel, cracking may result from the formation of CH_4 at high pressures in internal voids in the metal [Eq. (11.9)]. Chromium and molybdenum additions to a steel improve its

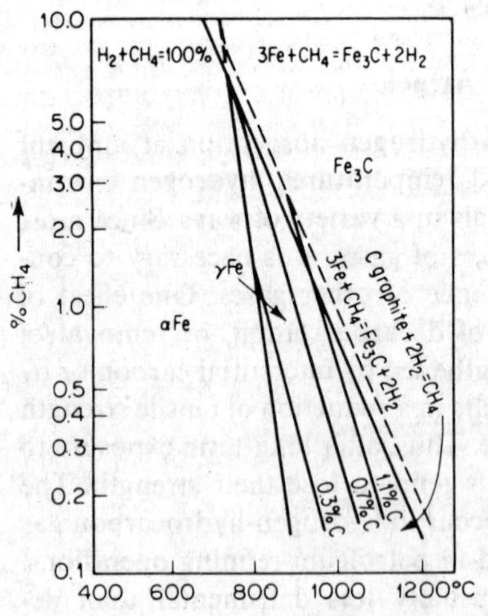

Figure 11-11 Equilibrium diagram of the Fe-CH_4-H_2 system. *Note:* Figure applies only for pure methane-hydrogen mixtures. Presence of inert gases (e.g., nitrogen) renders it inaccurate. [*W. D. Jones, Fundamental Principles of Powder Metallurgy, p. 543, Edward Arnold (Publishers) Ltd., London, 1960*]

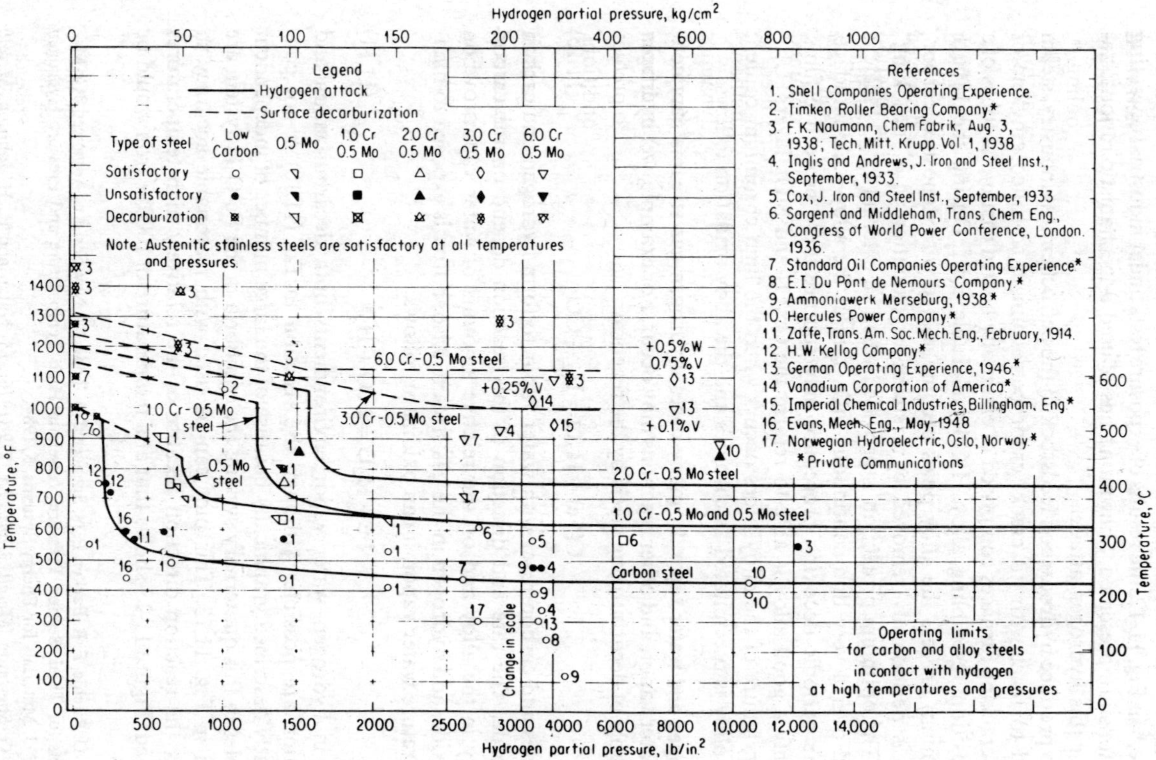

Figure 11-12 Operating limits for carbon and alloy steels in contact with H_2 at high temperatures and pressures. [*G. A. Nelson, Petrol. Refiner*, *29:104 (1950)*]

resistance to cracking and decarburization in hydrogen atmospheres by forming carbides which are more thermodynamically stable than iron carbide.* In Fig. 11-12, the results of exposure tests under industrial operating conditions are presented. Shewmon† has offered a mechanistic rationalization for this hydrogen attack.

To produce hydrocarbon feedstock in the petrochemical industry, steam is used to react with (to crack) crude petroleum in so-called ethylene or steam cracking furnaces comprised of externally heated cast metal tubes of a composition resembling alloy HK-40 of Table 5.3.‡ During the 1800°F (maximum) process the alloy pipes become coated with deposited carbon, which is periodically removed by a steam-air oxidation in a "decoking" cycle. The optimum protection is provided when an external Cr_2O_3 scale is formed over a very thin continuous SiO_2 film at the alloy-scale interface. But the periodic decoking procedure may rupture these protective films, and only a good alloy is able to reform the Cr_2O_3 and SiO_2 layers. Ultimately, failure of the alloy tubes occurs when the high chromium content of the alloy is precipitated as carbides and the alloy oxidation follows.

Hydrogen and water vapor In many cases, hydrogen gas streams may contain water vapor, and other reactions may occur. For example, wet hydrogen is capable of decarburizing steels by the reaction

$$\underline{C}(Fe) + H_2O = H_2 + CO \tag{11.10}$$

Carbides and carbon react with water vapor to form hydrogen and carbon monoxide. The rate and direction of this reaction depend on the carbon activity in the alloy and on the relative amounts of water vapor, carbon monoxide, and hydrogen in the gas stream. Also, if iron is exposed to high-temperature water vapor, it can oxidize as follows:

$$Fe + H_2O = FeO + H_2\text{¶} \tag{11.11}$$

Thus, in hydrogen–water vapor environments both decarburization and oxidation are possible. Although the reaction in Eq. (11.10) is not conveniently described graphically because of the large number of independent variables in this quaternary system, the equilibria in the Fe-O-H system are shown in Fig. 11.13. This plot can be used with reasonable accuracy to predict the reaction direction because of the relatively high temperatures involved. Figure 11-13 shows that at a temperature of 1000°F iron would be

*R. D. Ritchie, E. R. Parker, P. N. Spencer, J. A. Todd. "A New Series of Advanced 3Cr-Mo-Ni Steels for Thick Section Pressure Vessels in High Temperature and Pressure Hydrogen Service." J. Materials for Energy Systems, **5**, 1984.

†P. G. Shewmon, *Metall. Trans.*, **7A**, 279 (1976); M. Natan and H. H. Johnson, *Metall. Trans.*, **14A**, 963 (1983).

‡K. Ledjeff, A. Rahmel, and M. Schorr, *Werkst. Korros.*, **31**:83 (1980).

¶My Ph.D. dissertation studied this reaction at 1600°C.

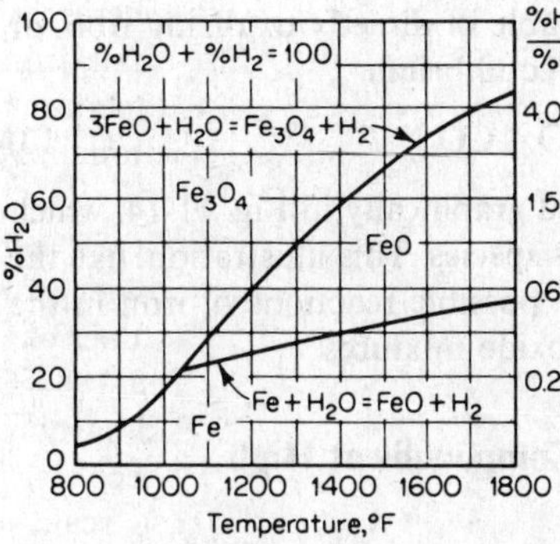

Figure 11-13 Equilibrium diagram of the Fe-H_2-H_2O system. *Note:* Figure applies only for pure H_2O-H_2 mixtures. *(Metals Handbook, vol. 2, p. 68, American Society for Metals, Metals Park, Ohio, 1964)*

oxidized in hydrogen containing 20% or more water vapor. In industrial practice, gas process streams containing carbon, hydrogen, and oxygen species do not maintain constant compositions, so that an alloy may be alternately subjected to oxidizing and reducing conditions; rapid deterioration of the metals is a common consequence, a reaction sometimes referred to as "metal dusting."

Carbon monoxide–carbon dioxide mixtures Carbon dioxide and other carbonaceous gases are frequently encountered in petroleum refining and during combustion of fuels. A CO-CO_2 atmosphere is capable of decarburizing or carburizing steels (and other alloys) by the forward and reverse directions of the reaction

$$C_{(Fe)} + CO_2 = 2CO \tag{11.12}$$

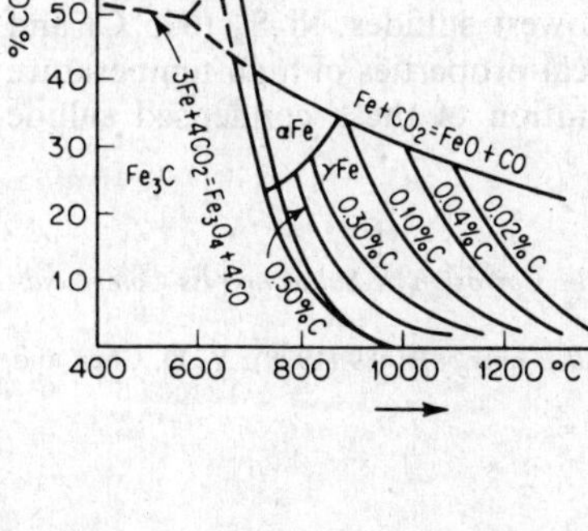

Figure 11-14 Equilibrium diagram of the Fe-CO-CO_2 system. [W. D. Jones, Fundamental Principles of Powder Metallurgy, p. 537, Edward Arnold (Publishers) Ltd., London, 1960]

Also, a CO-CO_2 atmosphere may be capable of directly oxidizing iron, or else reducing iron oxide, according to the equilibrium

$$Fe + CO_2 = FeO + CO \qquad (11.13)$$

Equations (11.12) and (11.13) are presented graphically in Fig. 11-14, which shows the region of stability of the various species. This illustration, like the previous ones, can be used to predict the possible reactions of iron in the presence of carbon monoxide–carbon dioxide mixtures.

11-12 Corrosion of Metals by Sulfur Compounds at High Temperatures

As the areas of energy conversion and environmental protection become increasingly concerned with the processing and cleanup of sulfur-bearing fossil fuels and their combustion products, the attack of metals by sulfur compounds at high temperatures has become a major corrosion problem. A symposium treats this subject.* The term "sulfidation" is used to describe the gaseous attack of metals by H_2S, S_2, SO_2, and other gaseous sulfur species. The term "hot corrosion" refers to the accelerated attack of metals where fused Na_2SO_4 condensed onto the metal surface. Although sulfides may be formed in hot corrosion also, the two basically different forms of high-temperature attack are treated separately here.

In connection with *gaseous sulfur attack*, the vapors in equilibrium with liquid sulfur comprise S_2, S_4, S_6, S_8 as principal molecular species, with S_2 the predominant vapor species at elevated temperatures and low sulfur activities. In reducing gases containing hydrogen, such as gasified coal, hydrolyzed refuse, or processed petroleum, H_2S is a major gaseous constituent or contaminant. In oxidizing gases, such as fossil fuel combustion products (automotive exhaust, incinerator gases, process effluent gases), molecular SO_2 may vary from a fraction to several percent. Although considerable SO_3 may be thermodynamically stable in excess oxygen at relatively low temperatures, SO_3 formation kinetics are generally quite slow in the absence of sufficient catalytic surface. Each of these sulfur-bearing gaseous compounds can lead to rapid scaling and to internal precipitation of stable sulfides, even when a more stable, protective oxide scale imperfectly covers the metal. The eutectic points† of the lowest sulfides Ni_3S_2 (645°C) and CoS (880°C) are so low that the mechanical properties of high-temperature alloys are endangered upon the precipitation of these condensed sulfide compounds.

*Z. A. Foroulis, ed., *High-Temperature Metallic Corrosion by Sulfur and Its Compounds*, The Electrochemical Society, Princeton, N.J. (1970).

†R. Y. Lin, D. C. Hu, and Y. A. Chang, *Metall. Trans.*, **9B**:489 (1978); Y. D. Chen and Y. A. Chang, *Metall. Trans.*, **9B**:61 (1978).

Hydrogen sulfide may act as an oxidizing agent in the formation of sulfide scales on metal substances at high temperatures. In general, nickel and nickel-rich alloys are usually rapidly attacked in the presence of hydrogen sulfide and other sulfur-bearing gases. Because of the 645°C eutectic temperature in the Ni-S binary system, attack is frequently catastrophic, with rapid intergranular penetration by the liquid sulfide product and subsequent disintegration of the metal. Iron-base alloys are often used to contain hydrogen sulfide environments because of their low cost and good chemical resistance. Figure 11-15 presents some experimental results obtained with chromium and chromium-nickel stainless steels. It is interesting to note that small alloy additions of nickel to iron-base materials are beneficial to the sulfide resistance of iron alloys. Table 11-7 presents a qualitative tabulation

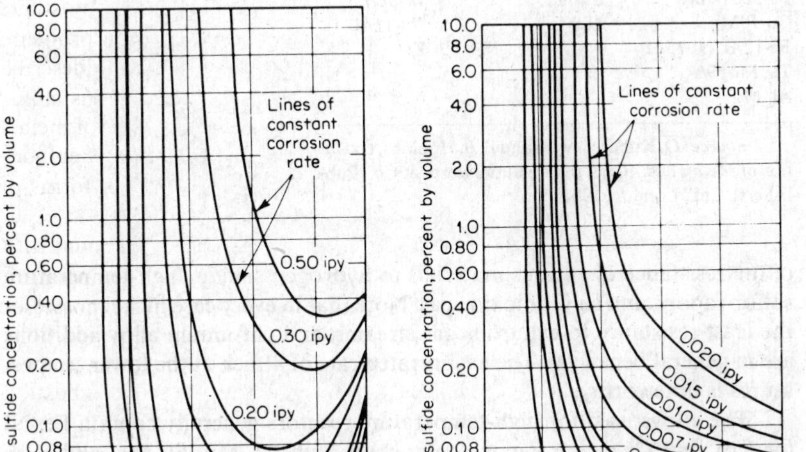

Figure 11-15 Resistance of Cr and Cr-Ni steels in H_2S at elevated temperatures. (a) Effect of temperature and hydrogen sulfide concentration on corrosion rate of 0 to 5% chromium steels. Graph is from laboratory data at 175 to 500 psig hydrogen pressure. (b) Effect of temperature and hydrogen sufide concentration on corrosion rate of chromium-nickel austenitic steels. Graph is from laboratory data at 175 to 500 psig hydrogen pressure. [*Technical Committee Report 59–10, Corrosion*, **15**:*125t (1959)*]

Table 11-7 Resistance of various metals and alloys to sulfur-containing gases

Materials listed in order of increasing corrosion resistance

H_2S	S	SO_2
Ni, Co	Ni, Cu	Ni
Mild steel	*Mild steel*	Fe
Fe	Fe	Cu-10Mg
Fe-Mn	Fe-14Cr	Fe-15Cr
Inconel	Cu-Mn	Ta
Cu	80Ni-13Cr-6.5Fe	Cu, *brass*
Fe-15Cr	Mn	Al *alloys*
Fe-25Cr	Cr	Mo, W
Cr	Fe-17Cr	Fe-30Cr
Fe-18Cr-8Ni	Fe-18Cr-8Ni	Fe-18Cr-8Ni
Fe-22Cr-10Al	*Hastelloy*	*Inconel*
Cu-10Mg	Al, Mg	Cu-12Al
Fe-12Al, Ni-15Al		Zr
Ta, Mo, W		
Al, Mg		

Source: O. Kubaschewski and B. E. Hopkins, *Oxidation of Metals and Alloys*, p. 277, Butterworth & Co. (Publishers), Ltd., London, 1962.

of the resistance of various materials to hydrogen sulfide, high-temperature sulfur vapors, and sulfur dioxide gas. Note that in every case nickel possesses the least resistance to sulfurous gas streams, and chromium alloy additions are in general beneficial. The exaggerated rate of attack at the lower temperatures is noteworthy.

The alloys used for high-temperature reactors generally contain Fe, Ni (or Co), and Cr as the major components with Al, Mn, Si, Mo, and C as important minor constituents. High-temperature scaling resistance in oxidizing gases is generally provided by the formation and maintenance of a slow-growing, protective layer of Cr_2O_3 or Al_2O_3. The diffusional growth rates of the sulfides or nickel, iron, cobalt, and chromium are all unacceptably high. Only the exclusion of sulfide formation offers a satisfactory condition for long service lives for components exposed to environments with the mixed oxidants: sulfur plus oxygen, or sulfur plus carbon plus oxygen.

The thermodynamics of phase stabilities for a pure metal in several oxidants is treated by log-log plots (sometimes called Ellingham diagrams) of two principal gaseous components as shown in Fig. 11-16 for the Cr-O-S system, in Fig. 11-17 for the Ni-O-S system, and in Fig. 11-18 for the Fe-O-S system at 1000°K. These plots assume that the gas phase is in internal equilibrium with respect to every gaseous species. Obviously, both the oxide and lowest sulfide of chromium are more stable than those for nickel or iron.

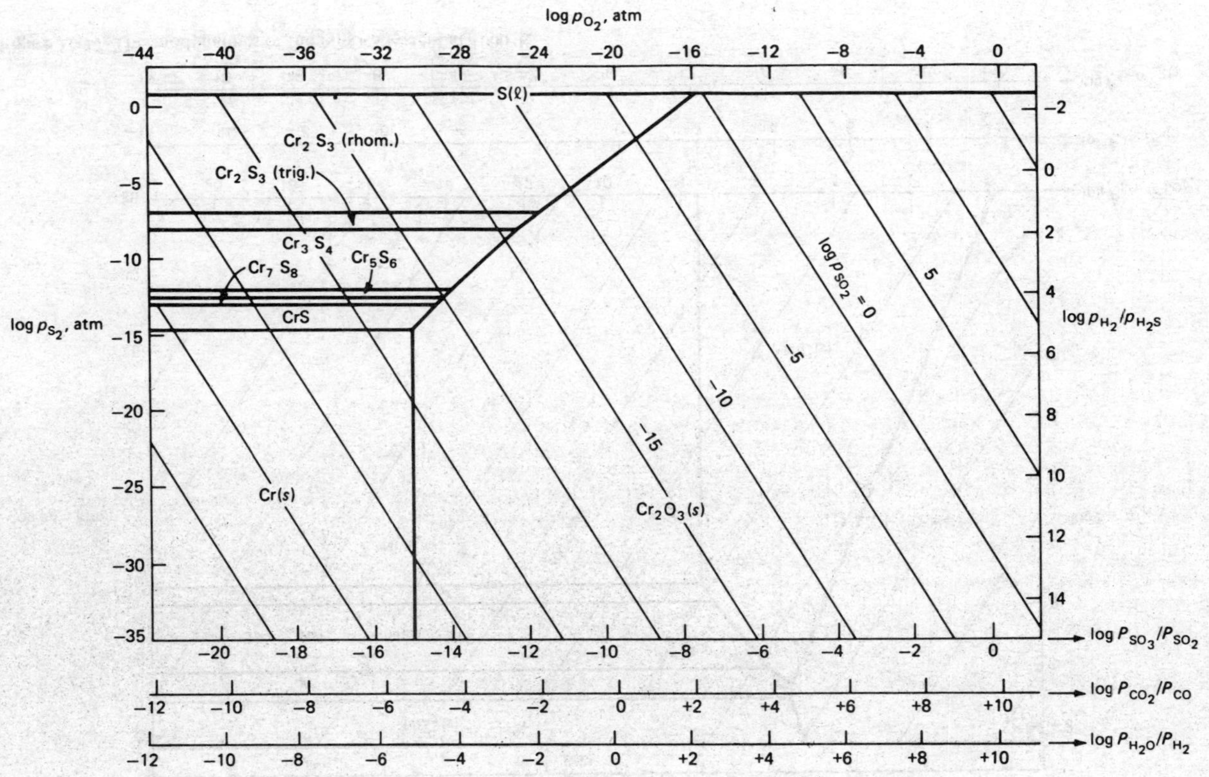

Figure 11-16 Thermodynamics of the Cr-O-S system at 1000°K.

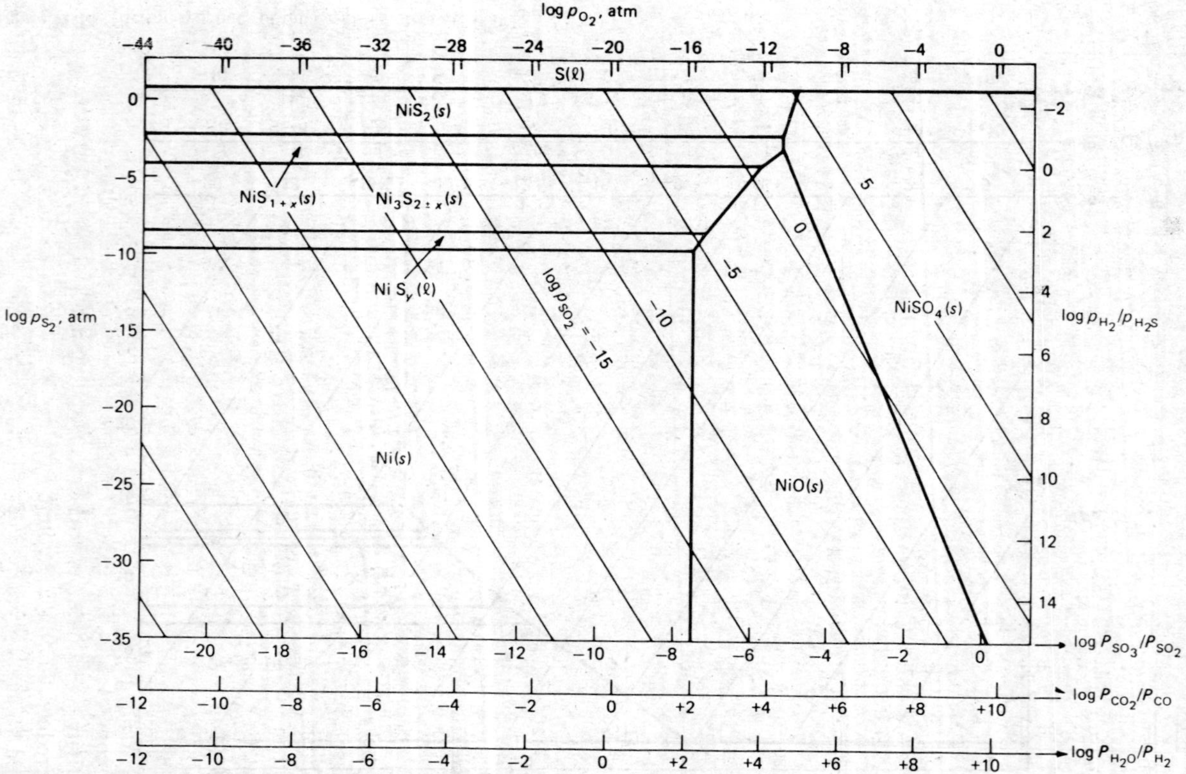

Figure 11-17 Thermodynamics of the Ni-O-S system at 1000°K.

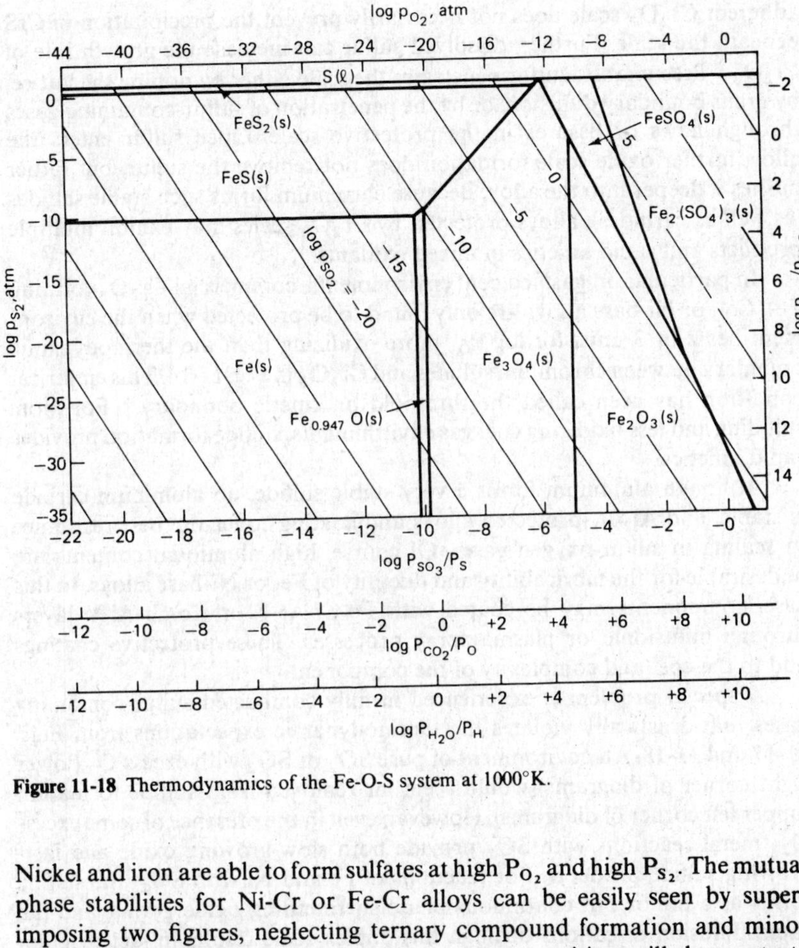

Figure 11-18 Thermodynamics of the Fe-O-S system at 1000°K.

Nickel and iron are able to form sulfates at high P_{O_2} and high P_{S_2}. The mutual phase stabilities for Ni-Cr or Fe-Cr alloys can be easily seen by superimposing two figures, neglecting ternary compound formation and minor shift in alloy-component activities. Such a superposition will demonstrate that Cr_2O_3 is frequently the most stable phase in common gas mixtures except in very reducing environments of relatively low P_{H_2}/P_{H_2S} and low P_{H_2O}/P_{H_2}. But most of the coal-gasification environments verge on these stability limits.

While the thermodynamic Ellingham diagrams specify the stable phases in contact with the gas mixture, in fact, at the metal-scale interface, the activity of the predominant nonmetal component forming the scale may be so reduced that a compound product for the second oxidant can be stabilized.* Then, for an Fe-25Cr alloy or a Ni-20Cr alloy, the formation of an

*F. S. Pettit, J. A. Goebel, and G. W. Goward, *Corr. Sci.*, **9**:903–913 (1969).

adherent Cr_2O_3 scale does not necessarily prevent the precipitation of CrS beneath the scale. Further, dissolved sulfur can increase the growth rate of Cr_2O_3.* Presumably, sulfur penetrates the scale either by doping the lattice, by grain-boundary diffusion, or by the penetration of sulfur-containing gases through flaws or fissures in the protective scale. Once sulfur enters the alloy, further oxide scale formation does not remove the sulfur, but rather pushes it deeper into the alloy. Because chromium forms such stable sulfides (as well as carbides), alloys protected by Cr_2O_3 scales may exhibit multiple products and rapid kinetics in mixed oxidants.

In particular, in gasified coal environments, commercial Cr_2O_3-forming Fe-, Co-, or Ni-base alloys are only found to be protected when the environment lies 2 or 3 units for log P_{O_2} more oxidizing than the thermodynamic boundary between chromium sulfides and Cr_2O_3 in Fig. 11-16. This empirical condition has been called the threshold or kinetic boundary.† For more sulfiding and less oxidizing coal gas environments, sulfide formation provides rapid kinetics.

Although aluminum forms a very stable sulfide, no aluminum carbide is stable, and Al_2O_3-protected alloys and coatings offer the best resistance to scaling in sulfur-oxygen gases. Of course, high aluminum contents are undesirable for the fabricability and ductility of Fe- or Ni-base alloys. In this case, components may be coated with Fe-Cr-Al-Y or Co-Cr-Al-Y layers through diffusional or plasma spray processes. These protective coatings add to the cost and complexity of the component.

A special problem is experienced in fully combusted sulfur-containing gases that drastically violates the thermodynamic expectations from Figs. 11-17 and 11-18. An environment of pure SO_2 or SO_2 with excess O_2 (lower right corner of diagrams) would seem far removed from sulfide formation (upper left corner of diagrams). However, even in the presence of some excess O_2, metal reactions with SO_2 provide both slow-growing oxide and fast-growing sulfide products. For reactions of Fe and Ni with SO_2, the sulfide forms as a minor but continuous phase surrounding oxide grains, and the rapid diffusion of cations through this duplex scale results in fact reaction rates. At 604°C, where nickel sulfide is still solid, the ratio of parabolic reaction constant for pure nickel in pure SO_2 is about 10 times higher than the rate constant for Ni oxidation in air.‡ In alloy reactions, the SO_2 species may penetrate flaws or fissures in a protective Cr_2O_3 or Al_2O_3 scale to form sulfides at the scale-gas interface. Clearly, special attention must be paid to SO_2 gases, even in the presence of excess O_2.

*G. Romeo, H. S. Spacil, and J. W. Pasko, *J. Electrochem. Soc.*, **122**:1329–1333 (1975).

†M. Danielewski and K. Nateson, *Oxid. Metals*, **12**:227 (1978); K. Nateson, NACE Preprint 72, CORROSION 84 (1984).

‡K. L. Luthra and W. L. Worrell, *Metall. Trans.*, **9A**:1055 (1978); **10A**:621 (1979).

11-13 Hot Corrosion of Alloys

With the increasing interest to combust impure residual fuels, considerable attention has been focused on the accelerated oxidation of alloy components beneath a thin surface coating of a molten ionic salt phase. "Hot corrosion" by a condensed fused salt is not observed above the boiling temperature or below the liquidus (eutectic) temperature of the salt (at the process pressure), and therefore the attack is frequently maximized at an intermediate temperature.*

In gas turbines used to power ships or planes over or near the ocean, ingested Na_2SO_4 plus NaCl from seawater leads to corrosive degradation of turbine hardware (e.g., blades and vanes). The observation of sulfides in the microstructure of corroded alloys initially led investigators erroneously to blame gaseous sulfur compounds for the accelerated attack. However, the oxidation of presulfidized alloys (which generate Cr_2O_3 or Al_2O_3 protective scales) is not greatly accelerated, and does not simulate hot corrosion.† Today, "hot corrosion" attack by thin fused salt films is recognized to encompass accelerated oxidation much more generally than just the Na_2SO_4 attack in gas turbines.

In fast breeder reactors under development, fission products from the fuel condense onto the stainless steel cladding as complex salts based on CsI and Cs_2O. A hot corrosion attack at the grain boundaries results. In boilers and turbines burning high-ash coal or residual fuel oil, condensed molten salts accelerate alloy oxidation. In heat exchangers, the condensates cause "fireside" corrosion; low-melting sodium vanadate ($T_M = 570°C$) is particularly damaging. In developmental fused alkali carbonate fuel cells, and for prospective carbonate energy storage systems, accelerated hot corrosion is a problem. In future coal conversion plants metallic internals are threatened by slag condensation. In future magnetohydrodynamic (MHD) channels, fused salt seed promises significant attack of refractories. Thus, the scope and mechanisms of hot corrosion by fused salts extend far beyond the initially observed, apparent sulfur attack.

Both salt chemistry‡ and electrochemistry¶ must contribute importantly

*M. A. De Crescente and N. S. Bornstein, Formation and Reactivity Thermodynamics of Sodium Sulfate With Gas Turbine Alloys, *Corrosion*, **24**:127–133 (1968).

†J. A. Goebel and F. S. Pettit, The Influence of Sulfides on the Oxidation Behavior of Nickel-Base Alloys, *Met. Trans.*, **1**:3421–3429 (1970); N. S. Bornstein and M. A. De Crescente, The Role of Sodium and Sulfur in the Accelerated Oxidation Phenomena—Sulfidation, *Corrosion*, **26**:209–214 (1970).

‡D. Inman and N. S. Wrench, Corrosion in Fused Salts, *Brit. Corr. J.*, **1**:246–250 (1966); J. A. Goebel, F. S. Pettit, and G. W. Goward, *Metal. Trans.*, **4**:261 (1973).

¶A. Rahmel and E. Tatar-Moisescu, Electrochemical Behavior of Alkali and Alkali Earth Sulfates at 800°C, *Electrochem. Acta*, **20**:479–484 (1975); D. A. Shores, Use of Anodic Polarization in Fused Na_2SO_4 for Estimating Hot Corrosion Rates, *Corrosion*, **24**:434–440 (1975).

to any comprehensive mechanism for hot corrosion. The thermodynamics and kinetics of hot corrosion in a thin-film salt electrolyte share many common features with atmospheric corrosion by aqueous solutions. In particular, a high-temperature Pourbaix-type diagram as shown in Fig. 11-19 represents the phase stabilities for the Al-S-O system superimposed on a field in the Na-S-O system where Na_2SO_4 is stable throughout. The ordinate may be considered as the redox or electrode potential relative to a standard oxygen electrode, and SO_3 is considered as the acid component of the salt, with oxide ions the basic component. For an alkali carbonate salt, CO_2 is the acid component.

The oxide Al_2O_3 may dissolve into a fused salt, such as Na_2SO_4 or Na_2CO_3, either by the reaction:

$$Al_2O_3 \rightarrow 2\,Al^{3+} + 3O^{2-} \qquad \text{(acid dissolution)} \qquad (11.14)$$

or else by complexing with an oxygen ion:

$$Al_2O_3 + O^{2-} \rightarrow 2\,AlO_{2^-} \qquad \text{(basic dissolution)} \qquad (11.15)$$

depending upon the local acid-base chemistry at the oxide-salt interface.

The analogous dissolution behavior of NiO in Na_2SO_4 to form one acid and one basic ionic solute species provided the solubility plot of Fig.

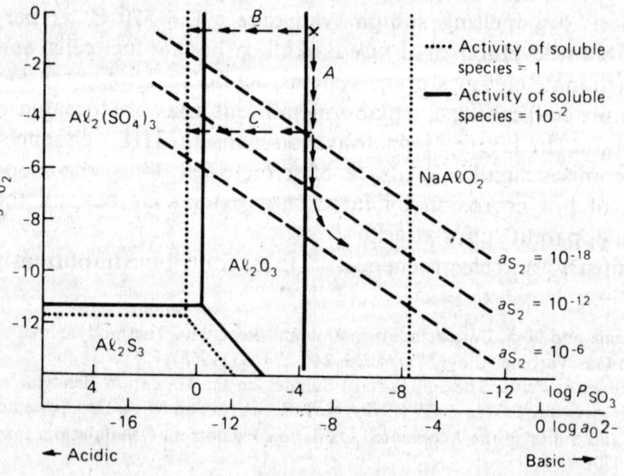

Figure 11-19 Phase stability diagram for Al in Na_2SO_4 at 906°C. Sulfur isobars are shown as dashed lines. The solubility of Al_2SO_3 is represented by solid lines for an activity of soluble species equal to 0.01 and by dotted lines for an activity of soluble species equal to 1. [*From D. A. Shores, Use of Anodic Polarization in Fused Na_2SO_4 for Estimating Hot Corrosion Rates, Corrosion*, **24**:*434–440 (1975)*]

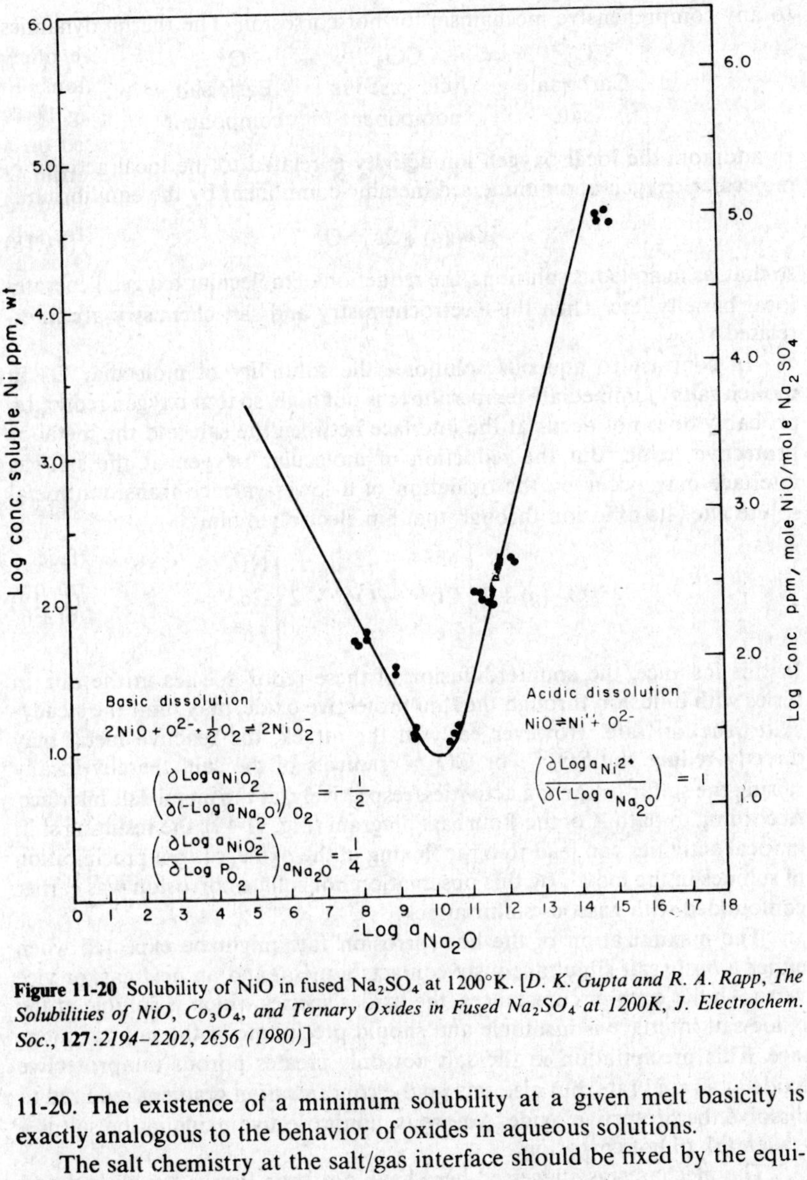

Figure 11-20 Solubility of NiO in fused Na_2SO_4 at 1200°K. [*D. K. Gupta and R. A. Rapp, The Solubilities of NiO, Co_3O_4, and Ternary Oxides in Fused Na_2SO_4 at 1200K, J. Electrochem. Soc.,* **127**:*2194–2202, 2656 (1980)*]

11-20. The existence of a minimum solubility at a given melt basicity is exactly analogous to the behavior of oxides in aqueous solutions.

The salt chemistry at the salt/gas interface should be fixed by the equilibrium:

$$\underset{\text{Sulfate salt}}{SO_4^{2-}} \rightleftharpoons \underset{\text{Acid gaseous component}}{SO_3(g)} + \underset{\text{Basic salt component}}{O^{2-}} \tag{11.16}$$

or

$$\underset{\text{Carbonate salt}}{CO_3^{2-}} \rightleftharpoons \underset{\text{Acid gaseous component}}{CO_2} + \underset{\text{Basic salt component}}{O^{2-}} \tag{11.17}$$

In addition, the local oxygen ion activity is related to the local activity of molecular oxygen at an immersed metallic component by the equilibrium.

$$\tfrac{1}{2}O_2\,(g) + 2e \rightleftharpoons O^{2-} \tag{11.18}$$

so that, as in aqueous solutions, the reduction of molecular oxygen generates local basicity, etc. Then the electrochemistry and salt chemistry are inter-related.

In contrast to aqueous solutions, the solubility of molecular O_2 in molten salts at immediate temperatures is not high, so that oxygen reduction probably does not occur at the interface between the salt and the metal or protective oxide. But the reduction of molecular oxygen at the salt-gas interface may occur by the oxidation of a lower-valence transition metal solute after its diffusion through the thin electrolyte film:

$$\tfrac{1}{2}O_2\,(g) + 2\begin{cases} Ni^{2+} \\ Co^{2+} \\ Fe^{2+} \end{cases} \rightarrow O^{2+} + 2\begin{cases} Ni^{3+} \\ Co^{3+} \\ Fe^{3+} \end{cases} \tag{11.19}$$

In this instance, the counterdiffusion of these redox species in the salt, in series with diffusion through the thin protective oxide, may limit the steady-state reaction rate. However early in the attack, the reactive metal may directly reduce the SO_4^{2-} or CO_3^{2-} radicals of the salt, thereby locally raising the sulfur or carbon activities, respectively, at the metal-salt interface. According to path *A* of the Pourbaix diagram (Fig. 11-19), the resulting shift in local activities can lead to basic fluxing of the oxide, or to a precipitation of sulfides in the metal. By this observation hot sulfate corrosion was earlier confounded with gaseous sulfur attack.

The maximization of the hot-corrosion rate might be expected when either a *basic* salt simultaneously contacts a metal and an *acid* gas, or vice versa. Under such circumstances, the oxides species which is soluble at the oxide-salt interface is insoluble and should precipitate at the salt-gas interface. This precipitation in the salt not only creates porous nonprotective oxide at a rapid rate, but also sets up the concentration gradient required to dissolve the protective oxide. A porous, nonprotective oxide in the salt is a trademark of hot corrosion.

The mechanisms suggested here have not been firmly established and should not be generalized. Clearly, however, hot-corrosion kinetics will depend greatly on gas solubilities in the salt, the salt and gas chemistries, and the availabilities of redox ions. Further, experimentation in large crucibles of molten salt cannot simulate the hot-corrosion process if transport control through a thin salt film is important.

INDEX